# THE CONTINENTAL DRIFT CONTROVERSY

## Volume II: Paleomagnetism and Confirmation of Drift

Resolution of the sixty-year debate over continental drift, culminating in the triumph of plate tectonics, changed the very fabric of Earth science. Plate tectonics can be considered alongside the theories of evolution in the life sciences and of quantum mechanics in physics in terms of its fundamental importance to our scientific understanding of the world. This four-volume treatise on *The Continental Drift Controversy* is the first complete history of the origin, debate, and gradual acceptance of this revolutionary explanation of the structure and motion of the Earth's outer surface. Based on extensive interviews, archival papers, and original works, Frankel weaves together the lives and work of the scientists involved, producing an accessible narrative for scientists and non-scientists alike.

Beginning in the early 1950s, continental drift found new life from an unexpected source, paleomagnetism, which records the Earth's magnetic field in rocks and how its direction and intensity has changed over time. This second volume provides the first extensive account of the growing paleomagnetic case for continental drift and the development of apparent polar wander paths that showed how the continents had changed their positions relative to one another – more or less as Wegener had proposed. Paleomagnetism offered the first physical measure that continental drift had occurred, and helped determine the changing latitudes of the continents through geologic time.

Other volumes in *The Continental Drift Controversy*:

Volume I – Wegener and the Early Debate

Volume III – Introduction of Seafloor Spreading

Volume IV – Evolution into Plate Tectonics

HENRY R. FRANKEL was awarded a Ph.D. from Ohio State University in 1974 and then took a position at the University of Missouri–Kansas City, where he became Professor of Philosophy and Chair of the Philosophy Department (1999–2004). His interest in the continental drift controversy and the plate tectonics revolution began while teaching a course on conceptual issues in science during the late 1970s. The controversy provided him with an example of a recent and major scientific revolution to test philosophical accounts of scientific growth and change. Over the next thirty years, and with the support of the United States National Science Foundation, National Endowment for the Humanities, the American Philosophical Society, and his home institution, Professor Frankel's research went on to yield new and fascinating insights into the evolution of the most important theory in the Earth sciences.

"A well constructed and gripping narrative, which preserves the complex scientific detail, but invites one in to this fascinating world and helps the reader patiently to find a way through its labyrinth. Frankel is a wonderful guide and worthy of your trust."

MOTT GREENE, University of Puget Sound and University of Washington

"This volume… is a complete account and benefits from the fact that many of those who were principals in the drama are still alive… Fascinating and full of humor, but very serious. A better book on the subject will probably never be written."

NEIL D. OPDYKE, University of Florida

"Tracing an exhaustive and comprehensive history, Frankel illuminates how different were geological and geophysical perspectives on continental drift, providing fascinating insights on the erratic and complex fashion in which science advances."

JIM BRIDEN, University of Oxford

# THE CONTINENTAL DRIFT CONTROVERSY

## Volume II: Paleomagnetism and Confirmation of Drift

HENRY R. FRANKEL

*University of Missouri–Kansas City*

To Johanna

**CAMBRIDGE**
UNIVERSITY PRESS

# CAMBRIDGE
## UNIVERSITY PRESS

University Printing House, Cambridge CB2 8BS, United Kingdom

Cambridge University Press is part of the University of Cambridge.

It furthers the University's mission by disseminating knowledge in the pursuit of education, learning and research at the highest international levels of excellence.

www.cambridge.org
Information on this title: www.cambridge.org/9781316616062

© Henry R. Frankel 2012

This publication is in copyright. Subject to statutory exception and to the provisions of relevant collective licensing agreements, no reproduction of any part may take place without the written permission of Cambridge University Press.

First published 2012
First paperback edition 2016

*A catalogue record for this publication is available from the British Library*

ISBN 978-0-521-87505-9 Hardback
ISBN 978-1-316-61606-2 Paperback

Cambridge University Press has no responsibility for the persistence or accuracy of URLs for external or third-party internet websites referred to in this publication, and does not guarantee that any content on such websites is, or will remain, accurate or appropriate.

# Contents

# Foreword

This is the story of the formative years – the decade of the 1950s – of paleomagnetism as a scientific discipline in conjunction with a focus on the big questions of the day – the origin of the geomagnetic field, polar wander, continental mobility. The exposition is meticulously documented with referral to primary published literature and enlivened by extensive referral to real-time correspondence and retrospective views based on the author's interviews and written exchanges with many of the principals dating back to the early 1980s. Some of the themes that emerge from the account are the ever-importance of serendipidity and the ability of top scientists to identify tractable aspects of a big problem, adjust the scope and direction of the research as needed, and recognize applications to seemingly oblique problems. Paleomagnetism involved some of the major figures in physics of the post World War II era, including the Nobel laureate Blackett (who studied under Rutherford, another Nobelist), who spins up the story with an ingenious experiment to test whether the geomagnetic field is a fundamental property of a rotating body. The results were famously negative yet the theory and experiment had several notable positive outcomes, namely capturing the interest of his Ph.D. student, Keith Runcorn, to test the fundamental theory versus the competing dynamo theory by making measurements of the geomagnetic field in mine shafts, and the deployment of the sensitive magnetometer developed for the experiment for paleomagnetic research on rocks. Runcorn went on to assemble what became the leading group in paleomagnetism research (started at Cambridge but soon moved to Newcastle), whose students would emerge in the vanguard of the subject's most influential practitioners. The enterprise was graced with luck right at the outset with the arrival in late 1949 at Cambridge of Hospers, a student from Holland who came with his own scholarship and wanted to sample young lavas in Iceland hoping to correlate them by their intensity of magnetization. In the process, Hospers produced evidence for a global correlation tool, polarity reversal stratigraphy; for the exquisitely simple geometry for charting polar motions or continental mobility, the field of a geocentric axial dipole; and providing data that motivated the development of statistical methods on a sphere, Fisher statistics. These are pillars of paleomagnetism and they were basically established by 1953. Soon after Hospers arrived, Runcorn recruited Irving with his background in geology to look for evidence of geomagnetic secular variation in the Torridonian, thick sedimentary beds of Precambrian age. This was a wildly optimistic effort that nevertheless developed modern techniques and produced the first magnetic polarity reversal stratigraphy in sediments and oblique directions that indicated magnetic stability, which pointed to such fine-grained redbeds as key sampling targets for studies of the ancient geomagnetic field. The range of research expanded and was conducted at an exhilarating pace in a global network of information flow with sharp attention to publication priority. Creer, Runcorn's second student in paleomagnetism, built a sensitive astatic magnetometer at Cambridge after the design of Blackett's machine and only managed to start sampling and measuring a series of rock units half-way through his three-year fellowship; nevertheless, by 1954 he constructed an apparent polar wander path for Britain in 1954, the first such path and the conceptual basis for testing continental drift. Irving leaves for Australia in 1954, builds a lab from scratch with a new student, Green, and they had new results on Mesozoic dolerites in press within 2 years. And so forth.

By 1958, there were published results from young lavas from four continents in support of a geocentric axial dipole and the reality of polarity reversals, full results from the Deccan of India by the Blackett group, data from Australia by Irving's lab in Canberra, from South America by Creer, and paleoclimate evidence from proxies like Opdyke's analysis of wind directions in full support of the paleomagnetic assumptions: the evidence from the British schools (and from others like Gough in Africa and the brave Khramov in the Soviet Union) was decisively in favor of crustal mobility. In contrast, Graham at the Department of Terrestrial Magnetism at the Carnegie Institute came to very different conclusions. Graham actually had a head start with the availability of a sensitive spinner magnetometer at the Carnegie that allowed him to publish in 1949 a paleomagnetic survey of sedimentary formations from throughout North America and to develop seminal reliability tools like the fold test. Unfortunately, Graham was unable or unwilling to counter what the author describes as the prevailing fixist and anti-field reversal orthodoxy of the American community, and called upon cryptic strain effects and self-reversal (given credence by theoretical work of Nobel laureate Louis Néel, followed shortly thereafter by the chance discovery by Nagata's group in Japan of a self-reversing rock, now known to be an exceedingly rare occurrence) to explain otherwise straightforward evidence for crustal and/or polar mobility.

By 1959, the author points out that every major paleomagnetist with the notable exception of the American Graham (and Cox and Doell) favored crustal mobility, but despite this level of success, the paleomagnetists who advocated continental mobility were a beleaguered group. For one, the U.S. effort simply lacked a charismatic leader like Blackett to counter the negativism of the geologic community. And to cap this desultory period in the paleomagnetic case for crustal mobility, a lengthy critical review by Cox and Doell that appeared in the GSA Bulletin in 1960 reserved judgment, an opinion that tended to conform with general disbelief in crustal mobility expressed by the pillars of the American geological community (e.g., Bucher, Gilluly) as well as some of the high priests of theoretical geophysics (e.g., expressed in Jeffreys' *The Earth* and in Munk and MacDonald's *The Rotation of the Earth*). A great irony is that despite what is appropriately described as one of the greatest flukes in the history of testing continental drift (ranking right up there with the self-reversing rock from Japan) – Cox's report in 1957 of the aberrant direction from the Eocene Siletz volcanics from Oregon falling close to the Deccan pole from India with continents in the present position and ascribed to rapidly varying geomagnetic fields, which turned out to be due to local tectonic rotation of the Siletz – Cox's misjudgements were basically forgotten and his (with Doell's) reputation rested on their subsequent work on the timescale of polarity reversals (motivated in part by the self-reversing fluke), which was the basis of the Vine and Matthews hypothesis. The decade-long effort to make the case for continental mobility with land-based paleomagnetism was not in vain. It not only helped prepare the community to accept plate tectonics (the topic of the author's next volume), it eventually provided the natural paleogeo-graphic reference frame.

*Dennis V. Kent*
*Earth and Planetary Sciences, Rutgers University, and*
*Lamont-Doherty Earth Observatory of Columbia University*

# Acknowledgments

I could not have undertaken and completed this book without enormous help from those paleomagnetists whose work led to the measurement of drifting continents. Ken Creer, Jan Hospers (deceased), Edward Irving, and S. K. Runcorn (deceased) answered many questions over many years about their work and that of others. Creer, Hospers, and Runcorn critically reviewed earlier versions of several chapters, and Irving reviewed the entire manuscript and provided flash forward updates about the current status of various problems. John Clegg (with assistance of Mike Fuller), Colin Bull, Ernie Deutsch (deceased), Richard Doell (deceased), Ian Gough (deceased), Ron Green, Aleksei Khramov, A. E. M. Nairn (deceased), Neil Opdyke, and Don Tarling discussed their own work and that of others. Maurice Adams, Bill Bonini, Martin Bott, David Collision, M. E. Evans, Warren Hamilton, Raymond Hide, Donald Hitchcock, Leo Kristjansson, Frank Lowes, Jim Parry, Graeme Stevens, Gillian M. Turner, and Stanley Westoll (deceased) kindly provided particularly useful information about the work of others. Their help has been essential, especially because much of the work by these paleomagnetists was collaborative. It is a pleasure to acknowledge their considerable help.

I also thank S. Warren Carey (deceased) for answering questions about his work. I thank Mervyn Paterson for helping me understand Carey's work in rheology, and for critically examining my account of Carey. Robert Fisher, Ronald Green, Edward Irving, and Curt Teichert (deceased) also provided useful information about Carey, his standing and reception of his views among Australian Earth scientists.

I thank Dan McKenzie for critically reviewing an earlier version of this volume.

I thank Nancy V. Green and her digital imaging staff at Linda Hall Library, Kansas City, Missouri, for providing the vast majority of the images; Richard Franklin for color image of the *Time Magazine* representation of Creer's 1954 Oxford version of his APW path for Britain. I should also like to thank the reference librarians at Linda Hall Library, and the interlibrary staff at the Miller Nicholas Library, UMKC.

I owe much to Nanette Biersmith for serving as my longtime editor and proofreader.

I am indebted to the United States National Science Foundation, the National Endowment of the Humanities, and the American Philosophical Society for financial support. I also thank the University of Missouri Research Board and my own institution for timely grants to continue this project.

I wish to thank Susan Francis and her staff at Cambridge University Press for believing in this project and for their great assistance throughout its production.

# Abbreviations

| | |
|---|---|
| AAPG | American Association of Petroleum Geologists |
| AF | Alternating fields |
| *AG* | *Advances in Geophysics* |
| AGU | American Geophysical Union |
| APW | Apparent polar wander |
| ANU | Australian National University |
| BMR | Australian Bureau of Mineral Resources |
| BAAS | British Association for the Advancement of Science |
| FRS | Fellow of the Royal Society (London) |
| GAD | Geocentric axial dipole |
| GSA | Geological Society of America |
| IGY | International Geophysical Year |
| IUGG | International Union of Geodesy and Geophysics |
| *JGG* | *Journal of Geomagnetism and Geoelectricity* |
| *JGR* | *Journal of Geophysical Research* |
| Lamont | Lamont Geological Observatory |
| Ma | Million years |
| NRM | Natural remanent magnetization |
| NSF | National Science Foundation (USA) |
| ORS | Old Red Sandstone |
| RAS | Royal Astronomical Society |
| RS1 | Research Strategy 1 |
| RS2 | Research Strategy 2 |
| RS3 | Research Strategy 3 |
| SEPM | Society of Economic Paleontologists and Mineralogists |
| Scripps | Scripps Institution of Oceanography |
| UCLA | University of California, Los Angeles |
| USGS | United States Geological Survey |
| IUGG | International Union of Geodesy and Geophysics |
| VRM | Viscous remanent magnetization |

# Introduction

By the late 1940s and early 1950s, mobilism was at a low ebb, perhaps its lowest ever. Volume I has shown that regionalism, isthmian links, the failure to find a generally acceptable mechanism and a host of special objections had left mobilism in tatters. Fixism ruled. Globally, mobilism had few advocates and there was no sign that their numbers were increasing. The fixism/mobilism debate was moribund and something entirely new was needed, something astounding, to breathe new life into it, and break the impasse. In timely fashion during the early 1950s, the fortunes of mobilism were revived by the work begun by two British research groups studying the natural remanent magnetization of rocks, paleomagnetism.

These paleomagnetists found that the directions of magnetization in rocks that were less than about twenty million years old were not along the present geomagnetic field but were, on average, along the field of a dipole situated at the center of the Earth and directed along the axis of rotation, the geocentric axial dipole (GAD). The time average field, its average over several thousand years, had this simple form. Making paleomagnetic surveys in Britain, they found that rocks older than about twenty million years, rocks (and this was of crucial importance) that could be shown to be magnetically stable, had magnetizations that were systematically oblique to the present geocentric axial field, sometimes very strongly oblique, differing from it by as much as 90°! It was as if Britain had moved many thousands of kilometers relative to Earth's present axis of rotation and rotated many tens of degrees relative to the present meridian. A survey in peninsular India suggested that during the past sixty-five million years it had drifted 5000 kilometers northward and rotated almost 30° counterclockwise relative to the present meridian. Certain results were also obtained by a third, older paleomagnetic group, from the United States of America, some of which could have been interpreted in terms of comparable motions of North America but were not. Over the next half-dozen years, from research carried out in Europe, India, Australia, the Soviet Union, southern Africa, South America, and Antarctica, systematically varying, oblique magnetizations were observed to be grossly inconsistent from continent to continent; inconsistent in much the same way as expected from the paleoclimatic evidence and from the reconstructed movements of continents relative to each other and to the paleogeographic pole as proposed on entirely different grounds by

Wegener (I, §2.7, §2.8, §2.15, §3.2, §3.10, §3.13, §3.15), Köppen (I, §3.15), and du Toit (I, §6.5–§6.7). These astonishing paleomagnetic results obtained between mid-1951 and 1959 provided the first solid physical evidence for continental drift and reversed the downward trend of mobilism's fortunes. Collectively they confirmed that continental drift had happened, and almost every paleomagnetist accepted them as evidence of drift. However, a few from the United States saw otherwise. Most fixists outside paleomagnetism also rejected the results as evidence of drift, while old-time mobilists welcomed them. Some opponents raised difficulties, often the same ones repeatedly, which pro-drift paleomagnetists showed to be either phantom difficulties or ones that had already been disposed of. How all this came about is the subject of this volume in which certain other topics of much concern at the time will also be addressed.

As in Volume I, I shall describe how researchers acted in accordance with what I have identified as three standard research strategies (I, §1.13). Workers did not recognize or say that they acted in this way; the three research strategies are my retrospective description of how they went about their tasks, how they addressed their problems. Research Strategy 1 (hereafter, RS1) was used by researchers to expand the problem-solving effectiveness of solutions and theories. Research Strategy 2 (hereafter, RS2) was used by them to diminish the effectiveness of competing solutions and theories; RS2 was an attacking strategy used to raise difficulties against opposing solutions, and to place all possible obstacles in their way. Workers used Research Strategy 3 (hereafter, RS3) to compare the effectiveness of competing solutions and theories, and to emphasize those aspects of a solution or theory that gave it a decided advantage over its competitors.

The development of paleomagnetism's case for mobilism is a story of how a small, disparate, often quarrelsome band of researchers working in Britain in the early 1950s took a backwater discipline in the Earth sciences and made it of central importance; how they found a way to measure, quantitatively, past movements of continents relative to the paleogeographic pole, and, less directly, to each other. Besides reviving the fortunes of mobilism, the work described in this volume has had a long-lasting and likely permanent legacy: the provision of a geographical frame of reference for mapping Earth's major features in the remote geological past, a frame of latitudes and longitudes analogous to that we have for the present world. This work began in the early 1970s with a synthesis between rock magnetization directions transformed into paleomagnetic poles and plate tectonics, which began in the early 1970s beyond the time frame of this book.[1]

## Note

1 It was Smith, Briden, and Drury (1973) who initiated this synthesis in a general way with their atlas of paleogeographic maps. A short history of the formative stages of this synthesis has been given by Irving (2005). Later developments, which became possible as data accumulated, have involved the construction of "composite" apparent polar wander paths (also variously called "world" or "synthetic" APW paths) in which all continental paleomagnetic data are combined into a single path (Phillips and Forsyth, 1972; Besse and Courtillot, 2002; Kent and Irving, 2010). At present, this synthesis can be made only for Late Triassic and later times, because there are no oceans, on which plate tectonic methods depend, older than this.

# 1

# Geomagnetism and paleomagnetism: 1946–1952

## 1.1 Breaking the impasse: the three main paleomagnetic groups

Three groups were primarily responsible for the developments in paleomagnetic work. Two were founded by S. K. (Keith) Runcorn and P. M. S. Blackett; both were physicists who became interested in paleomagnetism through their work in geomagnetism. Runcorn, formerly an assistant lecturer at Manchester University, where he worked for his Ph.D. under Blackett, began in mid-1951 recruiting to his group at Cambridge University. Blackett, Head of the Department of Physics at Manchester University, began forming his group in early 1952, placing John Clegg in charge; Clegg will be introduced in Chapter 2. The third group, at the Department of Terrestrial Magnetism at the Carnegie Institution in Washington, DC, began working in paleomagnetism much earlier, in the late 1930s; its efforts lapsed during World War II, and recommenced afterwards. I deal with the British groups first as it was they who made the startling discoveries, recognizing them as the key to what was to become the first physically based measure of mobilism.

In 1947 Blackett revived interest in fundamental or distributed theories of the origin of the geomagnetic field, arguing that all rotating bodies produce magnetic fields. He constructed an astatic magnetometer specifically to test such theories, a test that proved negative. Runcorn became interested in Blackett's ideas, and carried out, at Edward (Teddy) C. Bullard's (later Sir Edward) suggestion, a different test of the distributed theory for which he earned his Ph.D. This too was negative. Blackett and Runcorn recognized that from studies of the natural remanent magnetization (NRM) of rocks there was much to be learned about the long-term history of the geomagnetic field, and thus better understand its origin. They both realized that Blackett's magnetometer was, with adaptations, well suited for such paleomagnetic studies.

When Runcorn arrived at the Department of Geodesy and Geophysics in Cambridge in 1950, Jan Hospers was already there. Hospers, from the Netherlands, had just begun working on a Ph.D., and planned to undertake a paleomagnetic survey of Icelandic lavas. He had no interest in mobilism, thinking, as almost everyone in the Netherlands then did, that it was a dead issue. His plan was to use variations in the strength (intensity) of magnetization to correlate the lavas.

He quickly found reversals of magnetization, which became his main interest. The directions of magnetization he observed were somewhat dispersed and required statistical analysis. He explained the problem to Runcorn who told R. A. Fisher (recently knighted Sir Ronald), the great evolutionary biologist and statistician. Fisher supplied the statistical method which enabled Hospers to show that the average field in Iceland, regardless of sign, was close to that of a geocentric axial dipole. Runcorn, who was then working on the problem of secular variation, decided to try to use paleomagnetism to study ancient or paleosecular variation, the variation in the strength and direction of the geomagnetic field over hundreds or thousands of years. In June 1951 he hired Edward (Ted) Irving as a temporary assistant to collect oriented samples, and to measure them on Blackett's magnetometer, which he did. Irving also happened to be interested in continental drift, and he figured out how to use paleomagnetism to test it; later that year, with help from Fisher, he initiated the first such test. However, Hospers, Runcorn, and Irving did not immediately redirect their main research programs toward testing mobilism, but continued gathering samples, hoping to learn more about the long-term history of the geomagnetic field, its reversal, and paleosecular variation. Irving and Runcorn soon discovered that fine-grained red sandstones (red beds) recorded well the average direction of the geomagnetic field but not the details of its secular variation that Runcorn had hoped for. This early discovery was crucial because it allowed paleomagnetists to quickly locate reliable recorders of the long-term behavior of the field, records that they needed regardless of whether they were working on problems in geomagnetism or testing mobilism.

Researchers at the Carnegie Institute for Terrestrial Magnetism in Washington, DC, had earlier used paleomagnetism to address problems about secular variation and reversal of the geomagnetic field, and J. W. (John) Graham, a key member of the group, also learned (1949) that paleomagnetism could be used to test mobilism, but he did not act on it for half-a-dozen years, long after British paleomagnetists had begun to do so. Importantly, however, Graham also developed two field tests invaluable to determine the reliability of paleomagnetic data, dependable standbys throughout the mobilism debate.

## 1.2 Blackett and Runcorn begin their years together at the University of Manchester (1946–1949)

Stanley Keith Runcorn (1922–95) was born in Southport, Lancashire.[1] He had one sibling, a younger sister. He attended King George V Grammar School in Southport, where he excelled early in history and mathematics. Later his headmaster convinced him to study science, and he gained a State Scholarship. He was a prominent member of the school Debating Society where he learned skills he practiced effectively all his life. At school he became a very good swimmer. Later he, with more enthusiasm than skill, became an increasingly elderly rugby and squash player; it was at swimming that he excelled. His enthusiasm for sports is important because it led to his finding two of his best students, Irving and N. D. (Neil) Opdyke.

Runcorn became an undergraduate at Gonville and Caius College, Cambridge University, entering the Faculty of Engineering (then called Mechanical Sciences) in 1941. It was wartime, and in 1943 he took what was then the usual two-year degree, and was recruited into the war effort, into radar work. During this time, he had become more interested in physics than in engineering, and decided not to return to Cambridge. He applied unsuccessfully for a fellowship in the Department of Physics at the University of Manchester. However, Bernard Lovell, already at Manchester where he led the radio astronomy group, encouraged Runcorn to apply for an assistant lectureship in physics because he had heard that Blackett had been impressed with his application. He did so, and found himself (Runcorn, August, 1984 interview with author) "in October of 1946 as an Assistant Lecturer in Physics, having not done physics as an undergraduate subject and not having a Ph.D."

## 1.3 Blackett's fundamental or distributed theory of the origin of the geomagnetic field and Runcorn's introduction to it

Patrick Maynard Steward Blackett (1897–1974), later Lord Blackett, was a giant among experimental physicists, a charismatic personality and a prominent public figure. His strong support in the 1950s for paleomagnetic work in Britain was likely the principal reason why it prospered there.

Earmarked for a career in the Royal Navy, Blackett at age thirteen entered Osborne Naval College, and two years later the Royal Naval College at Dartmouth where he received thorough training in science and technology as well as in normal naval subjects. He was present at the first Battle of the Falkland Islands (1914), at the huge Battle of Jutland (1916) just off the coast of Denmark, and in several smaller engagements in the Channel and North Sea toward the end of World War I.

He resigned from the Royal Navy in 1919 and entered Cambridge University where he read Part I Mathematics and Part II Physics, wasting no time graduating in 1921. He entered the Cavendish Laboratory under Sir Ernest Rutherford. There Blackett improved on the original design of C. T. R. Wilson's cloud chamber, turning it into a powerful tool for research in nuclear physics and cosmic radiation. Because of this and the discoveries made through it, Blackett received the Nobel Prize in Physics in 1948. He became head of physics departments at Birkbeck College, London, and then at the University of Manchester. In World War II he became "a founder of wartime operational research and one of the heroes in the British triumph in the U-boat campaign" (Nye, 2004: 99). (Accounts of Blackett's scientific career and eventful life are by Lovell (1975), Bullard (1974), Butler *et al.* (1975) and Nye (1999, 2004).)

Blackett returned to the University of Manchester after World War II, where he played a strong role in the national debate on atomic energy and the bomb.

He expanded cosmic ray research, and also became interested in magnetism "while considering the possible influence of the magnetic field of stars in the galaxy on cosmic ray phenomena" (Blackett, 1947: 658). He soon developed a highly speculative "fundamental" or "distributed" theory of geomagnetism whose central idea was that any rotating body produces a magnetic field by virtue of its rotation.

While considering whether the magnetism of stars could influence cosmic rays, Blackett realized that the magnetic moments of the Earth and Sun are nearly proportional to their angular momenta; that

the magnetic moment P and the angular momentum U of the earth and sun are nearly proportional, and that the constant of proportionality is nearly the square root of the gravitational constant G divided by the velocity of light.

(Blackett, 1947: 658)

Blackett was excited, for besides thinking that he had an explanation of the geomagnetic field, he thought this theory of magnetism might "provide the long-sought connection between electromagnetic and gravitational phenomena" (Blackett, 1947: 658). He began reviewing the literature on the origin of the Sun and Earth's magnetic fields, and realized that he was not the first to think that rotating bodies might generate a magnetic field. "I found to my surprise that the essence of these facts had been known for some years, but had, for various reasons, dropped later out of notice" (Blackett, 1947: 658).

The Manchester physicist and mathematician Arthur Schuster had first suggested that the Sun and the Moon might possess a magnetic field by virtue of their rotation. In 1891, Schuster, while discussing the nature of the solar corona, whose luminosity he attributed to electrical discharges, proposed as the source of the discharges a solar magnetic field, which also would explain the shape of the corona.

If then, as is probable, electric discharges take place near the sun, there must be some cause which keeps up the difference in electrical potential between the sun and outside space. The form of the corona suggests a further hypothesis, which, extravagant as it may appear at present, may yet prove to be true. Is the sun a magnet? We know that a body at such a high temperature cannot be magnetisable, but may not a revolving body act like a magnet, and may not the earth's magnetism be similarly due to the earth's revolution about its axis?

(Schuster, 1891: 275)

Schuster carried out experiments, and in 1912 Blackett (1952: 310) learned that Schuster had been unable to detect a magnetic field near a rapidly rotating nonmagnetic body. In 1923, H. A. Wilson (who like Blackett had worked at the Cavendish but eventually moved to Rice Institute in Houston where he much influenced the young Maurice Ewing, who later became a strong anti-mobilist (III, §6.3)) further tested Schuster's idea, but detected no magnetic field (Wilson, 1923), as did W. F. G. Swann and A. Longacre (1928) with the same result. Blackett realized that their magnetometers were not sensitive enough to detect the field predicted by his theory (Blackett (1947: 665–666); he needed a more sensitive instrument.

The magnetic fields of stars were important for Blackett's theory, and he learned from his friend Subrahmanyan Chrandrasekhar, the Indian-born American astrophysicist, that Horace W. Babcock (1947a, b) of Mount Wilson Observatory had recently determined but not yet published the magnetic field of a rapidly rotating star, 78 Virginis, using the Zeeman effect. Babcock was also aware of the relevance of his observation for Schuster's theory. To Blackett's delight, the ratio of magnetic moment to angular momentum of 78 Virginis closely matched that of Earth and Sun.

Blackett presented his idea in November 1946 at the University of Manchester, and in May 1947 to the Royal Society coincident with its publication. Runcorn recalled Blackett's eagerness to publish.

Blackett was asked to give a talk at the Royal Society in London about his work. He wanted to publish it in the *Proceedings of the Royal Society*. He thought at this time that it was a very hot topic. The Royal Society didn't give any indication that they would publish it quickly so he had it published in *Nature*. He was very cross with Robinson, the President of the Royal Society, for refusing to expedite publication in the *Proceedings*.

(*Runcorn, August 1984 interview with author; last sentence added during August 1993 interview with author*)

Initially Runcorn had wanted to go to Manchester partly to study cosmic rays, and during his first term he helped G. D. Rochester and C. C. Butler with their cloud chamber (Butler *et al.*, 1947). Within a month after his arrival, he heard Blackett's presentation, which got him thinking.

I became very interested. So after this meeting I discussed with him whether there were any experiments to do. The obvious one was, of course, to rotate a mass in the laboratory and see if it had a magnetic field. We did talk about that, and eventually he did an experiment related to that idea.

(*Runcorn, 1984 interview with author*)

Runcorn wanted an independent test of Blackett's theory, and he soon learned of a way not only to test, but also to compare Blackett's theory with its chief rival, the self-exciting dynamo. (Nye (1999) and Nye (2004) give fuller accounts of Blackett's work.[2])

## 1.4 Elsasser develops a self-exciting dynamo in Earth's core as the source of the geomagnetic field

It struck me then that if one assumed the metallic core of the Earth to be in convective motion [one could] account in a qualitative way for the remarkable phenomenon of the geomagnetic secular variation, with its unusual time scale ... These studies [on secular variation] were interrupted by the War and it was not until right after the War that I was able to put the magnetohydrodynamics of a spherical conductor into mathematical form. This seemed at last to permit a quantitative approach to the secular variation ... It left the main problem, the possibility of a dynamo theory, still unsolved. I then realized that I had overlooked a fundamental mathematical fact, namely, the existence of two sets of aperiodic modes of the sphere, the poloidal and toroidal modes ... The discovery that there could be a toroidal field in the Earth's core at once led to the well-known amplifying mechanism of this field by non-uniform rotation and thus to the main step in the dynamo model, suggesting also that the dynamo mechanism consists in a playing back and forth of magnetic fields ... It appeared early in my studies that the Coriolis force must be the agent which orders the amplificatory processes of the magnetic field relative to the axis of rotation of the fluid. Apart from much geophysical evidence, a qualitative but strong confirmation of this conclusion has come from the fundamental observations of Babcock on magnetic stars.

*(Elsasser, 1959: 93; my bracketed additions)*

While Blackett was reinventing the fundamental theory for the origin of the geomagnetic field, Elsasser was busy reviving what became its chief competitor, the self-exciting dynamo, which locates the source of the field within the core. Common to all versions is the idea that the metallic and fluid outer core generates an electric field, which through a coupling process creates Earth's dipole magnetic field. The self-exciting dynamo was first invoked by Sir Joseph Larmor (1919) as an explanation of the magnetic fields of Sun and Earth. He cautioned that his theory would require a highly conducting liquid region deep within Earth for which at the time there was some support from seismology (Oldham, 1906). In 1936 the Danish seismologist Inge Lehmann discovered the solid inner core (Brush, 1996). Larmor's theory caught the attention of T. G. Cowling, an applied mathematician then at University College, Swansea in Wales, who showed (1934: 44) that Larmor's self-exciting dynamo theory failed because no dynamo could be maintained from an axially symmetrical field.

Elsasser (1939) attributed the main field to thermoelectric currents in the metallic core; currents arose from temperature variations brought about by thermal convection maintained by radioactive impurities in the core. Of fundamental importance to the interpretation of paleomagnetic results, he recognized the Coriolis force as responsible for the general correlation between the axis of rotation and that of the main geomagnetic field, and assigned a slight asymmetry to the thermoelectric currents to account for secular variation.

After World War II, Elsasser returned to the problem (Brush, 1996a). In October of 1945 he submitted Part I of a tripartite article (Elsasser, 1946a,b; 1947) in which he described a self-exciting dynamo in Earth's core that accounted for the main field and its secular variation. Bullard (1949: 434) characterized it as a work of "great generality and elegance." At first Elsasser was unable adequately to account for the coupling between electric and magnetic fields. However, he gave a solution in Part III, which appeared in timely fashion six months after the publication of Blackett's fundamental theory. Elsasser introduced his coupling mechanism in this way:

The analysis of Part I and Part II has led to an interpretation of the geomagnetic secular variation in terms of interactions between fluid motions in the earth's core that are the sources of the magnetic field. This analysis suffers from the shortcoming that the current modes which give rise to a magnetic field outside the metallic sphere do not represent a complete set of solutions of the electromagnetic field equations. There exists a second set of solutions, representing modes of the electric type, whose magnetic field is confined to the interior of the conducting sphere. In the preceding parts these models have been disregarded on the assumption that they cannot be excited. It has been found however, that in the theory of inductive coupling by fluid motion there appear definite couplings between the two types of modes and that, therefore, the electric modes are an integral part of the field as described by this theory. It will appear in the course of this paper that from this viewpoint inductive coupling between the magnetic and electric modes is by far the most important feature of the earth's magnetic field.

*(Elsasser, 1947: 821)*

Elsasser postulated that coupling between electric and magnetic fields within the core provided the needed feedback mechanism to produce the external dipole, main or poloidal field. Because of non-uniform rotation of the conducting fluid core in the presence of a poloidal field, a toroidal field is generated that is entirely internal, the lines of force are parallel to lines of latitude; the field is confined within Earth and cannot be detected outside. Elsasser estimated the strength of the toroidal field to be about ten times that of the external poloidal field.

He speculated about the power source needed to maintain the core motions. He proposed (1947: 831) that the "power is a by-product of the change in Earth's speed of rotation caused by the lunar tide." Noting that although much of the angular momentum released by a decrease in the rate of rotation is either transferred to the Moon as it recedes from Earth or is dissipated by tidal friction in the oceans, neither effect completely exhausts the energy released; enough energy is left over to drive motions within the outer core.

Although Blackett discussed Elsasser's work, he focused on Larmor's older, self-exciting dynamo model and on a new model proposed by the Soviet scientist J. Frenkel. Frenkel (1945) attributed the movement of the metallic core to the action of convection currents, and developed an axially symmetrical self-exciting dynamo model. Blackett (1947: 659) dismissed Frenkel's account, referring to Cowling's theorem, which apparently disallowed axially symmetrical models.

### 1.5 Runcorn and colleagues carry out the mine experiment and discriminate between fundamental and core theories

On May 15, 1947, Runcorn traveled to London with Blackett, who presented his theory to the Royal Society. Bullard and Sydney Chapman attended. Chapman was professor of natural philosophy at Oxford, and co-author of *Geomagnetism*, then the standard work. Bullard, already interested in geomagnetism, later developed a self-exciting dynamo model based on Elsasser's work. Runcorn recalled:

At this meeting Bullard and Chapman were there, and Bullard threw out an idea during the discussion that, possibly, Blackett's theory might give a different variation of the geomagnetic field with depth from other theories – Blackett's theory involved the whole Earth in generating its magnetic field while other theories placed its source in the core.[3]

*(Interview with author, 1984; revised 1993 interview)*

Runcorn described the differences Bullard had in mind:

Whereas on core theories both the horizontal and vertical intensities increase with depth according to an inverse cube law, on a distributed theory such as the one recently put forward by Blackett we find, with reasonable assumptions, that while the vertical intensity should increase for small depths as an inverse cube law, the horizontal intensity should decrease.

*(Runcorn, 1948: 373)*

Bullard had indicated how to test Blackett's theory. While Blackett was designing and constructing his highly sensitive magnetometer and carrying out his laboratory experiment, he could test Blackett's theory and earn his Ph.D. by descending into mines with magnetometers to determine how the intensity of the field changed with depth; did it do so in accordance with Blackett's or Elsasser's theory? Formulating concrete predictions was not easy. Runcorn recalled that at first he and Blackett "couldn't make much of it," but with Blackett's encouragement, he worked out the expected changes and Blackett sent them to Chapman to check.

When we got back [to Manchester], of course, I talked to Blackett about the possibility of an experiment to test the theory by going down in mines with magnetometers, and I undertook to try and work out what one should expect. What one should expect is not very obvious from the very speculative idea that Blackett had talked about. But I did calculations, which Blackett sent to Sydney Chapman. Sydney Chapman wrote back and said that he thought I was wrong, and he would do them himself. He did them himself by a different method involving vector potentials. So Blackett said that I should go to Oxford to see Chapman about this vital question. In the end, Chapman agreed that my method was essentially correct, if unorthodox, though I had made a slight approximation: it concerned how density should be brought into the calculations. By this time I had got really interested in the idea of doing the experiment. I remember Chapman inviting me to lunch. He was a very austere person – indeed, rather frightening to a young person. I always remember Chapman saying, "Well your calculation is not exact." And, I said, "Well, I make an approximation because, there is no chance that we can go down to the center of the Earth with this formula. It is just a question of what the first few kilometers will give." We discussed the difference between our two formulas, and I remember making the terribly brash statement to him, "Well, the trouble is that you are thinking of this problem as a mathematician and I am thinking of it as a physicist." I always remember Chapman's gentle reply: "Well you know I sometimes think of myself as a physicist as well!" Anyway, as he said goodbye he said that we now understand each other. After that we became very friendly, and he was very helpful.

*(1984 interview author; revised 1993; my bracketed addition)*

With Chapman's help, Runcorn had turned Bullard's idea into a testable prediction. What is more, he had found himself able to hold his own in debate with the top workers in the field. Blackett reported Runcorn's results (Runcorn, 1948) to the Physical Society of London in April 1948.

Deriving predictions, however, was not the only obstacle; securing reliable data was also not easy. Success depended on determining the changes in intensity with depth of the horizontal and vertical components of the geomagnetic field freed of local anomalies due to geology or to human activities. As Runcorn *et al.* put it:

> The essential problem of the experiment is to find conditions in which measured differences of the geomagnetic components between a point on the Earth's surface and a point underground may be attributed to the main field and not to magnetic anomalies arising locally.

> *(Runcorn* et al., *1950: 784)*

Runcorn chose deep coal mines in Lancashire where the rocks were too weakly magnetized to affect the geomagnetic field. Recruiting undergraduate students to take the many measurements, he launched the mine experiments near the end of 1947. He presented the first results orally to the Royal Astronomical Society on February 27, 1948 (Chapman, 1948a), and with colleagues submitted the final paper in May 1951 (Runcorn *et al.*, 1951). The results from the first mine favored the fundamental, not the core theory; however, the mine "was too near the outskirts of the town of Leigh for a surface survey of adequate size to be made." His team eventually obtained reliable results from five other mines avoiding magnetic disturbances from towns, and concluded: "the experiments must be regarded as decisive evidence against a fundamental origin of the main geomagnetic field" (Runcorn *et al.*, 1951: 148, 150).

Runcorn was not the only one to encounter difficulties with local magnetic anomalies when doing such mine experiments. Anton Hales, who had years before examined Holmes' hypothesis of convection currents (I, §5.6), and D. I. (Ian) Gough, from the Bernard Price Institute of Geophysical Research at the University of Witwatersrand in Johannesburg, performed a similar experiment. At first they, like Runcorn, thought their results favored fundamental rather than core theories (Hales and Gough, 1947). They corresponded with Runcorn, and questioned whether his formula was strictly applicable to their situation, because their surface measurements were done near Johannesburg, 5200 feet (1585 m) above Earth's mean surface level, whereas Runcorn's formula applies strictly to depths below that level. They also suggested that there might be some unknown geological effects. Chapman examined their findings. Besides pointing out a further difficulty with Runcorn's formula and suggesting a more general one, for which he was thanked by Runcorn, Chapman (1948b), thinking like a physicist, correctly noted that more work was needed in mines less disturbed magnetically than the mine chosen by Hales and Gough.[4] Hales and Gough returned to the mine, made additional measurements, and realized, by July 1949, that their results had been affected by the abundance of nearby strongly magnetized intrusive igneous rock; Hales and Gough's results "gave no useful information with regard to the radial variation of the earth's field" (Runcorn *et al.*, 1951: 148).

## 1.6  Blackett and Runcorn become interested in paleomagnetism; Runcorn accepts a position at the University of Cambridge

Both Blackett and I went to a meeting of the Royal Astronomical Society in which Professor Bruckshaw of Imperial College discussed ... reversed magnetization. This was a reversed magnetization of Cleveland dykes of Tertiary age in northern England. He did magnetic surveys across them, and demonstrated that they were reversely magnetized. I think that this was Blackett and my first exposure to paleomagnetism.

*(Runcorn, 1984 interview with author)*[5]

Much went on at this meeting (February 27, 1948); Runcorn presented his first, erroneous, mine results, and Bullard talked about secular variation, attributing it to changes in electric currents induced by the movement of the conducting core material through the main magnetic field. Although Bullard did not offer his account of the main field until later that year, he located its origin within the core.[6] Bruckshaw followed.

J. McG. Bruckshaw was reader in geophysics at the Royal School of Mines, Imperial College, London. He described the work that he and E. I. Robertson, a research student from New Zealand, had begun during the summer of 1946 on a system of Early Tertiary dykes, extending southeastward from the Isle of Mull in western Scotland to the northeast coast of England. These dolerite dykes had been intruded into much older strata, and had since remained undisturbed. They made magnetic surveys across them, and much to their surprise, the dykes were magnetized in a direction nearly opposite to that of the present geomagnetic field. Noting that such rocks acquire a "residual magnetism" as they cool down "through the Curie temperature of the magnetic material within the dyke" in a direction parallel to the ambient magnetic field, they proposed:

Since there has been no significant earth movement in this region during the 30 million years this dyke system has existed [now known to be ~50 million years], the Earth's field in the area at the time of cooling through the transition temperature would appear to have been approximately in opposition to its present direction.

*(Bruckshaw and Robertson, 1949: 316; my bracketed addition)*

They continued:

Thus such characteristics (igneous intrusions of inverted polarity) are fairly common in both surface distribution and in age. There can be no doubt of the changed direction of the magnetic field necessary to produce the observed polarization. Whether inverted fields were widespread over the Earth's surface in the past, or whether they were an abnormal, but local, state associated with the conditions necessary for the invasion of the crust by molten magmas cannot yet be decided.

*(Bruckshaw and Robertson, 1949: 318)*

In his summary of the meeting, Chapman speculated about the possible causes of inverted magnetizations – many were cited in *Geomagnetism*, his book with Bartels

(1940) – and although Chapman himself expressed little faith in any of them, he described an open and vigorous discussion. There was the possibility that the geomagnetic field as a whole had once been of opposite polarity, which "would, of course, be quite inconsistent with any fundamental theory associating magnetization with rotation in the case of large rotating masses, such as Prof. P. M. S. Blackett has advocated" (Chapman, 1948b: 464); rotational theory would require Earth to reverse its direction of rotation. Perhaps the inverted polarity of the dykes had been caused by a local inversion of the geomagnetic field due to abnormal secular variation. But Chapman noted, "It is difficult also to see how Dr. Bullard's explanation of the secular variation could account for a local reversal of the field." Or, if the dykes were located near the equator, as the Pilansberg dykes of South Africa are today, a reversal of the vertical component of their magnetization might be explained by postulating a slight shift in the position of the magnetic equator. However, a slight shifting in the magnetic equator could not account for the inverted mid-latitude samples such as those discovered by Bruckshaw and Robertson. Chapman even considered continental drift, but thought it "unattractive."

A very extensive continental drift carrying land masses across the equator is another possible but unattractive hypothesis that would obviate the conclusion that the earth's field as a whole was once of opposite polarity.

*(Chapman, 1948b: 464)*

Although, according to Runcorn (1984 interview with author), Blackett "was rather upset" about the possibility of field reversals, and everyone "was quite agnostic as to whether or not it was the Earth's field that had imprinted itself on these rocks," they immediately saw the necessity of further paleomagnetic studies. Blackett recalled:

Such [paleomagnetic] information would not only be of great importance for its own sake but would be of immense value in an attempt to understand the physical mechanism giving rise to the field. For one of the main difficulties of finding a plausible theory of the origin of the earth's field lies in the fact that we have direct measurements of the earth's magnetism by means of the compass needle only over the last 400 years, compared with the more than 500 million years of reliable geological history ... Without the study of rock magnetism we had no possibility of knowing whether the field might not have been vastly different in the distant past, perhaps, a thousand or more times greater or smaller. Definite facts about the past history of the earth's magnetic field were essential in order to have a reliable foundation for theorizing about the mechanism of its origin.

*(Blackett, 1956: 5; my bracketed addition)*

By 1948 Blackett and Runcorn, although themselves not yet involved, were well aware of paleomagnetism and that it was needed for an understanding of the origin of the geomagnetic field and its secular variation. Indeed, concurrently with the mine experiments, Runcorn had begun work on secular variation. Their initial attraction to paleomagnetism had everything to do with the origin of the geomagnetic field and nothing whatsoever to do with continental drift.

In 1948 Runcorn was elected to a fellowship at Gonville and Caius, his undergraduate college at Cambridge. He debated about whether to take it up. He had been promoted to lecturer in physics at Manchester, and as he (1984 interview with author) put it, "A fellowship was high in prestige but it didn't pay very much." So, he decided to remain in Manchester, at least for another year, continuing his mine experiments and teaching geophysics, thereby broadening his knowledge of the subject. The following year, the Department of Geodesy and Geophysics at Cambridge was permitted to replace Bullard who had left in 1948. Runcorn was asked, and with the extra stipend that came with the university appointment, he, noting his unplanned good fortune, added, "I couldn't refuse and returned to Cambridge in January of 1950" as Assistant Director of Research with a fellowship at Caius (Runcorn, 1984 interview with author). Serendipitously when he arrived, Jan Hospers was there as a new graduate student already planning to begin paleo-magnetic work that summer in Iceland (§1.12).

### 1.7 Work at the Carnegie Institution in Washington and the case for a geomagnetic field without gross changes

According to Blackett (1956: 5), his interest in paleomagnetism was aroused by reading "Pre-history of the Earth's magnetic field," a paper by E. A. Johnson, Thomas Murphy, and O. E. Torreson (1948), of the Department of Terrestrial Magnetism at the Carnegie Institution. The authors stated their wish to test Blackett's fundamental and Elsasser's core theories. At the time, the physicist Merle A. Tuve was director of the department, where paleomagnetic research had begun in 1937, lapsed during World War II, and resumed in 1946. This was shortly before Runcorn arrived in Manchester.

Using a highly sensitive spinner magnetometer, developed by Johnson, Carnegie workers measured many different kinds of sedimentary samples including glacial varved clays, folded and undisturbed sandstones from across the United States, and cores of ocean floor sediments from the Pacific. Working on increasingly older samples, and finding them commonly magnetized along the present field, they had, by 1949, concluded that the average direction of Earth's magnetic field had remained substantially constant for the past 400 million years. In addition, they found certain Silurian sedimentary samples that had magnetizations oblique to the meridian, and inclined upward above the horizontal, which they took to be reversed polarity. Fully confident that these magnetizations were stable, they began to wonder if they indicated that the Earth's field had in the past been reversed in polarity and if continental drift had occurred. They rejected both, and for five years they ceased to consider them to be reasonable possibilities. In the meantime, as we shall see, British workers had discovered good evidence that both had occurred. I shall consider at some length this remarkable clash of results and ideas.

Carnegie workers set out to study the record of the geomagnetic field in sediment-ary rocks and they quickly asked the fundamental questions: Did sedimentary rocks retain their directions of magnetization over geological time? Were they magnetically stable? Over the following decade, paleomagnetists developed two sorts of answers to these questions. First they developed *field stability tests*, which made use of some readily observable geological phenomena. To this end, Graham (1949), who became the most prominent paleomagnetist at the Carnegie Institution, developed the important fold and conglomerate field tests. About nine years later, paleomagnetists developed *laboratory stability tests* based on various demagnetization procedures, which I shall introduce in Chapter 5.

John W. Graham (1918–71) was born in Boston.[7] His family moved to Albany, New York, where he spent his boyhood. He became attracted to the Earth sciences during his early teens, spending much time at the state science museum. Graham received an A.B. degree in geology from Johns Hopkins University in 1940. During his senior year he studied inorganic chemistry and spent his first year in graduate school doing chemistry. During World War II, he served as a commissioned naval officer working on radar, proximity fuses, and guided missiles. After the war, he returned to graduate school at Johns Hopkins to do geology. In his second year he obtained a fellowship at the Carnegie Institution to work at the Department of Terrestrial Magnetism for his Ph.D., which he was awarded in 1949. In his seminal paper (Graham, 1949), "The stability and significance of magnetism in sedimentary rocks," he described his two field tests. Despite the paper's significance, he almost did not receive his Ph.D. As Richard Doell remarked in his memorial:

In view of the presently recognized importance and success of this work, it may be surprising to some to learn that his thesis referees divided on awarding his degree, and only a favorable note by an outside examiner led to his being awarded the Ph.D. in 1949.

*(Doell, 1974: 106)*

Graham remained at Carnegie for a number of years, but, as support for paleomagnet-ism lessened, he left in 1957 to join the staff at Woods Hole Oceanographic Institution. Tuve had come to think that paleomagnetism had little to offer. This was not a personal matter. Glen (1982: 119) reports that Tuve told Doell when at MIT in 1957 that "if he wanted to remain on the staff of MIT he should eschew paleomagnetism and 'get into some serious geophysics, like seismology or gravity'." Tuve provided none of the encouragement that Runcorn and Blackett gave their associates. But, as we shall see, Graham did not give Tuve much cause for optimism (Le Grand, 1989).

The first post-war study by the Carnegie group was of Pleistocene glacial varved clays from New England, and it was this that caught Blackett's eye. It had a simple message.

The data presented in this paper indicate that in the case of the glacial clays the polarization of these clays has remained constant in direction and intensity since these clays were deposited 10 000 to 20 000 years ago.

*(Johnson et al., 1948: 371)*

They also examined five cores of ocean sediments that had been collected by J. L. Hough during the 1946–7 United States Antarctic Expedition. Finding no appreciable variation within them, they

concluded that one million years ago the Earth had a magnetic field similar in direction and size to the present field and that the south pole was in approximately the same position as at the present time.

*(Johnson et al., 1948: 370)*

The following summer they collected samples of older flat-lying sedimentary deposits from the western United States to extend "knowledge of the changes of Earth's magnetic field farther back in geologic time" (Tuve, 1949: 58).

The data here discussed, though representing a very small exploratory sampling of the sedimentary rocks of the earth's crust, are consistent with the idea that, for the past 50 million years or so, the polarity of the earth's magnetic field has been the same as now, and the magnetic axis has had an average orientation that coincides with the earth's geographic axis.

*(Torreson et al., 1949b: 129)*

They maintained that the geomagnetic field had generally remained unchanged for 50 million years, but that during the Tertiary the declination of the field had been axial and had moreover not reversed in polarity, flatly contradicting the results of Bruckshaw and Robertson in Britain (§1.6).

They [our results] do not support the contention, recently reiterated on the basis of measurements of the polarizations of 50-million year old igneous dykes in England, that the earth's field in the past has had a reverse polarity.

*(Torreson et al., 1949a: 209; my bracketed addition)*

### 1.8 Graham develops field tests of stability

Field tests of stability make use of some geological (or archeological) feature and were first developed by Folgerhaiter (1899), David (1904), and Brunhes (1906) (reviewed in Irving, 1964: 6–7). Graham extended their ideas, and in the geological context, formulated them as the bedding tilt (or fold) and conglomerate tests. In the conglomerate test, the directions of magnetization of large clasts are measured, and if they are all the same, their magnetizations and that of nearby parent rock were acquired after deposition of the conglomerate. However, if clast directions are random, their magnetization and likely that of their parent predate the formation of the conglomerate. For the tilt (fold) test, if magnetization directions are all parallel irrespective of the orientation of the strata, they postdate deformation; if they are the same relative to the bedding plane, then they predate folding. In the absence of laboratory stability test in the early years, field stability tests were of fundamental importance in the paleomagnetic contribution to the forthcoming mobilism debate attack on the global test of continental drift. As the global study progressed, three other tests were developed: the *igneous baked contact test*, the *reversal test*, and the *consistency test*. These will be explained shortly.

Although Torreson, Murphy, and Graham (1949a: 209) mentioned both tests, it was Graham who was responsible for them. He developed the conglomerate test in the Miocene Ellensburg Formation near Selah, Washington, USA, comprising flat-lying, fine-grained lake sediments with interbedded conglomerate beds. The magnetization directions of the clasts were scattered, in contrast to the uniform directions in fine-grained strata. He argued (1949: 145) that the magnetization of the clast, and hence that of the interbedded strata, had remained unchanged since its formation. Several conglomerate tests in Paleozoic rocks were negative.

Graham attempted four fold tests: in slumps in Pleistocene varved clays at Middletown, Connecticut; in coarse-grained, gray, Eocene sandstones near Gardner, Colorado; in Late Paleozoic red sandstones near Glenwood Springs, Colorado; and in Silurian gray-colored siderites (now called ferroan dolomite) from the Rose Hill Formation near Pinto, Maryland. The Pleistocene clays were stable, the directions when corrected for deformation, grouped together, differing slightly from the present field. The Eocene rocks were unstable. His study of the Late Paleozoic rocks was inconclusive; their magnetization lay close to the strike of bedding and hence did not respond significantly to bedding correction. Most interesting was the Silurian Rose Hill Formation, which has since become a standard illustration, and to which Graham repeatedly returned. Graham found that the twenty samples across a fold had scattered directions which became well grouped when corrected for folding (Figure 1.1). Graham concluded that their magnetizations had been acquired before they were folded near the end of the Paleozoic Era (Appalachian Orogeny), and had remained unchanged (stable) in direction.

The close relation between the bedding orientation and directions of magnetization is presented as satisfactory evidence that this rock has retained its direction of magnetization *since a time prior to its folding* 200 million years ago.

*(Graham, 1949: 151; emphasis added)*

Although Graham had clearly demonstrated that the magnetization of the Rose Hill Formation was stable, he was puzzled because the magnetizations had upward inclination which he regarded as reversed polarity. One possibility, he thought, was that the rock had been struck by lightning before folding. To check, Graham measured the magnetizations of the same formation at Cacapon Mountain, some thirty to fifty miles away; they agreed roughly with Pinto, so lightning was an unlikely cause. Graham found his Pinto results "disconcerting"; his colleagues wondered if they could be explained by unusually large secular variation.

In spite of these inconsistencies, a tendency towards uniformity of polarizations in the bulk of the Rose Hill formation over long distances is clearly indicated. It might be remarked that, although this reversal of the polarizations is disconcerting, it may not be an unusual feature of old rocks, particularly for limited periods of time. My physicist colleagues at the Department

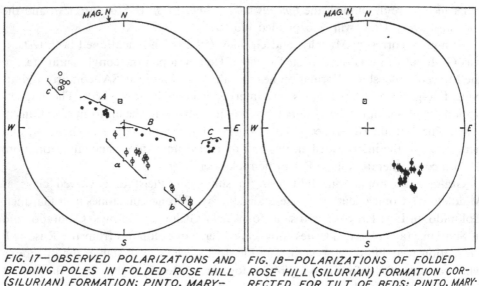

FIG. 17—OBSERVED POLARIZATIONS AND
BEDDING POLES IN FOLDED ROSE HILL
(SILURIAN) FORMATION; PINTO, MARY-
LAND, NOVEMBER 23, 1948
•=BEDDING POLES
♦=SOUTH SEEKING POLARIZATIONS
○=NORTH SEEKING POLARIZATIONS
▣=EARTH'S PRESENT FIELD

FIG. 18—POLARIZATIONS OF FOLDED
ROSE HILL (SILURIAN) FORMATION COR-
RECTED FOR TILT OF BEDS; PINTO, MARY-
LAND, NOVEMBER 23, 1948
♦=SOUTH SEEKING POLARIZATIONS (CORRECTED)
▣=EARTH'S PRESENT FIELD

Figure 1.1 Graham's Figures 17 and 18 (1949: 150) showing bedding poles and uncorrected
and corrected natural remanent magnetization in folded Rose Hill (Silurian) Formation, Pinto,
Maryland.

of Terrestrial Magnetism expect to find evidences, if the history of the earth's magnetic field
can ever be well established, for stronger secular-variation foci (disturbances of the general
magnetic field of the earth over local regions) in remote epochs when the interior of the earth
may have been hotter and the disturbing current-systems nearer the surface. These foci at
present are relatively localized and move slowly over the surface of the earth; the corresponding current-systems now appear to be roughly at 1200 km depth.

*(Graham, 1949: 153)*

These apparently reversed magnetizations disturbed Graham and his colleagues. He
continued:

The approximate agreement of the reconstructed compass-direction over a distance of roughly
50 miles is striking and difficult to account for on an accidental basis. With my physicist
colleagues at the Department of Terrestrial Magnetism, various efforts to reconstruct electrical
current systems in or near these folds to account for the observations have led to what they
regard as completely implausible hypothetical conditions. Agreement has been reached in our
discussions that even the present evidence ... is exceedingly difficult to understand on any basis
other than the assumption that the beds were magnetized either at the time of deposition or by
the growth of magnetizable minerals in a regionally uniform magnetic field prior to folding.
No mechanism is known ... by which a uniform magnetic field of this kind over a distance as

great as 50 miles can be produced by electrical current systems (for example, associated with volcanic action). The reconstructed residual-polarization is, therefore, accepted as a measure of the direction of the earth's magnetic field in this locality at some time prior to the Permian and very probably at the time of deposition.

*(Graham, 1949: 154)*

Graham thought his fold test was an excellent way of determining if rocks were reliable indicators of the past direction of the geomagnetic field. However, if his test truly had worked, then it seemed to him that Earth's magnetic field had once been reversed, and Graham, his colleagues, and Tuve, the director, were not pleased with that. Other explanations were continental drift and polar wandering, but these were not even mentioned at the time as possible interpretations. It is ironic that because of continental drift, North America having crossed the equator in the interim, the polarity of the Rose Hill Formation is actually normal and not reversed.[8]

## 1.9 Graham and others at the Carnegie Institution abandon the fold test

Two years after describing his Maryland fold test, Graham changed his mind and came to regard it as a mixed blessing. Throughout 1949 and 1950, Graham and others at the Carnegie Institution, hoping to settle the uncertainty about the orientation of the Paleozoic geomagnetic field, undertook an extensive survey sampling along the Appalachians from Alabama to Maine from rocks ranging in age from Ordovician through Permian. They again found what they regarded as reversely magnetized samples, but always from deformed beds in the Appalachian Orogenic Belt, while samples from contemporaneous flat-lying, undeformed beds outside the orogen had normal magnetizations approximately aligned close to the present field. Graham and Torreson presented their results at the spring 1951 AGU meeting. They claimed that it was the magnetizations from the flat-lying beds that truly recorded the ancient field while those of folded strata from within the orogenic belt (notably from the Rose Hill Formation) did not.

Contrasting Magnetizations of Flat-Lying and Folded Paleozoic Sediments – The effort has been continued to trace the history of the Earth's magnetic field in geologic time by making determinations of the directions of permanent magnetism possessed by many sediments. These magnetizations apparently were acquired originally at the time of deposition and are retained faithfully for millions of years. Determinations on 170 rock samples taken from horizontal strata at 14 exposures of Ordovician through Permian age are contrasted with a comparable number of determinations on samples, chiefly of Silurian age, from folded [Rose Hill Formations from] Appalachian sections. The data from east and west flanks of anticlines differ from each other and differ from the observation in the flat-lying beds. *These differences are attributed to deformation.* Inverse magnetizations are present in the deformed geosynclinal sections but have not been found in contemporaneous flat-lying beds. It is believed that these anomalous magnetizations resulted from the physical and chemical conditions peculiar to the geosynclinal sections, and that they do not have any direct bearing on the past directions of the Earth's magnetic field. It is proposed that the Earth's field in North America has remained substantially constant for the past 400 million years.

          *(Graham and Torreson, 1951: 336; my bracketed addition; emphasis added)*

So the flat-lying strata gave what Graham and Torreson regarded as "good" results and the folded rocks gave "anomalous" results. But what about the positive fold test? They abandoned it. The test no longer applied to the Rose Hill Formation, the very formation which two years before had served (and still does) as a most convincing example. They no longer thought that its pre-folding magnetizations necessarily recorded anything about the orientation of the Paleozoic geomagnetic field. Instead they maintained that certain unspecified chemical and physical processes during the Appalachian orogenesis had magnetized the folded sedimentary rocks in scattered directions, the fanning of directions due not to folding but to remagnetization.

     In his report of the activities of the Department of Terrestrial Magnetism for 1948–9, Tuve, its director, had accepted Graham's fold test, arguing that the magnetization of the Rose Hill fold was stable.

*Stability of magnetization.* Workers in the field of rock magnetism have long recognized the necessity for demonstrating that rocks are capable of retaining their initial directions of magnetization from the time of their origin to the present ... It is possible in many localities to locate rock exposures where sedimentary beds were long ago squeezed into contorted arches by mountain-building forces. Observations in these squeezed and folded beds have yielded knowledge of the permanence of the directions of magnetization. One particular series of observations on Silurian rocks (350 million years old) [in the Rose Hill Formation] that are well exposed at Hancock and Pinto, Maryland is of special interest ... The directions [of magnetization] were found to vary systematically throughout the fold in a manner that has an intimate relation to the attitude, or position in space, of each part of the bed where an observation was made. The closeness of this relation is clearly brought out when a graphic reconstruction is made to restore the bed to its initial flat condition. Although the directions of magnetization in the fold are fanned out, and show differences in direction of as much as 127°, when the bed is made flat by a graphical transformation the magnetizations are all brought into general alignment so that these differences are usually less than 15°.

          *(Tuve, 1949: 58–60; my bracketed additions)*

In the next annual report, Tuve made no mention of the fold test. Although he did not refer to the previous work at Hancock and Pinto by name, he noted that work there had ostensibly revealed reversed magnetizations, and that this

striking observation stirred the Department's investigators into an extension of these measurements throughout the Appalachians. Further work was carried out in about eight localities from Birmingham, Alabama, to Eastport, Maine, and it was found that there were wide fluctuations in the remanent magnetizations. It was thus clear that no immediate deduction could be made regarding the direction of the earth's magnetic field at the time of deposition of these Silurian deposits some 350 million years ago.

          *(Tuve, 1950: 62)*

Tuve (1950) then turned to results from flat-lying Silurian strata at Clinton, New York. Their magnetizations were approximately along the present field, which he took to indicate the true orientation of the Silurian field. He dismissed the oblique and apparently reversed directions as "anomalous" and endorsed the opinion of Graham and Torreson that they had been caused as a consequence of deformation.

One clue to this puzzle [about the wide fluctuations in remanent magnetizations] is that a Silurian deposit at Clinton, New York has residual magnetization in a direction consistent with that of the earth's present magnetic field. The Clinton Silurian deposits have never been involved in mountain-building deformations, in contrast with the sediments at the other localities, which are in the Appalachian Mountains. It is possible that the extreme folding processes that have gone on in the mountains may have in some way affected the magnetic polarization at the time of folding, and that since the deformation the polarizations have remained unchanged.

*(Tuve, 1950: 62–63; my bracketed addition)*

Tuve then went one step further, arguing that a certain fossil established the contemporaneity of the two sequences.

At both sites [two sites in West Virginia where the beds were folded and the magnetizations reversed but enclosed above and below by beds having normal magnetizations] there is found throughout the whole series of beds a rare fossil ostracod, *Matigobolibina typus*. This fossil is considered to be a good time marker because of its sudden appearance and disappearance, and it is probable that it existed during only a small fraction of geologic time, perhaps 100 000 years. The flat-lying deposits at Clinton, New York, bearing the same unusual fossil and thus presumably of the same age as the West Virginia deposits, show no anomalous magnetization.

*(Tuve, 1950: 63; my bracketed addition)*

So Tuve, the director, like the workers themselves, Graham and Torreson (1951), argued that the flat-lying and folded beds were contemporaneous and only the former recorded the ancient field; the folded beds in West Virginia were, by implication, "anomalous." Geomagnetic field reversals and drifting continents could be dismissed.

These facts lead us to believe that the most straightforward hypotheses for the observed anomalous magnetization – a reversal of the earth's magnetic field or a drifting of the continents – are not satisfactory.

*(Tuve, 1950: 63)*

There was no doubt about Tuve's dissatisfaction with mobilism. I also believe that in 1950 Graham himself found continental drift and polar wander unpalatable; had he considered them, he could have saved his fold test and avoided field reversals. However, his attitude toward drift did not become explicit until the middle 1950s when, as I shall describe, he raised contentious unreliability difficulties with the paleomagnetic case for continental drift (§7.4). What is remarkable about this episode is the unanimity among Graham, Torreson, and Tuve. There was no dissent within the Institution and no one else in the United States presented alternatives; the Carnegie group unanimously set aside both field reversals and continental drift, and there was no one in the United States to argue with them. In their eyes, what they regarded as reversed samples from Maryland and West Virginia were the exception not the rule, and hence they provided no interpretable record of the Paleozoic geomagnetic field. Confident in their belief, they saw no reason to pursue Graham's earlier question: "Will the study of rock magnetism be useful in throwing light on questions regarding large-scale movements of the crust, such as continental drift and polar migration?" (Graham, 1949: 160).

When Graham lost confidence in his fold test, he abandoned one of the basic procedures that British paleomagnetists would soon successfully adopt. It was not for another five years that Graham became interested in continental drift and polar wandering, long after the British groups had actively begun to promote them, and he still remained skeptical about reversals of the geomagnetic field.

## 1.10  Graham opts for self-reversals rather than field reversals

I was about to be thrown out of the Department for proposing the self-reversal.
*(John Graham Doell, 1974: 106)*

Graham was so unhappy about either continental drift or geomagnetic field reversals that he had essentially given up his fold test. He also wondered if rocks could become magnetized in the direction opposite to the ambient field (self-reversed). He could then maintain that the Rose Hill Formation had undergone a self-reversal before folding, and reclaim his fold test without accepting field reversals. In 1949, there appeared a paper of seminal importance by Louis Néel of the University of Grenoble on "traînage magnétique," and Graham wrote to him inquiring about the possibility of self-reversal.[9] This prompted Néel (1951) to propose four ways in which rocks could, theoretically, become self-reversed. Soon afterwards Seiya Uyeda, of the University of Tokyo, found a self-reversing rock. Uyeda and Takeshi Nagata, his supervisor, did not at the time know of Néel's paper (Nagata *et al.*, 1951), but Nagata (1952) later referred to it. Uyeda's rock was a dacite lava from Haruna volcano in Japan.

Graham (1952), impressed with Néel's work and the Japanese discovery of a naturally occurring self-reversing rock (albeit of uncommon type), assessed the relative merits of self-reversals and geomagnetic field reversals. While admitting that Hospers' very new study of Icelandic lavas (§1.13) offered support for the idea of field reversals, that they were next to impossible was "a matter of well-founded faith" for Graham.

For many decades there has existed a fascinating dilemma among students of the earth's magnetic field and students of rock magnetism. Intuitively, and often as a matter of well-founded faith, one regards the earth's magnetic field as an inherent property of such grandeur that reversals of its sense are difficult to grant, and yet there are so many examples of rocks having a magnetic polarity so close to the opposite of the present field that one's faith in this constancy may become shaken (see, for example, Hospers). It thus remains important to understand fully the mechanism by which these "inverse" rocks become magnetized.

*(Graham, 1952: 429; Graham's reference is to Hospers, 1951)*

He believed that geomagnetic reversals faced a serious, ostensibly fatal, theoretical difficulty (RS2). His concern was more serious than the fact that nobody knew how reversals of the field occurred; it was highly unlikely because of what was known about the geomagnetic field. Perhaps he felt that to suppose field reversals was to create a new problem that would be harder to solve than the original one (RS2); the cure worse than the disease. Furthermore, I suspect that he wrote to Néel because he thought the chance of understanding how field reversals occurred within the framework of what was then known about the field was remote. To him anything seemed preferable to global field reversals, even abandoning his fold test.

But Néel gave Graham the chance to salvage his fold test. Perhaps the Rose Hill Formation had undergone self-reversal. Néel had proposed that certain self-reversing rocks may have two iron mineral phases that interact magnetically, and his Silurian strata, Graham contended, showed behavior uncharacteristic of a single-phase system.

Néel has worked out in some detail various theoretical mechanisms by which the polarity of a rock during cooling may become the opposite of the impressed magnetic field. In these, it is required that there be two magnetic phases that interact on one another, whether it be on the scale of adjacent lattice planes, or on a scale of minutely intermingled different crystalline species ... In some Silurian sediments of the Appalachian mountains, stable inverse polarizations are to be found. Magnetization experiments, the details of which will be published later, show for these inverse rocks an anomalous behavior that is not characteristic of a single-phase system.

*(Graham, 1952: 430)*

Several months later, Graham began vacillating between self-reversals and a new phenomenon – short-term regional reversals or "excursions" as he called them. He gave a talk on August 1, 1952, at the Institute of Geophysics, University of California, which was not published until two years later (Graham, 1954: 215). The editor wanted publication "without revision," so Graham's comments reflect his changing views in 1952. At the close of his paper, Graham returned to the correlation of reversely magnetized folded strata of the Appalachian orogenic belt and the normally magnetized (and, like Tuve, he appealed to the same fossil ostracods) apparently contemporaneous flat-lying strata outside the orogenic belt, and tentatively suggested that the geomagnetic field might have experienced a very short period reversal over a relatively small area.

In the Appalachian Mountains, we have encountered a number of widely separated sections where many dozens of feet of Silurian strata with inverse magnetizations are interbedded with others of identical appearance having normal polarizations. Through these sections, there are a number of well-established faunal zones characterized by ostracods having a very restricted occurrence in the geologic time scale. Outside the deformed geosynclinal area, the flat-lying contemporaneous beds containing the same faunal assemblages have normal polarizations exclusively. Granting the stability of all these polarizations and taking them at their face value, these findings would seem to indicate that such reversals as the earth's field may have undergone in Silurian time took place over a relatively small area of the earth that was nevertheless still a large area when expressed in square miles; they took place very rapidly; they lasted for a short time; they took place without yielding transitional directions between the two senses; and, finally, they took place in a zone that later was destined to become first deeply buried by additional sedimentation and then later became folded and deformed during a mountain building orogeny. *In view of the complexity of this problem, it seems well not to believe that these findings prove a reversal of the earth's field during Silurian time.*

*(Graham, 1954: 221; emphasis added)*

He could now keep his fold test without invoking continental drift or field reversals. But he was not very confident about his excursions, however brief they might be.

The next year Graham speculated further about what he thought to be the reversed magnetizations of the Silurian sediments from the Rose Hill Formation. He (1953: 255) supposed that the Rose Hill strata had formed through the erosion of igneous rocks whose magnetic mineral fragments had the capacity to undergo self-reversal as proposed by Néel but at first they did not do so. Then, long after deposition the strata could self-reverse as a result of "an appropriate temporary change of chemical environment" adding:

A process of this general character appears to have taken place in some Silurian sediments of the Appalachian geosyncline, but not in sediments outside the geosyncline.

*(Graham, 1953: 255)*

Graham's speculations and actions were not the only ones possible. He could have argued that his fold test showed that rocks from the Rose Hill Formation were magnetically stable and invoked polar wandering or continental drift because the magnetizations' directions were strongly oblique to the present field. He could with good reason have argued on the basis of Néel's "traînage" paper (1949) that the flat-lying sedimentary strata had been remagnetized recently because their magnet-izations were roughly grouped not far from the present field. Instead of writing to Néel, he could have written to Bullard or Elsasser asking them if they could come up with a theoretical mechanism for geomagnetic field reversals. But to have done so would have meant flirting with field reversals and going against the prevailing fixist and anti-field-reversal orthodoxy at Carnegie and in North America in general. Graham was undoubtedly strongly influenced by Tuve and his geophysics colleagues who did not favor field reversals, polar wandering, or continental drift. Perhaps he felt pressured by them. If so, his abandonment of his fold test study as providing a meaningful determination of the Paleozoic geomagnetic field might have been a momentary loss of courage. Graham later told Doell, another

US paleomagnetist who resisted mobilism during the 1950s (§3.8, §8.2–§8.5), that he had little support at Carnegie.

[Graham] felt so strongly about such lack of support that he later left a fund at Johns Hopkins to award fellowships to scholars with "intriguing ideas not yet sufficiently formulated or developed to attract consideration by more conservative fund-granting agencies."

*(Doell, 1974: 106; my bracketed addition)*

However, I do not think Graham's refusal to act individually on the idea that paleomagnetism could be used to test continental drift and polar wandering was due entirely to pressure. I shall describe how (§7.4) Graham (1955) later became somewhat inclined toward polar wandering, but still rejected continental drift; how in 1957 he argued against Irving's defense of mobilism at the Toronto IUGG meeting (§8.12); how he seriously maintained in 1956 that removal of stress from rocks could alter their remanent magnetization rendering them unreliable indicators of the ancient geomagnetic field and useless for testing mobilism (§7.4). There was no need for anyone to tell or pressure him not to seriously entertain ideas about continental drift; he just seems to have avoided them like the plague.

This sad and inconclusive discussion at Carnegie about the Paleozoic geomagnetic field had begun in 1949 with the very successful Rose Hill fold test, which had held out bright prospects for obtaining accurate records of the ancient geomagnetic field. Graham's initial, simple interpretation based on a straight reading of the evidence had, by 1953, degenerated into a number of unresolved, complex possibilities, some of dubious relevance and none very clearly defined. Graham dithered, and Director Tuve's annual reports showed that eventually none of these possibilities commanded his confidence (Le Grand, 1989). The opportunity via paleomagnetism of testing continental drift and polar wandering in a new way and of understanding more about reversals of the geomagnetic field was acknowledged by early Carnegie workers. These topics, although not yet center stage, were certainly hovering in the wings. But Carnegie workers did not take the cue. These possibilities did not attract them. They made none of the vital connections (hinted at times but never firmly made) that were necessary to link paleomagnetism in a practical way to mobilism or geology generally: linking paleolatitude and paleoclimates for instance, or linking reversals to stratigraphy to determine the order and timing of strata. Back across the Atlantic attitudes were very different, and things had already begun to change.

### 1.11 Igneous baked contact test of stability

To be believed paleomagnetists had to show that the directions of magnetization of their samples had remained little changed over time – they needed tests to determine the stability of these directions. *Conglomerate* and *bedding tilt* (or *fold*) tests have been described above (§1.8). Another is the *consistency* or *dispersion* test that is more commonly thought of, not as a test, but as a statistical requirement. The test is that regardless of polarity and when allowance is made for secular variation of the ancient geomagnetic field, results from the same formation should be in good agreement; specifically, the directions of magnetization of specimens cut from the same rock sample should agree, and so should results from different samples from the same outcrop, and from outcrops in the same rock formation. In this way one can then be assured that a good time average of the field had been obtained. The estimation of dispersion will be described shortly (§1.13).

There is a third field test of stability, the *igneous baked contact* test, the essence of which is that the country rock immediately adjacent to an igneous body is metamorphosed (mineralogically reconstituted) and its original magnetization destroyed; as the contact rock and igneous body cool, they become magnetized together in the same geomagnetic field, acquiring the same magnetization directions. Rock that has undergone contact metamorphism is readily recognized in the field so that if, millions of years later, it is found to have magnetization directions identical with that of the igneous body, the likelihood is that both were acquired at the time of intrusion. Contact metamorphism (or baking as it is usually called) increases the intensity of magnetization dramatically, often by orders of magnitude. This is not only a useful stability test, the concern here, it is also, as will be explained in IV, §5.5, a principal means of detecting self-reversals of magnetization, the possibility of which has just been discussed (§1.10).

Early attempts to study the ancient geomagnetic field were made using lavas and also bricks and pottery, human artifacts that had been fired to red heat and become magnetized along the ambient magnetic field according to the position they had within the kiln. If lava is erupted from a volcano and flows over a deposit of clay, then the clay immediately beneath is baked forming what Bernard Brunhes described as "la brique naturelle." This not infrequently happened during the Cenozoic in the Massif Central of France, then an active volcanic region. Brunhes (1867–1910) was professor at the University of Clermont-Ferrand and director of the nearby Puy de Dôme Observatory. Using a quote from Laj and colleagues (2002: 381), he "had been very impressed with the work of Giuseppe Folgerhaiter who had showed that ordinary bricks and pottery carried a particularly strong and stable remanent magnetization that is aligned with the magnetic field in which they were baked." In the translation of Laj *et al.*, Brunhes recognized, "if the direction of the magnetization in beds of naturally baked clay is well defined and different from that of the present field, then we are entitled to admit that the direction of the magnetization is that of the geomagnetic field existing when the volcanic flow baked the clays." Brunhes, a learned man, would know of the local outcrops of massive basalt underlain by "brique naturelle," and he and assistant Pierre David set out to sample, first the baked clay and later the overlying lava – "brique naturelle" was their first objective. These massive plateau basalts, with their hexagonal columnar cooling joints underlain by baked horizons, along with other famous examples such as the Antrim basalts (Giant's Causeway) of Northern Ireland and similar occurrences in the

Scottish Hebrides, had been favorably cited 130 years earlier by Nicholas Desmarest (1725–1815) as evidence for their volcanic origin. This was utterly contrary to followers of Braham Gottlob Werner (1749–1817) who believed all rocks had been deposited through water (Neptunism). Brunhes and David would doubtless have been aware that they were sampling geologically hallowed ground.

As expected on Vulcanist theory, they found in several instances that the baked clays were strongly magnetic and had the same directions of magnetization as the lavas overlying them. Evidently, heat from the lava had metamorphosed the subjacent clay into "brique naturelle" which, upon cooling along with the lava, had become magnetized in the ambient field. Near the village of Pontfarein about 90 km south of Clermont-Ferrand, they found magnetizations of baked clay and lava in excellent agreement and reversed in direction from the present Earth's field – the first igneous contact study and the first observation of reversed magnetization! Fossils in sediments intercalated with the lavas were Miocene in age. Subsequently, Laj and colleagues (2002: 387) have obtained a radiometric date of $6.16 \pm 0.08$ million years for the Pontfarein flow, uppermost Miocene.

This situation, in which very hot, liquid basalt flowed over lake sediments that had been derived from earlier volcanic rocks and therefore contained readily magnetized minerals, is ideal. Equally suitable are subhorizontal contacts above and below intruded sills, and subvertical contacts on either side of dykes and around small plutons – indeed any place where hot liquid magma has come into direct contact with cool country rock. During the paleomagnetic test of continental drift, very many investigations of such occurrences were carried out in the manner of Brunhes and David. Although the rocks were older and the geological contexts more complex than those studied by them, workers interpreted their results in much the same manner as these French pioneers did.

### 1.12 Hospers arrives in Cambridge, 1949: his early education and commencement of Iceland surveys

When Runcorn returned to Cambridge, to the Department of Geodesy and Geophysics in January 1950, the mine experiments and his work with F. J. Lowes on secular variation remained unfinished. He wanted to set up a group working on paleomagnetism, but he himself knew nothing of the technicalities of such work, and he needed to obtain funding, to recruit assistants, and generally to establish himself. On his arrival there was a welcome surprise. The previous September (1949), Hospers, an energetic Dutch graduate student, had arrived in the department and was already planning, with his former supervisors in Holland, to undertake a paleomagnetic survey of Icelandic lava flows during the coming summer.

Jan Hospers (1925–2004) was born in Groningen. He spent his teenage years under Nazi occupation. He entered the University of Groningen in 1945. The professor who most influenced him there was Ph. H. Kuenen, who taught sedimentology, and he obtained a B.S. degree in geology and physics in 1948 which "was a somewhat unusual combination at the time" (Hospers, March 1983 interview with author). He enrolled in the University of Utrecht for a master's degree in geology and physics (mainly geophysics) where his professors were Vening Meinesz (geophysics), M. G. Rutten (general geology), R. W. van Bemmelen (economic geology) and W. Nieuwenkamp (mineralogy and petrology) – a very distinguished group! Reflecting on his training, Hospers said:

I am not a fully-fledged physicist, for I have very little knowledge of atomic and quantum physics – theoretical physics. Am I a geologist? Partly. Though I am sadly lacking in knowledge of mineralogy and petrology – I have had very little opportunity to look at rocks under the microscope, for example. So I suppose I have to call myself a geophysicist/geologist.

*(Hospers, 1983 interview with author)*

Unlike Blackett and Runcorn, Hospers had been among scientists who were aware of continental drift and polar wandering, but who placed little stock in them.

I knew, of course, about continental drift because it had been a very lively issue, I think, in the 1920s. I knew about polar wandering. But I think that I can say that from what I heard and read about continental drift particularly, I thought they were more or less relics of past thinking. Here, I'm talking, of course, about the late 1940s.

*(Hospers, 1983 interview with author)*

Of the classical arguments for continental drift, Hospers (1983 interview with author) recalled that what "impressed me most was the fit of the continents, particularly South America and Africa." He was less impressed with drift's solution to the Permo-Carboniferous glaciation because "I was aware from my own observations of tills in the Netherlands; that it is often very, very difficult to establish that a till is indeed a till – a deposit by glacial ice." His reading of Holmes, Wegener, and du Toit did not change his mind.

I was on a geological field trip to Britain and I bought Holmes in London – the book couldn't be bought in Holland at the time – and read it. I remember thinking it interesting that Holmes still had continental drift. And, again, I think I considered it a relic of the past. Wegener and du Toit I bought second hand at the market store in Cambridge. I was very much impressed and intrigued but I did not think that this was still a living issue.

*(Hospers, 1983 interview with author)*

Continental drift and polar wandering were discussed at Groningen and Utrecht.

Kuenen devoted quite some time – quite a few hours – to discussing continental drift, which he did not believe in. He had worked in the East Indies before the War – what was then the Dutch East Indies and now is Indonesia – and had interpreted quite a lot of marine geology without taking into account continental drift, and it was difficult for him to accept that continental drift had occurred. At Utrecht, continental drift was discussed. But, for example, van Bemmelen had very strong views that what one saw in the way of tectonics was due to vertical movements in the Earth's crust, and he did not believe in horizontal movement.

And only Vening Meinesz – if you are familiar with his work you know that he did gravity work over the oceans including Indonesia – and his interpretation required the large scale movement of large tracts of the earth [polar wandering but not continental drift].

*(Hospers, 1983 interview with author; my bracketed addition)*

Kuenen and van Bemmelen did not favor drift and Vening Meinesz supported only polar wandering (I, §8.14).

Hospers recalled that his decision to study for a Ph.D. at Cambridge was a purely financial matter. I was a poor student. I had no private income, no private resources, and had started with about $250 in 1945 when I began as a student and had to live on grants and stipends to complete my studies. So when I competed for the Royal Dutch Shell Studentship in geophysics at Cambridge and got it, I was very pleased because that solved my financial problems – the problem of how to finance my studies for a Ph.D. and that is how I went to Cambridge in 1949.

*(Hospers, 1983 interview with author)*

His decision to work in Iceland during summer of 1950 on paleomagnetism was equally fortuitous.

When at Utrecht (1948–9), Rutten planned an expedition to Iceland with a view to study the volcanics and tectonics of Iceland. And he invited van Bemmelen. He also had an interest in the study of volcanism. They decided to take along one student. They invited me, and I said, "Yes." And so it was decided in the summer of 1950 we would – the three of us – work in Iceland.

*(Hospers, 1983 interview with author)*

Once it was determined that Hospers would go to Iceland, the next question (1983 interview with author) was "What would young Hospers do in Iceland?" Van Bemmelen's father (1883) had worked on geomagnetic secular variation, and knew about geomagnetism and paleomagnetism. The monotonous Icelandic lava flows were difficult to correlate, and van Bemmelen suggested that Hospers should measure the intensity of their remanent magnetization and use it for correlation.[10] As Hospers succinctly put it:

The prime concern was: the Earth's field varies in strength, and therefore lava flows may have different magnetizations. How one would get the strength of the field from the strength of the rock itself was an unsolved problem. But, anyway, one could hope that by measuring the strengths of the natural remanent magnetization in different flows, one might be able to correlate them.

*(Hospers, 1983 interview with author)*

Hospers had no plans to work on reversals of the geomagnetic field; he did not even know about them before arriving in Cambridge. Nor did he intend to use paleomagnetism as a means for testing continental drift and/or polar wandering; he thought drift a dead issue (Hospers, 1983 interview with author). Indeed, he had no idea that continental drift could be tested by means of paleomagnetism until Irving told him about it when they met in late summer 1951 (Hospers, 1981 letter to Irving, §2.9).

B. C. Browne, then Head of the Department of Geodesy and Geophysics, became and remained Hospers' official supervisor, even though he had never worked in paleomagnetism. Hospers educated himself before leaving for Iceland. He took up Browne's suggestion to attend lectures in physics, studied for his M.Sc. exams (still needed for his Utrecht degree), read about the geology of Iceland, and worked on two gravity surveys (Hospers, May 28, 1985 letter to author).[11] However, the two most important things that happened to him before leaving for Iceland were reading a popular article on paleomagnetism by H. Manley (1949) and becoming friends with Runcorn.

Manley's paper, entitled "Palaeomagnetism," was published in the July 1949 issue of *Science News*, and Hospers thinks (May 28, 1985 letter to author) that he bought a copy in October or November of that year. Manley, who had been a technical journalist, was then a student of Bruckshaw's at Imperial College, London, studying the thermomagnetic properties of rocks. Manley gave an excellent summary of the discovery by Bruckshaw and Robertson of the reversed magnetization of dolerite dykes in northern England (§1.6). Manley also cited the existence of other reversely magnetized formations (such as the Precambrian Pilansberg dykes of southern Africa), reviewed the use of paleomagnetism as a dating tool in archeology, and (I believe) was the first to speak of "reversed sample" in place of "inverted sample."[12] Although reading this paper reinforced Hospers' decision to work on paleomagnetism, it did not inspire him to search for reversals when he went to Iceland:

When I read [Manley's article], I was not impressed, as I had earlier seen some of the complexities of geology. Accordingly, I did not think about the possibility of finding reversals in the lava flows of Iceland before I had collected and measured the samples. What I did think about was the possibility of correlating lava flows from different sections (exposures) by means of the intensity of their natural remanent magnetization. I did not consider, at the time, the possible importance of directions of magnetization (secular change or reversals).

*(Hospers, May 28, 1985 letter to author)*

Nor, at this stage, did Hospers begin an extensive review of the paleomagnetic literature.

I did not read very much else on paleomagnetism before going to Iceland. I was engaged in a number of other things. As I recall it, my main concerns in relation to the planned fieldwork in Iceland in the summer of 1950 were the financing, the transport to Iceland for myself and my motorcycle (I got there in 1950 as a paying passenger on an empty Icelandic fishing boat returning from Hull [UK] to Iceland) and, most of all, the reading of the relevant geological literature and planning in detail of the geological fieldwork in Iceland. My reasoning was that I should collect my paleomagnetic samples from a succession of lava flows about the geology of which I wanted to know as much as possible. The actual measurements and their interpretation could, I thought, later be done in the laboratory.

As I recall it, I did not read any of the references cited by Manley before going to Iceland for the first time though, of course, I did later on read some of them. I certainly did not read *Geomagnetism* by Chapman and Bartels before going to Iceland in 1950. I did read it though soon after my return. The only papers on paleomagnetism which I think I may have read [other than Manley's article] before the summer of 1950 were Mercanton's (1926) papers and possibly Chevallier's 1925 paper.

*(Hospers, May 28, 1985 letter to author; my bracketed additions)*

And then in January 1950 Keith Runcorn arrived. Runcorn became Hospers' friend, mentor, and unofficial supervisor, and without question it was he who influenced Hospers most at Cambridge.[13] However, it would be a mistake to think that Runcorn actually directed Hospers' research or fueled him with key arguments. Rather the opposite, it was Hospers' work that was soon to convince Runcorn of the reality of geomagnetic reversals.[14] What was crucially important was that Runcorn offered continual encouragement, scientific companionship, and provided Hospers with key contacts, greatly facilitating his work. Hospers recalled:

Runcorn encouraged me very much. There were the samples collected in 1950, and he very much encouraged me to return on my own to Iceland in 1951, and collect more samples. And, of course, he took care of the contacts.

*(Hospers, 1984 interview with author)*

### 1.13 Hospers' first results from Iceland, 1950–1951, and genesis of Fisher's statistics

Hospers' work on his first lava collection from Iceland had important consequences: it led to his defense of the idea that Earth's magnetic field had undergone at least two reversals of polarity, it sparked Fisher's interest, and it began to provide key new information about the time-averaged form of the geomagnetic field, all of which spurred Runcorn into starting his own paleomagnetic group at Cambridge.

In summer 1950, Hospers accompanied Rutten and van Bemmelen to Iceland as planned. He returned with twenty-two oriented samples spanning a substantial stratigraphic thickness. Runcorn made arrangements for him to measure them on the spinner magnetometer recently constructed by S. A. Vincenz, a graduate student of Bruckshaw's at Imperial College. Two things quickly became apparent: the opposing polarities of samples and the substantial dispersion of their directions of magnetization.

When I came back from Iceland in 1950 and measured my samples, one could see that there were two opposite directions – what we would call normal and reversed now – and there was a large scatter in these directions. It was obvious that if I wanted to extract some more accurate information from the data, one had to apply some kind of statistical method to find out how one would calculate an average direction and how reliable that direction would be.

*(Hospers, 1984 interview with author)*

Hospers explained to Runcorn the patent need for a statistical method. Runcorn knew just the person who could provide it, R. A. Fisher. Runcorn recalled:

By that time I had become friendly with R. A. Fisher, who was a fellow of Caius. I had breakfast with him almost every morning in the senior common room at Cambridge in my college. He was one of these people who was interested in everything. He found the various geophysical studies we were doing very interesting, and he used to come to the colloquium which I organized – as did Jeffreys. I began to talk to Jan Hospers about, you know, how one would deal with the statistical problem. It looked to me as though these normal directions and reversed directions were exactly at 180°, which I thought could not be due to random processes. I told Fisher about the problem with Hospers' data, and I asked Fisher if there was a way to generalize Gaussian error theory to directions in space that I could find in a textbook. He said, "There was not; it had not been done." Then, he wrote his famous paper.

*(Runcorn, 1984 interview with author; revised by Runcorn, August 1993)*

Fisher had done preliminary work on the problem "in the early 1920s" (N. L. Fisher *et al.*, 1987: 12) while working on a different application, a test of significance in harmonic analysis. Runcorn consulted Fisher, who apparently pulled out some old notes, which he reworked to apply to paleomagnetic data; he modeled paleomagnetic data by what has come to be known as Fisher's distribution.[15] In fact, Fisher himself even did the actual calculations for Hospers' first paper, deriving estimates of the concentration parameter (Fisher's precision), the mean direction, and the radius of the 95% cone of confidence around the mean: these are the essential statistical requirements. His method was not published for two years (Fisher, 1953), but by 1951 it was available to Cambridge workers, and it is difficult to overestimate the importance of its timely, indeed essential, assistance to them and for the development of paleomagnetic studies generally.

In the Fisher distribution, individual NRM directions are regarded as points dispersed on the unit sphere and have a frequency density proportional to $\exp(\kappa \cos \theta)$, where $\theta$ is the angular departure from the mean and $\kappa$ describes how the points are distributed about the mean. Fisher (1953) explained how to obtain an estimate $k$ of $\kappa$. If $k$ is 3 or less, the directions are very widely scattered, essentially random. If $k$ is high, say $\sim 100$, the dispersion is low and the points are concentrated closely about the mean. Paleomagnetic data sets generally fall between these extremes. Fisher called $k$ the *precision*, and more recently the term *concentration* has been introduced for this fundamental parameter (N. L. Fisher *et al.* (1987), where spherical statistics are fully described). Fisher also showed how to determine the *mean* of the group of directions and the radius of its *circle of confidence* $\alpha$, which is usually quoted at 95% ($P = 0.05$); $\alpha_{95}°$ is the radius of the circle centered around the estimated mean that contains the true mean with a probability of 95%. The concentration parameter $k$ is inversely related to the square of the *dispersion*; for example the radius, $\theta_{63}°$, of the circle described around the mean containing 63% of the observed directions is given by $81k^{-1/2}$ (Watson and Irving, 1957); $\theta_{63}°$ is called the *spherical standard deviation* and is the analogue in spherical statistics of the standard deviation in linear Gaussian statistics. For the first few years of using Fisher's statistics, the only measure of dispersion workers had was Fisher's $k$.[16]

Hospers submitted his first paper on his Iceland work to *Nature* in June of 1951. He had become convinced of field reversals, or as he (1951) initially referred to them, "inversions" of Earth's magnetic field. He appealed to Bruckshaw and Roberston's work (§1.6) and that of others on reversals. He (1951: 1112) found no lithological differences between normal and reversed samples. As the cause of reversals, he eliminated "pressure, reheating, recrystallization, lightning, etc., because it is not conceivable that these processes should only have affected the reversely magnetized Plio-Pleistocene lavas and not also the overlying normal Quaternary flows." He also ruled out tectonic explanations.

> An inversion due to tectonic dislocations is out of the question, as top and bottom in the series can definitely be determined. Moreover, an inversion of the whole series would only mean that before the inversion the group which is now normal would have been reversed; so the problem remains the same.
>
> *(Hospers, 1951: 1112)*

In addition, he (1951: 1112) noted that regardless of sign the difference between the mean direction of magnetization of his samples and the geocentric axial dipole (hereafter, GAD) field was "not significant." His mean directions were predominately axial (north/south), none were strongly oblique to the GAD field; there was nothing about them that suggested significant (i.e., greater than 1000 km) continental drift or polar wandering.

## 1.14 Consistency or dispersion as a test of paleomagnetic stability

*Bedding tilt* (or *fold*), *conglomerate* (§1.8), and *igneous baked contact* (§1.11) tests have been described above. There are two further field stability tests – tests that have the field relationships of samples to one another as their basis. Both make great use of the Fisher statistics just described.

The *consistency* or *dispersion* test may be more commonly thought of as a statistical requirement rather than a stability test, but – and I am anticipating developments that were at a time far in the future – when executed to its fullest extent it is the most powerful and general of them all, and it formed the very basis of the paleomagnetic global mobilism test. At the first stage, magnetization directions from a single region can be analyzed using Fisher's method carried out through several ascending hierarchical levels. To pass the test, the requirement is that, regardless of polarity, and when due allowance is made for secular variation of the ancient geomagnetic field, results from the same geological rock formation should be in good agreement, the dispersion should be low and errors (half-angles of cone of confidence) small; specifically the directions of magnetization of specimens cut from the same rock sample, the directions of magnetization from different samples from the same outcrop, and directions from outcrops in the same rock formation. Fisher's method allowed all this to be quantified. To make comparison over larger areas, account must be taken of the variation of the geomagnetic field over the Earth's surface using Hospers', time-averaged, GAD model of the field just described. This is done by transforming directions of magnetization into corresponding poles (§3.6) and analyzing them using Fisher's method. Comparisons can then be made across continents and between continents, tailor-made for the global mobilism test; the development of very large-scale intra- and intercontinental consistency tests then became a reality when Ken Creer in 1954 drew the first apparent polar wander path (§3.6). Many examples of such comparisons will be described.

The fifth and final field-based stability test makes use of reversals of magnetization. The presence of a statistically exact reversal (180°) means that substantial secondary magnetizations (overprints) are absent. Often, however, reversals are not exact because of the systematic biasing effect of overprinting; but when normal and reversed are averaged regardless of polarity, the sign of the secondary magnetizations is inverted and their effect thereby removed to a first approximation. This powerful procedure (it might be thought of as a "procedure" rather than a test) was crucial before the development of magnetic cleaning methods (§5.5); it worked, regardless of whether reversals were caused by field reversal or by self-reversal. By 1954 it had been used, often implicitly, by Hospers, Clegg and colleagues, Creer, and Irving, and was so obvious that at first its usefulness was not explicitly spelled out nor its efficiency perhaps fully realized. Later it was named the "consistency-of-reversals-test for stability" by Cox and Doell (1960: 656)[17] and "the reversals test for secondary components" (Irving, 1964: 74). More simply it may be called *the reversal test of stability*.

### 1.15  Runcorn arrives in Cambridge, 1950, decides to work on paleomagnetism, and hires Ted (E.) Irving, 1951

We [in 1951] viewed the possibility of looking at the paleomagnetic record in rocks, really, purely from the point of view of looking at the behavior of the Earth's magnetic field: the secular variation and the issue of reversals.

*(Runcorn, 1984 interview with author; my bracketed addition)*

Knowing of Hospers' success in the spring of 1951, Runcorn decided to involve himself directly in paleomagnetism. Although Hospers' work was the immediate stimulus, Runcorn had recognized while working with Blackett at Manchester that paleomagnetic work could provide needed information about the history of Earth's magnetic field (§1.6), and there is no reason to doubt that this was the general reason for his decision. He recognized two problems: neither had anything to do with continental drift or polar wandering.

First was the origin of secular variation. Lowes, a former graduate student of Blackett's and Runcorn's at Manchester, who joined Runcorn at Cambridge in 1952, wrote a paper with him on secular variation in which they criticized Bullard's account and offered their own.[18] In addition, Runcorn had encouraged Raymond Hide, one of his students at Cambridge and a Manchester graduate, to do experiments on convection in rotating fluids that might have relevance for secular changes.[19] Second was the problem of reversals. Hospers' work persuaded him that reversed magnetizations were caused by reversals of the field and he became interested in them.

I remember that I had been skeptical about the origin of reversed magnetization, but when I saw Hospers' plot – I can almost see it now – I said to him, "Well, you know, you have demonstrated that there is no possibility of explaining reversed magnetization by tectonic movements or speculative geological processes not otherwise evident. If you turn the reversed ones upside down, you turn the normal ones upside down." At that time, geophysicists and even geologists might have entertained very speculative geological processes of a local character. Other geologists might have said, "Look you cannot do that, but we do not believe that the magnetics tells you anything about the field." So, it was Hospers' work that convinced me of the reality of reversals.

*(Runcorn, 1984 interview with author)*

Hospers gave several talks on his work. According to Runcorn, he gave one "soon after Hospers measured the collection of rocks he made in summer of 1950." Hide (1984 communication to author) recalled that Runcorn announced his belief in field reversals while they walked home after Hospers' talk. Commenting on Hide's recollection, Runcorn stated:

I certainly remember that this [Hospers' talk] would be soon after Hospers measured the collection of the rocks he made in the summer of 1950. I certainly remember detailed discussions with him about the results and I certainly remember being very struck by the fact that the mean directions of the normal and reversed rocks were nearly opposite and this seemed to me to dispose of the possibility that "reversed magnetizations" were due to peculiar geological movements or anomalous magnetization processes.

*(Runcorn, letter to author, September 19, 1984; my bracketed addition)*

There is much uncertainty about just when and where Hospers gave this talk.[20] In any event, it is likely that Runcorn became sympathetic to field reversals because of Hospers' work and Hospers' talk spurred him to work himself in paleomagnetism.

As a start, Runcorn approached the Department of Geology at Cambridge, probably W. B. R. King the chairman, to find a student to collect samples for him. He received no recommendations. However, Brian Harland, a lecturer in the department, volunteered (see III, §1.11 for Harland's later contribution). Runcorn recalled:

I suppose it was the spring of 1951 after I had already got Hide and two other people at Cambridge that I decided to collect some specimens to see whether rocks in England had got this remanent magnetization. Now I did not know anything about geology. I knew Harland because he was in my college. So, I said to Harland, "Would you like to cooperate?" And, so he said, "Yes." I said, "We'll collect some rocks." So he went away on some sort of field trip, and they brought back a lot of little chippings from some formation which, you know, they said was a pretty important formation – It was the graywackes of Wales. After I saw this, I knew that I wouldn't get anywhere unless I went and did it myself.

*(Runcorn, 1984 interview with author)*

So Runcorn decided to collect himself, but realized that he needed geological help. The Department of Geology at Cambridge had recommended no one, but Runcorn found out about Ted Irving in another way, and hired him as a personal assistant in June 1951.

When I decided to go and collect samples for myself, I thought I needed a geologist. I asked the geologists at Cambridge if they had any graduate students who would come out with me. And the reply was: "All our graduate students are doing much more important work." So then I asked, "Have you got any students who have just graduated?" (It was probably Harland, but, perhaps, other people in the department too, I asked.) And the general view – it hadn't been a good year – was that there was only one of them, and that was Martin Bott, now Professor of Geophysics at Durham. Ted Irving had, in fact, taken geology in that year, but he was not recommended to me. I played rugby. I was friendly with a person in Cambridge who was doing research in agriculture. We played on the same rugby team, and we used to go out and drink beer together. His name is Maurice Adams, he is a farmer now. He used to say to me, "You know, you have a lot of interests in philosophy and religion and politics. You should meet Ted Irving." But these things happen, he never got around to introducing me. So, one day, rather depressed, I went around to see Maurice. We went out to have a drink, and I told him of the completely uncooperative attitude of the Geology Department at Cambridge. (Petrologists would have been more helpful, but I was, obviously, asking for a stratigrapher.) Maurice said, "This chap, I keep talking about is a geologist." So I said, "Well it really sounds as though he would be just the person." I then found that Ted had gone home – I suppose it would have been June. I rang him up. I said, "Would you like to come and help me as a research assistant?" He said, "Yes, sounds interesting." It wasn't long before I realized that he was a very able chap.[21]

*(Runcorn, 1984 interview with author)*

Maurice Adams confirmed Runcorn's remembrance and gave further background.

In 1951 I was studying graduate Agriculture at Cambridge and through a mutual interest in rugby football became friendly with Keith Runcorn. We spent much time together drinking beer and discussing politics, religion and philosophy, and generally putting the world right. Toward the end of the summer term of 1951 Keith said that he was having difficulty in finding a geologist to help in his research in geomagnetism. I said that I had a friend in Fitzwilliam in College, and incidentally another rugby player, who I thought had a very original and broad thinking mind. They subsequently met and you know the rest of the story. Keith has occasionally mentioned Ted Irving and how well he was doing over the intervening years, and I am pleased to hear that he is now near the top of his particular line.

*(Maurice Adams, February 23, 1987 letter to author)*

### 1.16  Irving's early education and undergraduate years

> You say potato and I say po-tah-to
> You say tomato and I say to-mah-to ...
> Let's call the whole thing off!
>
> *(George and Ira Gershwin, 1937)*

Edward (Ted) Irving was born on May 27, 1927 in Colne, Lancashire, England, a small industrial town in the Pennine Hills of northern England. He went to the local council school at age five and progressed to class 9, where there were two other pupils who were selected to take certain lessons with class 10 in preparation for sitting the Junior County Scholarship examinations. A scholarship would pay the fees at Colne Grammar School. Having marks as good as the other two, he told his father about the scholarship. His father went to see the headmaster who agreed that, although not yet ten years old and younger than usual (11+ years), he should be given a chance. He sat the exam in spring 1937 and passed. This was important because looming ahead, although not known at the time, was conscription into the armed forces at age eighteen. Irving's early entry into grammar school in 1937 at just turned ten meant that throughout his school years he was a year younger than most in his class and this allowed him a third year in the sixth form before being "called up." It was in that third year that he was to gain a County Major Scholarship, which became his ticket to Cambridge.

In the middle years at grammar school he was "streamed" into the "humanities" and took only "general science," not any science separately. As a result, in the sixth form he took a mixed bag of geography, biology, mathematics, and chemistry. He represented the school at cricket and rugby, was a house captain and Head Prefect.

In first-year sixth form (1942–3), as part of physical geography, he had lessons on isostasy and Wegener's theory of continental drift. Irving is sure about this. In the class Irving asked for further explains, but in an attempt to pronounce Wegener's name, which the teacher had written on the blackboard, Irving uttered something that sounded more like "vagina" than Wegener."

I remember the lesson quite clearly for the very Freudian reason that in class I mispronounced the middle part of Wegener, saying a soft "g" as in Wage (which is sound English practice when "e" follows "g") and this caused gales of laughter and a temporary break-down in discipline in the mixed class! One doesn't forget things like this, although one may exaggerate a little.[22]

*(Irving, December 8, 1980 letter to Runcorn; copy sent to author)*

The teacher called off the discussion.

The school library had a book on local geology written by a previous headmaster, Dr Wilmore. It also had the *Textbook of Geology*, authored by Lake and Rastall, which he studied. Colne had an excellent, Carnegie-endowed, municipal library and Irving read books of all sorts, among them Julian Huxley's newly published *Evolution: The Modern Synthesis*, and large sections of Parker and Haswell's massive *Text-book of Zoology*. In 1944 he bought a copy of Darwin's *Origin of Species*. His interests were settling on biology and geology.

Irving had become a self-mover. At an early age, he learned to be self-sufficient and to work with others.

I learnt how to work with other people – building Meccano models with pals or going with them on long hikes and bike rides for weeks on end, youth hostelling. Always I had a project, not always successful. I never used a bought plan or assembly kits. I looked down on that as an inferior activity;

it was just copying. Mercifully my parents could not afford them anyway. My parents were wonderful. My father explained to me that he "did not want any bother." I can't remember him ever making clear to me what that meant. Not creating bother for others in the family is of course a very wise and simple idea. It is the essence of family life. My parents let me do more or less what I wanted; I guess they trusted me not to create bother. I was never given an allowance – pocket money. I began earning at age eight delivering milk door to door for a local farmer. Then in my early teens, I took up gardening, selling produce excess to family needs. Now I simply give excess away to friends. That and odd jobs during school breaks is how I financed hiking and biking holidays.

*(Irving, November 10, 2011 email to author)*

At age eighteen, Irving was conscripted into the British Army (1945–8) initially as an infantryman. World War II was over. He spent almost two years in Palestine, north and east Africa. Abroad, he transferred to the Education Corps. By late 1947 it became likely that he would be demobilized the following year and it was time to think about university. At the time he was stationed in Asmara, Eritrea, and visited an office downtown, probably the British Council. It had information on British universities including Oxbridge colleges about which he knew nothing. Armed with a major scholarship, he wrote enquiring letters to many colleges about a place in fall of 1948. Several suggested he see them when back to Britain, which was in May 1948. His ignorance of Oxbridge was such that he applied to some theological and, what were then, women's colleges The upshot was that, despite the short notice, he was lucky enough to get a place at Fitzwilliam House, Cambridge, to read natural sciences in the autumn of 1948. Fitzwilliam, a mid-nineteenth century establishment, had all the normal clubs and societies of a college but no residences. He went with very little idea of what to expect. At school he had read far less about Cambridge than he had about the Middle East before going there. In Part I he read geology, mineralogy, and zoology, in Part II, geology, specializing in stratigraphy and paleontology. He graduated in 1951 with geology as a special subject, receiving a Lower Second Class Honors degree.[23]

Irving found first-year courses solid and formal, lacking zip. In zoology there was no course on evolution and no prospect of one in the second year. Much of first-year geology was taught by Professor W. B. R. King whom we have just met. Irving has no recollection of his mentioning continental drift, which is not surprising when we recall that King had reviewed unfavorably Holmes' account of drift in *Principles of Physical Geology* (I, §5.9). It was only in the third year that mobilism was mentioned by staff and not as a central concern. Most teachers did not mention mobilism, but two instructors did in the third year. Brian Harland discussed it favorably in supervisions. Maurice Black discussed the glacial evidence for continental drift in lectures on the Gondwana System. According to Irving, "Black leaned strongly toward drift, as I can show from my lecture notes" (Irving, December 8, 1980 letter to Runcorn; Irving sent copy to author).

Glaciation must be regarded as due to cooling of the S. Hemisphere only as ("tropical") corals continue to flourish in N. Hemisphere. Glaciation cannot be explained on supposition that Poles shifted only – some regions would still be distant from it. So far drift has been the only adequate explanation.

*(Irving's class notes, p. 99)*

Nor can Irving recall any instructor arguing that continental drift was impossible. There were lectures on exploration geophysics on Saturday mornings, but Jeffreys' criticisms of mobilism were not discussed.

As an undergraduate Irving read about continental drift, and discussed it with fellow students.

Several of us had read Wegener (1924, English Edition), myself and several of the group that talked together in Part II – Bob Stoneley Junior, Alan Wells and Martin Bott, in particular. I had also read Argand – it was, I think, on the reading list for Part II, summer term 1950. There was no doubt that in that group some of us took a very favorable attitude to drift.

*(Irving, December 8, 1980 letter to Runcorn)*

He read Jeffreys' *Earthquakes and Mountains* (1950) which devoted one sentence (p. 173) to continental drift. But he was much more impressed with Argand's 1924 *La tectonique de l'Asie.*

I found [Jeffreys'] book rather incomprehensible. It was certainly less attractive to me than ... Argand's enormously prophetic essay on the tectonics of Asia – certainly it was that rather than Jeffreys that made an impression on me.

*(Irving, 1981 interview with author; my bracketed addition)*

So Irving was sympathetic to drift before Runcorn hired him in June 1951. As he put it:

I came away from both school and undergraduate courses at Cambridge knowing *very* well what continental drift was, and without any particular prejudices for or against it.

*(Irving, 1981 interview with author)*

Little did he know that he was soon to work on it.

## 1.17 Irving and Runcorn's first work, July through December 1951:
### only red beds give coherent results

If it is something in paleomagnetism, Ted Irving did it first.

*(Neil Opdyke[24])*

During the remainder of 1951, Irving's work was exceedingly fruitful. He and Runcorn collected samples of several types of old sedimentary rocks, and, using Blackett's magnetometer, their samples of fine-grained red beds gave coherent data. Their directions of magnetization were strongly oblique to the present field, and did not record the ancient secular variation as Runcorn originally hoped. Irving realized that paleomagnetism could be used to test continental drift, and initiated the first such test. In a remarkable turn in Irving's fortune, both Runcorn and Blackett were sufficiently impressed by him to ask him if he wanted to pursue a Ph.D.

Runcorn was interested in secular variation and in the possibility of finding reversals, especially in light of Hospers' work. But secular variation was his priority. He wanted

to see whether one could get any suggestion about what the parameters were – whether they were similar to the kind of secular variation one saw in recent times. Because so little was known about the Earth's magnetic field, if you show there was secular variation way back in the geological past as there was in recent times, this might show that the same mechanism of core motions was responsible for the field over the whole geological time. It may seem rather odd now that one would be a bit bothered by wanting to see if the secular variation was there. But so little was known about the field, in particular the role of fluid motions, there was very much a question of whether motions are going on in the core throughout the geological past.

*(Runcorn, August 1984 interview with author)*

To study secular variation Runcorn needed samples from strata that had been deposited as uninterruptedly as possible. Igneous rocks are formed sporadically, and do not track secular variation step-by-step. He knew that the intensities of remanent magnetism of sedimentary rocks are very much less than that of igneous rock, but thought that with Blackett's high-sensitivity magnetometer it would be possible to study them. Acting on a suggestion of T. C. Phemister of the University of Aberdeen, he planned to start with the Torridonian Sandstone Series of northwestern Scotland.

I did not know much about geology but I spent a nice holiday in the northwest highlands, and I'd seen all these red sandstones of the Pre-Cambrian, and I kind of realized that one needed to make collections in a geologically simple environment where there had not been much tectonics, and that was one reason I wanted to collect there. And another reason was that it was so far back that there was a contrast with the Tertiary. Looking back on it, it was perhaps a silly choice because it was known to be Late Pre-Cambrian but there was no radioactive age – but then there weren't too many accurate radioactive ages available.[25]

*(Runcorn, 1984 interview with author)*

Runcorn and Irving did not know at the time what sort of sedimentary rocks would provide them with reliable paleomagnetic data. They did not know if any really old sedimentary rocks would even yield consistent remanent magnetization directions. There was only one way to find out: study different old lithologies.

They made their initial collections during the first two weeks of July 1951 from three very different lithologies: Silurian deepwater mudstone from Cumbria in northwestern England at the suggestion of W. B. R. King of Cambridge University, fluviatile red beds from the Torridonian Sandstone Series from northwestern Scotland at Phemister's suggestion as just noted, and gray and buff siltstones of Devonian age from the lacustrine Caithness Flags in northeastern Scotland on Irving's initiative (Irving, field notes).[26] Later that summer, Irving sampled massive shelf carbonates of the Mam Tor limestone, and the overlying estuarine Edale shales of Derbyshire. Large block samples were collected, each spanning thicknesses of about 20–30 cm stratigraphically at each locality, so that there were many individual beds in each sample.

Blackett having just finished using his magnetometer to test his fundamental theory of the geomagnetic fields (§1.3), Runcorn made arrangements for Irving to use it. Irving's measurements of their first collection were made between August 21 and September 15, 1951.[27] Jim Pickering, Blackett's laboratory assistant, instructed him (Irving, *Synopsis of Records*, p. 4).

Irving first had to learn how best to cut and measure samples. Runcorn and he wondered about optimum specimen size and shape, and soon discovered that inhomogeneity of magnetization within specimens affected measurements. They settled for taking cores sliced into thin discs. These were easier to cut and apply corrections than the cubes that had until then been generally used in Britain. The discs were 1.35 cm in diameter (a few millimeters thick) because that was the size of cores produced by the drill press already in the department, left there by Bullard who had used it to prepare rock cylinders for heat conductivity measurements. Also, orientation marks could be easily and accurately transferred from block to core to disc. They found they had to measure their discs in both upright and inverted positions and at many azimuths, directly beneath and offset to either side of the magnetometer axis, the latter a procedure devised by Blackett (1952).

We (that is Runcorn and I) hadn't yet worked out the full theory of measurement to take account of inhomogeneity. We were beginning to realize the necessity of taking readings both upright and inverted to fully describe the magnetization.

*(Irving, Synopsis, p. 5)*

Runcorn wanted to observe secular variation, so Irving took two parallel cores perpendicular to bedding plane from each block. If secular variation was being observed, both should show the same regular variations. Quickly they focused their attention on samples from a small bluff (called locally "Craigriannan" or "Sunny Rock") on the north shore of Loch Torridon.

At the time the Torridonian was estimated to be about 5000 m thick, and was divided into three: the Diabaig, the oldest (~1000 m), Applecross (~3000 m), and Aultbea (~1000 m) groups. Craigriannan lay near the base of the Applecross. They also took very coarse-grained samples from the Isle of Ewe in the Aultbea Group. It was only in blocks of fine-grained red beds that the directions of discs were well grouped. The coarse-grained specimens gave scattered directions. Not only were the fine-grained specimens well grouped, they were all directed to the southeast and down, far removed from the present field. In all, Irving had thirty-nine fine-grained discs from Craigriannan, arranged in two parallel columns. There were no common shifts in magnetization directions common to both columns, and both columns gave the same average direction about which individual disc directions were randomly scattered (Irving, *Synopsis*, p. 5). There were no regular fluctuations as expected from secular variation.

Runcorn did not give up and decided that a second more detailed study of Craigriannan be made. In October 1951, they collected twelve blocks spanning ~130 cm continuously, mainly fine-grained, red sandstones with occasional thin coarser layers. Parallel cores were sliced into 224 discs. Of these, 135 were siltstones or very fine sandstones with directions well grouped to the southeast and down, far from the present field, reproducing within narrow statistical limits the first results. Again there were no discernible bed-by-bed variations in parallel cores and Runcorn's search for secular variation experiment died.

At this stage Irving had realized that the question, "What samples yield the most coherent results?" was not reducible to "What kind of sedimentary rock yields the most coherent results?" because some types of red beds give more coherent results than others. There was therefore a great advantage to be gained by sampling specific lithologies.

Although they did not realize it at the time, their discovery that fine-grained red beds were generally suitable for paleomagnetic study was very important for the development of the paleomagnetic case for mobilism in the next few years. Irving, I think correctly, believes that it ranks high among his contributions.

It was from the work on [the two Craigriannan] collections ... and from collections I made on a weekend trip to Derbyshire of Carboniferous limestone and black shales, that we decided that red beds were the best type of sediment for paleomagnetic work. In 1951 I also began work on determining which type of red bed was best and established this fairly quickly – they had to be coarse silts or fine sandstones. There were thus two problems – first what general rock type (red beds) and then what particular variety within that broad type (fine-grained). So far as I knew this systematic search, begun in 1951 for the optimum sedimentary rock for paleomagnetic work, was the first ever undertaken. I believe this to be one of my most important contributions to paleomagnetism.

*(Irving, Synopsis, p. 2; my bracketed addition)*

Finally, there were Craigriannan's oblique magnetization directions. What were they to make of them? And what were Irving and Runcorn to do next? Should they persist with the search for ancient secular variation? Their study was at a rudimentary stage. The strata at Craigriannan were laid down in an active fluviatile environment perhaps not ideal for highly accurate recording of the geomagnetic field. Lake deposits laid down under more tranquil conditions might prove more suitable. Should they set out on a search elsewhere for fine-grained lacustrine red beds of other ages? The Caithness Flags might have been deposited under more suitable conditions but they were not red and had not yielded coherent results, so redness and the implied presence of hematite appeared important. Was it the oblique direction itself that was of most interest, and should it become the focus of their attention? Should Irving and Runcorn remain with the Torridonian and determine the lateral and vertical (stratigraphic) extent of the oblique magnetization as a first step in determining its meaning? Should they move to other rock formations? These questions remained undecided until early the following year (§2.5).

There was also a general analytical question. Some fine-grained samples preserved the southeast down oblique magnetization; some coarse-grained sample discs did too but with much higher dispersion. Irving noticed also that some samples had a component of magnetization close to the present Earth's magnetic field. Thus there were many oblique directions and some present Earth's field directions, and there was scattering endemic to both: all required statistical treatment. What good fortune it was that Fisher, who had just developed procedures for rigorously analyzing directions distributed on a sphere, should be right there in Cambridge and only too willing to give Irving a tutorial. Irving thinks the tutorial was in late 1951 or early 1952. I suspect it was the latter, for Fisher was busy going to India and Irving probably was not very confident about the legitimacy of his results until after he and Runcorn had finally worked out the theory and practice of measurement. Regardless, Fisher showed Irving how to use his statistics, which he already had created to analyze Hospers' data.

Neither Runcorn nor Irving, as one might retrospectively think, immediately related the oblique directions of magnetization at Craigriannan to continental drift or polar wandering. Nor did they provide the key for Irving's realizing that paleomagnetism could be used to test continental drift. Irving is quite emphatic about this.

The results that we got from the Torridonian sandstones in 1951 showed a direction that was strongly oblique from the present field ... There was no question of connecting this to drift or polar wandering at the time. I don't think there were any discussions of that sort. There had been no suggestion ever made about continental drift in the Pre-Cambrian [The age of the Torridonian]. Wegener's theory was about the very recent geological past. Even now it is extremely difficult to have ideas about Pre-Cambrian drift that have any sort of basis in reality at all. So the answer to the first part of your question [Were the oblique directions of your results from the Torridonian what led you to think about using paleomagnetism to test drift?] is that we didn't really talk about these results from the Torridonian in terms of drift.

*(Irving, 1981 interview with author; my bracketed addition)*

How, then, did Irving come to realize that paleomagnetism could be used to test continental drift?

### 1.18 Irving devises a paleomagnetism test of continental drift, autumn 1951

The point at which I made the connection between paleomagnetism and continental drift is really quite clear to me, and it's pretty well documented. In October of 1951, I was looking around for somewhere to live in Cambridge. I was then at Trinity Hall. I couldn't live in College because as a research student you didn't get a room. So I took residence at Westminster College which is just at the end of the Cambridge Backs. It's a college for English Presbyterians, and they had space there and I took a room. One of the first things I did was to reread Wegener and to read Du Toit, which I had not read before – I bought a copy at that time. I also bought Chapman's *Geomagnetism* monograph, which was on Runcorn's recommendation. Chapman told me that the inclination of the geomagnetic field provided you with an estimate of latitude, and that is written down in my notes. Then, looking at the maps Köppen and Wegener had produced – the timetable of continental drift – India was the one that had moved the greatest amount in latitude. It had moved from the Southern to the Northern Hemisphere. These maps were reproduced in du Toit's book. And so [I realized] that by simply applying the simple idea of the Earth as a large magnet, that one should be able to determine the very large change of latitude readily in India.

*(Irving, 1981 interview with author, revised 2003; my bracketed addition)*

As soon as Irving learned of the relationship between the inclination of the field and latitude, he, already favorably disposed toward continental drift and familiar with the proposed dramatic northward movement of India across the equator, realized how he could use paleomagnetism to test drift, and why India was critical. The test requires reliable paleomagnetic inclinational data and would work only if the time-averaged geomagnetic field had been a GAD.

Irving came up with the idea during or before October 1951 (Hospers, 1981 letter to Irving, §2.9). He recorded his thoughts in the first of his "Notes" notebooks, which contains 206 pages of notes. He wrote "Begun Oct 1951." on the first page, which also includes a table of contents. Almost every entry deals with geomagnetism, paleomagnetism, or continental drift. The remainder summarizes papers mainly about various aspects of the ocean floor. The second entry (pp. 3 to 24) summarizes du Toit's *Our Wandering Continents*. The third, dated, Bristol, Nov. 1951, is a summary of Irving and Runcorn's discussion with Sir Lewis Fermor, former Director of the Indian Geological Survey (for his pro-mobilist views see I, §6.14). They stopped to see Fermor in Bristol during their trip to the West Country to inspect rock-cutting equipment.

Fermor lived in retirement in an apartment full of tropical plants, and Runcorn and I visited him to obtain information on Indian geology as part of our trip to the west country, where we also visited "Impregnated Diamond Company" at Gloucester to select suitable rock-cutting machinery.

*(Irving, Synopsis, p. 10)*

Since, as Irving has retrospectively remarked, there was no reason to have visited Fermor other than to discuss Indian geology, and since Irving's notes list ages and lithologies of various possible Indian formations (Notes, Notebook I), he must have formulated his India test by November 1951. The Fermor entry is followed by a one-page discussion of Quaternary glacial deposits, which might have tied in with Irving's reading of du Toit on glacial evidence. Regardless, the next entry contains notes on Chapman's *Geomagnetism*, including the formula giving the relationship of co-latitude ($\theta$) to magnetic inclination ($I$): $\tan I = 2 \cot \theta$.

Although Irving was first among the British to realize that paleomagnetism could be used to test continental drift, he was not the first to see the general connection: the idea was already in the literature. Mercanton (1926a) suggested it, but did nothing with it. Wegmann repeated it in 1949 (I, §8.10) as did Graham (1949), as we have just seen. Beno Gutenberg (1951b: 204), swift to recognize a good idea, also repeated Mercanton's suggestion in a review article on global theories of Earth: surely, he was hoping that some paleomagnetist would act on it (for Gutenberg's pro-mobilist views see III, §2.3–§2.4). He had not long to wait.

### 1.19  Realization in 1943 by Sahni that paleomagnetism could be used to test continental drift

The above were not the only ones to note the connection. Birbal Sahni, the Indian paleobotanist who argued for mobilism because it explained the juxtaposition of the *Glossopteris* and *Gigantopteris* floras of peninsular India and of Asia (I, §3.5, §6.13), saw the possibility of using paleomagnetism to test mobilism. His suggestion has so far gone little noticed and is further evidence of his adventurous mind.[28]

Sahni (1944: xv) "first publicly expressed" his thoughts in "December 1944 during a symposium on Magnetism in Relation to Structure, over which Professor Krishnan was presiding." He got the idea the year before after hearing the Indian physicist Kariamanikkam Srinivasa Krishnan (1898–1961) talk about magnetic properties of crystals.

The idea arose from Professor Krishnan's well known work on the magnetic properties of crystals, on which he had given a series of lectures at Lucknow a few years ago. He was demonstrating the way in which crystals of certain mineral substances, when freely suspended in an artificial magnetic field, align themselves in that field. It occurred to me that the principle underlying this simple experiment might be applied to stratigraphical and tectonic problems in the following way.

I started with the assumption that just before coming to rest the mineral particles of a sediment laid down in quiet waters are freely enough suspended to orientate themselves approximately in the earth's magnetic field at the time and place of deposition. Not all mineral substances are equally susceptible to magnetisation when placed in a field, and some are neutral. Moreover, during sedimentation many of the lighter particles would tend to be disturbed out of their magnetic alignment by the impact of others happening to settle upon them. But even after allowing for these aberrant particles, and for those that are quite neutral, we could expect that at least in some kinds of sediments there would be a sufficient number of susceptible particles to impart a magnetic polarity to the sediment as a whole ... If these premises are correct, the deposit as a whole when it sets into rock might be expected to possess, and retain permanently (unless subject to metamorphic forces), a magnetic polarity more or less marked according to the nature of the component minerals. In other words the rock would behave, in the mass, like a magnet with its axis along the direction of the earth's magnetic field at the place during the epoch of sedimentation. If therefore, we could experimentally determine the magnetic orientations of a close series of rock samples, taken from below upwards through a stratigraphical sequence, we should be able to visualise the changes that have taken place in the magnetic field of the earth at the given locality during the course of geological time.

*(Sahni, 1944: xv–xvi)*

He discussed his ideas with K. S. Krishnan and others, read papers by workers at the Carnegie Institution, and about the magnetization of igneous rocks.

Paleomagnetism could be used to help solve two riddles. The first, a major conundrum of Indian geology, was whether the Saline Series of the Salt Range of the northern Punjab is pre- or post-Cambrian. Sahni and his co-workers had discovered Eocene microfossils throughout the Series. But the Saline Series *underlies* Paleozoic beds of purple sandstone, suggesting that it is Precambrian. Sahni therefore argued that the Paleozoic beds had been overthrust on top of the Saline Series. Sahni saw that paleomagnetic survey might be able to solve the problem.

If the rock sequence is a conformable one we should expect to find only a gradual series of changes in orientation, reflecting secular variations in the [magnetic] field of the earth. But if the geological section conceals a large stratigraphical break, such as the suspected overthrust in the Khewra gorge, we should find a sudden discordance between the readings just below and just above the tectonic junction. Even if the thrust involved no torsion in the strata, the enormous time gap which the microfossils indicate between the Saline Series and the Purple Sandstone may be marked by a larger discordance in the magnetic readings than could be explained by secular variation alone.

*(Sahni, 1944: xvi; my bracketed addition)*

The second riddle was continental drift. Sahni proposed the idea in a section entitled "Magnetic Orientations and Continental Drift."

The Gondwanas offer such a vast and impressive series of freshwater sediments, largely undisturbed by folding, that they may be eminently suited for testing Wegener's theory of continental drift by the magnetic method. If the present Gondwana blocks have really drifted apart by the disruption of a Pangea it might be possible, with the aid of magnetic readings taken from above downwards through the stratigraphical sequence in different parts of Gondwanaland, to work backwards and reconstruct Pangea. At least we might be able to visualise the location of some of the blocks as they stood in their magnetic meridia round the South Pole of successive epochs since the Carboniferous glaciation, when they are supposed to have been juxtaposed.

*(Sahni, 1944: xxi)*

Sahni did not describe a model of Earth's magnetic field, but he certainly envisioned the use of paleomagnetism to test mobilism. He wanted to trace magnetizations of the Gondwana fragments backward in time through the Carboniferous; he thought their high southerly latitude ought to be detectable, but never pursued his idea. Sahni died in 1949.

### 1.20  Irving initiates his test of motion of India

What is important about Irving's linking continental drift and paleomagnetism, and which distinguishes him from others, is that he immediately acted on it. As Runcorn nicely put it:

Graham lists, of course, a lot of possible applications, and he obviously had been asked by someone, perhaps Tuve, to write down the possible applications. But this was a kind of idle notion, which, of course, had no meaning really until it was understood, really by us, that the Earth's field had to be along the axis of rotation.

*(Runcorn, 1984 interview with author)*

Runcorn, upon learning of Irving's specific test told Browne, who had a habit of asking him whether any of his expanding number of students had come up with any good ideas. Having got one, Runcorn took Irving specifically to see Browne. It is very much to Runcorn's credit that he recognized the fecundity of Irving's idea and encouraged him to follow it up. At the time, Runcorn was very far (five years) from being a proponent of continental drift; this may have been the first time that he had thought seriously about drift, and was likely the first occasion that he heard of a practical way of testing drift paleomagnetically. Although Irving cannot recall Browne's reaction, Runcorn remembers (1984 interview with author) that Browne was not terribly excited. More important, however, was the reaction of R. A. Fisher, which Irving recalled:

Runcorn knew that Fisher was going to India at about that time. And, so he took me to dinner with Fisher. Fisher was absolutely delighted. Of course, he thought this was a tremendous idea and he immediately said he would do everything he could to get me a few samples ... I think he was overjoyed because he had begun to read widely in drift. I think he saw that paleomagnetism was going to add a lot to the discussion and, I think, perhaps, he may have had a puckish delight in teasing Harold Jeffreys in this way. Because he had had a difference of opinion with Harold Jeffreys about statistics and because of Harold's well-known views about continental drift, Fisher may have taken some delight in taking the opposite point of view.[29]

*(Irving, 1981 interview with author)*

Fisher went to India near the end of 1951 as a guest of the Indian Statistical Institute. He went to see M. S. Krishnan, then Director of the Geological Survey of India (I, §6.15). Krishnan agreed to help, and gave Fisher a copy of his newly published *Geology of India* as a gift for Irving (Irving, 1981 interview with author; Irving, 1980 letter to Runcorn).[30]

In his Notes, Notebook I (p. 52), following mention of co-latitude and inclination, Irving summarized much of Krishnan's book, and followed it with three pages of notes entitled "Indian Requirements" in which he listed twenty-one horizons in India he thought suitable for testing continental drift. Irving remembers:

I noted the book [Krishnan's] in considerable detail, set out a list of horizons and wrote a very long letter to Krishnan about all this. My notes still exist, but unfortunately, not the letter. I set out all the key horizons, and the letter went to the Geological Survey of India, and subsequently, samples were sent to us.

*(Irving, 1981 interview with author; my bracketed addition)*

Irving now had to wait. He could not advance his test until samples arrived from India. However, he continued to read and to think about continental drift.

### 1.21 Why Runcorn and Irving did not immediately redirect all their work to test continental drift

Although Irving saw the paleomagnetism–drift connection, and initiated his "India test," neither he nor Runcorn redirected their main efforts toward testing continental drift. Although Runcorn had recognized the potential of Irving's Indian idea, told the head of the department about it, and taken Irving to see Fisher who had strong Indian connections, he did not drop what he was doing to test continental drift; he continued to pursue his own interests in geomagnetism, and to be interested in the results Irving was getting from the Torridonian. Nor, for that matter, did the non-axial or strongly oblique paleomagnetic directions of their first Torridonian samples affect their immediate thinking about continental drift. As already noted, even Irving did not think of the Torridonian oblique directions in terms of continental drift because the time frame was wrong; Wegener had not talked about Precambrian continental drift. Although later, in his thesis (1954: 109), he came to recognize continental drift as one possible explanation of the oblique directions that he eventually observed throughout the Torridonian, he did not, in late 1951, think of the data from Craigriannan in this way: it was, after all, from one locality in a huge sequence more than 5000 m thick; at that early stage it would have been presumptuous to have done so. Being an experimentalist, Irving became obsessed with finding lithologies that would provide internally consistent data, and later hunting for them elsewhere in the Torridonian. In addition, the obliqueness of their Cambridge findings ran counter to the American experience that oblique directions were rare and the exception not the rule (§1.9, §1.10). Finally, at the time when Irving came up with his India idea, many of the analytical procedures needed to test continental drift and polar wandering paleomagnetically had yet to be developed. Irving's test was just a hunch, albeit an educated one.

Although the oblique directions of Torridonian samples provoked no discussion of testing continental drift, doing so paleomagnetically had by 1953 become a general topic of conversation in Cambridge once Irving begun finding oblique directions more generally in the Torridonian; even though Wegener had not talked about drift during the Precambrian, the strongly oblique Precambrian directions nevertheless suggested that old, large-scale effects were present, and, as I shall describe, they prompted Clegg and colleagues and Creer, who were aware of Irving's results, to collect younger samples in order to track oblique directions forward to the present day (§2.4, §2.12).

## 1.22 Reaction of Blackett and Runcorn to Irving's work

Runcorn and Blackett were impressed with Irving and his early work. As a consequence, Runcorn decided, as part of his plan to expand Cambridge's research program in paleomagnetism, to take steps to have him registered as a Ph.D. student. Runcorn recalled:

It wasn't long before I realized that he was a very able chap. Of course, I introduced him to Blackett, because we did our first measurements on the magnetometer that Blackett had made at Manchester – in the summer he was using Blackett's magnetometer. And I said to Blackett, "Don't you think he is rather a good chap?" And, Blackett said, "Yes." And, I said, "Well, the geologists only gave him a bottom 2nd class degree." But, by that time I had got a grant to build a "Blackett" magnetometer at Cambridge and start this work ... So, after a while I said to Ted, "Would you like to do research for a Ph.D.?" And, so he said he would.

*(Runcorn, 1984 interview with author)*

Runcorn spoke with Browne about admitting Irving as a Ph.D. student. Browne was not in favor. Irving's poor degree troubled him, and he had not taken university courses in mathematics and physics; as far as Browne was concerned, he was doubly unimpressive. Not only was he a geologist, but he appeared on paper not to be a particularly good one. Nor was Browne very sanguine about Runcorn's desire to get yet another student for his rapidly expanding geomagnetism group. Browne and Runcorn had very different attitudes about how to do research. Runcorn addressed both issues:

I then said to [Ben Browne] the head of the department at Cambridge, who was beginning to get a little hostile to my empire building, because I had got quite a lot of students, "Would you be agreeable [to Irving's become a Ph.D. student]?" He had the idea, his philosophy, of doing research was to have one person doing paleomagnetism, one person doing something else, another doing something else. So, of course, we got at loggerheads because I wanted to build up a group entirely concerned with geomagnetism. So, Ted entered. Ben was not a very forceful person, and anyway, with Blackett's backing he accepted Ted as a research student. But he was always embarrassed with the fact of letting in someone with a bottom second class degree, you know, because he had the very highest standards. So he was always skeptical about Ted, even after he began to get very interesting results.

*(Runcorn, 1984 interview with author; Runcorn's comment about Browne's "very highest standards" was sarcastic; my bracketed additions)*

Irving officially became a Ph.D. research student in the Department of Geodesy and Geophysics in the autumn of 1952. Irving later summarized his status in the department:

During the academic year 1951–1952 I was employed in a temporary position as assistant to Runcorn, but I cannot remember the exact title of the position. In July 1951 it was too late to apply for a research position for 1951–1952 so Runcorn arranged it this way, very sensibly. I was registered as a research student in the autumn of 1952, I think. I understand that he had difficulty persuading B. C. Browne, head of the department, of my suitability as a research student – lack of physics, etc.

*(Irving, August 20, 1985 letter to author)*

In retrospect, Browne turned out to be wrong on both counts. Runcorn's idea of getting several students together to work on a related cluster of problems proved very fruitful, for he, Hospers, and the students became more responsible than any other group for the rise of paleomagnetism in the early 1950s. Runcorn's idea of hiring a geologist was not only sensible, it was necessary. His group was interdisciplinary, which had much to do with its success. Hospers' and Irving's backgrounds in geology were essential to their work. Fisher's contributions to the Cambridge group were enormous, and Ken Creer, whom Runcorn was soon to recruit, although primarily a physicist, had had two years of geology courses. Runcorn, I believe, was one of the first Earth scientists to actively recruit geologists, physicists, and geophysicists to work on the same group of problems. He recognized that he would need a varied group to tackle problems that cut across established disciplines. Runcorn's acting on this realization is one of his many important contributions to the development of the Earth sciences. Runcorn recognized that Irving had energy, and ideas to boot. The Cambridge Geology Department's and Browne's assessments of Irving were wrong.

## 1.23 Summary

At this embryonic state of the revival of paleomagnetism I want to emphasize several points. Workers had no prior commitment to continental drift. In Britain, Runcorn and Blackett knew little about it when introduced to paleomagnetism. Before starting research, Hospers knew about continental drift but thought it was a dead issue, and he had not heard of reversals of the geomagnetic field. Irving, although well aware of drift, was not committed to it; he thought it an intriguing idea, and understandably was the first British paleomagnetist to make the connection between mobilism and paleomagnetism and initiate its first paleomagnetic test. At the Carnegie Institution, Tuve and Graham and their colleagues were unsympathetic or hostile to both mobilism and field reversals. Although Graham recognized that paleomagnetism could be used to test mobilism, he did not explicitly act on the idea until March 1955 when he began collecting samples in the southwestern United States (§7.4), at least six months after Runcorn had collected rocks from the same general area (§3.8), and a couple of years after other British paleomagnetists had begun testing drift paleomagnetically.

Second, the possibility of using paleomagnetism to test continental drift was not why paleomagnetic research at the Carnegie Institution and in Britain began. Workers there did so to learn about the history of the geomagnetic field, its origin and behavior. The exception was Hospers who went to Iceland to use the intensity of magnetism as a stratigraphic tool; only after getting his first results did he turn to reversals. Runcorn was interested in secular variation and, after seeing Hospers' results, he became interested in reversals. Researchers at Carnegie knew that paleomagnetism might be used to test mobilism, but did not begin to do so until 1955, six years after Graham had noted this possibility.

Third, although British paleomagnetists came to realize the connection between paleomagnetism and mobilism, and Irving, with encouragement and practical help from Runcorn and Fisher, had even launched the first test, they did not suddenly redirect their work toward testing Wegener's theory of continental drift; they could only do so much at once. Hospers continued to concentrate on reversals. Runcorn learned how paleomagnetism could be used to test continental drift from Irving;

Runcorn liked the idea, but did not redirect their paleomagnetic research. He did not direct Irving to stop collecting Precambrian rocks and begin working with younger rocks that had formed when Britain supposedly drifted to its present position. Similarly, although Graham and his colleagues at Carnegie knew that paleomagnetism could be used to test mobilism, they made no move to do so.

Fourth, early paleomagnetists, Irving excepted, directed none of their attention to testing mobilism specifically. They were busy on other things, notably getting reliable (reproducible) data which became an end in itself, paleomagnetism being a relatively new field. There was no point in wondering about its global implications until they knew how to obtain coherent and reliable results. Later, they could worry seriously about what the data meant – perhaps in six months or a year, but not just then.

Fifth, sedimentary rocks became the focus of attention because they were, at the time, generally better dated and could be expected to contain a more continuous record than igneous rocks. Although their remanent magnetism was weak, Blackett's magnetometer built to test his fundamental theory of the Earth's magnetism was for the next decade sufficiently sensitive for this new task. British paleomagnetists had the prototype magnetometer before they began working on reversals or secular variation, and before turning to drift. Blackett let Hospers and Irving use his magnetometer, wrote a lucid account of it, and it was soon duplicated elsewhere and modified specifically for paleomagnetic studies. Paleomagnetists at the Carnegie Institution had Johnson's spinner magnetometer, designed in the late 1930s, which they continued to improve after World War II, and which was fully capable of measuring the weak magnetization of many sedimentary lithologies.

Sixth, interest in the ancient geomagnetic field required field tests to determine when remanent magnetism was acquired. Building on earlier work, Graham developed such tests as were available before drift-testing began in earnest. Before they could test mobilism, paleomagnetists would need statistical methods for analyzing dispersed data, and these had to be created by Fisher before drift-testing could begin. Next, paleomagnetists needed to know what kind of sedimentary rocks gave reliable records of the geomagnetic field. Eventually several lithologies (including many igneous rocks) would prove useful, but at this early stage Irving and Runcorn's demonstration that fine-grained red beds gave reliable results was, as I shall describe, crucial to drift-testing.

Finally, this study of the development of paleomagnetism after World War II shows that it flourished in Britain while declining in the United States, a situation which, as I shall show, continued throughout the remainder of the 1950s. Work continued at the Carnegie Institution but there was no surge of progressive activity as there was when Runcorn found Hospers, hired Irving, and soon Creer, and built up paleomagnetic work at Cambridge University, and when Blackett formed a group at the University of Manchester, later moving it to Imperial College, London. Moreover, everyone at the Carnegie Institution of Washington was dead set against geomagnetic field reversals and mobilism, the central problems that paleomagnetists in Britain would address in the years to come; the negative attitude of those at the Carnegie Institution, especially to mobilism, was, as I showed in I, §7.10, common to most North American Earth scientists, and it is this resistance to mobilism, I believe, which more than anything else explains why paleomagnetism stalled in the United States. Because the mandarins of North American Earth science viewed mobilism as a relic of bad geological thinking, there was no reason to encourage young workers and to fund their paleomagnetic research. Although mobilism had had a mixed reception in Britain (I, §8.13), from the start paleomagnetists there were open to the possibility of geomagnetic field reversals or mobilism, or both, and once they saw their relevance to the fixism/mobilism debate, they pursued them vigorously. Britain also had Blackett, a charismatic national figure whose enthusiasm for paleomagnetic work was a great support. In the United States, no one of his stature supported paleomagnetic work.

## Notes

1 This account of Runcorn's early life is drawn from Collinson (1996, 1998).
2 Nye (1999) details Blackett's development of his fundamental theory, and its eventual demise. She covers some of the theoretical objections to Blackett's view, and brings out the media attention Blackett enjoyed upon proposing his theory. She also provides a view of Blackett as someone willing to propose bold hypotheses, but demanding that they be testable. As Nye (1999: 91) puts it, "In short, Blackett expressed the need for boldness in theoretical conjectures or in experimental strategies if they could be tied rigorously to what he called phenomenological statements." I add that this approach made Blackett a good Popperian. Nye's 2004 highly readable *Blackett* provides the most thorough account to date of Blackett's scientific and political life. She provides excellent accounts of his experimental work in particle physics, later work on cosmic rays, development of his fundamental theory, and its testing with the construction of his astatic magnetometer at Jodrell Bank. She also discusses Blackett's contributions to paleomagnetism, and his support of continental drift.
3 That Bullard made this suggestion is well documented. Runcorn (1948) and Chapman (1948b) both credited Bullard with it.
4 E. C. Bullard also suggested (correctly) that Hales and Gough's result and Runcorn's result from Leigh might be unreliable due to local effects.

> The theory [Bullard's self-exciting dynamo theory], like all theories that ascribe a deep internal origin to the field, is inconsistent with the experiments of Hales & Gough (1947) and Runcorn (Chapman, 1948) on the variation of the field with depth. It is not impossible that their results are due to local anomalies, and it is of great importance that the doubt should be removed by measurements at other places and especially at sea.
>
> *(Bullard, 1949: 452; my bracketed addition; Chapman 1948 is my Chapman 1948a)*

5 Apparently, Blackett either found Bruckshaw's presentation of little interest or simply forgot about it, for he later recalled:

> My own interest in rock magnetism was aroused initially by the publication in 1948 of a paper entitled "Pre-history of the Earth's Magnetic Field" by Johnson, Murphy and Torreson (1948) of the Department of Terrestrial Magnetism, Washington. These workers showed that it was possible to trace back the history of the earth's magnetic field far beyond historic times by the measurement of the weak magnetism of certain sedimentary rocks.

Thus stimulated I explored some of the extensive literature of the last fifty years dealing with the magnetism of both igneous sedimentary rocks, and of the degree to which they could give a record of the earth's field during geological times.

*(Blackett, 1956: 5)*

Nor is there any doubt as to Blackett's attendance at this meeting because Chapman reports that Blackett made a remark there (see, Chapman, 1948b: 464).

6 Bullard quickly wrote up his talk and submitted it (Bullard, 1948) for publication on March 20, 1948. He located the origin of the geomagnetic field within the Earth's core. His self-exciting dynamo model of the geomagnetic field appeared the following year (Bullard, 1949). He developed his account of secular variation before learning of Elsasser's work. Apparently Bullard found out about Elsasser's self-exciting dynamo model shortly before submitting his own article (see Bullard, 1948: 249). Perhaps Chapman told him of Elsasser's work at the meeting. A difficulty which he recognized with Bullard's account of secular variation was that it required an extremely large magnetic field within the interior of the Earth's core; he found that he could remove it by adopting a self-exciting dynamo model for the generation of the Earth's mean magnetic field. The needed internal field, following Elsasser, was extremely strong, contained within the core, and toroidal in shape. Beginning with Elsasser's work, Bullard expanded one of Elsasser's feedback mechanisms for generating a self-exciting dynamo. The particular mechanism, rejected by Elsasser as "so complicated and artificial that it would hardly seem convincing," was greatly simplified by Bullard. For comments about the complication of Elsasser's original feedback mechanism, see Elsasser (1947: 827) and Bullard (1949: 446); for comments about the simplifying aspect of Bullard's analysis see Bullard (1949) and Elsasser (1950: 31).

7 Most of this biographical information about Graham is from R. R. Doell (1974).

8 Because Graham (and his colleagues) did not consider seriously the possibility of continental drift, he mistakenly assumed that the Rose Hill magnetizations are reversed. I make two comments. First, they are far from being reversed with respect to the steeply down-northerly directions that he had observed from flat-lying beds outside the Appalachian orogenic belt and which he believed recorded the mid-Paleozoic geomagnetic field; the declinations of the Rose Hill are northerly and for their polarity to truly be reversed they would have to be southerly; the inclinations were about $-40°$ whereas to be reversed they would have to be much steeper around $-70°$. Second, the alternative and now generally accepted interpretation of its $40°$ inclination is that the Rose Hill Formation was deposited in low southerly latitudes and the polarity was normal. This was explained in Creer, Irving, and Runcorn (1957: 152) in which they show that the Rose Hill paleolatitude was $25°$ S, as later confirmed by French and Van der Voo (1979). Although, in his 1949 paper, Graham thought paleomagnetism offered the possibility of testing continental drift, it is ironic that he did not even consider that possibility in interpreting his Rose Hill results and as a consequence mistook their polarity; thus his request to L. Néel, asking if rocks could undergo self-reversals in an attempt to avoid admitting field reversals (§1.10), rested on a mistake.

9 Néel was awarded a Nobel Prize in 1970 for his work on antiferromagnetism and ferrimagnetism.

10 Van Bemmelen's father (W. van Bemmelen, 1883) had worked on the problem of secular variation; in 1893, he helped establish a baseline for the measurement of secular variation by surveying navigators' records made from 1540 to 1680. For a brief discussion of the older van Bemmelen's work, see Elsasser (1950).

11 These gravity surveys resulted in two publications: A. H. Cook et al. (1952) and J. Hospers and P. L. Willmore (1953).

12 Hospers initially suggested to me that he thought he had been the first to use the expression "reversed sample," but, as far as I have been able to discover, the precedence goes to Manley. My guess is that Hospers got it from Manley. Hospers, in responding to such a suggestion, stated:

As I explained before, reading Manley's article on paleomagnetism in *Science News*, was important to me. I still possess this little book, and have it now before me. Penguin

Books published it in July of 1949. I agree with you that my use of "reverse" may have come from this article. Also, in looking at this article again after more than thirty years, I find it a very good overview, much better than I had remembered it.

*(Hospers, May 28, 1985 letter to author)*

13  Hospers addressed the question of Browne's official and Runcorn's unofficial statuses.

You may remember that I was a research student at the Department of Geodesy and Geophysics of Cambridge University during the four academic years of 1949–1953. B. C. Browne was then Head of the Department. He was my official supervisor during this entire period. Keith Runcorn joined the Department (January 1950) as Assistant Director of Research. As Browne was principally interested in gravity and Runcorn was interested in geomagnetism, I turned to Runcorn whenever I had any questions concerning my research. However, Runcorn was never my official supervisor but my unofficial advisor.

There was never any clash between Browne and Runcorn about this matter. I think this was partly due to the fact that I actually formally asked Browne's permission to change from some vaguely formulated gravity research to paleomagnetic research. This happened late 1949 or early 1950, before Runcorn arrived, as I was already interested in paleomagnetism (on the basis of hints from my professors at Utrecht and because I had read Manley's paper late in 1949).

*(Hospers, May 28, 1985 letter to author)*

14  Runcorn is in general agreement with these claims about the independence of Hospers' work and its importance in convincing him of the reality of geomagnetic reversals in polarity. However, Runcorn thinks that his own theoretical work in geomagnetism, work which offered a model consistent with the occurrence of field reversals, influenced Hospers in his own acceptance of the idea of field reversals. Hospers firmly denies this. I agree with Hospers, and I shall explain why when considering differences between Hospers' and Runcorn's methodology (§2.14) (Runcorn, 1984 interview with author; Hospers, July 8, 1985 letter to author).

15  This story about Fisher pulling out a set of old notes is recounted by Joan Fisher Box, one of Fisher's six daughters, in her biography of her father (Box, 1978: 439). Hospers also recalled how Runcorn told him the same story in the early 1950s (Hospers, May 28, 1985 letter to author). See also N. L. Fisher *et al.* (1987: 1–14).

16  I thank E. Irving for this technical fast forward.

17  Cox and Doell described the "consistency-of-reversals test" as follows:

Parallelism [regardless of sign] between tightly grouped mean directions of magnetization in two groups of samples which are reversely magnetized with respect to each other is a much stronger test than simple consistency of directions without reversals. This test applied to reversals due either to field or self-reversal, since in both cases the mean directions of magnetization are 180° apart. If subsequent to the original magnetization, both groups acquire an additional component of magnetization ... the two resultant groups will no longer be 180° apart.

*(Cox and Doell, 1960: 656; my bracketed addition)*

18  They submitted their work on December 28, 1950 and revised it in January 1951.

19  Hide was one of several undergraduate students at Manchester who helped Runcorn with the mine experiments. Hide came to Cambridge as a Ph.D. student of Runcorn's in 1950. Because Hide's work became so important for meteorology, it is worth stressing the fact that the original motivation came from geomagnetism. Runcorn recalled:

We talked about projects, and I had got the idea that it would be interesting to look at convection in connection with the Earth's core, because by then Bullard had the first paper on how an eddy in the Earth's core could generate a field. I thought of how one could simulate the Earth's gravity and do a model experiment of convection, and the nearest, of course, you can get to it in the laboratory is a rotating cylinder. And, so he set up his experiment, which has since become very famous, in which you have a cylinder, a fluid, and you have a heat source outside and a cold source inside, and you made use of the centrifugal force to produce this convection. And, of course, he did his Ph.D. on that. But, when in setting him on this, I was thinking that it might throw some light on the core, but, in fact, it interested the meteorologists more, and that was how Hide started.

*(Runcorn, 1984 interview with author)*

20 Where and precisely when Hospers gave this talk that convinced Runcorn of field reversals is unknown. Some suggested either the $\Delta^2V$ Club or the Kapitza Club; however, neither is correct. Daphne Sulston, formerly of the Department of Applied Mathematics and Theoretical Physics at Cambridge University, at Irving's request, kindly examined the records of both clubs at the Churchill College Archives Centre. She found that Runcorn and Hospers gave a joint talk on geomagnetic reversals to the Kapitza Club on April 14, 1953. However, this was much later and clearly not the talk by Hospers that inspired Runcorn and convinced him of reversals. She also found that there is no record of Hospers having given a talk to the $\Delta^2V$ Club from March 14, 1950 through May 19, 1953. Hospers recalled (Hospers, February 17, 1981 letter to Irving sent to author) that he gave talks to the Sedgwick Club but in the winter 1952/3, at a symposium at the Department of Geodesy and Geophysics, and at a joint session of the Royal Astronomical Society and Geological Society on January 23, 1953, and at the Department of Geology, University of Birmingham in January 1954. All these talks are far too late to be the talk Hide referred to. The talk in question might have been given to the Geophysics Club in King's that Maurice Hill organized and whose meetings were unrecorded.

*(Irving, December 20, 2007 email to author)*

21 According to Irving (January 16, 2007 note to author), Runcorn actually wrote him a note, and Irving then telephoned Runcorn by public phone; his parents never had a phone.

22 Nor, for that matter, has one of Irving's fellow students forgotten the incident. Donald Hitchcock, who was a year ahead of Irving but also in Miss Whitaker's class, recalled Irving's Freudian slip:

Wegener came into the story when, in my case I was in the sixth form, Miss Whitaker had certainly been taught about either sliding continents [Daly's theory] or the convection theory [Holmes' theory] when at Manchester University ... Your story and personal involvement is not a figment of the imagination. The lower and upper sixth did the same course year about, as far as possible, sometimes the upper sixth mainly me, would be absent in the library. Your Freudian slip was on such an occasion. The subject arose from a discussion of earth tectonics and the class consisted of Land, Yourself and the girls, I was out of the room at the time. [Hitchcock was out of the room because he was in the upper sixth form; Irving was still in the lower sixth form.] Apparently you simply announced, as was not uncommon, that there was a theory put forward to explain the movement of the continents by a man called "Vagina". The girls tittered and you asked me later why did Miss Whitaker not follow up a normal observation or something to that effect. As I said earlier we were rather naïve.

*(Donald Hitchcock, July 22, 1987 letter to Irving, copy sent by Irving to author; my bracketed additions)*

23 Actually, nobody in Irving's class at Cambridge did very well, and it is perhaps understandable why the Geology Department did not recommend anyone to Runcorn. However, all that Runcorn was asking for was somebody with enough expertise in stratigraphy to help him collect rocks. Irving, in remarking about his and his classmates' lack of success on their exams, said:

Yes, I did only averagingly (lower second) in my final exams. In Cambridge there are 4 grades: first, upper second, lower second, third. In my year in Geology (if I remember correctly) all the class got thirds or lower seconds except for one upper second.

*(Irving, August 20, 1985 letter to author)*

24 Conversation between Neil Opdyke and author.

25 Looking back, Runcorn's choice to study Pre-Cambrian rocks was excellent because it eventually anchored the old end of Creer's APW path (§3.6).
26 Much of this detailed analysis of Irving's work comes from a synopsis of records that Irving kept from 1951 through early 1955. Irving sent the synopsis, which he prepared in May–June 1985 to me at my request in a letter to Irving of March 12, 1985.

These records cover the period when I began research in the summer of 1951 until the time I left England for Australia in November 1954 ... Minor entries are not given, for brevity's sake. Dates given in the records themselves are underlined. Comments are given in brackets and labeled "COMMENTS"; these are not historical facts but my opinions 34 years later.

*(E. Irving, Synopsis of records kept by E. Irving 1951 to early 1955;*
*letter to author June 1985).*

Irving recorded his activities in three types of notebooks: field, research, and notebooks (which I shall refer to as the *Note notebooks*) containing comments on books and papers he read, lectures he attended, ideas and discussions he had, etc. All are hard-backed books. The field notebooks cover the period from July 1951 to September 1953. There are eight research notebooks. They are chronological, but often contain overlapping dates since they are somewhat subject specific. The first research notebook, entitled *Helman Head Series Devonian of Caithness* contains dated entries from August 21, 1951 to September 15, 1951. The latest dated entry in all of the research notebooks is July 1954. The *Note notebooks* contain dated entries of October 1951 through undated ones about work Irving was doing in 1955 after he had taken a position in Australia. There are four *Note notebooks*. Unless stated otherwise, all dates in this section are from data-entries in Irving's notebooks. These dates are secure.

27 Irving recorded his results in two research notebooks entitled *Helman Head Series* and *Torridonian*. The appropriate pages of the notebooks are dated August 16 – September 3 (*Torridonian*), August 21 – September 15 (*Helman*), and September 13–16, and August, 1951 (*Torridonian*).
28 It is understandable that Sahni's connection between paleomagnetism and continental drift went almost entirely unnoticed. He made the suggestion during his 1944 Presidential Address before the Indian National Academy of Sciences. The title of the published version of his address, "Microfossils and Problems of Salt Range Geology," would not have caught the attention of either paleomagnetists or mobilists. Moreover, the talk served as an introduction to a symposium on the age of the Saline Series in the Salt Range of the Punjab. Indeed, the only reference I've seen to Sahni's connection between paleomagnetism and continental drift is by J. S. Lee. Lee, a pro-mobilist Chinese tectonicist contributed a paper to the posthumous memorial volume honoring Sahni for his many contributions to the Earth sciences. Lee began his talk by recalling a much-appreciated letter he had received from Sahni just after the end of World War II.

In the fall of 1945 when the writer, like many others, was lying in Chungking seriously ill and exhausted under the strain of the recent war, a letter from Professor Birbal Sahni came in one morning through the thin line of communications of those days. That letter not only conveyed the warmth of friendly sympathy, but also contained illuminating hints on certain promising scientific undertakings which imparted, in no small measure, revitalizing interest to a life in distress. Fresh hope shone in the almost failing eyes of the afflicted. Professor Sahni alluded to, among other things, subjects so remote from orthodox geology as fossil magnetism, in the hope of securing possible evidence for the successive stages of diastrophic development leading to such extraordinary distortion of the crust of the earth as the hair-pin bend in northeastern India. This is just one example of how his searching and restless intelligence was probing the dark corners of vast, and as yet unexplored fields.

*(Lee, 1952: 298)*

Even Lee did not refer to Sahni's Presidential Address. Perhaps he knew about Sahni's insight only from the paleobotanist's letter, and did not know that Sahni had publicly raised the idea in his Presidential Address.

29  Jeffreys and Fisher had a long-standing feud about the nature of statistical inferences. Jeffreys defended Bayesianism while Fisher espoused the classical view of probability. Box (1978: 422) has an amusing story from Bullard about their disagreement. Jeffreys and Fisher attended a lecture by Eddington on scientific inference, "and were so horrified that they shook hands and promised not to write any more rude things about each other."

30  Apparently, Fisher returned to England before the year's end because he gave to Irving this copy of Krishnan's book which Irving dated 1951.

# 2

# British paleomagnetists begin shifting their research toward testing mobilism: summer 1951 to fall 1953

## 2.1 Outline

This chapter describes the paleomagnetic work in the Manchester/London and Cambridge groups from 1951 to the end of 1953, during which most British paleomagnetists directed much of their attention toward testing mobilism.

Toward the end of 1951, Blackett invited John Clegg to start a paleomagnetic group at Manchester. Clegg constructed a new magnetometer, and recruited Mary Almond and Peter Stubbs. They began collecting rocks in October 1952, discovering, as Irving had, that fine-grained redbeds give the most coherent results, and shortly after began a survey across England of red marls of Triassic age. Wasting no time, they submitted what became the first published paper describing paleomagnetic evidence in support of drift and/or polar wandering (Clegg, Almond, and Stubbs, 1954a); it was seminal work. Meanwhile, Blackett accepted a position at Imperial College, London, and in 1953 Clegg and colleagues went with him.

During the summer of 1952 Hospers made a second field survey in Iceland. The results of this survey together with his previous work demonstrated for the first time that reversals were not random but occurred sequentially in stratigraphic succession, greatly strengthening his earlier claim that they were caused by reversals of the geomagnetic field; he disagreed with American workers at the Carnegie Institution who had claimed that self-reversal was likely responsible (§1.10). Hospers also provided empirical support for the GAD hypothesis, which became fundamental to the paleomagnetic case for mobilism. However, his own test of mobilism, based on Miocene and younger Icelandic results, gave within statistical errors no support for polar wandering or lateral drift of Iceland for that interval. In the fall of 1953, he joined Royal Dutch Shell and left paleomagnetism for fifteen years.

Irving continued working on the Torridonian in Manchester and later at Cambridge. By 1954 he had finished a three-year study of this thick Precambrian sequence. Just as Hospers had observed in Iceland lavas, he found normal and reversed strata in stratigraphic succession, although the axis of magnetization was strongly oblique to Earth's present axis; his were the first sequential reversals to be observed in sedimentary strata. To explain the obliqueness of his results he invoked polar wander or drift of "that part of the crust containing NW Scotland" (Irving, 1954: 109). He also worked on samples from India that indicated it had drifted northward.

In spring 1952, Runcorn recruited Ken Creer, who designed and constructed a Blackett-style magnetometer at Cambridge. Earlier, Creer had worked without much success on a range of lithologies. Like Irving, he was led by a process of elimination to focus on fine-grained redbeds. By late 1953, he too began finding support for continental drift and/or polar wandering.

In 1953, Runcorn developed a model of Earth's magnetic field, providing theoretical support for the GAD hypothesis. Leaving Great Britain to his students, in the summer of 1953 Runcorn began his own fieldwork; inspired by Hospers' work on the Late Cenozoic basalts of Iceland, he collected rocks of similar age from Oregon and Idaho – the Columbia River basalts. He was interested in reversals and, in contrast to his students, his intention was not to test continental drift, which he did not favor at the time (§2.13). Fisher continued to be interested in the work of Cambridge paleomagnetists, and gave a talk in late 1952 or early 1953 in favor of continental drift.

## 2.2 Blackett initiates and Clegg leads the paleomagnetic group at Manchester

Although Blackett did not finish testing his distributed theory of the geomagnetic field until April 1952, he decided sometime before the end of 1951 to initiate a research program in paleomagnetism. His ambitious experimental test of his theory (§1.3) was winding down, and he was, as Runcorn reported (§1.22), impressed with Irving's initial paleomagnetic results.[1] He certainly was impressed with Irving because he asked him if he wanted to pursue a Ph.D. at Manchester in paleomagnetism. Irving thinks that Blackett probably asked him during

July or August 1951. It was during lunch in a pub near Jodrell Bank and Runcorn was there. Almost certainly it was after Runcorn had mentioned to me the possibility of returning to Cambridge but before he had arranged the financing.

*(Irving, March 13, 1987 letter to author)*

Blackett also wanted to learn geology and with Runcorn and Irving visited the Sedgwick Museum at Cambridge. Runcorn recounted Blackett's attempts to educate himself:

In 1951 Blackett visited me for a few days at Caius to concert plans and, as neither of us had had any training in geology, I took him to the Sedgwick Museum where we were jocularly greeted by Professor W. B. R. King, who said, "Have you come to learn all about geology this morning?" Well Blackett had. It was one of his remarkable characteristics that he was able, by vigorous questioning of a chosen expert (who often felt less of an expert afterwards) and by a judicious choice of literature (in this case Arthur Holmes' *Physical Geology*) to get up a new subject with remarkable speed and effectiveness.

*(Butler et al., 1975: 156–157)*

Earlier, in Manchester, Blackett had asked Irving to suggest a geology text. Irving lent him the copy of Holmes' *Principles of Physical Geology* that he had borrowed from Manchester University Library. Irving had not read Holmes' textbook as an undergraduate; he does not remember it being on the Cambridge reading list, after all King had reviewed it very critically (see I, §5.9 for a discussion of the review).

I can date Blackett's reading of Holmes as October, 1951 + or − three months because he borrowed from me the copy that I had borrowed from the Manchester University Library whilst living temporarily at Donner House.

*(Irving, 1980 letter to Runcorn)*

Blackett returned the book to him a few days later. Because Blackett made no comment, Irving, a temporary research assistant, thought it was not his place to ask and assumed that Blackett must not have thought very much of it. However, upon opening the book, Irving noticed several loose pages of notes in which Blackett had outlined the principles of geology as presented by Holmes. He read the notes and returned them to Blackett. They showed that Blackett, within a few days, had managed to work through Holmes' textbook, and apparently was quite impressed with its discussion of continental drift (Irving, 1981 interview with author). Blackett later described his reaction to Holmes' defense of mobilism:

I myself am convinced that the case for continental drift was rather strong quite apart from the new evidence from rock magnetism and oceanography. For instance, the survey by Arthur Holmes in *The Principles of Physical Geology* certainly played a valuable part in convincing the early workers in rock magnetism that the probability that the continents had drifted was high enough to justify an intensive study of the directions of magnetizations of ancient rocks. To a physicist like myself one of the most convincing single pieces of evidence was the Permo-Carboniferous glaciation of the Southern Hemisphere at about the same time that the great coal deposits of Northern Hemisphere were being laid down. I remember the impression made on me by Holmes's remark that there was just not enough water in the world to produce a large enough ice-cap if the continents were then in the same relative position to each other as they are today.

*(Blackett, 1965: ix)*

Blackett, unlike Runcorn, had quickly realized what a good idea it would be to use paleomagnetism to test mobilism. Just when and how he realized that paleomagnetism could be used to test continental drift is not clear. He could have come across it through reading Graham, Mercanton, or Gutenberg, or he could have come up with it on his own. Blackett also may have learned about it from Irving or Runcorn. Irving is not sure whether he told Blackett about his intended India test. Both Irving and Runcorn knew about Blackett's strong interest in India and his support of Indian science, and may have told Blackett of the planned attempt to obtain paleomagnetic samples from the Deccan Traps. Regardless, Blackett soon decided that the Manchester group would test continental drift.

As explained above, Blackett began by recruiting John Clegg. Clegg (1913–87) attended Queen Elizabeth Grammar School in Middleton, Lancashire. His sixth form physics and mathematics teacher, George Matthews, had received a degree at Manchester under Rutherford, and served as a science instructor in the Navy during World War I. Clegg was inspired by Matthews, whose course was wide ranging and included topics probably not on the syllabus (Clegg/Fuller 1987 interview with author[2]).

Following his old teacher, Clegg attended the University of Manchester where he received a B.Sc. degree in 1935, majoring in physics. During World War II, Clegg again followed his teacher and chose the Navy, where he was assigned to radar research. Stationed at Swanage, Dorset, in 1941, his group of eighteen learned about radar, working on aerials. Clegg eventually had the responsibility of checking radar receivers throughout Great Britain. He enjoyed the work because he was especially fond of hiking and mountain climbing and the receivers were usually located off the beaten track.

After the war he returned to the University of Manchester to earn a Ph.D. in physics, entering the Department of Physics in October of 1946 as a lecturer. There he met Keith Runcorn, another new Ph.D. student and assistant lecturer in physics. Clegg, because of his wartime radar research, decided to study radio astronomy with Lovell.[3] It was a good choice: the field was wide open, and Lovell was one of its leaders. Clegg spent much of his time during the remainder of the 1940s at Jodrell Bank helping to construct and fine-tune radio telescopes.[4] He used them to investigate meteor showers. Receiving his Ph.D. in 1949, he authored and co-authored over ten radio astronomy papers, published from 1947 through 1952.

While Clegg worked in radio astronomy, Blackett was perfecting his astatic magnetometer and testing his fundamental theory (§1.3). The magnetometer was in a farmer's field 200 yards away from the new Jodrell Bank observatory. As already noted, Blackett had decided by late 1951 to start a paleomagnetics group; Irving was then measuring his early rock collection on the magnetometer. Clegg had begun to have difficulty continuing to work with Lovell (Clegg/Fuller, 1987 interview with author),[5] and was asked by Blackett to head up the Manchester paleomagnetics group. During his undergraduate days, Clegg knew of Blackett, who was the external examiner when he took his undergraduate degree, Blackett then being at Birkbeck College, London. Clegg quickly accepted, and "felt that it was a great privilege to work with Blackett" (Clegg/Fuller, 1987 interview with author).

## 2.3 Clegg builds a new magnetometer at Manchester

Just when, in late 1951 or early 1952, Clegg actually began working in paleomagnetism is unclear. His last singly authored article in radio astronomy (Clegg, 1952) was received for publication in February of 1952. By April 1952, he was familiar with Blackett's work in geomagnetism, because he found and corrected several errors in the draft of Blackett's monumental paper on his negative experiment for which Blackett (1952: 369) thanked him; the paper was received for publication April 24, 1952.

One of Clegg's first tasks as operational head of the new paleomagnetic group was to design and construct a magnetometer better suited than Blackett's for paleomagnetic investigations. He must have begun early enough in late 1951 or early 1952 to have completed his task by, at the very latest, April 1952, because Blackett (1952: 329) makes reference to the recent setting up of a magnetometer "in a basement room of the Physical Laboratories of the University" of Manchester, and it is evident from other sources that the magnetometer in question was designed and built by Clegg.[6] A major advantage of Clegg's magnetometer was that it could measure samples more quickly than Blackett's, having a shorter period. It was somewhat less sensitive than Blackett's, but could measure the magnetization of many sedimentary rocks. Irving commented on this new magnetometer, and although his guess as to why it came to be built is, I believe, incorrect, I quote his remarks in full because they not only confirm Clegg's involvement, but reveal in a rather interesting way Blackett's kindness toward young researchers, his joy in experimental work and, perhaps, the last use of the Jodrell Bank magnetometer.

Yes I remember about the magnetometer in the Physics Department at Manchester. It had a shorter period but lower sensitivity than the Jodrell Bank magnetometer. I cannot remember when I first saw it, but Clegg was deeply involved and probably did much of the building of it. The switch to the town instrument [the one in the basement of the Physics Department at Manchester] may have been precipitated by me because I broke the suspension on the Jodrell Bank instrument sometime in early 1952. I did this by forgetting to adjust the stop to the large piston, which had been designed to carry the heavy gold cylinder not tiny rock specimens. The piston hit the magnetometer case and broke the quartz suspension. Only Blackett was allowed to change suspensions. So I went penitently into Manchester and confessed to the great man. He was very decent about my error. The following day (it was almost certainly Saturday) Blackett drove me out to Jodrell Bank from the city. He removed the glass-stemmed astatic system, but found that he could not see properly because he had brought his wrong glasses. He then borrowed mine and added them in front of his own. It was disastrous – he now couldn't focus at all, he dropped the system. It broke. It was never made to work again. Blackett was very mad, not at me but at himself. He was a great classical experimentalist with great pride in his manual skills (he had beautiful hands) and because of the bizarre circumstances his skill had failed him. This ended my measurements in Manchester. It is very unlikely that the short period astatic magnetometer was completed by then because otherwise I would have continued to work in Manchester, which I didn't. All this probably happened about March or April 1952.[7]

(*Irving, March 13, 1987 letter to author; my bracketed addition*)

## 2.4 The Manchester group expands and focuses on the Triassic redbeds

Mary Almond and Peter Stubbs were recruited to the Manchester paleomagnetic group in 1952. Almond had been a student of Clegg's, working with him and Lovell in radio astronomy. She had already written two articles on radio astronomy (Almond *et al.*, 1951, 1952), and had helped Clegg (1952) with computational work in one of his articles. Stubbs, a geology graduate, was a protégé of Blackett; their families were friendly (Mike Fuller, March 30, 1987 letter to author).

Neither Clegg nor Almond had any prior training in geology when they began collecting rocks in about October 1952. They found redbeds to be the most suitable (Clegg et al., 1954a: 583). Clegg recalled what happened:

In 1951, when Blackett began to build up a research group in rock magnetism, little was known about the magnetic properties of ferromagnetic materials. He and his team, like other British workers, began to collect rocks almost at random in the hope that some might turn out to have a measurable remanent magnetization. Samples were taken from handy road cuttings, quarries, river beds and banks, sea cliffs, and the walls of underground mine workings. The samples consisted of small slabs of rock which were orientated *in situ* before being knocked out of their parent formation. They were later cut up in the laboratory into cubic and cylindrical specimens and the direction and intensity of magnetization of each specimen was measured by means of the astatic magnetometer.[8]

*(Clegg, 1975: 9)*

However, these samples were too weakly magnetized for Clegg's new magnetometer.

The earliest rocks to be sampled were from the Carboniferous limestone, millstone grit, and coal measures occurring close to the Manchester area. These, disappointingly, turned out to be too weakly magnetized for measurement.

*(Clegg, 1975: 9)*

Finally during the first few months of 1953 they hit on the right type of sedimentary rock, namely very fine-grained redbeds. Clegg recalled:

Early in 1953, the Manchester group had its first major success. Samples were collected of New Red Sandstones of the Triassic period, from two sites in Cheshire 35 miles apart. The rocks from both sites were found to be consistently magnetized along an axis whose direction differed significantly from that of the present Earth's field.

*(Clegg, 1975: 9)*

Clegg found these beds at Frodsham, Cheshire, while taking one of his regular walks from Jodrell Bank.

With Clegg's recognition that fine-grained redbeds provide coherent results, the question arises as to whether or not he tested them because of Irving's and Runcorn's success with the Torridonian. Irving recalled that he probably

first met Clegg in the January-March 1952 period. He visited Jodrell Bank only very rarely because he had fallen out with Lovell and had lost his interest in radar research. His main research in radio astronomy was done before 1951 I think.

*(Irving, March 13, 1987 letter to author)*

There is no question that Irving knew by January 1952 that some very fine-grained redbeds yielded magnetization with low dispersion because he and Runcorn made their second collections from the Torridonian in October 1951, and by the end of 1951

Irving had measured them and confirmed their earlier result. Did Irving tell Clegg of this discovery? Irving thinks that he probably told Clegg. But he is not certain.

I cannot say whether I actually mentioned my predilections for red sandstones to Clegg. I remember being very open with him – he was not in paleomagnetism at the time I first met him, so I had no competitive reasons not to tell him. Probability is that I did mention it to Clegg, but since he was not a geologist he probably didn't regard what I said as very important. Our friendship was social and generally intellectual ... I cannot remember Clegg ever asking me specifically about my data. I remember being rather close to him and his wife Marjorie who were very hospitable to me and not feeling in any way inhibited in what I said. Later we drifted apart.
*(Irving, March 13, 1987 letter to author)*

To this day, Irving still does not know if Clegg tested redbeds because of what he might have told him, or if he did it on his own. Irving and Clegg did some sampling together but they collected grey sandstones from an Oldham coal mine.

So our discovery of this [that red sandstones give coherent results] predated Clegg's entry into the subject in 1952. He could have then decided this for himself at that time and quite independently of us. However, it is unlikely to have been entirely independent because he and I talked a great deal and we sampled together from coal mines near Manchester.
*(Irving, July 4, 1988 letter to author)*

It is not possible to determine with any certainty whether Irving influenced Clegg in his use of fine-grained red sedimentary rocks. Clegg was unable to add anything in his interview with Fuller, and as Fuller has remarked:

I do not know whether Johnnie [Clegg] just stumbled on the New Red or whether he tried it by analogy with Torridonian. As one of his passions was walking – I am sure that he would have been aware of localities in the New Red Sandstone.
*(Fuller, March 30, 1987 letter to author)*

However, as Irving (August 20, 1985 letter to author) said, this "is by-the-by." The important things were that by early 1952 Irving and Runcorn had realized that Torridonian fine-grained red sandstones were good carriers of paleomagnetic information, that by the beginning of 1953 Clegg had come to the same realization regarding the Triassic redbeds, both groups had developed a certain expertise for distinguishing rocks that were good prospects from those that were not, and Clegg and Irving continued to collect fine-grained redbeds. The description of Clegg and company's work on Triassic redbeds is completed in §3.5.

## 2.5  Irving investigates the origin of magnetization of the Torridonian and begins magnetostratigraphic survey

The entire Torridonian job was not finished till early 1954 with a lot of help with measurements from David Collinson. Remember the Torridonian was far and away the most detailed study of any rock unit, and remained so for many years. I do not believe even now that any other single rock unit has had studies (by me and others) made of folds, conglomerates (of two ages), of slumps, grain-size effects, reversal sequence, although of course techniques have improved. It was a long 3-year slog because we began almost from scratch.

*(Irving, August 20, 1985 letter to author)*

I return now to Cambridge and begin with Irving's diverse activities during 1952 and 1953. He made three more collecting trips to the Torridonian (five in all), focusing, as he (*Synopsis*, 1985: 3) retrospectively remarked, "on field tests of magnetic stability and on determining which grain-size within the redbed type was the best. There was also a first attempt at stratigraphic coverage." Irving and Leslie Flavill, chief laboratory technician of the department, had installed rock cutting equipment by November 1952 (Irving, *Synopsis*, 1985: 3). The Cambridge magnetometer became operational in February 1953 (§2.11) and, with much help from Collinson, Irving measured and analyzed his large collections. He remeasured (June 1953) specimens that he had previously measured on Blackett's magnetometer and, finding the results in very good agreement, thus inter-calibrated the Cambridge magnetometer with Blackett's. He measured the samples of the Deccan Traps (§1.20, §3.4). Irving also confirmed from many horizons in the Torridonian that red silts or very fine-grained sandstones had well-grouped remanence directions always strongly and systematically oblique to the present Earth's field. He used Graham's conglomerate and fold tests and studied slump beds. He made a third unsuccessful attempt to trace secular variation. He enquired into the processes of sedimentation, and argued that the Precambrian geomagnetic field was the primary force controlling the alignment of its remanent magnetization; this enquiry also led to an explanation why fine-grained red sandstones yield more coherent results than coarse-grained ones, and why some slump beds gave internally coherent paleomagnetic directions whereas others did not. Finally, Irving presented many of his Torridonian results at the Birmingham meeting in January 1954. I describe in more detail his Torridonian work here and his India test elsewhere (§1.20, §3.4).

Irving's third collecting trip (April 1952) began with a visit by Irving and Runcorn to the University of Leeds where they consulted John Hemingway, a sedimentologist who had spent many years working on the Torridonian. Remember that in early 1952 after completing their work at Craigriannan they had made no decision as to how to proceed (§1.17). However, by April 1952, the time of their Leeds meeting, Irving had proposed to Runcorn that he sample through the entire ~5000 m Torridonian (Irving, 1988), and consequently asked Hemingway for advice about the location of sections that would provide stratigraphic coverage. Following Hemingway's suggestions, Irving sampled those localities on his third and fourth (November 1952) collecting trips. Together with the Loch Torridon section that Irving and Runcorn had begun to sample at Craigriannan, Irving had, by late 1952, samples from four thick sections spaced through the Torridonian but with large gaps between them.[9] After meeting with Hemingway, Runcorn headed back to

Cambridge while Irving drove to northwest Scotland. Irving eventually made a fifth collecting trip before completing his study of the Torridonian.

Irving recognized the importance of Graham's field tests and he searched for places to apply them. Many conglomerates with Torridonian clasts were truly of New Red Sandstone age (Permo-Triassic) as Peach *et al.* (1907) said; they passed conformably into well-dated Jurassic beds and had an ill-consolidated matrix. But Irving found others intimately caught up in the Torridonian itself whose matrix had continuous quartz cement; these he described as of "indeterminate age"; later they proved to be of Torridonian age (Stewart, 2002). Finding that the clasts in both instances had random directions of magnetization, he reasoned, following Graham, that the magnetization of the parent rock (i.e., the Torridonian) had to be at least pre-Triassic and possibly much older. Irving also sampled a large amplitude fold formed during the mid-Paleozoic Caledonian orogeny, long after the deposition of the Torridonian. He found scattered directions at seven sites along the fold. However, because the scatter largely disappeared once he corrected for the tilt of the beds, he concluded that the magnetizations predated folding (<400 Ma).

Alerted by Hemingway to the abundance of intra-formational slump structures, Irving studied and sampled many examples hoping to determine if the remanence was acquired before or after slumping. Slumping occurred before lithification and the soft strata became folded or contorted or even broken. Slumped horizons can be several meters thick with folds meters in amplitude or as small as a few centimeters. They are now believed to have been induced by earthquakes, the Torridonian having accumulated in a rapidly subsiding fault-bounded trough (Stewart, 2002). Irving argued that if the magnetic grains held their alignment during slumping then the magnetization directions would be highly scattered, and if this were observed then the magnetization had to be Torridonian. If, on the other hand, magnetizations are uniformly aligned and parallel to those in adjacent flat-bedded unslumped strata, the magnetization in the slumps post-dated slumping. He found instances of both. In the former he knew that the magnetization was Torridonian and therefore Precambrian in age. In the latter he was able to show that although the magnetization had been acquired after slumping it had very probably done so between deposition and lithification, and it too was therefore very likely Precambrian.

Of special interest was an eminence called Stac Fada on the north shore of Stoer Bay in Wester Ross. Irving described it simply as a "slump." Later it was considered to be a volcaniclastic debris flow within the Stoer Group (Stewart, 2002). It contains, at its base, rip-up clasts meters in length derived from the underlying red sandstones. More recently it has been described as the product of a 1200 million-year-old meteoritic impact that spread debris over at least 50 square kilometers (Amor *et al.*, 2008). Irving observed that within each clast the magnetizations are well clustered but they are highly scattered between clasts, showing that they acquired magnetization before disruption. The disruption is penecontemporaneous and the horizon is within the oldest Torridon strata, which have therefore retained their original Precambrian magnetization. At the time Irving (personal communication) did not understand that the Stac Fada Member might have been a volcanic debris flow or a meteoritic impact, but he did understand that the field evidence required that the deformation was intra-formational and argued correctly that the magnetization of the Stoer sedimentary rocks had to be Precambrian. Rocks could retain their magnetization since the Precambrian.

How might the Torridonian rocks have acquired their magnetization? The use of sedimentary rocks as a record of the geomagnetic field and as a way of testing continental drift and polar wandering assumes that the ambient geomagnetic field was the major factor controlling their magnetization. For this to happen, the magnetization of their iron mineral grains must magnetically become aligned in the then existing geomagnetic field. However, Irving recognized there were controlling factors other than the geomagnetic field: the alignment of non-spherical magnetized grains in a sedimentary deposit could, for example, have been brought about by water currents driven by fluvial conditions, or indirectly by prevailing winds, or by mechanical stress during burial long after deposition. In such circumstances, no information about past orientations of Earth's magnetic field could be inferred, even though well-defined magnetizations were observed. Consequently, Irving turned his attention to the processes of sedimentation and to processes occurring after deposition, and was able to argue on the basis of extensive field tests that action of the geomagnetic field was the dominant aligning force.

Just precisely when he came up with his explanation is unclear. He must have worked out much of it by late 1953 because he summarized it at the January 1954 Birmingham meeting (§3.3). He likely was working on the problem during 1952 or early 1953, because he then was reading books on sedimentation by Ph. H. Kuenen and Francis Shepard, presumably to help him understand how the Torridonian became magnetized.[10]

Irving reasoned:

Stable current systems induced by prevailing winds are conceivable, but a sub-aqueous environment of constant depth and configuration would be necessary. The rapid alternation of fine and coarse beds, sun cracks, and the truncation of slump and cross-bedding by erosion surfaces are universal characteristics of the Torridonian. They indicate rapid changes of conditions during deposition and it is hardly likely that unified and constant current systems could become established.

(*Irving, 1957c: 104*)

He appealed (1957c: 103) to his study of the Caledonian fold in order to eliminate the possibility that magnetizations were stress related, pointing out that the magnetization of the folded beds retained the same angle relative to the original bedding plane "in spite of compressive forces sufficient to produce vertical tilts."

Irving associated fine-grained sediments with deposition in relatively still water, which allowed for close alignment of magnetic particles to the geomagnetic field, and coarse-grained sediments with turbulence, which did not – a natural explanation of why fine-grained red sandstones had the far more coherent magnetizations.

During the deposition of the coarse arkoses, in which the water turbulence is considerable, the alinement of the magnetic grains in the geomagnetic field will be poor. However, in finer sandstones where the overhead water turbulence is much less and the bottom smoother, less scatter in the alinement will result and the directions measured in adjacent samples should not differ greatly.

*(Irving, 1957c: 105)*

This solution explained other findings. Irving noticed that black-banded fine-grained red sandstones showed very little dispersion, even less than massive, non-banded beds of the same very fine grain size at the same locality. He found through X-ray studies that the dark bands had a high concentration of tiny grains of specularite (crystalline hematite) which were responsible for their magnetization. The bands, he reasoned, had been produced by the gentle winnowing effect of water currents on freshly deposited sediments; the non-magnetic grains, being larger and less dense than the specularite grains, were preferentially removed and the specularite concentrated. He then added:

During winnowing the specularite particles would be in to-and-fro motion induced by the slight turbulence overhead, and continuous adjustment to the earth's magnetic field may have been possible over long periods of time.

*(Irving, 1957c: 106)*

Irving then fleshed out his account of slump beds, classifying them into larger and smaller amplitude. The larger, with amplitudes of a meter or more, were uniformly magnetized (post-slumping) in the same direction as adjacent undisturbed sediments. He proposed that they had become remagnetized after slumping by rotation of the (smaller) specularite grains. As the sediment slumped it became quasi-liquid, thus allowing for magnetized grains to swiftly realign themselves in the direction of the geomagnetic field.[11] Smaller slumps gave two different results. Some, like the larger slumps, were uniformly magnetized in the same direction as the overlying flat-bedded layers, and for these Irving again invoked remagnetization through grain rotation. Other small amplitude slumps showed wide internal scatter with directions lying far from those displayed by adjacent flat-bedded layers; the degree of scatter in these slumps was much greater than that displayed by unslumped sandstones of the same grain size. Why were they so scattered? He suggested that dewatering was likely very sudden so that individual grains had no time to rotate into the direction of the geomagnetic field; they became fixed in the random positions acquired during the slumping process.

Irving concluded his defense of the view that the geomagnetic field was the main cause for the systematic alignment of magnetizations by saying:

These arguments lend support to the view that the mean directions [of magnetization in the Torridonian] may be those of the geomagnetic field in the late Pre-Cambrian times.

*(Irving, 1957c: 110; my bracketed addition)*

He claimed that unless they were deposited under turbulent conditions (coarse-grained) or were part of a quickly dewatered slump, the action of Earth's magnetic field was responsible for the good alignment (low dispersion) of magnetized red sedimentary rocks he observed in the Torridonian. There could be other situations, but because his study was so extensive, the chances were that these were exceptional, not the rule.

By identifying just how and when the normal action of the geomagnetic field may be over-ridden by other factors during sedimentation, Irving gave weight to the idea that the ancient orientation of Earth's magnetic field is recorded by certain red sedimentary rocks during or soon after deposition. The Torridonian result was notable because it meant that certain identifiable lithologies could retain their directions of magnetization over hundreds of millions of years, even since the Precambrian, and that Earth possessed a magnetic field at that time. Consequently, studying the geomagnetic field throughout the Phanerozoic and back into the late Precambrian became a very feasible objective: it was not just a dream. It warranted the use of paleomagnetism as a tool for testing mobilism during the Phanerozoic.

## 2.6 Irving completes magnetostratigraphic survey of the Torridonian

By mid-1953 Irving had established that for the most part the remanent magnetization of the Torridonian was Precambrian in age, and that in the four stratigraphic sections he had sampled it was always strongly oblique to the present field. There were, however, large gaps between sections, and later that year, in a very extensive fifth collection trip made with Collinson, an attempt was made to fill these gaps and obtain a representative coverage of the entire Torridonian sequence.

Typically, in the upper three-quarters of the sequence he found reversed and normal magnetizations, respectively, toward the southeast and downward, as at Craigriannan, and to the northwest and upward. There were at least twenty polarity zones (Griffiths and King, 1954; Irving, 1954; Irving and Runcorn, 1957). This was the earliest description of stratigraphically sequential reversals in sedimentary rocks. Irving and Runcorn argued that they were best explained by reversal of the geomagnetic field during the Precambrian.

By contrast, the lower quarter of the sequence had no reversals, only normal magnetizations aligned along a very different axis, northwest and downward. It was a surprise to Irving that the magnetic boundary between what he called the Lower and Upper Torridonian did not correspond to any geological boundary recognized at the time; the boundary lay just below the top of the Diabaig "Group," the lowermost unit recognized by the British Geological Survey (Peach *et al.*, 1907: 107; see also §1.17). This was not the last of many instances when Irving's results were to clash with orthodoxy.[12]

## 2.7  Fisher defends mobilism

R. A. Fisher gave a talk in favor of continental drift at a meeting of the Kapitza Club, a Cambridge science club, at Gonville and Caius College. Raymond Hide, who set up the meeting, believes this was between October 1952 and March 1953. Hide also asked Jeffreys to speak, but he declined. Fisher talked about the biological and geological evidence in favor of mobilism. He also considered the mechanism difficulty that had plagued mobilism, and suggested, reminiscent of Wegener, that small forces acting over great lengths of time might be sufficient to cause drift. He discussed the Great Glen Fault in Scotland, arguing that, by the cumulative effect of horizontal displacement along it and other collinear transcurrent faults, movements amounting ultimately to continental drift could occur. Irving recalled Fisher's argument and the style with which he proposed it.

The thing that sticks in my memory about this talk was Fisher talking about the Great Glen Fault. There are granite masses on both sides of the fault, called the Strontian on one side, and the Foyers on the other side near Inverness. W. Q. Kennedy (1946) who was a professor at the University of Leeds, I think, at the time, had suggested that this is evidence for a transcurrent fault – the matching granite masses are displaced by about 65 miles. Fisher's argument was that if you had a fault like the Great Glen Fault, which may have been a very long one extending out into the Atlantic and God knows where in the east, if it had been moving by this relatively small amount, if it had continued moving or if companion faults nearby had moved, the total motion on either side of the fault or fault zone could be very considerable and this would amount to continental drift. It was this accumulation of small effects over long time spans that recurred in Fisher's discussion about continental drift and its viability. I can remember when he explained this on the blackboard, and showed Kennedy's maps, he stood as tall as he could and fixed the audience with his beady eyes, and declared that this was his answer to MacDuff's question, "Stands Scotland where it did?"

*(Irving, January 2001 conversation with author)*

Runcorn also remembered the meeting, and has his own story about what happened.

There were two societies in Cambridge – one in experimental physics called the Kapitsa Society, and the $\Delta^2 V$ club, which was theoretical. Meetings were held in college after dinner, and everybody in Cambridge came to them. The $\Delta^2 V$ club was apparently where Crick first talked about his discovery. (Oddly enough I used to go to the Kapitsa club regularly, and I don't recall it.) Kapitsa founded it … before the War. The $\Delta^2 V$ club was traditionally organized by graduates, and Hide was an organizer, and by this time we were interested in continental drift in connection with the interpretation of our results, and I said to Hide, "let's ask Fisher to give a talk." [Hide believes that Fisher actually gave the talk at the Kapitsa not the $\Delta^2 V$ club, and I tend to agree with him since he organized the meeting.] Fisher was by that time, you know, talking a lot about continental drift. And so, they asked Fisher, and Fisher was just a little bit reluctant because he said, "There are all these distinguished physicists in the department, and I don't know much about continental drift." I don't believe any geologists came because they would not have thought of coming to listen to anybody else talk about geology. Fisher gave the talk.

By that time Blackett was interested, and I must have told him, and he said, "I'll come to Cambridge and listen." He was still at Manchester. Fisher had a particularly difficult way of lecturing, which was, you see, to assume that you knew all about the subject and make penetrating comments about it. He was witty, and he used to throw in little, very academic type jokes. Anyway, Blackett came. I suppose he must have stayed with me in College. He had met Fisher. They didn't hit it off too well because Fisher was very rightwing politically. Fisher was always very friendly to me at Caius. But when I came, you see, and he knew that Blackett had been my professor, he used to be very critical. Blackett had written a book about the atomic bomb. I used to defend Blackett. But Fisher was very distrustful of these leftwingers. Of course, Blackett came and had dinner with me many times in Caius, and I always remember Fisher saying to me afterwards, "I've revised my view about Blackett because he has such boyish enthusiasm about science." That was very characteristic about Blackett. Blackett on the other hand – his one great defect was that he had no sense of humor whatever. The fact that Fisher was humorous upset Blackett. And, I always remember, after this talk, which was full of animated discussion, you see, Blackett said to me, "I don't think that was a very good lecture. You or I could have read up the subject and given a straightforward account."

But, we were talking about Ray Littleton, Hoyle, Gold, and Bondi, they always used to come to geophysics colloquia which I used to organize. They came to this meeting. I always remember – they were very skeptical about continental drift – and Fisher tried to say, "Well there are the rift valleys, and they show extension." And, while blowing out air, he then said, "if you have fifty km extension parting of the continental blocks, why cannot you have a thousand?" Littleton, who was also an amusing chap, got up and said, "Does Professor Fisher really want to say that because he can blow out his chest two inches he can blow it out twenty?" Undoubtedly that was a time when we had in the department talked a lot about continental drift.

*(Runcorn, August 1984 interview with author; my bracketed addition)*

Fisher had become favorably inclined toward mobilism, but many in the audience did not agree with him.

## 2.8 Hospers returns to Iceland, builds an "igneous" magnetometer, and develops his case for reversals of the geomagnetic field

Runcorn encouraged Hospers to return to Iceland during the summer of 1951 to extend his survey of Icelandic lavas. Hospers collected over 600 samples[13] between May and September (May 28, 1985 letter to author) while Irving and Runcorn were beginning their work. His samples, mainly lavas, dated back to the Miocene. He began measuring them during the winter of 1951/2, finishing in summer 1952. Given an old instrument housing by Runcorn, Hospers constructed a magnetometer sufficiently sensitive to measure his strongly magnetized basalts. He used Blackett's more sensitive astatic magnetometer at Jodrell Bank for his smaller collection of sedimentary samples. Hospers performed many laboratory experiments during 1952. He wrote his doctoral thesis "Paleomagnetic studies of Icelandic Rocks," during the winter of 1952/3, submitted it on April 20, 1953 (Hospers, 1953c), and took his orals on June 8. O. T. Jones and Blackett were his examiners. He passed, and left Cambridge for Holland on September 29 (Hospers, 1985 interview with author).

Hospers wrote seven papers on the results from his second Icelandic collection, and an additional paper on Eocene basalts from Northern Ireland, which at Runcorn's suggestion he got H. A. K. Charlesworth to collect in 1952. Hospers began writing the first paper immediately after submitting his thesis. It was published in three parts, the first two (Hospers, 1953a, b) appeared before the end of 1953, and the third (Hospers, 1954a) the following year. He completed the other papers (Hospers, 1954b, c, d, 1955; Hospers and Charlesworth, 1954) by October of 1953, while staying with his mother in Holland before beginning work with Royal Dutch Shell on November 1.

The first part of Hospers' three-part paper (1953a) was seminal. Making two first-order contributions to the rise of paleomagnetism, he offered extensive support for polarity reversals and, in the process of analyzing his data, as a bonus found strong support for the GAD hypothesis (§2.9), which also showed no significant latitudinal drift of Iceland since the Miocene. His was the first treatment of the reversal problem that took into account all the field and stratigraphic evidence, and Néel's self-reversal mechanisms. He was also the first to construct a reversal timescale for the last few million years. His marshaling of empirical support in favor of the GAD hypothesis provided the base on which paleomagnetists would later build their case for mobilism. His own Icelandic samples were geologically too young to record any latitude change.

Hospers' work on reversals is elegantly introduced in this letter by Fisher to T. Einarsson, an Icelandic scientist who with his countryman T. Sigurgeirsson continued Hospers' work.

21st May, 1955

Dear Dr. Einarsson,

I have this morning read with the greatest pleasure your letter with Dr. Sigurgeirsson published in *Nature* on May 21st.

When he was in Cambridge, I had the pleasure of frequent discussions with Dr. Hospers on his results and plans for his expeditions to Iceland. It is, therefore, with great satisfaction that I see how thoroughly you have confirmed his conclusions as to the successive inversions of the earth's magnetic field, adding to his evidence the extensive and cogent observations of your own. The resistance felt to the acceptance of a theory of successive reversals seems largely to have been centered in a small group at the Carnegie Institute in Washington, who have shown the greatest ingenuity in devising every sort of alternative explanation of facts which from all sources have been becoming increasingly decisive.

It was in connection with the statistical examination of data of the kind obtainable by geophysicists that I was led to the mathematical investigation of the enclosed paper [Fisher's "Dispersion on a Sphere"], which, as you see, leads to a tolerably expeditious method by which the observations can be combined and summarized.

Yours sincerely,
Signed: Ronald A. Fisher[14]

The problem of reversals was fraught with difficulties (§1.10), which Hospers proceeded to discount one by one. In particular, he raised a reliability difficulty against the study of sedimentary rocks by Torreson, Murphy, and Graham (1949b), which had led them to claim that there had been no reversals during the last 50 million years; he argued that they had not generally shown their samples to be magnetically stable. He noted that non-red sedimentary samples (as most of them were) were often unstable (RS2), appealing to Graham's own work and to Irving's unpublished work on the Torridonian sandstones.

However, more recent work in this field shows that the permanent magnetization of a sediment, acquired on deposition, is often unstable (personal communication of Mr. IRVING, Cambridge University; cf. also GRAHAM 1949 ...) The stability of the magnetization must therefore be established before definite conclusions can be drawn and this has not been done in the case of the sediments in question.

*(Hospers, 1954a: 114–115)*

He noted (1954a: 115) in the 1949 Carnegie study that "of the 99 samples on which the above conclusion is based, seven were actually found to possess R [reversed] magnetization." He then raised a reliability difficulty attacking Tuve's suggestion that the apparently reversely magnetized Silurian samples from Maryland and West Virginia could be discounted because they were inconsistent with supposedly contemporaneous, normally magnetized strata from Clinton, New York (RS2); the correlation was weak and the Clinton rocks had not been shown to be stable.

This author [Tuve], however, believes that the most straight forward explanations of the observed anomalous magnetization – a reversal of the earth's magnetic field or a drifting of the continents – are not satisfactory, as at Clinton, New York, beds bearing the same unusual fossil (a rare ostracod) as the West Virginia deposits and thus presumably of the same age as these beds, do not show anomalous magnetization. In the present author's opinion the validity of this argument cannot be admitted, as the figure of 100 000 years quoted as the period of time this rare ostracod could have lived is quite arbitrary. If this period of time is assumed to be 500 000 or 1 000 000 years, sufficient time is available for one period of N (normal) and one of R (reversed) polarity of the earth's magnetization. Moreover, there is no evidence that the normal magnetization of the Clinton beds was acquired on deposition.

*(Hospers, 1954a: 115; my bracketed addition)*

Hospers (1954a: 116) raised a similar reliability difficulty with the assessment by Johnson, Murphy, and Torreson (1948) of five Pacific Ocean cores, noting that they had not shown them to be magnetically stable.

Hospers considered Néel's self-reversal proposals in the second of his three-part paper (1953b), a masterful demonstration of how to determine which of two competing hypotheses is the more worthy. Hospers listed the three Néel self-reversal mechanisms relevant to the Iceland problem.

The various explanations of the observed reverse magnetization of lava flows and other igneous rocks can therefore be listed as follows: N I comprises all explanations which consider reverse magnetization as due to an inherent property of certain self-reversing ore minerals. N III explains reverse magnetization as due to the interaction between two different ferromagnetic substances with different Curie points. N II and N IV account for reverse magnetization, both of igneous rocks and of sediments, by a combination of later selective chemical changes with imperfect results of the mechanisms N I and N III respectively. The above explanations are thought to be able to produce reverse magnetization in a normal field, and have in fact been advocated as an adequate explanation of the reverse magnetization found in Iceland (Hospers, 1951) by others (Néel, 1952; Graham, 1952). The fifth possible explanation is that by a repeatedly reversing field.

_(Hospers, 1953b: 482)_

His general strategy was to raise anomaly difficulties against each of Néel's solutions by showing that field and laboratory data were inconsistent with them (RS2), and to maintain that the hypothesis of geomagnetic field reversals explained more than Néel's solutions (RS3). He argued that because both N II and N IV required the chemical substitution of a mineral in the reversed lava flows _after_ they had been laid down, it would be surprising to find sequences of reversed samples _underneath_ normal ones (RS2); his results showed that is just what commonly happens, reversed samples occur below normal ones, and normal and reversed samples are each associated with definite stratigraphic horizons or zones. N I and N III also faced other adverse field relationships.

If R magnetization were due to the presence of a self-reversing mineral (N I), alone or in combination with more normal ore-minerals, one would hardly expect it to manifest itself in a small number of zones in an otherwise very uniform series of lava flows as is actually observed. Similarly, if R magnetizations were due to interaction between two different minerals (N III) one would expect a more random distribution than observed.

_(Hospers, 1953b: 483)_

Raising another anomaly difficulty (RS2), Hospers commented (1953b: 483), "It seems strange that all of these mechanisms ... should produce an approximately equal number of N and R specimens, as is observed in the Early Quaternary–Tertiary sections." And there was a further difficulty (RS2).

The nearly perfect agreement between corresponding N and R series of flows in intensity of permanent magnetization and susceptibility is difficult to reconcile with any of the explanations N I – N IV of the R magnetizations.

_(Hospers, 1953b: 484)_

He cited observations from France:

In 1953, Roche, who also argued for geopolarity reversals, found that clay, baked by overlying reversely magnetized lava flows was invariably reversely magnetized in the same direction.

*(Hospers, 1954u. 114)*

A. Roche (1953) quite rightly took this as strong evidence for polarity reversals. Roche argued as David (1904) and Brunhes (1906) had done, fifty years earlier, that the underlying clay was magnetized by the molten lava in the reversed magnetic field at the time of extrusion (§1.11) (Laj *et al.*, 2002; Courtillot and Le Mouel, 2007). Hospers agreed and concluded:

The agreement between comparable N and R series of lava flows [with respect to their magnetic intensity and susceptibility] therefore suggests strongly that we are dealing with series of flows similar in all respects, except in that they cooled in fields of opposite directions (but comparable strength).

*(Hospers, 1953b: 484; my bracketed addition)*

Field reversal offered a more complete explanation than self-reversal (RS3); it was not plagued by anomaly difficulties, and it explained field data that posed serious anomaly difficulties for self-reversal (RS2).

Hospers then described laboratory experiments, which raised further difficulties for self-reversal (RS2). He heated normally and reversely magnetized Icelandic lavas above their Curie point and let them cool in Earth's magnetic field. Without exception they acquired a magnetization in the direction of that field, indicating that "mechanisms N I and N III are not responsible for the reverse magnetization" because every sample upon cooling acquired a permanent magnetization in the direction of Earth's magnetic field (1953b: 485). But the "experiment … is not conclusive as it is not impossible that the reheating causes chemical or physical changes in the rock" (1953b: 485–486). Hospers also obtained a sample of Uyeda and Nagata's Haruna self-reversing dacite (§1.10), repeated and confirmed their experiment, finding that it truly was self-reversing, acquiring, after heating and cooling, a magnetization in a direction opposite to that of the ambient field, very different from his Icelandic basalts among which Hospers had found no comparable example. He concluded that whatever caused the self-reversal of the Haruna dacite could not account for the reversed magnetism of his own Icelandic basalts.

… [It] may be said that the regularities in the horizontal and vertical distribution of reverse magnetization in the lava flows practically exclude the possibility of later chemical or physical changes (mechanism N II and IV) being responsible for the reversed magnetization. On the other hand, the presence of reversely magnetized sediments can only be accounted for by these chemical or physical changes. The laboratory experiments show that mechanism N III is definitely not responsible for the reverse magnetization of the lava flows and that mechanism N IV is most unlikely to be so. It therefore appears that only the assumption of a repeatedly

TABLE VI.   Stratigraphic Column of Iceland

| | |
|---|---|
| Historic lava flows: from A.D. 1729 to 1947–8 . . . . . . . . . | N |
| Postglacial lava flows: from 2,000–7,000 years old . . . . . . . . | N |
| Interglacial lava flows and sediments, approx. 150,000 years old . | N |
| Palagonite Formation: sampled part 500,000 years old or less . . | N |
| Early Quaternary lava flows and sediments: 500,000–1,000,000 years old . . . . . . . . . . . . . . . . . . . . . . . . . . . . . . | R |
| . . . . . . . . . . . . . . . Break . . . . . . . . . . . . . | |
| Pliocene sediments and lava flows: 2,–12,000,000 years old . . . . | { N<br>{ R |
| . . . . . . . . . . . . . . Break . . . . . . . . . . . . . | |
| Tertiary lava flows (and some sediments); those sampled are probably Miocene: 12–26,000,000 years old . . . . . . . . . . . . . . | ( R<br>) N<br>) R<br>( N |

Figure 2.1 Hospers' Table IV (1953b: 475) showing the sequence of normal and reversed groups based on his study of Icelandic lavas.

reversing magnetic field can satisfactorily account for the observations, an assumption which is confirmed by the striking similarities between N and R magnetization.

*(Hospers, 1953b: 490–491; my bracketed addition)*

Reversals of magnetization were best accounted for by reversals of the geomagnetic field, although he was not entirely certain.

It is therefore suggested, though not without some reserve, that the Icelandic rocks have preserved a record of a repeatedly reversing main geomagnetic field. As the reversal is through a 180° it seems that this reversal is worldwide and is, in fact, a reversal of the polarity of the earth's magnetization. The manner in which the reversal takes place is still unknown, but it appears that the field changes to the opposite direction within one-fifteenth of the period over which a normal or a reversed field persists.

*(Hospers, 1954a: 119; see also Hospers, 1953b: 491 for a similar passage)*

The empirical support for field reversals was strong, and the lack of a known mechanism for field reversals was not reason to doubt them. It was up to theoreticians to explain how they occurred. Indeed, after summarizing his work on reversals at the January 1953 joint meeting of the Royal Astronomical Society and the Geological Society of London, Hospers "said that he tried to keep clear of geomagnetic theories while reaching conclusions from his experimental results" (Anonymous, 1953: 69). Graham thought very differently; for him this lack posed a serious theoretical difficulty for field reversal (§1.10).[15]

Hospers recognized the importance of erecting a reversal timescale. If the observed reversals truly reflect reversals of the field, then each would be worldwide, and a sequence of them would provide a global timescale. If he could date the reversals in the Icelandic lava flows, he could develop a reversal timescale. His attempt to do this back to the Miocene in light of the then existing stratigraphic record is shown in Figure 2.1.

Although far from complete, Hospers' and also Roche's (1953) timescales were significant developments;[16] they were far-sighted precursors of the quantitative reversal timescale developed a decade later.[17]

## 2.9 Hospers develops the geocentric axial dipole hypothesis and tests polar wandering and continental drift

Hospers' grounding of the GAD hypothesis was foreshadowed in his first Iceland collection and confirmed by the second. He recalled:

As to the axial geocentric dipole concept, the earliest published discussion is in *Nature* (Hospers, 1951). However, the results were inconclusive in this respect. Pinpointing the demonstration is not difficult. The specimens used were obtained in the summer of 1951, the measurements made winter 1951/52 and summer 1952. My thesis was written winter 1952–1953, the thesis was submitted spring 1953 and I had my exam summer of 1953 with Blackett and O. T. Jones as examiners. Accordingly, this concept must have been established autumn 1952 because I worked backwards in geological time (Recent – Quaternary – Tertiary) and the effect is already evident from the postglacial lava flows. As to the published record, you know of course that this work is described in my thesis [1954c] and in Hospers 1953–54.

*(Hospers, November 24, 1980 letter to Irving; my bracketed addition)*

Using Fisher's statistics, Hospers calculated the mean direction of the NRM of volcanic rocks from Iceland and elsewhere, and compared them with the directions of the present field and that of the GAD. He found that results from Miocene and later rocks agreed well with the GAD field, better than with the present field. These sequences from youngest to oldest were: Early Quaternary lava flows, western Iceland (Hospers, 1953a); Plio-Pleistocene and Mio-Pliocene lavas, Massif Central, France (Roche, 1951); Miocene lavas, northern and southwestern Iceland (Hospers, 1953a). Hospers concluded:

The measurements show that taken over periods of several thousands of years the magnetic pole centres on the geographic pole. This has been so since Miocene times (approximately $20 \times 10^6$ years ago).

*(Hospers, 1954a: 119)*

Hospers had actually reached this conclusion by autumn of 1952, and his results quickly became, as I shall show, common knowledge among Cambridge paleomagnetists (Hospers, March 1983 interview with author).

Hospers then turned to the question of continental drift and polar wandering in the third part (1954a) of his three-part article. He first learned about the application of paleomagnetism to mobilism from Irving.

I now want to amplify ... about your joining the Department at Cambridge, even though I am not sure about the time (autumn 1951?). However, I still remember very clearly meeting you for the first time in the workshop of the Department's laboratory on Madingley Road. I also remember clearly that on that occasion you had fully realized the potential value of palaeo-magnetism in studies of continental drift. In other words, as I remember it, you knew what the theory of continental drift was about, what palaeomagnetism could do to test the theory and how to go about testing it, namely by getting palaeomagnetic data from continents alleged to have drifted apart.

*(Hospers, February 17, 1981 letter to Irving)*

Hospers explained what would be expected if continental drift, polar wandering only, or neither had occurred.

Polar wandering and continental drift can be distinguished from each other if data from several continents are available for the same geological formation. If polar wandering has occurred (it is assumed that in that case the magnetic pole wanders with the geographic pole, as for the last thousands of years the magnetic pole has centred on the geographic pole) a certain geological formation gives in each continent a mean direction of magnetization which is significantly different from the present and the dipole field, and these mean directions will agree as to the position of the pole. If continental drift has occurred, the mean direction of magnetization of a certain geological formation will in each continent be significantly different from present and dipole field, but these mean directions will not agree as to the position of the pole.

*(Hospers, 1954a: 117)*

Because of the good agreement between the theoretical GAD and the paleomagnetic directions as just described, he concluded (1954a: 118): "Iceland has suffered no rotation or changes in latitude since Tertiary (Miocene) times, i.e., since at least 20 million years." Hospers claimed (1954a: 118) that polar wandering, to the extent indicated by Köppen, "is definitely outside the range of possible positions." He also claimed no support for continental drift, but recognized Iceland was not the crucial testing ground.

These measurements therefore contradict the [Köppen's] theory of polar wandering, and do not support the theory of continental drift as proposed by Wegener, though it must be said that Iceland plays no significant role in Wegener's theory and is not suitable as a testing ground (cf. Wegener, 1924).

*(Hospers, 1954a: 118; my bracketed addition)*

Hospers considered polar wandering and continental drift in two other papers: a letter to *Nature* (Hospers, 1954c) in June 1954, and a paper (Hospers, 1955) that was not published until 1955, but received by the editors a year earlier in January 1954. His paper contains a diagram with a "locus" for each paleopole; by locus he meant "the somewhat deformed small circle on Earth's surface which corresponds to the circle of confidence (for P = 5 per cent)" (Hospers, 1955: 65). I conclude therefore that Hospers was correct in claiming that he was the first to calculate paleomagnetic poles (Hospers, 1983 interview with author). He explained how he calculated loci.

On a stereographic projection the (opposite of the) calculated mean direction of magnetization is plotted, with its circle of confidence for P = 5 per cent. For four points on this circle the declination and inclination are determined. The inclinations are used to calculate the corresponding co-latitudes (i.e., the distances on the globe from our sections in western Iceland to the corresponding position of the magnetic pole by using the formula $\tan I = 2 \cot \varphi$, where $I$ is the inclination and $\varphi$ the co-latitude (cf. Chapman [and Bartels], 1951, p 20). For each point chosen on the circle of confidence, a spherical triangle on the earth's surface can be drawn, the corners of which are constituted by (1) the mean position of our sections in western Iceland, (2) the present geographic north pole, (3) the required position of the magnetic pole. Two sides (the present geographic co-latitude of western Iceland and the calculated magnetic co-latitude) and the included angle (the declination found for each point chosen on the circle of confidence) are known in these triangles. The third side can therefore be calculated: it is the angular distance from the present geographic pole to the required position of the magnetic pole. As the angular distance from western Iceland to the required position of the magnetic pole is also known, the position of the latter can be determined by geometric construction. A nearly circular closed curve is drawn through the four points thus found on a stereographic projection of the Northern Hemisphere, this is our locus. The position of the geographic pole in early Quaternary times is now defined within narrow limits. The probability is only 5 per cent that the geographic pole was situated outside the locus.

*(Hospers, 1955: 65; my bracketed addition)*

Hospers calculated eight such loci including Early Tertiary results (Figure 2.2). None were older than Eocene. He did not give the mean position, or paleopole, within each locus. He did not intend to give an apparent polar wander (APW) path nor did he recognize that such a path could be determined paleomagnetically.

Hospers (1955: 70) was concerned to show that paleoclimatically based polar wander paths as then known were inconsistent with paleomagnetism. He dismissed all four of them. *Kreichgauer's* Tertiary pole was outside Tertiary paleomagnetic loci, his more recent poles were outside loci for the early Quaternary of Iceland and Plio-Pleistocene of France. He rejected the Pre-Quaternary part of *Köppen and Wegener's* path because both its Eocene and Oligocene poles (which Hospers interpolated between the Eocene and Miocene poles) were outside corresponding paleomagnetic loci, and its Miocene pole, even though it was within the loci of the Mio-Pliocene from France and Oligocene–Miocene from northern England, was outside the Miocene paleomagnetic locus from Iceland. He also rejected the younger part of Köppen and Wegener's curve as well as *Köppen's* later improvements to it because their early Quaternary poles were outside the loci for the early Quaternary locus for Iceland and the Plio-Pleistocene for France. *Milankovitch's* path, Hospers claimed, failed for similar reasons.

Consequently, Hospers rejected appreciable Tertiary and post-Tertiary polar wandering and continental drift.

In conclusion it may therefore be said that the amount of polar wandering in Tertiary and Quaternary times suggested by the authors of various polar-wandering hypotheses is far too large to be reconciled with our data. However this does not necessarily mean that no polar wandering or continental drift has taken place at all.

*(Hospers, 1955: 72)*

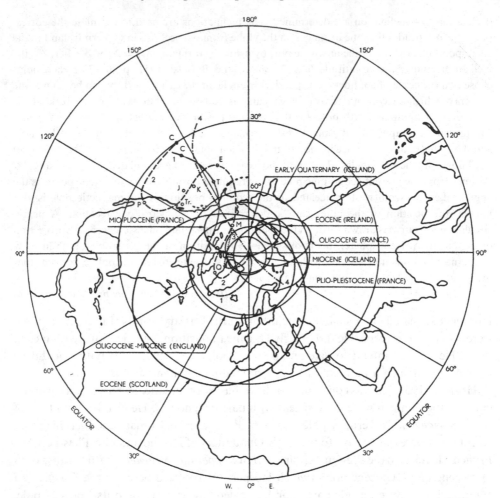

Figure 2.2 Hospers' Figure 1 (1955: 67) showing the loci of eight paleopoles calculated from rock magnetic data. From most recent to oldest, he used his own data from Early Quaternary lava flows from western Iceland (Hospers, 1953a, b, 1954a), Roche's data from Plio-Pleistocene lava flows from the Plateau Central, France (Roche, 1951), Roche's data from Mio-Pliocene lava flows from Plateau Central (Roche, 1951), his own data from Miocene lava flows from northern and southwestern Iceland (Hospers 1953a, b, 1954a), Bruckshaw and Robertson's data from tholeiite dykes of northern England of Oligocene or Miocene age (Bruckshaw and Robertson, 1949), Roche's data from Oligocene dykes, necks, and sills from Plateau Central (Roche, 1950), his and Charlesworth's data from Eocene lava flows from Northern Ireland (Hospers and Charlesworth, 1954), and Vincenz's data from Eocene lava flows from Scotland (Vincenz, 1952). He tested polar wander paths of Kreichgauer (1902), Köppen and Wegener (1924), Köppen (and Wegener) (1940), and Milankovitch (1933). Kreichgauer's curve is 1, Köppen and Wegener's is 2, Köppen's slightly altered curve is 3, and Milankovitch's curve is 4. C, Carboniferous; P, Permian; Tr, Triassic; J, Jurassic; K, Cretaceous; T, Tertiary; E, Eocene; M, Miocene; Q, early Quaternary position of the geographic north pole.

He tempered his rejection, pointing out that Iceland, according to Wegener and other mobilists, had not drifted relative to adjacent lands since the Miocene. Then, after correctly identifying (1955: 71) Iceland "as a piece of ocean floor above sea-level," he turned to results from Ireland:

It is shown that even for P = 0.01 per cent the mean direction of magnetization of the Eocene basalts of Northern Ireland is still outside the circle of confidence ... There is therefore some reason to believe that the position of the geographic pole in Eocene time relative to Ireland was somewhat different from its present position.

*(Hospers, 1955: 71)*

Did this mean that Ireland had drifted relative to Europe?

One may wonder whether there is any difference between the results from continental localities and those from Iceland. The available information does not suggest any such difference, but the question could be decided only if much more information were available and particularly if Eocene data from Iceland could be used. If any such difference could be found, it might indicate that the apparent small shift of the geographic pole deduced from the Eocene data from Ireland is not due to a wandering of the poles but to a drifting of the European continent relative to the ocean floor.

*(Hospers, 1955: 72)*

This was an interesting speculation.

Hospers stressed the importance of obtaining magnetic data from older rocks and other continents. However, he was not very sanguine that old rocks could retain their original remanent magnetization.

The idea that rock magnetic studies might eventually help to solve the problem of continental drift was first expressed by Mercanton (1926) and in later years by Gutenberg (1940, p. 448; 1951, p. 204) and Graham (1949, p. 160). To this end we shall have to possess much more information than we do at present. We need rock magnetic data from as many continents as possible and from as many different ages as possible, going back as far as the Carboniferous. It may be expected that igneous rocks will be more suitable than sedimentary rocks for this purpose; but even for igneous rocks it seems doubtful whether they will have been able to preserve their original magnetization over such enormous periods of time.

*(Hospers, 1955: 72. Reference to Graham is the same as mine; Gutenberg (1940)*
*is also the same; Gutenberg (1951) is my Gutenberg (1951b); Mercanton (1926) is*
*my Mercanton (1926b).)*

Hospers suggested a specific test of mobilism, based on Lester King's speculation that the black rocks he saw while aboard ship off the coast of Queen Maud Land might correspond with the Jurassic Drakensberg basalts of South Africa (I, §6.10). Once again, however, he had his doubts about whether the test would work because the rocks were old.

King (1953) has recently announced the possibility of a new test of the hypothesis of continental drift, consisting of a stratigraphic comparison of South Africa and Queen Maud's Land (Antarctica). The present author would suggest that a rock magnetic comparison of the

Drakensberg plateau basalts and the plateau basalts of Queen Maud's Land (which are thought to be stratigraphically equivalent) might give valuable information. The difference in latitude between the two series before drift occurred cannot have been more than a few degrees, whereas at present it amounts to approximately 30°. However, it does not seem certain that the method can be applied, as the age of these plateau basalts is considerable.

*(Hospers, 1955: 72; King reference is the same as mine)*

Thus, although Hospers was not willing to reject mobilism entirely, and even suggested specific tests, he doubted that they would ever be successful because of long-term instability in old rocks. Hospers was strangely resistant to Irving's ample demonstration from field tests of the very long-term stability of sedimentary rocks (§2.5, §2.6). Hospers remained at Royal Dutch Shell until the early 1960s, his next papers on paleomagnetism did not appear until the late 1960s.

Because general polar wander had been mooted but individual motion of Iceland had not, Hospers' negative results were relevant to the discussion of polar wander but not to drift. His doubts about using old rocks had no effect on the other British paleomagnetists, and were surprising because of Irving's results as just noted. Likewise Clegg and colleagues for the Mesozoic (§2.4), and Creer (§2.12) was actively gathering promising data from Paleozoic rocks. Although Hospers left Britain in September 1953, a year before Irving completed his thesis, he was fully informed of all these new results at the Birmingham meeting of January 1954, which he attended (§3.3). Nevertheless, Hospers' justification of the GAD hypothesis in Iceland was an essential starting point for paleomagnetic testing of mobilism. It was paradoxical but nevertheless helpful to the paleomagnetic case that the lack of Iceland's drift advanced the paleomagnetic test of drift; if Iceland's latitude had changed perceptibly during the Late Cenozoic, Hospers would not have observed magnetizations directed along the GAD field.

## 2.10 Creer, his education and initiation in research

Runcorn, pleased with Hospers' Iceland work and Irving's initial results, proceeded to expand paleomagnetic work at Cambridge. Wanting a high-sensitivity Blackett-style magnetometer at Cambridge, in the fall of 1951 he invited Ken Creer, a new graduate student at the Department of Geodesy and Geophysics, to undertake the task. Creer did not accept immediately, but did so the following year.

Kenneth (Ken) M. Creer was born on the Isle of Man on October 18, 1925. He attended Douglas High School, where he became "aware of the drift hypothesis, from background reading, as a schoolboy" (Creer, April 9, 1984 letter to author). He was awarded a scholarship endowed by Henry Bloom Noble. He served in the British Army (1944–8) in infantry regiments, much of the time in Africa. During his last year of service he was in charge of the Science Department at the Middle East College where courses were given to servicemen to prepare them for return to civilian life. He left the army in April 1948 with the rank of captain.

Creer entered Cambridge in the fall of 1948, the same time as Irving. They met in geology and mineralogy/petrology classes. Creer was a model student. On his first year's work he won a Shell Scholarship in Geology, and at the end of his second year, in May 1950, he was awarded First Class in Part I of the Tripos (physics, geology, and mineralogy) and won a Foundation Scholarship and a Munro Studentship at Queens' College. As an undergraduate he bought Holmes' *Principles of Physical Geology*, but recalls (Creer, April 9, 1984 letter to author) that in Part I "we were concerned mainly with basic principles – Sedimentology, Stratigraphy, etc." According to Creer (April 9, 1984 letter to author) none of his undergraduate teachers in geology "aroused any particular interest" in continental drift, and he, unlike Irving, had

no special interest in continental drift until, on joining the Department of Geodesy and Geophysics as a research student it became apparent that palaeomagnetism offered a new method which held the potential of contributing to proving whether or not drift had occurred.

*(Creer, November 24, 1983 letter to author, revised January 5, 1999)*

After considerable anguish, Creer decided to read physics rather than geology during his third year and gained a First in Part II of the Tripos.

With such a stellar undergraduate career, Creer had no trouble being admitted to graduate studies in the Department of Geodesy and Geophysics. He did so in September 1951, receiving (as Hospers had) a three-year Royal Dutch Shell Studentship. He was offered a choice of research projects.

When I was received into the Department, I was offered the choice of several research projects. One of these was to work on paleomagnetism with Keith Runcorn's new group, but I decided against this because the grant had not then come through to fund this work, and being a physicist I wanted to build and use my own magnetometer – it was a different matter for Ted [Irving] because he already had access to the Blackett magnetometer. Another project proposed to me was in marine seismics with Maurice Hill's group, but I ruled that out immediately because I am prone to seasickness. I spent about two months on a project to develop a portable seismograph to detect local anomalies for use in urban areas, the signal source being hammer blows rather than explosives. The prototype I built (before transistors and magnetic tape recorders were available) used thermionic valves and proved too cumbersome and fragile for field-use. I then chose a project involving laboratory compaction measurements on muds from marine sediment cores that had been collected a year or so earlier by Teddy Bullard from the Pacific. The relationship between applied pressure and void ratio is linear (Terzaghi's Law). Through this technique the maximum pressure that the muds have been subjected to *in situ* on the geological timescale can be deduced, and thence it can be inferred whether some of the overlying sediments have since been eroded or, for example, deeply buried or covered by an ice cap. Unfortunately for me Bullard was no longer at Cambridge, having taken up an appointment at the University of Toronto, and there was no one in the Department who had any direct connection with the overall project for which the cores had been collected. And so

I found that I was becoming more and more interested in compaction processes at very low pressures and this was leading me into reading about processes of random fracture of solids. I designed a new apparatus to be able better to carry out such investigations.

*(Creer, February 1987 interview with author)*

In designing his new apparatus, Creer consulted with staff and students in the department, but with Bullard gone, got little help. The suggestion was made, "Why don't you go and ask the engineers for some advice?" Creer went over to the Department of Engineering and was directed to an undergraduate laboratory set up for soil mechanics studies.

I went into the laboratory and was immediately shocked by the sight of a row of apparatus very similar to the one I had designed for my research. There were some differences in size and proportions but the basic principles were the same. On the one hand I was quite pleased to see that my design was good, but on the other hand I was shocked by the realization that I had lost six months of the three years funded by my research scholarship.

*(Creer, February 1987 interview with author)*

The realization that he must abandon the compaction project took some time to sink in, for he was reluctant to throw away months of work. Creer's immediate response was to go to a movie, though he was not a big fan of the cinema. He is confident that he did not watch much but he needed a place to think.

And so I went to the cinema by myself, and sat by myself. The cinema was on my way home. I don't think I watched the film, what I wanted was isolation to sit and think out what I should do. Part of me was thinking: "You should really stick in with this project and make a go of it." On the other hand I thought, "Better transfer to another project while there is still time."

*(Creer, February 1987 interview with author)*

Runcorn's offer was still open, and his grant was about to come through. Before he left the cinema and after two false starts, Creer had decided to transfer to paleomagnetism.

## 2.11  Creer constructs the Cambridge magnetometer

Creer accepted Runcorn's offer around Easter 1952.

At about Easter time, 1952, Runcorn finally persuaded me to join his group with the initial task of constructing an astatic magnetometer. I agreed because Runcorn told me that he had just then received news that his research grant application had been successful.

*(Creer, November 24, 1983 letter to author; slightly altered January 5, 1999)*

Creer still had some doubts about the wisdom of his decision – would he be employable after choosing to work in such an esoteric area? In fact, Ben Browne cautioned him about the slim chances of future employment. Creer later addressed his uneasiness and Ben Browne's lack of enthusiasm.

I had a certain feeling that I wanted to do research in something more applied that would prepare me better for working, for example, with an oil company or in commercial geophysics. And Ben Browne was not too pleased when I told him of my decision. He said, "Well look, you'll never get a job; you're probably wasting your time doing this [i.e., working in paleo-magnetism]." But, anyway I did.

*(Creer, 1987 interview with author; my bracketed addition)*

Creer designed the magnet system. Like Clegg, he wanted an instrument that would measure more quickly than Blackett's. He aimed for a magnetometer that would measure specimens of weakly magnetized sedimentary rock in about eight minutes, several times faster than the Jodrell Bank magnetometer, which had been designed for other purposes; on it one could measure "12 to 15 specimens per very long day" (Irving, June 29, 1983 letter to Runcorn). Thus Creer and Clegg did not simply copy Blackett's magnetometer but made instruments better suited to measuring routinely the remanent magnetism of small sedimentary samples rather than much larger ones.

Creer was the major force behind the design and Leslie Flavill behind the construction. David Collinson and Irving helped. Collinson, a physics graduate from Manchester who was hired by Runcorn as a research assistant, helped Irving and Creer with fieldwork and laboratory measurements. Flavill's contribution was crucial. Leslie Flavill had a very high reputation as an outstanding instrument maker. Creer recalled:

Then in Cambridge we had a very fine technician, Leslie Flavill. He was very good at making small things. The whole key to making this magnetometer was that the magnet system had to be small and delicate. And he was just the person to do this. You had to keep the whole thing as light as possible with the minimum moment of inertia. Of course, the whole thing was light. It was hung on a suspension system, which was also very light. Ideally, it should have been weightless. He was able to make the suspension system. It was really a beautiful piece of work. Since then, I don't think I've met another technician in another laboratory who could have done this. So he played quite a key part in making the magnetometer so good ...

*(Creer, 1987 interview with author)*

Taking over six months, Creer got the magnetometer operational in February of 1953. There were delays in the construction of the building for it, which was not finished until the summer 1952. It had to be distant from magnetic and mechanical disturbances, to be non-magnetic and held at an even temperature. It was made of wood, copper nailed, and thermally insulated. It was in a mechanically and magnetically quiet place, five hundred yards from the nearest main road (Creer, 1955: 4). It still stands in the department site. When returning from his Christmas break at the beginning of 1953, Creer ran into difficulties.

Well, I thought we had the magnetometer going at the time of the Christmas–New Year break. When I got back from the holidays I left it with Ted and David Collinson to work with, since I was in sick bay in Queens' with flu. I remember them coming to see me (I was taking my pulse) telling me that when they switched on the newly installed Helmholtz coils the magnet-ometer twisted right around and became locked in a certain direction. Having been nicely astatisized, it should have swung quite freely in the "zero" field which was down to a few gammas at the center.

*(Creer, 1987 interview with author)*

Creer took his medicine and designed a new coil system.

The bible was then Chapman and Bartels' textbook on Geomagnetism, and I got this out and looked up their calculations of the degree of homogeneity of the field produced by the Helmholtz coil systems. I soon realized that our Helmholtz coils were just too small, and there was a problem that the laboratory could not take larger ones. Obviously, the bigger the coils, the smaller the field gradients. When the coils were switched on, each magnet [of the suspended astatic pair] saw a slightly different field, and this caused a strong controlling force that could not be asticized away. In Chapman and Bartels I found that other four-coil systems had been designed that, for a certain size, made a much more homogeneous field. I opted for the "Fanselau" design, which I had constructed when I returned to the lab. It was quite a lot more work. But lo and behold, it worked. It was a challenge for me personally because the others were pretty pessimistic at the time, and I was wondering whether I could cure the problem.

*(Creer, 1987 interview with author)*

Had Creer not come up with this quick, neat solution, the Cambridge effort would have been greatly delayed, a high sensitivity magnetometer being essential to the program.

## 2.12 Creer begins fieldwork

With the calibration and testing of the magnetometer in February of 1953, Creer was finally able to begin his own research project, half-way through the three years allotted to his graduate program. He was very pressed for time. However, beginning late allowed him to make use of the findings of Hospers, Irving, and Clegg. He chose to test continental drift by investigating very long-term variation in the ancient geomagnetic field. Rather than a detailed study of closely associated rocks he needed samples spread throughout the Phanerozoic. A gamble but, as it turned out, a brilliant strategic choice, leading as it later did in 1954 to his constructing the APW path for Great Britain, the first such path ever, and the key to the paleomagnetic test of continental drift.[18] He also examined reversals, secular variation, and the question of which lithologies truly recorded the ancient field. Creer, like Clegg, did not initially choose fine-grained red sandstones, but did so by a process of elimination after finding them to be the most productive.

Creer knew that Clegg and company had been sampling the Triassic New Red Sandstone that transformed the Phanerozoic from one huge blank slate into two lesser but still huge blank slates between Hospers' Eocene and the Triassic and between the Triassic and Irving's Late Precambrian Torridonian. He began with the former.

I was getting a bit pessimistic because all this time was passing. By that Christmas (1952), almost a year and a half of my three years had gone, and I hadn't got a single result. I wanted to finish in three years. It was my target to finish on time. And so, it must have been about, I would say, February 1953 before the magnetometer was going. I first thought I would like to fill in the space between Hospers' Tertiary and Eocene and Clegg's Triassic.

*(Creer, 1987 interview with author)*

His first collections were from the classic, very well dated Jurassic sedimentary rocks. But he found them too weakly magnetized to measure or magnetized close to the present Earth's field and they could have been remagnetized recently. Given the advantage of hindsight, he later recalled:

I measured many Jurassic clays and found their direction of magnetization was not very different from the direction of the present Earth's field. With hindsight, this result is understandable now that we know much more about how Britain has moved with respect to the poles. But I was then disappointed to find that I wasn't getting the big differences that Clegg had found in the Triassic.

*(Creer, 1987 interview with author)*

He also studied Early Tertiary brown sandstones, which were not red like Irving's Torridonian. They too proved to be unsuitable.

I then did some Tertiary sandstones – rather coarse grained. They were not red but they had bags of iron, far more iron than the Triassic red beds or the Torridonian. But, the iron was sort of brownish and yellowish. They were very strongly magnetized but again I took it to be along the present Earth's field. They were not giving oblique directions indicative of continental drift.

*(Creer, 1987 interview with author)*

His first collections were unpromising.

Creer also collected from the Triassic red Keuper Marls at Sidmouth on the south coast of England, which were some improvement but not ideal. He found that many samples had two components of magnetization, which led him to the problem of overprinting, and he applied Néel's (1949) theory of magnetic viscosity.

When I was down in the south [of England] ... I went to Sidmouth where the Keuper Marls were exposed on the coast. Since they are Triassic, I thought, "Well, let me try them to see if they confirm Clegg's results." It is really a very nice exposure. These were Marls, kind of windblown, not siltstones like Clegg had measured. They were red colored. I found that when I came to measure them that instead of getting a circular group of directions, they were "streaked" along a great circle. I interpreted this later as due to their having two components of magnetization: one a fossil component and a secondary component along the present Earth's field. I think that all along I was as much interested in the development of the paleomagnetic techniques as in the application of the results as relevant to continental drift. And, surely this is natural for a physicist.

*(Creer, 1987 interview with author; my bracketed addition)*

Also at Sidmouth, he determined by geometrical construction what he regarded as two primary magnetizations, normal and reversed, imperfectly defined but consistent with what Clegg and company were finding in Keuper sediments elsewhere. Normal was found high in the cliff and reversed below. During summer 1953 he returned and sampled between them: but there was no transition, only an abrupt change.

And then when I looked at the samples, I realized that the reversed ones came from the bottom of the cliff and the normal ones from the top. So then I went back with David Collinson who, by then, was helping me in the field. The object was to find the horizon between normal and the reversed samples because, I thought, if the field is reversing, we will be able to see *how* it reverses. We never found a record of the polarity transition: below a certain level they were reversed and above that they were normal. One of the prime objectives of present day paleomagnetic researchers is to find out more about how the field actually reverses polarity. You have to find sediments that were deposited fairly fast at a particular time when a polarity reversal happened. You know the field spends only about 1% of the time actually reversing, so looking for suitable exposures is rather like looking for a needle in a haystack. One needs some luck, and if we had had more luck then it would have been very nice, and pertinent to the then lively controversy over self-reversals.

*(Creer, 1987 interview with author)*

He had not yet found coherent, unambiguous, single-component magnetizations, as required for the continental drift test, and the possibility of being able, within the compass of British geology, to "fill in the space between Hospers' Tertiary and Eocene and Clegg's Triassic" was receding. He needed to consider trying to fill the earlier gap. Accordingly he sampled Permian basalts, the Exeter Volcanic Traps of Devon. He had also begun thinking more about Irving's finding that the best Torridonian results came from red and reddish-purple siltstones or fine sandstones, well cemented. Could this be generally true and could it be used to select samples for future study? Creer later summarized his thoughts.

And so, this is when I thought: Well what sort of rocks should I now collect because I knew (from Irving's work and also from Clegg's) that it was possible to get coherent results. It was then that I realized that one had to get red beds – it was not sufficient to select rock beds that were rich in iron. I did not reason why as I had to get on with my collecting while the favorable summer and autumn weather lasted. Clearly, Ted had developed an "eye" for selecting rocks that were good for paleomagnetic study. We all talked together and were familiar with the "look" of one another's rock samples – the "good" ones and the "bad" ones. It was not easy to describe criteria entirely verbally, though in general terms dark red/purple siltstone to fine-grained sandstone were the key words. We didn't know exactly why at this time, and in later years David Collinson applied a major effort in investigating the origin of remanent magnetization in "red" beds. Ted had also noticed that his drab (greenish) colored Torridonian samples were much less strongly magnetized than the "red" ones. Could this observation be extended to rock formations of different ages? My negative results for the Jurassic also suggested to me that this could be a reasonable working assumption to make in planning my future work.

*(Creer, 1987 interview with author)*

Creer recalled that samples of the Devonian Old Red Sandstone (ORS) he had seen in the Sedgwick Museum were similar in color and grain size to Irving's Torridonian sandstones. Prompted by these thoughts and being in Devon, he decided not to go directly back to Cambridge but to take a detour to South Wales, and have a quick look at the Devonian ORS. He liked what he saw and decided to return with the relevant geological information and get some samples as soon as possible. It was a good move. They and the Exeter Permian lavas would begin to fill in the older temporal gap, that between Clegg's Triassic and Irving's Pre-Cambrian.

I think it was probably the beginning of November [1953]. I still had not got anything good and solid. I had these partially unstable rocks from Sidmouth that made quite a nice study in its own right. But I really wanted to get some rocks that would yield data relevant to the continental drift problem. And the answer came: "You've got to get dark red beds that look like the Torridonian." And samples of Old Red Sandstone that I saw in the museum seemed to me to hold great potential. But it was November 1953 before I made my first collection, in South Wales, with David Collinson's help. We brought them back in December, and before Christmas had cut some up and measured them. It was immediately apparent that they were going to yield good data, and that with hard work I would be able to complete my thesis on time. I spent a very happy Christmas holiday!

*(Creer, 1987 interview with author; my bracketed addition)*

After the first few measurements Creer knew that he had made an excellent choice. The intensity of magnetization was comparable to the Torridonian and readily measurable. The directions were well grouped around a direction oblique to the present geomagnetic field, but different from both Irving's Torridonian and also Clegg's Triassic. All samples had reversed magnetization. So also had the stably magnetized Permian Exeter basalts. From this early work he was able to present preliminary results at the important January 1954 meeting in Birmingham (§3.3).

From the first few measurements I got highly promising results. From the way in which the directions grouped, I could tell that they were comparable in quality to Ted's Torridonian data, and with John Clegg's Triassic data. So, at the end of that year [1953] I saw the way clear to finishing my thesis. I knew all I had to do was to complete the Old Red Sandstone study with some additional fieldwork, and to collect from other "red" bed formations that I knew were to be found in the South Wales area. At the January meeting I reported the first results from the Old Red.

*(Creer, 1987 interview with author)*

Eventually Creer collected from 117 sites in the ORS of the Anglo-Welsh Curvette. Except for a few Silurian samples from the Ludlow Series at the base of the ORS, which proved to be overprinted, all were Devonian. After the meeting, Creer and Collinson also sampled the Early Cambrian Caerbwdy Sandstones in West Wales and the Late Precambrian Wentnor series of the Long Mynd in Shropshire. Both are red-purple in color and lithologically similar to the Torridonian. The samples were measured early in 1954 and had well-grouped magnetizations strongly oblique to the present field.

Creer now knew from his own results, and those of Clegg and colleagues from the New Red Sandstones and Irving's from the Torridonian (both at the time unpublished), that he could reasonably expect to make a first attempt in his thesis to trace the movement of Great Britain relative to Earth's rotational axis since the late Proterozoic (the last ~600 million years); to get a quick "look" at the entire Phanerozoic it likely would not be necessary, at least initially, to study all the rock formations as intensely as that of the Torridonian. At this stage (early 1954) Creer had not expressed results as pole positions, only as directions of magnetization. These old oblique magnetizations from Britain had been obtained by half a dozen British workers.

## 2.13 Runcorn and his research

During the early 1950s at Cambridge, Runcorn devoted much enthusiasm and energy to recruiting students to work on paleomagnetism and geomagnetism. He soon however began travelling widely, especially in the United States, and his extensive absences from Cambridge gave his research students and associates ample opportunity to develop their own projects, and they did. Although his students were largely self-motivated – self-starters, carving out their own areas of research, and making their own original contributions – Runcorn was the motivating force behind the early development of paleomagnetism at Cambridge. It was he who got the whole effort going. He recognized the need to get researchers with different backgrounds working on related problems; an idea, contrary, as already noted, to Browne's way of doing things in the Department of Geodesy and Geophysics (§1.22). Runcorn recognized and attracted good students, knew how to get them started and set up contacts for them, and then, as he was perpetually busy doing all manner of other things, more or less left them alone. He also learned from those around him, from Hospers whose work convinced him of the reality of geomagnetic reversals and the GAD hypothesis (§1.13), and from Irving (§3.4) and Creer (§3.7) who were arguing in favor of continental drift several years before he was willing to accept it (§4.3).

While traveling widely, Runcorn did his own research. Beginning in the summer of 1952, during which he worked on high-pressure studies, he spent several months of the following four summers in the United States. In May 1953 he attended the spring meeting of the American Geophysical Union (AGU), where he presented a paper on the origin of the geomagnetic field and described the Cambridge work in paleomagnetism. Invited by L. B. Slichter, Director of the Institute of Geophysics at the University of California, Los Angeles (UCLA), he spent the summers of 1953 and 1954 there as a visiting researcher. After working at the institute in 1953, he teamed up with C. D. Campbell, a geologist at the State College of Washington, and collected samples from flow in the Late Tertiary Columbia River basalts. Collinson measured them back in Cambridge. Runcorn soon gave a preliminary account at the 1954 Birmingham meeting (§3.3) but the results were not published for two years

(Campbell and Runcorn, 1956). Much as Hospers had found in Iceland, half the lavas were normal and half reversed, and they argued that this reflected reversals of the geomagnetic field. Also as in Iceland, regardless of polarity, the mean field during their eruption corresponded closely to that of the geocentric axial dipole – the first evidence for the GAD outside Iceland and Europe.

Runcorn worked on other aspects of geomagnetism. Following Elsasser (1939), he regarded the Coriolis force as the main factor in controlling the axial orientation of the geomagnetic field (§1.4), It was, he argued, the reason why the axis of the geomagnetic field, when averaged over periods of several thousand years, coincided with the axis of rotation (Runcorn, 1955a), as Hospers' work on Icelandic lavas completed in late 1952, and the Columbia River basalts work demonstrated. Runcorn also benefited from the work of Raymond Hide, another of his highly successful Cambridge students, who had constructed an apparatus analogous to Earth's core, which allowed him to study the effect of various forces including the Coriolis force on convection in a rotating fluid system. From the work of Hospers and Roche, and now his own, Runcorn became convinced of field reversals, and argued that his overall model could account for them, for the westward drift of secular variation, and for variations in the length of the day.[19]

Runcorn believed he had provided a theoretical framework to bolster Hospers' GAD hypothesis regarding the time-averaged form of the geomagnetic field, and was patently pleased to have done so. Hospers, Irving, and Creer gave much more weight to empirical arguments; he was more theoretically inclined.

## 2.14 Rationality in deciding to launch a paleomagnetic test for continental drift

Except for Runcorn, who as just shown worked on other problems, British paleomagnetists, increasingly conscious of the tectonic potential of paleomagnetism, saw testing mobilism as the long-term goal. Entering the debate when they did was wise, because by 1953 they had good reason to think that they could determine whether or not drift had happened. They had the instruments to do it. They knew that Blackett-type magnetometers were sufficiently sensitive and although slow could be modified to work more quickly. They intercalibrated their new magnetometers. They had growing evidence that redbeds (siltstones and fine sandstones) and certain igneous rocks yielded coherent results; their magnetizations were not only stable but likely recorded the orientation of the geomagnetic field at or very soon after deposition and could do so even as far back as the Late Precambrian. For the GAD hypothesis, Hospers had provided empirical and Runcorn, theoretical support. Although when doing so neither had continental drift in mind, indeed Hospers opposed it and Runcorn did not take it seriously for another three years (Chapter 4). The growing preoccupation of most paleomagnetists in Britain with tectonic applications did not stop them attending to other problems: they continued to work on reversals, notably the reversal timescale, and Hospers made a first attempt at magnetostratigraphy. Above all they developed a growing awareness of the fruitfulness of paleomagnetic work; the same samples from the same sequences simultaneously provided data for both tectonics and for geomagnetic reversals; the same data were a rich source of information for investigating both phenomena; between the two, there were fruitful feedbacks.

## Notes

1 During the second half of 1951 and early 1952 I worked at Jodrell Bank for weeks on end. Occasionally I would get in to Manchester and call in at the Physics Department. Blackett would invite me into his office and enquire about my progress. I would explain what I had done with my rock discs and his magnetometer. There were lots of drawings on graph paper and Blackett would work always with a small slide rule he kept in the breast pocket of his jacket. Then he would ask what I was planning to do next, and I would tell him. Then he would ask how long would it take me, and I would tell him. He would then enunciate in the best quarter-deck style what he referred to as Blackett's Law – "Irving," he ordered "multiply that by phi."

*(Irving, October 24, 2010 note to author)*

2 Mike Fuller, at the author's request, kindly conducted an interview with John Clegg. Fuller, a geophysicist and paleomagnetist, is Clegg's nephew. In addition, as a young man he often accompanied Clegg and other members of the Manchester–Imperial College paleomagnetic group on field trips. Because Clegg was in poor health and had difficulty speaking, Fuller conducted the interview, asking Clegg a number of questions prepared by the author. Fuller's answers constitute summary statements of Clegg's remarks. They also include some editorial remarks by Fuller.

3 Sir Bernard Lovell (1968) wrote an entertaining history of radio astronomy at Jodrell Bank. He discusses Clegg's work on aerials, and his help in building various radio telescopes.

4 Fuller has recalled that he used to go out to Jodrell Bank with Clegg when Clegg "worked on some of the early stuff." Fuller reported that Clegg was involved with some of the actual construction of radio telescopes but was unsure as to whether Clegg had anything to do with the design or construction of the big dish. Lovell was the main figure behind the big dish and the research program in radio astronomy at Jodrell Bank. However, Clegg got into radio astronomy at Manchester early and, as demonstrated by his publications, played an important role in the study of meteor showers (Clegg/Fuller, 1987 interview with author).

5 No reasons were given for Clegg deciding he no longer wanted to continue working with Lovell. Perhaps the dynamics of their relationship changed once Clegg earned his Ph.D. Irving had a bit of a run-in with Lovell, although, given the circumstances, it would be quite unfair to suggest that Lovell was to blame. Irving (August 20, 1985 letter to author) recalled his encounter with the following:

The golf ball story is that I found such a ball by the road outside Lovell's office at Jodrell Bank. Lovell was head of the radio-astronomy observatory there. In my delight, and being fond of ball games, I threw it down hard to bounce up against a blank part of the office wall intending to catch it, and then perhaps throw it again. Instead it hit a hole in the road and shot sideways through Lovell's office window. He was inside at the time and got quite mad. It was all so comic that I smiled when he reprimanded me, which made it worse. He told Blackett about it and he reprimanded me mildly. You see, professors really were "in loco parentis." Both Lovell and Blackett were very serious people, and expected seriousness in others.

6 Both Irving and Clegg/Fuller interviews with author confirm that Clegg played a central role in the construction of the magnetometer housed in the basement of Manchester's Physics Laboratories.

7 Either this episode occurred after April 22, 1952, or Blackett fixed the magnetometer and used it again, since Blackett states how he made sets of measurements on the Jodrell Bank magnetometer on the evenings of April 21 and 22, 1952 (see Blackett, 1952: 361). In addition, Irving made his third collecting trip to the Torridonian in April of 1952. Unfortunately, his field notes are often not dated. After consulting with J. E. Hemingway, a sedimentologist at the University of Leeds, about the Torridonian on April 4, he drove north to Scotland in a Land Rover. But there is no record as to how long he spent in the field or when he began measuring the samples on the Jodrell Bank magnetometer. Most likely, the magnetometer was damaged at the very end of April or in May. Irving made his fourth collecting trip in November 1952. If he measured samples from his third trip before launching his fourth, then he would have had to have used the magnetometer at Jodrell Bank. He did not use the one in the Manchester physics laboratory and could not have used the Cambridge magnetometer until after February 1953 because Creer did not get it operating until then. However, Irving, in a retrospective comment (*Synopsis*, pp. 2–3) remarks that all rocks collected after the first two field trips were measured on the Cambridge magnetometer. If this retrospective comment of Irving's is correct, then perhaps he measured some of the samples from the second trip at Jodrell Bank after completing his third trip. Regardless of details, it is clear that Clegg had his new magnetometer in working order by the end of April 1952.

8 This twenty-two page unpublished essay by Clegg, entitled "Blackett and rock magnetism," was used by Lovell as a source for his Royal Society biographical memoir of Blackett. Clegg sent an undated copy of his essay to the author, and some of the passages from it are quoted by Lovell (1975).

9 Actually, I am not sure in what detail Hemingway spoke of the stratigraphy of the Torridonian sandstones to them at this time. However, Hemingway certainly answered a number of questions Irving posed at this time about the conditions of deposition of the Torridonian. Hemingway was helpful; for example, Irving's (1957c) paper based on his dissertation in which he acknowledges his help. Certainly, at this early stage, he directed Irving to three important sections, giving him an excellent start on the task of achieving stratigraphic coverage of the Torridonian. Much of Irving's information, however, came from the classic 1907 memoir of the Geological Survey of Great Britain, "The Geological Structure of the North-West Highlands of Scotland," by Peach, Horne, Gunn, Clough, Hinxman, and Teall, and from his own extensive reconnaissance of prospective sections. I might also note that Irving recorded his sampling on Ordnance Survey Maps at the scale of 6 inches to a mile and these are in his files (Irving, note to author, July 2007).

10 Irving's notes on Kuenen and Shepard were definitely taken before March 1953 since they precede notes he took at a symposium during March at Cambridge on the geology of Cyprus and Oman. Because they occur after his reading of several papers that were published late in December 1951, he probably read Kuenen and Shepard during 1952 or early 1953.

11 This process is now widely recognized and called post-depositional detrital remanent magnetization (Dunlop and Özdemir, 2001).

12 This work on Torridonian was important for the early development of paleomagnetic work in Britain and for Creer's seminal polar wander path (§3.6) which was the starting point for the global paleomagnetic test of mobilism. Although outside the time-frame of this book, knowing how his results have fared is of historical interest, hence this note, which has been kindly supplied by E. Irving (12/15/2011 email to author): Following Irving's thesis work (1954), nothing further was done on Torridonian rocks for almost 20 years. Stewart and Irving (1974) resampled the Stoer Group and the base of the Torridon Group applying thermal demagnetization. By then, plate tectonic reconstructions had confirmed that the Lewisian Platform on which the Torridonian rested was part of the Laurentian (Canadian) Shield, and they related Torridonian pole positions to the Proterozoic apparent polar wander path of Laurentia by correcting them for the opening of the Northern Atlantic. This practice has been followed by Smith *et al.* (1983) and later workers. From the 1970s,

there have been half a dozen papers on particular and usually local problems. Recently, the entire sequence has been sampled by M. H. Darabi and J. D. A. Piper (2004); they resampled and greatly extended Irving's sections, reviewed previous work and obtained important new results concerning the origin of the primary magnetization and subsequent overprinting. Comment here is concern mainly their revisions.

*Stoer Group*. Thickness ~2 km. Stewart and Irving (1974) resampled across the newly recognized stratigraphic boundary between the Stoer and Torridonian Groups and confirmed that it coincided with the paleomagnetic boundary recognised by Irving. Darabi and Piper (2004) have reviewed the age information on the Stoer Group, noting a Pb-Pb isochron of 1199+/−70 Ma by Turnbull *et al.* (1996) in agreement with the discovery in rare limestones of nannofosils of middle Riphean age (Cloud and Germs, 1971).

The original NRM studies of Irving (1954, 1957) and Irving and Runcorn (1957) and the demagnetization studies by Stewart and Irving (1974) of clasts in basal Torridon conglomerate of the Torridon Group and of the main Stoer section with it's Stac Fada member (see §2.5) demonstrated that the magnetization of the Stoer Group generally was not greatly affected by later overprints. This has now been abundantly confirmed by Darabi and Piper (2004). The mean directions for the Stoer Group given by these three studies are very close (Creer and Irving, 2012) and hence the pole for the Lower Torridonian (Stoer) in Creer's figs. 3.4 and 3.5 of chapter 3 has barely changed. The NRM of the red silts and fine sandstones of the Stoer Group is among the most magnetically stable every observed.

*Torridon Group*. Thickness ~5 km. Piper and Darabi (2004) review age information on the Torridon Group. The youngest concordant detrital zircon from the Applecross Formation gives an age of 1060+/−18 Ma (Rainbird *et al.*, 2001), consistent with an age of 1039-1025 Ma obtained by comparison of Torridon Group poles with the age-calibrated Keeweenawan poles on the Laurentian pole path. Piper and Darabi (2004) have shown that the Torridon Group has a widespread secondary magnetization of Caledonian age, steeply inclined to the ESE. The overprint is of one polarity only. It is absent in the underlying Stoer Group. This discovery by Piper and Darabi enabled them to provided improved estimates of the directions of primary magnetization (see endnote # 2 Ch 3) and explains many features of Irving's early data, notably the strong polarity bias; of his 81 sampling horizons, 53 were directed to the SE with downward inclination and only 28 to the NW with upward inclination.

13  The majority were small chips with an arrow pointing up on which he measured polarity only. The rest were fully oriented and cored.

14  I want to thank Leo Kristjansson, Science Institute, University of Iceland, for sending me Fisher's letter. Kristjansson, a paleomagnetist who has done extensive work in Iceland, was given the letter by the late Dr. Einarsson. Kristjansson has written three excellent reviews of the history of paleomagnetic research in Iceland from its initiation until around 1981. See Kristjansson (1982, 1984, 1993). The "Fisher" letter appears in Kristjansson (1993).

15  This difference in attitude of Hospers and Graham toward the question of "What is the cause of field reversals?" is analogous to the difference in opinion of mobilists and fixists prior to the acceptance of seafloor spreading and the rise of plate tectonics. Wegener and many of his fellow mobilists viewed the problem of finding a mechanism for drift as an unsolved second-order problem. In contrast, most opponents of drift during the pre-seafloor spreading period viewed the issue of the driving mechanism for continental drift as a pressing theoretical difficulty facing the drift theory. Moreover, once the empirical support for plate tectonics became overwhelming, the vast majority of Earth scientists came to accept mobilism in the form of plate tectonics even though the mechanism issue regarding what drives the plates remained (and still remains) imperfectly answered.

16  If I were writing a history of post World War II paleomagnetism, Roche would play an important role, especially in the discovery of reversals and their interpretation as field reversals. Roche studied many igneous bodies in the Massif Central of France. Finding that all his upper Pleistocene lavas were normal while some lower Pleistocene

lavas were reversed, he estimated that the last reversal occurred during the middle of the lower Pleistocene. He also identified reversals during the upper and lower Pliocene, an upper Miocene reversal, and reversals during the early and middle Oligocene (Roche, 1950, 1951, 1953).

17 Hospers' and Roche's timescales are similar to the first potassium–argon polarity reversal timescale developed by M. G. Rutten (1959). Rutten, recall, was with Hospers during his first trip to Iceland, and presumably was introduced to paleomagnetism by him. Hospers' and Rutten's timescales offered similar estimates for the duration of the current (Brunhes) normal interval. For a discussion of Rutten's timescale see Glen (1982: 130–139).

18 Such paths later became known as apparent polar wander (APW) paths, which is how Creer and other paleomagnetists interpreted them: as evidence of continental drift, polar wandering in the classical sense, or their combination. Magnetization directions map unambiguously into poles, so to claim that an APW path is correct is not to make an ontological commitment that polar wandering in the classical sense has occurred. When Creer developed the convention of displaying paleomagnetic directions in terms of APW paths, they treated them as APW paths. Irving introduced the expression "apparent polar wander path" in his textbook (Irving, 1964: 130). Before that they were called simply "pole paths" or "paleomagnetic pole paths." I shall refer to paleomagnetically derived polar wander paths as apparent polar wander paths because it better expresses the use of such paths by these early paleomagnetists.

19 Runcorn was invited to the AGU meeting by Walter Bucher. Bucher, a very influential American geologist and a renowned foe of continental drift (I, §3.8, §5.11, §6.10, §7.3, §8.4, §8.8, §9.7), was interested in geophysics, believing that it had much to offer geology. Bucher met Runcorn in 1951 at an IUGG (International Union of Geodesy and Geophysics) meeting in Brussels. Bucher was elected President of AGU for a second time in 1953. Instead of giving a second Presidential Address, he decided to invite Runcorn to talk about Earth's core. Bucher introduced Runcorn with the following:

Three years ago your President, a geologist, gave a report on the largest problems of the Earth's dermatology. When you reelected him, he had to plan for another presidential address. If this were a Union of terrestrial dermatologists, he could have served up more than one acceptable dish of specialized knowledge. But we are proud of our function as a Union of men who get together at intervals to see the ultimate object of their studies in broad perspective. So your President decided upon a departure from custom, not to dodge his duty, but to fulfill it in a larger measure. Having himself spoken on the Earth's skin, he invited now a number of specialists to round out the picture of the Earth in a Symposium on the Interior of the Earth.

We start with the heart of the matter, the core of the Earth. We have with us some of our own great cardiologists of the Earth, who will then speak. For the opening address, your President has invited a leader among the younger generation of geophysicists of England, in conformity with the Union's basic purpose to help creative workers to overcome the handicaps of geographic separation while they are in the midst of their most fruitful years.

It gives me great pleasure to present to you the Assistant Director [of Research] of the Geophysical Laboratory of Cambridge University, Professor Stanley Keith Runcorn.
*(Walter H. Bucher (1954): 48; my bracketed addition)*

Bucher also invited Vestine and Graham from the Carnegie Institution of Washington, Walter Elsasser, and a young geophysicist working in paleomagnetism named L. W. Morley. Morley, then a Ph.D. student working under J. T. Wilson, briefly described his work in paleomagnetism. He came out in favor of polarity field reversals, citing Runcorn's discussion of Hospers' study of Icelandic lavas. Graham, in contrast, did not think that field reversals had been adequately substantiated. Elsasser welcomed Runcorn's appeal to the Coriolis force to account for the rotational symmetry of the Earth's magnetic field, but added (1954: 74): "A mathematical attack upon this difficult dynamical problem has not yet been made."

# 3

# Launching the global paleomagnetic test of continental drift: 1954–1956

## 3.1 Paleomagnetists on the move

The broad outline of the paleomagnetic case for mobilism emerged from early 1954 through much of 1956.[1] These were fruitful years for British paleomagnetists who traveled far and wide, and for Soviet and South African workers. During this interval surveys were initiated on all major continents except Antarctica. Blackett left Manchester University in fall 1953 to become head of the Department of Physics at Imperial College, London. Clegg followed him and was joined by Almond and Stubbs. Their paper on the paleomagnetism of Triassic red sandstones, received in March 1954 and published in June, was the first publication by paleomagnetists that contained a clear endorsement of mobilism. In Cambridge, Creer constructed the first paleomagnetic APW path, the path for Britain, in the second half of 1954.[2,3] He argued that his APW path could be explained by continental drift or polar wandering, or their combination, and that APW paths for other continents would be required to distinguish among them. Creer then left paleomagnetic research for two years. Runcorn returned to the United States during summer 1954 to collect rocks, this time to collect not as previously from Cenozoic lavas (§2.13) to study reversals, but from Mesozoic and Paleozoic rocks from which he constructed an APW path for North America. This he claimed was indistinguishable from that for Europe, reasserting his belief that polar wander but not continental drift had occurred. Irving moved to the Department of Geophysics at the new Australian National University (ANU) in Canberra, Australia, at the end of 1954, and by mid-1955 obtained results, and argued that Australia had moved ~5000 km relative to Britain. He initiated global intracontinental comparisons of paleoclimatic evidence with paleomagnetically determined latitudes, and intercontinental comparisons that were consistent with large-scale relative motions of continents.

During this period paleomagnetic surveys were begun in the USSR at the All Union Petroleum Institute in Leningrad, notably that of Alexei Khramov. His work on the late Cenozoic red beds of the south Caspian region became of interest because of the reversal sequences they contain. Their work on the Late Paleozoic and Early Mesozoic red strata of the eastern Russian Platform (especially the Permian System) was later used for the global test of drift. In this period Gough, Hales, and K. W. T. Graham at the Bernard Price Institute in Johannesburg obtained results from older dykes (Pilansberg) and the critical Jurassic Karroo dolerites of South Africa. A. E. M. Nairn, a Runcorn recruit, worked in Central Africa, obtaining results from the Karroo lavas of Zimbabwe. Clegg, with Blackett's strong support and encouragement, began a long-term research project in India, following Irving's early reconnaissance. He and co-workers began sampling in October 1954. Their work in India ended in 1960. Their first paper, submitted in December 1955, unequivocally supported continental drift. Importantly, Blackett endorsed drift. In December 1954, he gave the second in the series of Weizmann Memorial Lectures in Israel; published in 1956, they described Blackett's support of continental drift. In mid-1956 Runcorn left Cambridge to take the Chair of Physics at the University College of Newcastle upon Tyne (then part of the University of Durham). Moving with his group, he offered a lectureship to Creer, who promptly began a paleomagnetic survey of South America. This work in the USSR, South Africa, Zimbabwe, South America, and India, which began to bear fruit after 1956, will be described in Chapter 5.

## 3.2 Four stages in the paleomagnetic test for continental drift

The paleomagnetic test of mobilism evolved through four stages. Sometimes they overlapped in time and stages were very occasionally skipped. During the *first* stage (Stage I), paleomagnetists determined the orientation of the time-averaged geomagnetic field for a *single* landmass at a *single* geological interval of time. They found directions of magnetization strongly oblique to the present and to GAD fields, which they realized reflected polar wandering (motion of the axis of rotation relative to Earth's surface either by rotation of the whole body of the Earth through the axis of rotation, or by slippage of the entire outer shell (crust or lithosphere) relative to the interior), drifting of the landmass (continental drift), or a combination of the two. Clegg and colleagues' "England" test (§3.5) was a first stage paleomagnetic test of mobilism. First stage tests allowed the past latitude and orientation of that landmass relative to the paleomeridian to be determined for particular intervals.

During the *second* stage (Stage II), paleomagnetists collected samples of several different ages from the same landmass and so determined the changes in latitude and orientation it had undergone through time. This was first done in Britain by Creer, who expressed these changes as pole positions and constructed the first APW path, the successive positions of the poles of the mean geomagnetic axis relative to a certain landmass (§3.6). Pole paths determined in this way were later called APW paths. At this stage it was not known if APW paths were caused by polar wandering, continental drift, or some combination of them.

Paleomagnetists then moved to the *third* stage (Stage III). They obtained first single *paleopoles* or, somewhat later, whole new APW paths from *different* landmasses and compared them with Creer's British path; his path and progressive updates of it became the reference against which results from other regions were compared. If they coincided, then only polar wandering had occurred. If there were differences, then continental drift would be implied; alternatively, differences could be attributed to experimental error, uncertain geological correlations, very rapid polar wandering, failure of the geomagnetic field to remain dipolar, or some combination of these. Creer compared his APW path for Britain with Graham's Silurian data from Maryland, North America (§3.7). Irving compared his results from the Indian Deccan Traps and from the Tasmanian dolerites with Creer's APW path (§3.12). Runcorn compared his new North American APW path with Creer's British path, and found no significant difference (§3.10). Irving's reanalysis showed that there was a significant difference (§3.12).

Meanwhile, other paleomagnetists did further sampling in North America. In 1954, P. M. Du Bois, one of Runcorn's new students, collected Precambrian samples from Michigan's Keweenawan peninsula that were very roughly correlated with the Torridonian of Scotland, publishing his initial results the following year. In early 1955 J. W. Graham obtained Permian and Triassic samples from the Grand Canyon and surrounding area (§7.4). Doell and Cox began working in paleomagnetism at the University of California, Berkeley, under Verhoogen. Doell collected Paleozoic and Pre-Cambrian samples from the Grand Canyon in 1954, in the same summer as Runcorn, publishing his results the following year (§3.8). Cox collected Tertiary samples in Oregon in summer 1956, publishing in 1957 (§7.6).

The *fourth* stage (Stage IV) in the development of the paleomagnetic case for mobilism was markedly different; paleomagnetists now reached for independent paleoclimatologic checks to determine the reliability of their results and the validity of the GAD hypothesis prior to the Miocene. Although Hospers (1955) and Creer (1955) drew comparisons between paleomagnetic and paleoclimatological data, Irving (1956a) was the first to make globally significant comparisons. He came up with the idea of using paleoclimatic data in this way in his Ph.D. thesis (1954). However, it was not until his arrival at ANU that he was able to give it close attention, writing what he later described as his most creative paper (1956a) (§3.12). Meanwhile Runcorn with his new graduate students, Neil Opdyke, and others devised a new way to test paleomagnetic results, notably by comparing them with wind directions (§4.2, §5.12–§5.14). Blackett turned to paleoclimatology in the early 1960s (III, §1.7). Irving's India test was somewhat unique, it immediately moved straight from the first to the fourth stage (§3.4).

### 3.3 The January 1954 Birmingham meeting

D. H. Griffiths and R. F. King of the Department of Geology at the University of Birmingham hosted a meeting on January 8 and 9, 1954, which they summarized in *Nature*. They themselves had begun working on the paleomagnetism of

Quaternary sediments. Most British paleomagnetists attended; Hospers came from Holland, and the hosts gave presentations.

Historically the meeting is of first importance. For the first time strongly oblique magnetizations were described from rocks of widely different age – Precambrian, Paleozoic, and Mesozoic – which showed that major changes in the orientation of Earth's axis of rotation relative to Britain, or of Britain relative to the axis of rotation, likely have occurred. The meeting made clear that by early 1954 the use of paleomagnetism to test the old but hitherto unresolved ideas of polar wander and continental drift had become a major focus. Reversals were also a major concern and there was general agreement that reversals of the geomagnetic field had occurred; although Néel's mechanisms for self-reversals were discussed and participants recognized that self-reversals might conceivably happen, the prevailing view was that it was the geomagnetic field that had undergone reversals. Also center stage was the reliability of paleomagnetic data. Participants reported that some samples were unstable, others had both "soft" and "hard" components of magnetization; they reported results of field tests to determine magnetic stability, and described laboratory experiments on how magnetization of sediments was acquired on deposition. The upbeat mood regarding mobilism and reversals, and the realization that these problems could now be confidently addressed could hardly have been more different from that prevailing at the Carnegie (§3.15); this was the real significance of the Birmingham meeting.

G. D. Nichols described the mineralogy of naturally occurring magnetic reversals. Fisher outlined his statistical method. Hospers recounted how his study of Icelandic lavas provided no evidence for substantial polar wandering during the Neogene as suggested by Köppen. Bruckshaw described studies of reversely magnetized Tertiary lavas from the Isle of Mull, Scotland. Irving described results from the 3000 m thickness of Torridonian red sandstones, remarking that he had found sequential reversals about an oblique axis (northwest up/southeast down), had identified at least sixteen polarity zones, and had used folded bed and slump tests to determine when magnetization was acquired. Creer discussed early results from his surveys of Devonian, Triassic, and Eocene sediments, and Permian lavas. He reported that some samples displayed reversed magnetization and varying degrees of obliqueness. He noted the variability of the stability of Triassic samples, which he explained in terms of Néel's theory. Runcorn described the results of his and C. D. Campbell's work on Miocene lavas in the western United States, and argued in favor of field reversals (§2.13). He argued that his account of the origin of the geomagnetic field, which laid stress on the Coriolis force, provided theoretical support for the GAD hypothesis. Furthermore, Runcorn prematurely defended polar wandering, without giving a polar wander path (none had yet been constructed (§3.9)), as the "simplest" solution to the oblique directions in British rocks, and suggested that paleomagnetic surveys should be extended to other continents. Clegg described reversals of magnetization about a strongly oblique axis (northeast down/ southeast up) in Triassic red marls (§3.5). He

considered a number of possible reasons for this preferential direction of the magnetic polarization and concluded tentatively that it might be due to a movement of Britain relative to the earth's geographical axis.

*(Griffiths and King, 1954: 1117)*

Clegg and company were not favoring continental drift over polar wandering; both were possibilities; this remark appears to be the first publicly recorded statement by a British paleomagnetist interpreting paleomagnetic results in terms of either polar wander or continental drift. This contrasted with Runcorn who favored polar wander only; I suspect his students, Irving and Creer, agreed with Clegg. Griffiths and King summarized their work on very recent varved clays from Sweden. King described experiments he had performed on the varved clays, redepositing them in a magnetic field and finding that the magnetic inclination of the resulting sediment was systematically shallower than that of the applied field. Clegg reported the same effect.

Reports at the Birmingham meeting established the existence in pre-mid-Tertiary rocks of magnetizations strongly oblique to the present axis and this set the stage for a full-scale attack on the long-standing problems of polar wandering and continental drift. Before any APW paths had been constructed (the first was constructed later that year, §3.6), plans were being laid and quickly implemented to obtain paleomagnetic results from North America (§3.8), India (§1.17, §3.4, §3.13) and Australia (§3.12).

### 3.4 Irving's thesis, support of continental drift, and plans for Australia

By late 1953, Irving's Torridonian work had shown that the geomagnetic field had existed in the Precambrian and therefore could likely be studied over a substantial portion of Earth's history (§2.6). The Precambrian magnetic field in northwest Scotland was strongly and systematically oblique to the present field; during the accumulation of what he called the Lower Torridonian (now the Stoer Group) the field was aligned over 70° from that during the accumulation of the Upper Torridonian (now the Torridon Group), and during the former no reversals of the field occurred whereas during the latter there were many reversals. At the Birmingham meeting he had identified sixteen polarity zones (§3.3) and later in Figure 27 of his thesis he described twenty-six polarity zones in the Upper Torridonian.

But what happened to his India test, the test that jumped without pause from Stage I to Stage IV? Receiving from Dr. Sahasrabudhe of the Indian Geological Survey seven basalt hand samples from the Deccan Traps, Irving began coring and slicing them in November of 1952.[4] He measured them some time after June of 1953.[5] The seven samples were from the Western Ghats, which form the western edge of the Deccan, the most prominent feature of peninsular India. Here the traps are over 2400 m thick. At the time they were known to be uppermost Cretaceous or Paleocene with the possibility of extending into the

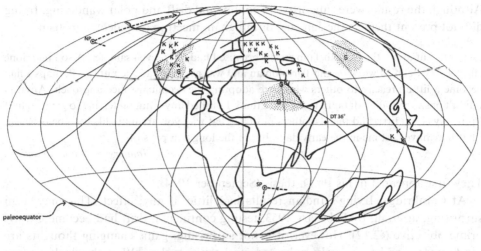

Fig. 14. Moore und Wüsten im Eozän
(K Kohle, S Salz, G Gips, punktierte Räume: Trockengebiete)

Figure 3.1   Eocene paleogeography of Köppen and Wegener; their Figure 8 (1924: 52). The present grid (fainter) is fixed relative to Africa. Paleolatitudes are in heavier lines. NP and SP are the north and south paleogeographic poles in the Eocene, and the dashed lines are the directions of motion of the pole before and after that time. Arid areas are stippled. K, coal; G, gypsum; S, salt. DT is the approximate position of the Deccan Traps as determined by mean latitude by Irving (1954, 1956a) at 36° and by Clegg, Deutsch, and Griffiths (1956) at 34°; their close fit to Köppen and Wegener's map is remarkable.

Eocene (Krishnan, 1949: 433).[6] The samples were taken one per locality, spaced at ~300 km intervals from Poona to Belgaum. Their magnetizations were aligned along a northwest up (329°, −5°), (329°, −56°), down (149°, +56°), (149°, +56°), two localities normal and five reversed, respectively, error 10°. Averaging them regardless of sign substantially corrected for later overprints, the axis differed from the axis of the present field by more than 70°. The reversals indicated a significant span of time. The corresponding pole is shown in Figure 7.1 (DT 1). The inclination indicated a paleolatitude of 36° south. (In principle this could have been a northern paleolatitude but this possibility was readily discounted (Figure 5.18).) This implied a northward motion of 5500 ± 1000 km. The declination signifies 30° counterclockwise rotation of the Indian Peninsula. These motions are remarkably similar to those required by the Eocene reconstruction by Köppen and Wegener (1924) (Figure 3.1). Or as he proposed, briefly and without elaboration, in his thesis, "India has moved from the Southern Hemisphere through 53° of latitude and has rotated counterclockwise through 28°, a motion required by Wegener's reconstruction" (Irving, 1954: 113). However, he was unwilling to single this out as the correct interpretation; there were others consistent with the data. He again suggested:

It is possible that geomagnetic pole wandering and continental drift may have both contributed in some complex way. This matter could almost certainly be settled by collecting many more similar results from the Southern Continents.

*(Irving, 1954: 113)*

Although the results were supportive of continental drift and polar wandering, Irving did not present them at the January Birmingham meeting on paleomagnetism.

I was reticent about the Deccan Trap results because I had only seven samples. Also I had done all this very detailed work on the Torridonian and was aware of all the pitfalls. Perhaps I had become a little preachy to others who were adopting the reconnaissance approach. Anyway I didn't push the Indian data. In fact I don't think I pushed any data, preferring to get it right if I possibly could first, and then let others push it. That way you are more likely to be believed. That may have cost me in the early days, but in the long run pays off.

(*Irving, January 2000 note to author*)

They first appeared in his Ph.D. thesis (September 1954).

At Cambridge, Irving's Indian results had little, if any, effect. This may seem surprising; after all, by February 1953, testing continental drift had become Creer's prime objective (§2.12). However, it is not surprising; game-changing thoughts are rarely widely or immediately embraced, and the samples, although widely spaced over 300 km and from different levels, were only seven. Hospers had left to work with Royal Dutch Shell before Irving obtained his Indian results. Creer, as will be seen (§3.7), made no appeal to them in his Ph.D. thesis of January 1955, even though he argued in favor of continental drift and/or polar wandering. Runcorn paid them little attention and so did not refer to them in print until 1956 (§4.11). Irving himself did not publish his Indian findings until 1956 (§3.12).

Irving continued to wonder more generally how paleomagnetism related to continental drift and polar wandering. Eventually he (1954) began to think of the oblique paleomagnetic directions of the Torridonian as possible evidence of continental drift, to speculate about the paleoclimatological aspects of the classical case for drift, and to relate them to the developing paleomagnetic case. Just when Irving began to think of his Torridonian results this way is unclear. It was not through the initial discovery of oblique directions in the Torridonian that he made the link between continental drift and paleomagnetism (§1.17); because Wegenerian accounts of continental drift did not posit drifting landmasses until the Mesozoic, Irving did not immediately think of the oblique directions in terms of the drift of that particular part of Scotland. However, things began to change for him, perhaps around the beginning of 1952. Argand (1924/78: 139) and Holmes (1929: 347) had stressed a long-continued movement of Britain from lower to higher latitudes, and he had read them. Moreover, the idea of using paleomagnetism to test mobilism had become a topic of discussion at Cambridge, and Blackett had set up the Manchester/London group apparently with the express intention of testing continental drift, which Clegg and his co-workers began to act on. Indeed, Clegg (§2.4) and Creer (§2.12) began to fill in the gap between Hospers' Tertiary and Irving's Pre-Cambrian data.

Irving had been fired up by Runcorn, who on Phemister's advice (§1.16) had selected the Torridonian on which Irving had served his apprenticeship, and which showed him how, by selective sampling of the best lithologies, good records could be

obtained from very old rocks. Besides being the first among the British paleomagnetists not only to note the connection but also to initiate the first paleomagnetic test
of continental drift (§1.18), his Torridonian results influenced the work of other
Cambridge and Manchester paleomagnetists. There was his (and Runcorn's)
double-checked discovery of the systematic oblique directions, which by late 1951
had demonstrated they were real. Although his oblique Torridonian directions did
not lead him directly to Wegener's continental drift, they literally became, during the
Royal Society Soirée in May 1954, the starting point of the first APW path, the
second stage in the global test (§3.6). This was an outcome neither Runcorn nor
Irving could have possibly imagined when they began sampling Craigriannan on the
north shore of Loch Torridon in June 1951.

A few weeks after the Birmingham meeting (February 17, 1954), Irving applied for
the position of research fellow in the Department of Geophysics at the Australian
National University; Runcorn showed him the advertisement in the *Cambridge
Reporter*. He was offered the position, receiving a letter dated April 22, 1954.
J. C. Jaeger was head of the department, and Irving was engaged to set up a paleomagnetics laboratory and carry out surveys. In May 1954, Jaeger, who was on
sabbatical leave in the UK, visited Irving in Cambridge. They set out detailed plans
for the new laboratory in Canberra. Jaeger also gave a seminar to the department on
the magnetism of Tasmanian dolerites in which he explained that their magnetization
was near vertical. Irving was listening: that was just what the Mesozoic maps of
Köppen and Wegener required for Tasmania, high latitude, steep inclination.

It was at this time, early summer 1954, that Irving began to compile his results
and write his thesis, which he submitted in September before leaving for the IUGG
meeting in Rome. In his thesis, he emphasized the central role of the GAD hypothesis,
and cited the work of Hospers and Runcorn in its support. Regarding the paleopoles
he had derived for the Lower and Upper Torridonian, he suggested (1954: 109) that
their difference from each other and from the position of the current pole "may have
resulted from" polar wandering, continental drift, or a combination of the two. In the
absence of paleomagnetic results from other regions, he was unwilling to accept any
particular hypothesis. He then recalled that Wegener and Köppen had appealed to
paleoclimatological evidence in support of both polar wandering and continental drift,
and suggested that it would be "most desirable to compare" paleoclimatic conditions
with his paleomagnetic results from the Torridonian. However, because the paleoclimatic conditions under which the Torridonian was laid down were at the time not well
established, he resorted, tenuously, to an indirect comparison between the paleomagnetically derived paleopoles from the Torridonian and the Precambrian climate of
North America.

Clearly it is most desirable to compare the results from the above evidence and from palaeomagnetism. Unfortunately reliable inferences concerning the positions of the geographic poles
in Early Paleozoic and Precambrian times cannot be made and comparison with the

Torridonian results is impossible. It may however be noted that tillites which are generally taken as indicative of high latitudes occur in the Precambrian of North America not very far from the emergence of the Diabaig axis [the paleomagnetic axis derived from Irving's Upper Torridonian].

*(Irving, 1954, p. 110; my bracketed addition)*

His Deccan samples were much younger than the Torridonian, and he made no comparison between them. He stressed the need for paleomagnetic surveying of the southern continents, and the need to compare paleomagnetic latitudes and geological evidence of past climates. Once in Australia, he set about doing it. Now with his detailed field and laboratory experience working alongside Hospers and Creer and wide reading, he was well equipped for the task.

Cambridge University did not award him a Ph.D., only an M.Sc. In retrospect this seems incredible. Fred Vine remembered that what had happened to Irving was still a topic of discussion among graduate students in the Department of Geodesy and Geophysics at Cambridge during the early sixties. Irving's contributions to the development of paleomagnetism were substantial. Why was his application for a Ph.D. rejected? Was it because he offered tentative support for continental drift? Irving does not think so. Was it because Irving did not invent or construct a new instrument? At the time it was customary in the department for students to construct a new instrument or invent a new procedure while working on their Ph.D. When writing his own dissertation a decade later, Vine deliberately emphasized his development of a computer program for analyzing marine magnetic anomalies because he did not want to suffer Irving's fate (IV, §2.9). Irving, however, has offered a very different explanation. His thesis was very short. There was almost no literature review, no novel instrumentation described. It was a straightforward presentation of his groundbreaking research; the examiners failed to understand its significance because they knew little or nothing about paleomagnetism, and he did not take the trouble to explain it to them. Irving himself failed to understand that the examiners would need much more background than he provided in order to place his work in context.

What happened to my Ph.D. thesis? Well this is an awfully long story. My Ph.D. thesis was very short. It was not written in the sort of bulky style, which was perhaps normal for Ph.D. theses. I was very anxious to get to Australia quickly. There was no way in which I was going to let this hang over. I started writing my Ph.D. thesis about May, and I finished it and submitted it in September – a very short interval of time for writing a thesis. I wrote it in great haste and it was also very short. And in some ways my case was not very well put … I think the point was that the examiners who were Ben Browne and Hemingway, he was a professor of geology at Leeds, hadn't really understood the importance of my work in generally establishing some of the important things that led to the establishment of red sandstones as being one of the key indicators, the establishment for the first time of reversals sequentially in sedimentary rocks, the establishment of stability since the Pre-Cambrian which, of course, showed that the Earth had a field at the time. On top of that, there was Indian work and the introduction of the idea of testing paleomagnetic results against the paleoclimatic evidence from geology with the sort of tentative start of that in the case of the Torridonian sandstone. The evidence for paleoclimates at that time was rather thin, but all of the ideas were there. And, I don't think this was really appreciated by the examiners because they themselves didn't really know the background to the subject. The choice of examiners was rather dictated by the rules at Cambridge. One had to have an internal examiner and the only other person on staff who really could have done it was the head of the department, Ben Browne, whose background was in gravity, and the other person – you needed to have an outside examiner – was a geologist who had been interested in the Torridonian Sandstone. But he had very little appreciation at all of the subject of paleomagnetism and what I had done and what was being done by other people at the time in the Cambridge group. And with another choice of people more deeply knowledgeable about this, another choice of an examiner, for example, if I had been examined, say by Blackett, Griffiths, Rutten, van Bemmelen, or someone like that then I think the outcome may have been quite different.

*(Irving, 1981 interview with author)*

Although Runcorn agreed that Irving's dissertation was too brief and hastily written, he placed the blame on Ben Browne. He recalled Browne's resistance to Irving's admission into the department as a Ph.D. student.

The Department of Geophysics, the Department of Geology, and the Department of [Mineralogy and] Petrology were always at loggerheads. The Head of the Department of [Mineralogy and] Petrology was [C. E.] Tilley, who had been at the Carnegie Institution, and worked on the origin of minerals. He had separated himself from Geology to found this Department, and he had the very highest standards. He was very friendly with me, but I must say that he was a frightful tyrant in this own Department. Arkell and the other people also were very friendly to me, and we often used to go and chat with the petrologists. Tilley had evidently got a rooted idea that the Geology Department was no good, and even W. B. R. King was not very good, and he obviously decided that Ben Browne was no good. Tilley was an important University administrator too. He sat on the general board and he was a power. And Ben Browne was afraid of him. Ben used to say to me, "I understand you want to have Irving in the Department, but it's very embarrassing. I have to go to the meeting of the Faculty Committee and say that this person should be registered for a degree." (They complicated business in Cambridge: you are not registered immediately for the Ph.D., you're registered first as a candidate and then you're accepted as a bonafide student for a Ph.D., and then you have to have your thesis topic accepted, and all this has to be done by the Faculty Committee.) And so he used to tell me each time what trouble I'd caused him because, you know, these geologists immediately say, "Oh Irving, of course, we wouldn't have him in our Department." And I told Ted bluntly that he would have to write a very good thesis, and one which emphasized the geophysics side. But, at the time Ted and I were a bit at cross purposes as supervisors sometimes are, and he had become, quite rightly, very confident of his own abilities. He wrote the thesis very much as a straightforward account of the work he had done on the Torridonian sandstone making rather less of the geophysical interpretations than I warned him he would have to do. Of course, my fears turned out to be correct, because in Cambridge the supervisor couldn't be an examiner. Ben Browne and someone outside failed him. Of course, this was a cause célèbre. I remember I was exceedingly angry. I talked to Fisher. I talked to Blackett. I always remember Blackett saying, "I fully agree with you, and Ben Browne just has an inferiority complex. But this is just a minor tragedy." Blackett said it was a tragedy, but he put it in perspective by saying it was a minor one. But, of course, by then Ted had gotten his job in Australia.

*(Runcorn, 1984 interview with author; my bracketed additions)*

Blackett's wise words turned out to be correct. A decade later (1965) the University awarded Irving a Doctorate of Science based on published work.

Irving did not learn about his Ph.D. failure until he got to Australia. Jaeger helped him get over his disappointment, and he buried himself in his research, and in the next year wrote three influential papers (one with G. S. Watson), one of them was also rejected by the journal to which he first sent it.

I didn't know about the rejection of my Ph.D. thesis until I was in Australia. In fact, I heard about it when I came back from Tasmania from collecting the Tasmanian dolerites – about February 1955. And, of course, this was rather disheartening news. Jaeger was very good about it. He explained to me that he had failed to get a Fellowship at Trinity and that had been the best thing that ever had happened to him because it was for that failure, when he had been working in Cambridge in theoretical physics in the late 1920s in the heyday of quantum theory, that he had decided to go back to his home country Australia and get back into classical physics, into heat conduction, and as a result, became the world authority on heat conduction – whereas it would have been extremely difficult for him to become the world authority in quantum mechanics. Of course, the failure to get a fellowship at Trinity in the late 1920s is a failure of an all-together different order from failing one's Ph.D. So to have failed to get a Fellowship at Trinity was really no disparagement. But, nevertheless, these were comforting words. So I simply went away and got drunk and had a good game of cricket, and got on with my work to try and make good this unfortunate happening. This is something of an aside, but it tells you something about the sort of person that Jaeger was and how he was able to help a person in the right way when in difficulties. This was sort of a critical time in my life and he helped me very greatly.

*(Irving, 1981 interview with author)*

### 3.5 Clegg and colleagues publish the first paleomagnetic support for continental drift: Stage I

Clegg, Almond, and Stubbs began sampling in October 1952. They submitted their elegant paper in March 1954, and it appeared in June. It contained the first fully described oblique paleomagnetic directions from Britain. In their seminal paper, they used Fisher's new statistics. They collected red sandstones and marls from ten sites in the Upper Triassic (Keuper 1). The localities were spread along 400 km, over much of England. Localities in the Cheshire Basin span a stratigraphic thickness of about 500 m (Clegg *et al.*, 1954a: 586) but collectively they spanned much more than that. The consistency of the magnetization directions and the presence of reversals strongly indicated stability of the magnetization. Correction for bedding tilt decreased scatter. They were aligned along a northeast and downward (five localities) and southwest and up (four localities) axis. This axis differed by about 90° from that which Irving had reported at the Birmingham meeting. Clearly there had been huge changes in the time-averaged geomagnetic field since the Precambrian. They rejected samples from one of the sites (Sidmouth, which Creer had also sampled (§2.12)) because their magnetizations "were markedly unstable in the laboratory, and the direction of magnetization could be substantially changed by changing the orientation relative to the earth's field for a period of a few hours" (Clegg *et al.*, 1954a: 591). They reported results from Devonian and Carboniferous rocks from two other sites. They tested magnetic stability by subjecting samples to steady and alternating magnetic fields and increasing temperatures.[7] Pulverizing some specimens, they suspended the powder in water, allowed it to settle in the Earth's field, and measured its magnetization. The declination agreed with that of the ambient field but the inclination (dip) was about 8° less. This discrepancy was similar to that reported by King (§3.3).

The mean direction of the Triassic samples deviated 34° in declination and 41° in inclination from the present GAD field, which they argued indicated polar wandering, continental drift, or their combination. They began with the deviation in declination.

... it seems therefore that the most likely explanation of the observed horizontal direction of magnetization of the sediments studied is that the whole of the land mass which now constitutes England has rotated clockwise through 34° relative to the earth's geographical axis. This movement must have occurred since the rocks acquired their magnetization, which was probably at or soon after the time of deposition (150 to 200 million years in the case of the Triassic sediments), but conceivably much later ... If such a rotation of England has occurred, it could have been a local movement of only part of the earth's crust, or alternatively the earth's mantle could have moved as a rigid whole relative to the geographical poles. The first hypothesis would consider the rotation either as a purely local movement or as part of a drift of large continental land masses. The second would adduce pole wandering as the operative mechanism.

*(Clegg et al., 1954a: 596)*

They noted that obtaining samples of other ages from other regions would allow them to test these hypotheses.

The deviation of inclination, they suggested, could be explained if England had moved from a lower to a higher latitude since Triassic time. They noted that the lower inclinations could be partly, but only partly, explained by inclination flattening during deposition or later compaction.

It has been shown that the rocks also have a mean magnetic dip markedly less than that of the GAD dipole field in the latitude of England (68°). Some part of this difference may be accounted for by deformation or compactions subsequent to deposition, but when due allowance has been made for these effects it still seems probable that the magnetic dip at the time of magnetization was appreciably less than today. If this is true, it would imply an increase in magnetic latitude, since magnetization and consequently, on the assumption of an average coincidence between the magnetic and rotational axes, an increase in geographical latitude.

*(Clegg et al., 1954a: 596–597)*

In a shortened account of their work, which appeared in December 1954, Clegg, Almond, and Stubbs excluded the possibility of a local movement of England.

Finally, the most likely possibility appears to be that the whole land mass of Britain has rotated through 34° relative to the Earth's geographical axis, at some period since the rocks acquired their remanent magnetisation. The mean magnetic dip of the rocks, which is appreciably less than that of the GAD dipole field in Britain (+65°) also suggests a Northward movement of Britain. The extent of the latter type of movement is, however, more questionable, since the present dip may have been influenced by deformation or compaction after deposition ... A displacement of England such as that postulated here could be due to either of two causes. On the one hand it may represent a movement of the Earth's mantle as a whole relative to the poles (pole wandering), or on the other hand it could be due to a movement of the continental land mass relative to the surface of the mantle (continental drift). These two effects might be mutually operative and interdependent. It is hoped that by examining rocks of other ages, and from a wider geographical range, it will be possible to distinguish between them.

*(Clegg et al., 1954b: 198)*

Either polar wandering, continental drift, or their combination was needed to explain the data. About half of their samples were reversely magnetized and half normal, which they thought reflected geomagnetic field reversals. They rejected self-reversals.

### 3.6 Creer's 1954 APW path for Britain evolves through three versions: move to Stage II

In the second half of 1954 Creer constructed three versions of the APW paths for Britain. They were snapshots of his progressive search for a British APW path as new results emerged from his own studies and those of Irving and Clegg and company, and as he established confidence limits of poles. His first version (Figure 3.2) was based on data available through spring 1954; it appeared in print later that year (Creer *et al.*, 1954: 165) in a special issue of the *Journal of Geomagnetism and Geoelectricity* (*JGG*) devoted to papers given at the 10th Assembly of the International Association of Terrestrial Magnetism and Electricity (IATME) as part of the IUGG meeting in Rome in September 1954; at this meeting Runcorn presented a selection of Cambridge work but not however an APW path. In preparation for the Rome meeting, Creer in June 1954 prepared, at Runcorn's request, a report of work in Cambridge; this was prior to Runcorn's departure for the United States (§3.8).[8] When Takesi Nagata, the convener, told Runcorn in September or October of his intention to have the papers from the meeting published, Runcorn asked Creer to revise the June report. This was Creer's first version, and it is important primarily because it was the first to appear in the scientific literature and became the most referenced APW path for several years. Creer presented his second version (Figure 3.3) in Oxford on September 8, 1954, at the annual meeting of the British Association for the Advancement of Science (BAAS). Although actually constructed before the *JGG* path, it is best regarded as his second version because it included results that had become available during the summer. Unlike the first, the "Oxford" version contained Irving's results from the Lower and

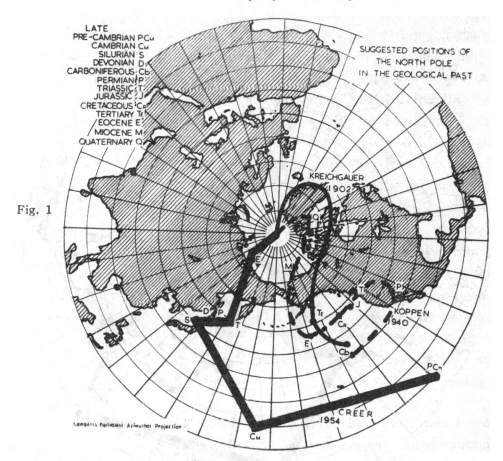

Fig. 1

LATE
PRE-CAMBRIAN PCₘ
CAMBRIAN Cₘ
SILURIAN S
DEVONIAN D
CARBONIFEROUS Cb
PERMIAN P
TRIASSIC T
JURASSIC J
CRETACEOUS Cr
TERTIARY Te
EOCENE E
MIOCENE M
QUATERNARY Q

SUGGESTED POSITIONS OF
THE NORTH POLE
IN THE GEOLOGICAL PAST

KREICHGAUER
1902

KOPPEN
1940

CREER
1954

Lamberts Equivalent Azimuthal Projection

Figure 3.2   Creer's first or "Rome" APW path based on NRM directions known in June, 1954. This APW path appeared as Figure 1 in Creer, Irving, and Runcorn (1954: 165). The late Pre-Cambrian pole was based on Creer's Longmyndian collection; the Cambrian on his Caerbwdy Sandstones; the Silurian on his Ludlow Sandstones; the Devonian on his Old Red Sandstones; the Permian on his volcanic Exeter Traps; and the Triassic on his work and that of Clegg and colleagues on the Keuper Marls. The Eocene paleopole was based on Hospers and Charlesworth's work on the Antrim Plateau basalts of Northern Ireland. Q, Quaternary; M, Miocene; E, Eocene; Cr, Cretaceous; J, Jurassic; T, Triassic; P, Permian; Cb, Carboniferous; D, Devonian; S, Silurian; Cm, Cambrian; PCm, Pre-Cambrian. (See endnote 2 for much later pre-Permian revisions to Creer's APW path, which did not affect the strong support it offered for Wegener's continental drift.)

Upper Torridonian. Version 2 appeared in the popular press before version 1. Creer constructed his third version (Figure 3.5) for his Ph.D. thesis. His thesis was based on the information he had in late 1954, and notably included ovals of confidence, and the seminal result of Clegg et al. (1954a) (§3.5). Although Creer submitted his dissertation early in the December of 1954, his oral examination did not take place (Professors O. T. Jones and P. M. S. Blackett) until January, which explains its reference date of 1955. This third version, with some additions, was published two years later (Creer

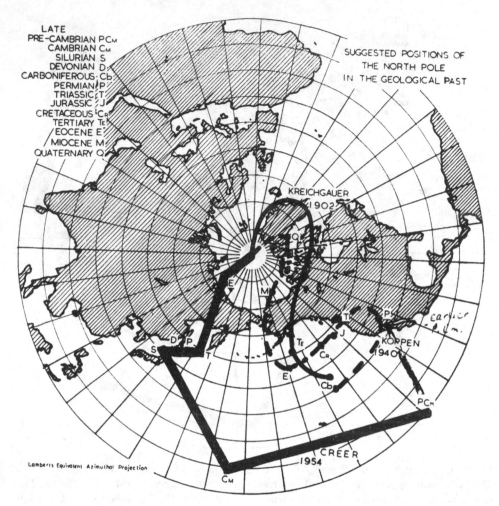

LATE
PRE-CAMBRIAN PCᴍ
CAMBRIAN Cᴍ
SILURIAN S
DEVONIAN D
CARBONIFEROUS Cb
PERMIAN P
TRIASSIC T
JURASSIC J
CRETACEOUS Cʀ
TERTIARY Tᴇ
EOCENE E
MIOCENE M
QUATERNARY Q

SUGGESTED POSITIONS OF
THE NORTH POLE
IN THE GEOLOGICAL PAST

KREICHGAUER
1902

KÖPPEN
1940

CREER
1954

Lamberts Equivalent Azimuthal Projection

Figure 3.3 The polar wander path that Creer showed at the Oxford meeting (Creer, 1954, "Rock magnetism and polar wandering," unpublished). It was his slide no. 12. The most important difference between this second APW path and the first one is the inclusion of an earlier Pre-Cambrian paleopole based on Irving's work from the Diabaig Group, the "Lower" Torridonian. Creer adapted his "Rome" APW path for his talk, printing "earlier P.Cm" to label the paleopole he derived from Irving's Torridonian work. (See endnote 2 for much later pre-Permian revisions to Creer's APW path, which did not affect the strong support it offered for Wegener's continental drift.)

*et al.*, 1957). These versions show the development of Creer's ideas. In his thesis he moved to Stage III of the test of mobilism.

Before examining development of Creer's paths, I want to dispel Runcorn's claim that there was an earlier path presented at the Royal Society's Conversazione on May 20, 1954. Referring to the Conversazione, Runcorn asserted:

This definitely was the occasion for which I got a globe to plot the pole positions from all the British strata and I distinctly remember getting Du Bois to do the spherical trigonometry and calculate the pole positions that we then put on the globe. I know we were puzzled about the Torridonian pole and finally reversed it to get the famous path circling around the North Pacific.

*(Runcorn, September 19, 1984 letter to author)*

However, Creer disagreed.

At the time of the Royal Society Conversazione in May 1954 there certainly was no polar wander curve. Our display consisted mainly of stereograms showing our results as obtained up to the date of the meeting ... much of our poster display was concerned with showing stereograms of individual sample directions from the different rock formations that we had studied: the respective mean directions and 95% circles of confidence were illustrated. There were several stereograms illustrating Hospers' Icelandic results and the point was made that, when averaged over long enough time intervals, secular variations of the non dipole field averaged zero. This observation formed the basis of our extension of the validity of the axial dipole field hypothesis from historic to geological time. Another theme of our display concerned geomagnetic reversals: Hospers' reversed and normal flow directions for Iceland were illustrated and reference was also made to possible self-reversal mechanisms.

We also showed stereograms of Irving's Torridonian results and of my own results from the Triassic Keuper Marls. The latter were of interest primarily because they illustrated the presence of secondary components of magnetization aligned along the present geomagnetic field direction: in this context the stable Triassic direction, later to be described in Clegg *et al.* (1954a, 1954b), was referred to in the captions to the diagrams forming the display. At this time preliminary results from the Old Red Sandstone of Wales were beginning to come through, and so they were displayed also, but I did not have any Cambrian, Ordovician or Silurian data then. I think I did show my Permian data from the Exeter traps, but this direction was similar to Clegg's Triassic direction.

Thus, at the time of the R.S. Conversazione, we did not have a spread of palaeomagnetic results through time. When we set up the display on the first day of the Conversazione, we had not plotted any pole positions, but later one of the visitors to our display (I can't remember whom) approached us to discuss the ADF [axial dipole field] hypothesis and, having seen Irving's Torridonian result, said "where would the pole have been in the pre-Cambrian?" This question led us to borrow a world globe and we were able to locate the pole position quickly using a piece of string and the ADF relation $\tan L = 0.5 \tan I$ ($I$ is the measured time-averaged palaeomagnetic inclination and $L$ is the calculated palaeo-latitude). Overnight, Phil Du Bois calculated the Torridonian pole position. (In those days there were no pocket calculators and the calculation took more time than it would nowadays.) We then marked the Icelandic and Torridonian pole positions on the globe which became part of our display. I'm quite sure that we had no polar wander curve at that time because my work on the Lower Palaeozoic, Permian and Triassic formations described in my thesis was not done at the time of the Conversazione: the path circling the N. Pacific was not drawn until September 1954 just prior to the BAAS meeting. However, the occurrence of reversals in the Torridonian result caused us to realize that there was a problem in deciding which of the two palaeomagnetic poles was the ancient north pole in very old rocks.

*(Creer, April 16, 1985 letter to author)*

Creer also noted in the letter to me that he had talked to Runcorn about this matter shortly before writing it.

Last week I was talking these early events over with Keith Runcorn at Newcastle. He says he thinks we had a primitive APW curve at the R.S. Conversazione, but his memory is not clear and he does not remember the episode of the borrowed globe. I feel that I trust my own memory better because I was actively occupied in obtaining the measurements and compiling them for my thesis.

*(Creer, April 16, 1985 letter to author)*

Creer and Runcorn agree that Du Bois calculated the pole for Irving's Upper Torridonian red sandstones. Creer is correct to point out that he had not completed analyses of his measured data for the Cambrian, Ordovician, or Silurian before the Conversazione. There is no evidence to support Runcorn's recollection thirty years after the event that there was an APW path at the Conversazione; no report that I have seen mentions an APW path at the Conversazione. Creer also wrote to Irving asking him if he remembered a path. However, Irving was unable to settle the issue; he had, he admitted, had "too many of the free drinks to remember clearly what happened."

Yes I was at the Conversazione ... and my recollections are not perhaps entirely to be trusted – too many free drinks to remember clearly what happened. I remember getting into an argument with Lees (chief geologist at BP) about continental drift which he was against. I also remember a globe and plotting things on it and at first I thought it was actually at the Conversazione but now I am not sure. It could have been back in Cambridge. I think you are right that we had no properly plotted polar wander curve at the Conversazione. The first time that I can be sure that I saw a curve was in your thesis which you sent me unbound to Australia sometime very early in 1955. I still have it. However that polar wander was being discussed amongst us before that for at least 18 months (probably over 2 years) in Cambridge [is clear]. See for example, *Nature*, 173, p. 1114, 1954. [This is a report of the Birmingham meeting, and referred to as Griffiths and King (1954) in my bibliography.] That paper makes incidentally a specific statement ... that Hospers had by Jan. 1954 actually calculated poles. This, his 1955 paper, and his own testimony leads me to believe that Hospers was the first to calculate poles. There is little doubt or no doubt in my mind that you were the first to construct a curve and have frequently said so.

*(Irving, letter to Creer, dated August 21, 1985; my bracketed additions)*

More importantly, Creer was the first to direct his research specifically toward obtaining an APW path. As described above, his first or "Rome" version, which he labeled "Creer 1954," appeared in late 1954 in a special issue of *JGG* (Creer *et al.*, 1954). As he noted:

I should stress that this polar wander curve [the "Rome" path] was exclusively my own effort and it was produced by myself during the weeks preceding the B.A. meeting in Oxford, after Runcorn had left for Rome. Realizing that there might be some dispute in future years as to its origin, I labeled the curve "Creer 1954."

*(Creer, November 24, 1983 letter to author; my bracketed addition)*

Following on from §2.12, I now consider Creer's somewhat frenzied activities leading up to and during the construction of the various versions of his APW path. Below (§3.9) I shall consider Creer's own discussion of his first version when I describe Runcorn's attitude toward mobilism, at this time just before he sampled from the Colorado Plateau in the summer of 1954. I begin by describing what Creer did after the Birmingham meeting in January 1954 (§3.3). Continuing to collect especially Paleozoic rocks, he recalled:

I had to press on, and as soon as the weather got better (no snow or ice), I went back to South Wales with David (Collinson). This would have been through the winter of [early] 1954 before Easter time. We made two or three more field collecting trips. And then I decided that I needed to get some older rocks – The Old Red Sandstone is mainly Lower Devonian, and extends up to the Middle Devonian and down into the Silurian. But, in the Silurian of that area, the beds I found were not so red; they were rather greenish (drab) in colour. I read ... and found that there were some Cambrian "red" beds (Caerbwdy Sandstones) exposed in the same geographical region. I collected them, and it worked out pretty well. But there was only one site. When David Collinson and I were not in the field, we were busy in the laboratory measuring. They were long days! Later I decided that I needed to extend my sampling further, and I found that there are some Pre-Cambrian rocks (called the Longmyndian) exposed in Shropshire, of the same age as Ted's Torridonian. So I went with David to collect these too. This would probably have been after Easter time.

*(Creer, 1987 interview with author; my bracketed addition)*

He now had to move very quickly. There was the upcoming BAAS meeting in September, and he wanted to finish his Ph.D. thesis before starting work in October with the Geological Survey of Britain.

To get the measurements done I was working long hours all through the summer. Through June I concentrated on measuring and sorting out the results for each site and rock formation. Then, for most of July and all of August I lived as a recluse in my lodgings, writing up. When they [Runcorn and Irving] went away to Rome, I had not finished. I still had a lot to do. I suppose it was towards the end of August – The British Association meeting came at the beginning of September – I remember that I just locked myself up for three to four weeks and I wrote more of my thesis. I just made a plan of what I had to do each day, and didn't go out of the house at all. I just wrote and wrote and wrote. Unfortunately, I didn't quite finish it by the beginning of October, and I had to go to the job. It took me two more months to polish it up.

*(Creer, 1987 interview with author; my bracketed addition)*

Blackett and O. T. Jones, retired Woodwardian Professor of Geology, were his examiners, and his Ph.D. dissertation was well received (Creer, letter to author, November 24, 1983).

By constructing his APW path for Britain, Creer moved to the second stage of the paleomagnetic test of continental drift. Creer has kept his notes for the BAAS lecture during which he actually described his second or "Oxford" version of the APW path.

From them I am able to construct an accurate view of his thinking at the time.[9] He began by discussing the intensity of the different types of remanent magnetism. He summarized Fisher's statistics, and explained how Hospers used them to support the GAD hypothesis. He addressed the issue of field reversals, arguing in their favor by appealing to Hospers' work on Icelandic lavas, Runcorn's on the Columbia River basalts, Irving's work on the Torridonian sandstones, and his own discovery of reversely magnetized Late Paleozoic rocks. He also noted that he had found indications of normal and reversed samples in the Triassic sandstones at Sidmouth, and had sampled through the level at which the reversal occurred, but had not succeeded in observing the manner in which the field reversed. He suggested that the geomagnetic field might have been reversed throughout the whole of the Devonian Period because all of the samples he had collected from the ORS had reversed magnetization. Finally, he addressed the question of the stability of remanent magnetization with special reference to the Sidmouth rocks, distinguishing between primary, secondary, and temporary components, and described how he had applied Graham's folded-bed and conglomerate tests.

Creer introduced his discussion of polar wandering by briefly mentioning that it had been suggested earlier from paleoclimatological evidence, notably by Kreichgauer, Wegener, and Köppen. He explained how paleomagnetism could be used to test the idea of polar wandering, and mentioned that Graham (1949) had made such a suggestion.

The biggest change between the first and second versions was the addition of an older Pre-Cambrian paleopole based on Irving's Diabaig (Lower Torridonian) work. Creer most likely added this to his path immediately before the Oxford meeting as indicated by his hand-printed label "earlier P.Cm" for Irving's Lower Torridonian paleopole (Figure 3.3), and his not adding "EARLY PRE-CAMBRIAN PC$_m$" to his key. Irving recalled with some confidence that he was graphically calculating poles for the Torridonian during summer 1954 when finishing his Ph.D. dissertation, and that before he left for the Rome meeting he bicycled over to see Creer at his lodgings, where Creer showed him how to calculate poles analytically. He likely showed him his final summary of his Torridonian results.

Returning to his BAAS lecture, Creer referred to the slide of his APW path, and explained that this was a convenient way of presenting paleomagnetic results and that he did not exclude the possibility of continental drift. He stressed that paleomagnetists should now begin gathering data from other landmasses.

Slide 12 [which showed his apparent polar wander path] shows the position of the north pole as calculated in this way [i.e., assuming the GAD hypothesis] for various geological epochs. Curves drawn up by Kreichgauer and Köppen are shown for comparison … The problem is complicated by the possibility of the relative displacement of the continents as was suggested by Wegener. The work carried out in Britain to date appears to support the pure polar wandering hypothesis but before the rock magnetician can say anything about Wegener's displacement hypothesis it will be necessary to make investigations similar to those described

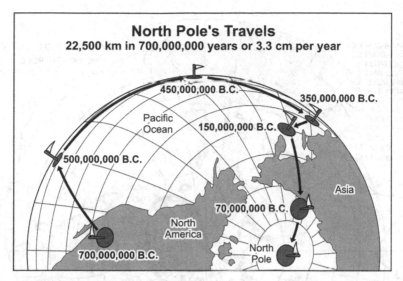

Figure 3.4 J. Donovan's plot of Creer's APW path from *Time Magazine*, September 27, 1954: 40. Donovan plotted the APW path that Creer showed at the Oxford meeting (Creer, 1954, "Rock magnetism and polar wandering," unpublished ). It was Creer's slide no. 12. The most important difference between this second APW path and the first one is the inclusion of an earlier Pre-Cambrian paleopole based on Irving's work from the Diabaig Group, the "Lower" Torridonian. (See endnote 2 for much later pre-Permian revisions to Creer's APW path, which did not affect the strong support it offered for Wegener's continental drift.)

on rocks of the same geological age from sites in different continents … It is intriguing to speculate on what the outcome will be.

(*Creer, 1954, "Rock magnetism and polar wandering," unpublished: 7; my bracketed additions*)

Creer presented the results in terms of polar wandering, but allowed for the possibility of continental drift, which members of the press also noted. *Science News* (Haslett, 1954) summarized Creer's talk. *The Times* also covered it (Special Correspondent, 1954) as did *Time Magazine* (Anonymous, 1954) and the *Manchester Guardian*. *The Times*, prematurely awarding his Ph.D., noted: "Dr. Creer mentioned also that the drift might be of continents, not of the crust as a whole." The author of the *Time* article noted:

Dr. Creer is not sure that the crust as a whole has moved. The continents may have drifted independently. By measuring the magnetism of more ancient rocks, he hopes to answer this question too.

(*Anonymous, 1954. 40*)

*Time Magazine* also showed a plot of Creer's APW path (Figure 3.4). Donovan, who worked for *Time*, talked to Creer after his talk, and plotted Creer's paleopoles on a large globe, which he then photographed. Although the second to be constructed, this was the first APW path constructed from paleomagnetic data to be *published*. Needless to say, Creer was interested in measuring rocks of the same age from

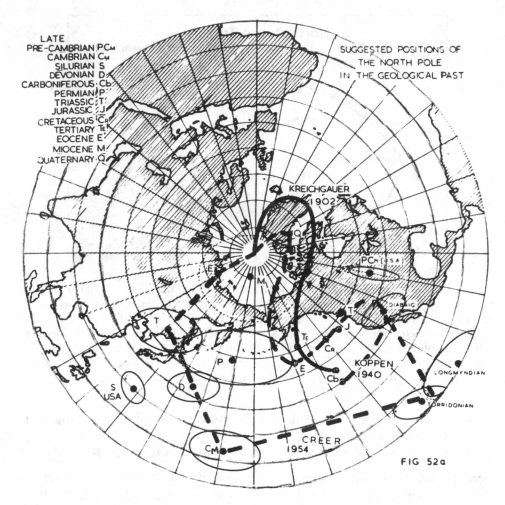

Figure 3.5 Figure 52a from Creer's Ph.D. thesis January 1955, opposite p. 177. This is the first polar wander path with ovals of confidence, and the first to compare results from Britain and the United States. (See endnote 2 for much later pre-Permian revisions to Creer's APW path, which did not affect the strong support it offered for Wegener's continental drift.)

different land masses.In this connection, *The Times* reported that Blackett, at the same BAAS meeting, talked about the work of Clegg's group and how the group had just gone to Spain to collect Triassic samples to check rotation of the Iberian Peninsula. Paleomagnetic studies of tectonics were certainly moving along.

Upon his return to Cambridge from the Oxford meeting, Creer began work on his third APW path, for his Ph.D. thesis (Figure 3.5). He located poles more accurately with improved documentation, and constructed ovals of confidence around them.

Ovals of confidence were not plotted on the first APW path that I showed at the B.A. Oxford meeting. I developed the method of calculating them the week after I returned to Cambridge having been led to do this because one of the audience at my Oxford talk asked why I had not drawn circles of confidence on the polar wander diagram, noting that circles of confidence about the average paleomagnetic directions had been illustrated by a slide. I discussed how this might be done at a tea break at the Department of Geodesy and Geophysics at Cambridge with Maurice Hill, the marine seismologist (now deceased) and it took me a few days to work out the formulae that I presented in my thesis. The first APW diagram showing ovals of confidence is that contained in my Ph.D. thesis and I did not draw this diagram until about the end of September 1954 at which time I left Cambridge to take up a post with the UK Geological Survey in London.

*(Creer, April 16, 1985 letter to author)*

In the "Rome" version he gave only rough geographical locations of poles; in the table accompanying the second and third versions he gave latitude and longitude (Creer, "Rock magnetism and polar wandering," 1954 unpublished; Creer 1955, Table 54, p. 181). In his first and second versions he included a Silurian pole, but he wisely omitted it in his third version, later giving (August 8, 1991 letter to author) as his reasons: the rocks were the wrong color, greenish rather than red or dark purple, he had only five samples, and no fold test (Creer, 1955: 57, 69). His samples from the ORS, which were located on two opposed limbs of a large structure, yielded substantial differences between limbs (Creer, 1955: 68–69).[10] He excluded highly deformed ORS (from Pembrokeshire) because of secondary overprinting, and recalculated the Devonian pole. He calculated the Triassic pole from the newly published paper by Clegg and company (§3.5), which had just become available, rather than using his own less complete and partially stable data from Sidmouth. Finally, Creer abandoned the Permian pole in constructing his APW path, reasoning as follows:

The fact that the Permian pole lies off the curve can probably be accounted for by the fact that insufficient results have been obtained for this period. Igneous rocks give spot readings of the ancient field direction and secular variation has probably not been averaged out in determining the axis of magnetisation.

*(Creer, 1955: 178)*

Creer's Permian lava samples may not have averaged out secular variation.

He stressed the continuity of poles and favored polar wandering.

A curve is drawn [in Figure 3.3 reproduced from Creer's Ph.D. thesis] connecting those pole positions determined from measurements made on rocks collected in Britain. The fact that a reasonably smooth curve can be drawn and that the pole positions do not lie over the whole hemisphere, justifies, to a certain extent, the attempt to account for the different axes of magnetization by the polar wandering hypothesis.

*(Creer, 1955: 177–178; my bracketed addition)*

He also noted (1955: 178) that further work was required to determine whether polar wandering has been gradual or intermittent.

Creer introduced continental drift by reviewing the classical arguments beginning with the paleoclimatological support, citing Wegener, du Toit, Gutenberg, and

Ting Ying Ma, a Professor of Geology at the National Taiwan University. Ma argued in favor of rapid polar wandering and continental drift, basing his work primarily on studies of corals. Creer noted:

Ma was by no means the first to put forward the hypothesis of polar wandering, but some of his ideas have been mentioned here because they are not so well known as those of many other authors, notably Wegener, Köppen and Du Toit.

*(Creer, 1955: 171)*

Creer later recalled that it was while working on his dissertation that he seriously began to review drift literature. He was influenced by Holmes.

While working on my thesis, I read parts of Wegener's and du Toit's books and also met Lester King on several occasions. Perhaps what attracted my interest most was the chapter [entitled "Continental drift"] in Holmes' textbook (Chapter XII, pp.487–509 of the 1947 reprint of the first edition). In my view, Holmes was the most perceptive geologist of his day. Consider, for example, his Fig. 262 on p. 506 [reproduced as Figure 5.6 in I, §5.8] and compare it with illustrations of the sea floor spreading mechanism. They are very similar indeed. Subsequently, I corresponded with Holmes' widow, Doris, prior to her preparation of the third edition.

*(Creer, April 9, 1984 letter to author; my bracketed additions)*

Creer cited drift's solutions to the problems of coastline congruencies and the distribution of Permo-Carboniferous glaciation, and thought the latter most important.

Perhaps the best known of Wegener's reasons is the similarity of the coastlines on either side of the Atlantic Ocean. Another more powerful reason is that by bringing together the continents affected by the Permo-Carboniferous glaciation the whole area covered by an ice cap in those times need have been no greater than that covered during the Pleistocene glaciation of the northern hemisphere.

*(Creer, 1955: 171–172)*

Mobilism faced difficulties, and although Creer singled out the lack of an acceptable mechanism, he wisely did not claim that solutions were not possible and stressed the lack of knowledge about Earth's interior – just the conclusion that many others had reached at the end of the 1940s and early 1950s. He found the overall range of problems solved by mobilism impressive.

The controversy as to whether the hypotheses of polar wandering or continental drift are plausible has been long and violent. A difficulty is that insufficient [information] is known about the physical and chemical properties of the crust and interior of the earth for the formulation of sound physical arguments as to whether forces could exist of sufficient magnitude to produce the required movements. However, they [mobilist arguments] have held the interest of geologists for over half a century, not only because of the numerous congruities and remarkable coincidences described by Wegener and Du Toit but because of the attractiveness of a unified theory (as envisaged by Ma, 1953) which could possibly be capable of explaining many outstanding geological problems.

*(Creer, 1955: 172; my bracketed additions)*

He characterized the paleomagnetic evidence as offering new and significant support for continental drift.

The importance of seeking fresh sources of more significant evidence is obvious. The possibility that the study of palaeomagnetism might provide such a source was suggested by Graham (1949).

*(Creer, 1955: 172)*

### 3.7 Creer compares his British APW path with a Silurian pole from North America: move to Stage III

At the Oxford meeting, Creer had remarked on the need to sample outside Britain to see if polar wandering alone was sufficient or had to be supplemented or supplanted by continental drift. After the meeting, he realized he could compare his APW path for Britain with Graham's 1949 Silurian data from the Rose Hill Formation of Maryland.

You raise the point of why I considered Graham's result from the Silurian. Well we had all been thinking about Graham's data from the beginning of our own work, after all it had been published several years earlier before we had even started. Whenever we discussed Graham's results in the light of our own, we all felt that most of the rock formations they [at the Carnegie] had studied must have somehow been remagnetized. And then, in September 1954 I wondered – well perhaps not all of the Carnegie data were affected by remagnetization! And so I came to reread through their Rose Hill Formation result – I thought – what if it is really his only "good" result? Let me calculate the pole and compare it with my own Silurian pole from Britain.

*(Creer, January 7, 1999 letter to author; my bracketed addition)*

Although this provided him with only a single point on the path that had yet to be obtained for North America, it was a start. *Indeed, this comparison by Creer was the first made between an APW path from one continent and a paleomagnetically determined paleopole from another.* Creer accepted that Graham's fold test on the Rose Hill Formation was positive and hence that the data from it were likely reliable. He read the directions of magnetization from Graham's stereograms, calculated their mean, its paleopole and oval of confidence, and plotted it alongside version three of his APW path for Britain (Figure 3.5).

In similar fashion, Creer also calculated a pole for Precambrian diabase dykes from Michigan based on Graham's work, but he realized that any comparison with the British path would be of little value because its age was imprecise.

The only test on these grounds of the drift hypothesis, which can be applied at the moment, is provided by Graham's results. The Pre-Cambrian results are valueless in this respect because the age of the diabase dykes cannot be compared accurately enough with the British Pre-Cambrian rocks measured. On the Wegener hypothesis the eastern coast of the Americas was close to the western European and African coasts. Maryland is now about 75° west of Britain. The position of the Silurian pole calculated from Graham's Rose Hill rocks lies to the correct side of the polar wandering curve to account for a subsequent relative westerly drift of America but only of about 30°: rather less than half the required amount.

*(Creer, 1955: 179)*

He did not argue firmly that this particular paleomagnetic result necessarily required continental drift. His path implied movement of Britain relative to Earth's pole, but with only one good and relevant result from North America, he rightfully was unwilling to accept continental drift.

The results so far obtained can fairly be said to support substantial movements of the pole relative to Britain. Evidence from other continents is essential before it is possible to suggest whether this movement is due to pure polar wandering or entirely to drift or perhaps to both.

*(Creer, 1955: 179)*

Reflecting thirty years later on his assessment of continental drift when preparing his thesis, he wrote:

At the time of writing my thesis no substantive palaeomagnetic evidence existed relevant to the drift hypothesis. So it would have been speculative, at that time, to have come out strongly in favor of drift or polar wander as the mechanism which had caused polar wander movement relative to UK, summarized in the polar wander curve in my thesis. I repeat: I presented the UK data in the form of a polar wander curve as being the most convenient and compact way of summarizing the palaeomagnetic results. It was not intended to imply that the results exclusively supported polar wander of a rigid outer shell of the earth or of the earth as a whole. How could this be when only two points (palaeomagnetic pole and UK) were plotted for any point of time?

The critical question at the time was: do polar wander [APW] curves for the different continents superimpose when plotted on the present globe? The first indication that they did not came from Irving's result from the Deccan Traps and then from his Australian results.

*(Creer, April 9, 1984 letter to author; my bracketed addition)*

Indeed, Creer did not accept continental drift until he saw the results from India and Australia.

I accepted continental drift when the first results from India and Australia became available. The reason was that the polar wander curves did not superpose when plotted on the present globe (world map). It remained to be shown that the Wegener/du Toit reconstructions were right, but at that stage it seemed to me that if drift had occurred, then it was most likely that their scenario was correct.

*(Creer, April 9, 1984 letter to author)*

His retrospective comments about becoming generally inclined toward continental drift accurately reflect what he wrote in his thesis; it is easy to understand why he did so so soon. His comment may also make us wonder why Creer did not use Irving's Indian results from the Deccan Traps when he compared his APW path for Britain with Graham's results from the Rose Hill Formation. There are two reasons. First, Creer likely did not know about Irving's Indian results until it was too late to include them meaningfully in his Ph.D. thesis. Earlier Irving occasionally talked about his Indian results in vague terms, but Creer never learned any of the particulars.

We didn't see much of each other through the summer of 1954 – I was in the field a lot, and measuring samples all the time when I was in Cambridge Ted was busy writing up, and then when I had finished measuring I locked myself up to concentrate on writing up too. I knew that Fisher had arranged for the collection of a few Deccan Trap samples for Ted, and earlier in the year I recollect that Ted said, informally, and on more than one occasion, that he was getting a Southern Hemisphere paleolatitude for India. But, he did not make a "big deal" of it. And I was not sufficiently interested to ask probing questions because I had plenty of my own work to do. I never asked him whether he would include the Deccan work in his thesis. In fact, neither of us showed one another a plan of his own thesis – both of us knew that we had done enough to be able to write up a very good thesis independently. Of course we shared the same overall knowledge of the ground that the others (including Clegg and Keith) had covered, from internal seminars and external conferences through the years 1952/54.

*(Creer, January 7, 1999 letter to author)*

Irving seconded Creer's recollection. Both spent most of their time alone writing away.

I vouch for this. You will recall that it was Ken who actually taught me how to calculate poles analytically (as opposed to graphically, which I figured out myself). I recall having to make a special call to Ken's lodgings to do this. I did not go much to the laboratory and so we did not see each other on a day to day basis. We both worked at our lodgings. I cannot recall seeing anyone during this period much, except a girlfriend who kept me sane.

*(Irving, March 3, 2001 note to author)*

Irving included the Indian results at the end of his Ph.D. thesis, and, as they had agreed to exchange copies, he gave one to Creer in London in mid-November 1954 when Creer saw him off on the boat-train to Tilbury docks, bound for Australia. By then Creer had finished most of his Ph.D. thesis, was already working for the Geological Survey of Great Britain, and was racing to finish so that it could still have a 1954 date.

At this time I was working in the British Geological Survey and learning about applied geophysics. This made heavy demands on my time. My thesis was then largely completed and well formed – I was tidying up some of the figures and the text was away with the typist. So Ted's thesis came into my hands too late to incorporate any more of his data ... I cannot recollect Ted presenting data, through the year 1954, from the Deccan samples that had been collected for him. These results were not available to me until he gave me a copy of his thesis, by which time I was too busy with the final preparation of my own thesis, getting it ready for submission (checking references, diagrams, chapter layout, etc.) even to have had time to scan through his. My target was to submit before the end of December 1954, and in fact I just failed to achieve this objective, getting caught up with the Christmas holidays – hence I did not get the thesis delivered to Cambridge University until early Jan 1955, and this greatly disappointed me because the thesis then came to carry a 1955 date rather than 1954.

*(Creer, January 7, 1999 letter to author; my bracketed addition)*

But even if Creer had seen Irving's results several months earlier, he still might not have used them. They were few, and to use them might be pushing data beyond reasonable limits. In fact, Creer recalled that he felt he was already pushing the paleomagnetic data to their limits.

The idea of attempting to introduce Graham's Rose Hill result into my discussion of the apparent polar wander curve came to me at the very end, after my presentation at the Oxford BA, and this (Rose Hill data) was the last new factor I added to my discussion before stopping to do new work in order to concentrate on finishing writing up [my Ph.D. thesis]. At that time I felt I was really pushing the data to their limits, and some colleagues in the lab (not paleomagnetists) cautioned me about letting myself be "carried away" by my convictions (which all did not share) and over (enthusiasm). I had forgotten about the existence of Ted's Deccan Traps data: the thought of using them never entered my head. Even if it had been suggested to me at the time (e.g., by Ted himself) I would not have had time to do so. Moreover, I felt there were too few samples to be able to average out the effects of secular variation: that is why the Blackett/Clegg group was planning to make a new and more thorough study of them. Actually it turned out that the result that Ted had obtained was not very different from the result that Clegg was to obtain. Importantly, Ted had carefully instructed the Indian Survey to collect his few samples from different and separated sites, rather than all from the same place, in order specifically to minimize the effects of secular variation. Even so, six samples (sites) would be considered even these days and with application of modern "cleaning" techniques to be too few to obtain a credible result.

*(Creer, January 7, 1999 letter to author)*

Creer was right. In contrast, Graham's result from the Rose Hill Formation had passed the fold test and, by the standards of the times, could be deemed more reliable than Irving's Deccan result, although it should be added that the reliability of the latter was not so bad because there was a reversal and lateral consistency over 300 km. Creer went ahead and accepted the Rose Hill result even though Graham had by then changed his mind; doubting the utility of his fold test, he no longer believed that his results reflected the Paleozoic field (§1.9).[11]

Once Creer got to the Survey he was unable to convince the administration to allow him to continue working in paleomagnetism (Creer, April 9, 1984 letter to author). He remained at the Survey for two years until September 1956, when he "was persuaded by Runcorn upon his appointment as professor of physics at Newcastle to join him there and start palaeomagnetic research again" (Creer, November 24, 1983 letter to author). He did not immediately settle there, but left to sample in South America, thereby extending his work on the third stage of the paleomagnetic test of mobilism (§5.6).

Creer constructed one more version of his APW path (Figure 3.6). This fourth version appeared in one of the five papers published in the *Philosophical Transactions of the Royal Society of London* that described Irving's and Creer's work at Cambridge under Runcorn's supervision. Runcorn, wanting their work to appear as a unit in a Royal Society (London) journal (III, §2.21), essentially lessened the impact of Creer's

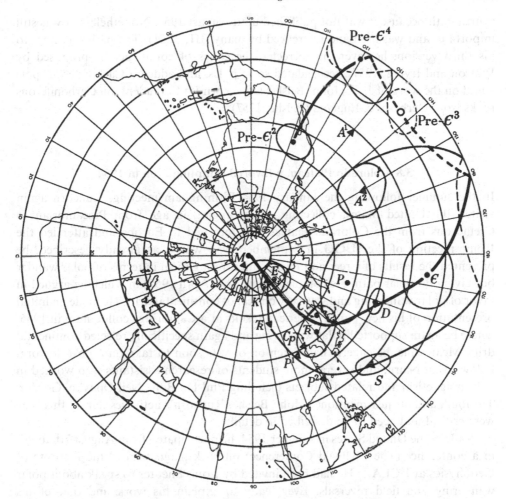

Figure 3.6 Creer, Irving, and Runcorn's Figure 1 (1957: 147). The bold APW path is based on rocks from Great Britain, Creer's fourth path. The finer APW path is based on American rocks. The solid part of each path is plotted in the Northern Hemisphere; the dashed, in the Southern Hemisphere. Northern Hemisphere British poles are solid circles; Southern Hemisphere ones are open circles; American, solid triangles. The British poles are as follows: Pre-$\mathcal{E}^2$ (Pre-Cambrian) is Irving's Lower Torridonian Sandstone; Pre-$\mathcal{E}^4$ is Creer's Longmyndian; Pre-$\mathcal{E}^3$ is Irving's Upper Torridonian Sandstone; $\mathcal{E}$ is Creer's Caerbwdy Sandstone; $D$ (Devonian) is Creer's Lower and Upper Old Red Sandstone; $C$ (Carbonifeous) is Belshé's Derbyshire Coal Measures; $P$ (Permian) is Creer's Exeter Volcanics; $TR$ (Triassic) is Clegg's Keuper Sandstones; $E$ (Eocene) is Hospers' Northern Ireland basalts; $M$ (Miocene) is Hospers' Icelandic basalts. The American poles are as follows: Pre-$\mathcal{E}^4$ is Graham's Michigan Diabase dykes; $A^2$ (Pre-Cambrian) is Runcorn's Hakatai Shales; $A^1$ is Doell's Hakatai Shales; $S$ (Silurian) is Graham's Rose Hill Fault; $Cp$ is Runcorn's Naco Sandstones; $p^2$ (Permian) is Runcorn's Supai Shales; $p^1$ is Doell's Supai Shales; $TR$ is Runcorn's Springdale Sandstones; $K$ (Cretaceous) is Runcorn's Dakota Sandstones; and $M$ is Campbell and Runcorn's Columbia River Lavas.

fourth path because it was not published until August 1957. Nonetheless, it was still important, and was used and referenced by many (III, §2.21). The path was close to his third version; however, he corrected the cones of confidence as proposed by Watson and Irving (1957), recalculated some poles, and added a Carboniferous pole based on the work of J. C. (John) Belshé, a Runcorn Ph.D. student, on Carboniferous rocks from Derbyshire, England (Belshé, 1957).

### 3.8  Colorado Plateau, a favored sampling area in 1954

In the summer following the January 1954 Birmingham meeting, Runcorn again visited the United States, collecting rocks ranging in age from Precambrian to Cretaceous from the Colorado Plateau. He returned to Europe and attended the Rome meeting of the IUGG in September 1954 where, as already described, he presented the results of Creer and Irving, interpreting them in terms of polar wander but giving no polar path. He completed a review (April 1955) in which he argued in favor of polar wandering and was very critical of continental drift. He made an initial presentation of his new North American results at a Cambridge colloquium in 1955, where he again supported polar wandering and again specifically rejected continental drift. Meanwhile, he eagerly had taken on or was soon to take on a post-doctoral fellow, Alan Nairn, and several more students or research assistants who worked in paleomagnetism and related subjects including Phil Du Bois, Neil Opdyke, Gordon Turnbull, and, as just mentioned, John Belshé. Runcorn's activities during this year were critical and need to be described in detail.

While in the United States in summer 1954, he participated from August 7th to 9th in a conference (The Idyllwild Conference) on rock magnetism at the Institute of Geophysics at UCLA.[12] He had been invited by Louis Slichter to speak about polar wandering and field reversals. Ever ready to explain his work and that of his students, Runcorn welcomed the invitation. At the meeting, he and John Graham differed about reversals. As Runcorn later recalled:

Slichter got interested in paleomagnetism, [and arranged a conference that was] funded by the National Science Foundation in the summer of 1954. And, I played quite a role in organizing it for him ... Slichter got this conference together. No proceedings were published. But, I remember quite a confrontation with Graham because Graham was arguing that these magnetizations differ [in direction] from the present [field], including reversals, [should not be interpreted in terms of polar wandering or geomagnetic reversals].

*(Runcorn, 1984 interview with author; my bracketed additions)*

However, Runcorn retrospectively remarked that he had another reason for accepting Slichter's invitation. He wanted to duplicate in the United States what his students and Clegg and colleagues had done in Britain. Having seen the Grand Canyon during summer 1952, he later realized that it had plenty of (structurally) little disturbed red beds covering a wide range of ages, an ideal place to collect. As Runcorn later claimed:

And, it was because of the opportunity of coming over in 1954 that I collected rocks in the Grand Canyon to make a comparison with Britain, because, of course, quite early in '54 or even in '53, the polar wandering story had become sufficiently debated by us to make it rather important to come to see whether one could get the same polar wandering curve from rocks in the United States. I'd seen the Grand Canyon in my first visit to Slichter's institute in '52, and I remember I went with some friends, and I remember seeing it. And, again this was very early in my geological education, and I thought, if our work in Britain goes well, this is obviously the next place to try – lots of red rocks, and lots of simple geology, where one would get good results. We were terribly afraid of going to an area where the tectonics would make the interpretation difficult. So, it was in '52 when I saw the Grand Canyon for the first time and I remember saying then that this was the next place. And when Slichter organized this meeting in '54 in the summer I came over earlier and went and collected rocks in the Grand Canyon from a variety of formations.

*(Runcorn, 1984, interview with author)*

There is no question that Runcorn saw the Grand Canyon in 1952, and that its red sandstones and their undisturbed nature may have reminded him of the Torridonian. However, I do not believe that in 1953 or early 1954 he thought of sampling there to compare with results from Britain. As I have just described, Creer's version of the British APW path was constructed by him in Cambridge in the late summer of 1954, a least half a year later, while Runcorn was in the United States. Before then it was not even known to anyone if paleomagnetic APW paths existed. There is no evidence that he told Creer or Irving or anyone that this was his intention, nor did he attempt to sample the Grand Canyon in 1953, even though he was in the vicinity that summer collecting Late Tertiary Columbia River basalts not far away in Oregon. Given Runcorn's resourcefulness, he surely would have found a way, or at least recalled trying to find a way, to collect Late Paleozoic and Mesozoic rocks from the Colorado Plateau, if he had actually thought about getting an APW path for North America as early as 1952 or 1953. Notwithstanding and despite the inaccuracies of his retrospective account, he certainly decided to sample the Grand Canyon and surrounding area because of Clegg's, Creer's, and Irving's success finding oblique directions, their wondering about continental drift and polar wandering, and certainly his work had very important consequences.

Securing, with the support of Walter Bucher, a grant from the Geological Society of America (GSA), and an invitation to use the Museum of Northern Arizona in Flagstaff as a field base, he left for the United States well before the upcoming conference at UCLA to ensure enough time for fieldwork. Runcorn completed his fieldwork, attended the conference, and sent his samples to Cambridge, returning there via Rome in late September or early October. He submitted two papers on the results in June 1955 (Runcorn, 1956a, 1956b), and discussed them at the Cambridge colloquium in early 1955 (Day and Runcorn, 1955). He interpreted the results as indicative of polar wandering without continental drift. To understand why he rejected drift, it is necessary to assess his overall attitude toward mobilism prior to these events. Before doing so, I want to mention Richard Doell's work in the Grand Canyon.

Doell received his Ph.D. in 1955 from the University of California, Berkeley, under Verhoogen (see Chapter 8 for discussion of Doell's later work). Verhoogen had an early interest in paleomagnetism, and in 1952–3 he gave a seminar (Glen, 1982: 144), which Doell has briefly described.

I can't be entirely certain about this, but I think it is essentially correct. That first grad seminar of Verhoogen's was, as I recall mostly – or entirely – about remanent magnetism. We reviewed all the Louis Néel papers and perhaps some of Koenigsberger and Nagata, as I recall. If we did do anything on drift or wandering it would have been very trivial, I'm pretty sure.

*(Doell, January 26, 2001 email to author)*

Verhoogen (Glen, 1982: 144–145) had spent fall 1953 at Cambridge University where he learned of Irving's work on the Torridonian. The next year he suggested Doell sample the Grand Canyon as part of his Ph.D. work. Doell recalled what happened.

That was Verhoogen's suggestion – more like a request, in a way. As you've noted, John came home from his sabbatical with knowledge of the English P.W. work. The Grand Canyon Rocks being the fantastic well-known, easily accessible sequence they are, it is not surprising that more than one person thought of collecting there. J. V. got the money for collection (and measuring) from Louis Slichter, who was "head" of UC's "Geophysics Institute." My guess is that these two, and perhaps others, got together after John's sabbatical and conceived this study, and that John volunteered my "services." They also arranged for me to do the measurements at Stanolind Oil and Gas Co.'s magnetometer in Tulsa. Ours at Berkeley was nowhere near sensitive enough to do seds [sedimentary rocks]. I think I spent 2–3 weeks (with my then wife Ruth) collecting samples in the Canyon, and a like time in Tulsa doing the measurements. The work did lead to my first publication and may have had some significance, but it occupied only a very small portion of my graduate studies research.

*(Doell, January 26, 2001 email to author; my bracketed addition)*

Like Runcorn, Doell collected from the Supai Formation (Permian) and Hakatai Shale (Pre-Cambrian), and his results agreed with Runcorn's. He stressed the independence of their results and their agreement.

It is important to note that these measurements were made entirely independently [of Runcorn's]. Runcorn restricted his collection to the Kaibab Trail, whereas my collections were taken equally from the Kaibab and Bright Angel Trails (separated by about 2½ miles). Further, Dr. Runcorn's measurements were made on an astatic type instrument, whereas mine were made on a "rock generator" type magnetometer … The comparison is very encouraging, in that the number of samples measured in each determination is of the order of ten to twenty.

*(Doell, 1955b: 1167; my bracketed addition)*

Doell said nothing about polar wandering or continental drift.

### 3.9 Runcorn's attitude to mobilism before his first
### North American survey

I definitely had begun to think that polar wander was a simpler explanation than continental drift, and, of course, it involves fewer parameters. So, it is the first hypothesis to try. Again, if I had a philosophy of science, it was that one should try the simplest explanations first. I heard the various arguments for continental drift, and, of course, discussed them. I thought the paleoclimatic evidence could be explained by rather quick polar wandering. I would have thought – probably, a lot of people I had contact with like Walter Bucher and O. T. Jones would have agreed with me – that polar wandering was perhaps a good bet at explaining that kind of evidence. And, the other kind of evidence for drift, even the close fit of Africa and South America, didn't mean that drift occurred as Wegener wanted it to occur in the last 100 million years. It was not impossible that this went right back to the beginning of the Earth. And, many people who now say that the fit of the continents proves continental drift illustrates the difficulty geologists have with the time scale. I remember Hess laughing one day [in the 1960s after he had come up with seafloor spreading] and saying to me, you didn't need to do paleomagnetic work to prove continental drift because anyone looking at the fit of the continents would say it occurred. Even Jeffreys wouldn't have objected to have drift in the breaking apart of Africa and South America in the early Pre-Cambrian, when he might have been quite happy of thinking of the crust shifting around an entirely molten mantle. With regard to the other kinds of evidence, you know, the tectonic evidence, I remember hearing about the fact that the Caledonian orogeny and the Appalachians fitted together. I didn't find that convincing. Why should mountain belts wind up in one long line along the globe?

*(Runcorn, 1984 interview with author; my bracketed addition)*

When Runcorn first saw Creer's APW path for Britain, unlike Creer and Irving, he was not disposed toward continental drift. Moreover, Runcorn's comments show that he, despite the new paleomagnetic work, had at the time not seriously studied and certainly had not grasped the significance of the evidence he cited (see, for example, I, §8.6–§8.13).

I begin by examining the paper, Creer, Irving, and Runcorn (1954), in the proceedings of the Rome meeting. This is historically significant in its own right for it included, as already explained, the first publication in a scientific journal of a paleomagnetically determined APW path. It also nicely pinpoints the differing attitudes of Runcorn and Creer to continental drift. Creer essentially wrote the paper, and Runcorn edited it. Irving had by then left Cambridge. Creer recounted the general situation.

In the fall of 1954, I had taken up my post at the Geological Survey in London. My office, in the South Kensington district was located just round the corner from the Blackett/Clegg laboratories at Imperial College, so I saw them fairly often, and Keith used to drop in when he came down to London. So Keith and I were in contact on occasions during the fairly short time while the typeset was being finalized for submission – we were keen to get a 1954 date on it. Ted may have had some discussion with Keith about the paper during his last weeks in Cambridge, though perhaps he went down to spend most of the time he had left in UK with his parents in Lancashire. In November he left England and it took a while to establish himself in Australia (the boat journey took many weeks).

*(Creer, January 7, 1999 letter to author)*

## Creer is correct, Irving was not in Cambridge.

I submitted my thesis in September and then very quickly went home to Lancashire to my parents. I spent probably about six to eight weeks with them. I was flat broke. I wanted to go home, and also I had to eat. I had previously (about August) borrowed money from my college to go to Rome, and from my father to buy a suit for my new job in Australia. I was married in it and still have it. Both debts I paid off from my first pay cheque in Australia.

*(Irving, March 3, 2001 note to author)*

## Irving believes that Creer was primarily responsible for the paper.

The APW curve was Creer's work. I don't know who wrote the paper but it has Ken "written" all over it – with perhaps some editing by Keith. I took no part in actually preparing the paper because I was gone to Australia. The paper neatly summarizes much of the Cambridge work including mine. I think the order of the authors [Creer, Irving, and Runcorn] is quite right.

*(Irving, March 4, 1999 letter to author)*

## Irving's assessment is correct. Creer recalled:

I rewrote the paper in the form published after I had given my BAAS talk, and (with Keith's approval) I added the polar wander curve, though Keith did not have it to present at Rome. I did not finish the *JG&G* [*Journal of Geomagnetism and Geoelectricity*] paper until after I had taken up my new job in London, but I did not add any new data.

*(Creer, August 24, 2007 letter to author; my bracketed addition)*

## Creer and Runcorn disagreed about continental drift. After presenting the results, they concluded with the following:

The oblique mean axes of magnetization found in Pre-Tertiary times are therefore simply explained by the polar wandering hypothesis. Since the rocks have been collected from Britain only, it is impossible yet to exclude continental drift as the cause of the change in the various ancient pole positions. When measurements have been made on rocks from several continents it should be possible to test whether or not polar wandering is a sufficient explanation.

*(Creer et al., 1954: 168)*

## Creer's assessment is that the first sentence is Runcorn's, the rest is his.

The [first sentence] was Runcorn's contribution. The remainder is what I insisted should go in following my thesis and BAAS Oxford talk. I wrote the sentences on continental drift, and persuaded him to include them.

*(Creer, January 7, 1999 letter to author; my bracketed addition)*

Runcorn was not at all in favor of continental drift, he thought the APW path indicated polar wandering, not continental drift or their combination. Creer (and Irving from the evidence of his September 1954 thesis) thought that the paleomagnetic results could be interpreted at least in part in terms of continental drift, and refused to exclude it. This was not to be the last time that Creer argued with Runcorn over continental drift.

Runcorn made clear his preference for polar wandering and his dislike of continental drift in a lengthy review article on paleomagnetism that he was asked to write for *Advances in Physics*, and which he completed before he had obtained results from the Colorado Plateau. His paper contains an extensive review of the results in unpublished theses of Irving (1954) and Creer (1955). Unlike them, he reviewed continental drift unfavorably. After noting that Wegener proposed continental drift as an alternative to landbridges to explain biotic disjuncts, Runcorn (1955a: 283) suggested that the theoretical difficulty raised against landbridges, their incompatibility with the principle of isostasy, was no longer "regarded as a valid objection." He then raised the standard objections to Wegener's mechanism problem, noting (1955a: 283) that the forces he invoked "are now known to have negligible effects"; he appealed to Jeffreys' objection about the greater strength of oceanic as compared to continental rock.

Nevertheless the concept of the continents as a granitic scum (Wegener's sial) floating in a denser basaltic medium which also forms the floor of the oceans has had a powerful effect in suggesting such drift might be feasible. Jeffreys (1952) points out that however plastic the basaltic material below the continents may be because of the moderately higher temperatures, the basaltic shell under the oceans must have greater strength than the continental blocks. Drift of the continental masses through the floor of the oceans is therefore unimpressive as a geophysical hypothesis.

(*Runcorn, 1955a: 285–286; the reference is to the 1952 edition of Jeffreys'* The Earth)

He carelessly repeated Jeffreys' mistake regarding the match of the Atlantic coastlines of Africa and South America, which, he added, "figures prominently in popular expositions of continental drift is a poor fit, even allowing that the continental shelf rather than the coast is the most permanent boundary of the continental masses" (Runcorn, 1955a: 286).[13]

Making clear his preference for polar wandering only, and sounding very much the confident classical fixist, he suggested that it could also provide a solution to the origin of Permo-Carboniferous glaciation. He (1955a: 286) cautioned that the "meteorology of remote geological epochs is not a subject on which it would be profitable to dogmatize," and argued that polar wandering alone might be able to account for the glaciation because the glaciation in the southern continents and India probably had not been contemporaneous. He cited du Toit for support.

If is often said that because the Permo-Carboniferous glaciation is contemporaneous in these four widely separated continents of the Southern Hemisphere, polar wandering cannot alone be the explanation. This objection is based on the same misunderstanding of time scales in geophysics and palaeontology, as the objection to reversals of the field popularly held. Geological correlation in the Palaeozoic may be inaccurate and supposed contemporaneity may allow a difference of even tens of millions of years in age, during which the Pole might have moved fast enough to shift from one continent to another. In fact the South American glaciation is known to be earlier than the Australian glaciation and later than those of South Africa and India (du Toit 1954).

*(Runcorn, 1955a: 286; the du Toit reference is to the third edition of his* Geology of South Africa*)*

Runcorn went on to argue that while the suggested causes of continental drift were unimpressive, those for polar wandering were promising; T. Gold and Vening Meinesz both had suggested how it might have occurred and thirty years later he recalled their work and its influence on him.

I may have been influenced by Tommy Gold, I may also have been influenced by Vening Meinesz because I was invited over to Holland quite a bit, and he thought of polar wandering, but he hadn't thought of a particular mechanism, but I certainly, by this time recognized that if you had sufficient movement in the mantle ... I do remember thinking that if you could have the equatorial bulge move around, then you could move the axis of rotation. And Gold, of course, develops that a lot by talking about how heat pushes up mountains on the surface to produce it. I talked about convection, and obviously I must have got that idea, not necessarily from Vening Meinesz, but I knew that he was talking about convection. I began to think that might be the motive force of polar wandering. I could see, you know, no simple principles on which one could get continental drift.

*(Runcorn, 1984 interview with author)*

In discussing Vening Meinesz's ideas, Runcorn (1955a: 287) acknowledged that "widespread convection is not definitely known to occur," but still found his hypothesis "a very attractive explanation of polar wandering." He then turned to Gold's hypothesis. Gold was well aware of the paleomagnetic work of Runcorn's group, having heard lectures at Cambridge and at the September 1954 Rome meeting. In fact, Gold presented his thoughts at that meeting, and his paper appeared in March 1955, a few months before Runcorn's review. Gold (1955) argued that rapid polar motion could occur through vertical shifts of continents. Runcorn claimed that Gold's hypothesis could explain the paleomagnetic data.[14]

Gold suggests that tectonic movements, representing the addition of an excess mass between the pole and equator would be effective in producing polar wandering. Gold estimates that if a continent of the size of South America was suddenly raised by 30 meters ... polar wandering at a rate of 0.001°/ year would occur until the additional mass was situated on the equator. Gold estimates that large angles of polar wandering of the type envisaged above could occur in a million years.

*(Runcorn, 1955a: 288)*

Runcorn clearly favored polar wandering over continental drift. He even noted that the amount of paleomagnetic data from regions other than Britain was too incomplete to draw any conclusions about whether there had been continental drift. Thus, at this time he was unwilling to take seriously Irving's Indian work. As he retrospectively put it:

The Indian data was very interesting because we all believed after Ted's work on the Torridonian that you didn't need to collect a vast amount of material to get a good value for the average value of the direction of magnetization ... Ted measured about six [actually seven] specimens. Ted was very impressed by the results that he obtained. But, I would have, you know, I suppose, if anyone asked me at the time, I would have hesitated in coming out in support. I didn't pin a lot of faith on six samples from the Deccan Traps. I was probably a bit more exposed than Ted or Ken were to criticisms by geologists, and, you see, here again, if I had a philosophy it was that only when you demonstrated that there was consistency of magnetization over a reasonably big area of strata and when you showed that the next age gave something different that I would pin a lot of faith on it. In my review article I suppose that I must have made a lot of the fact that the pole positions from Britain lay in serial relationship in time. So, you know, the Indian work, at that stage would not have impacted strongly on me. But, of course, Ted had done the work, he was very enthusiastic, and he must have thought that it fit in with Wegener. Of course, this is what influenced Blackett to do a proper study.

*(Runcorn, 1984 interview with author; my bracketed addition)*

Runcorn's initial dismissive attitude toward Irving's 1954 Deccan results changed two years later (1957: 176) when he compared them with results by Clegg *et al.* (1956) obtained from two localities several hundred kilometers apart, each with several lava flows from which they collected 209 hand samples; all were reversed. Irving's result was based on the reversal (§1.11) and consistency (§1.11) tests of stability; from a single sample for each of seven localities spread over 300 km, five were reversed, two were normal, and averaging them regardless of sign went some way to correct for recent viscous overprints. The mean directions of the two studies were only 3° apart; their respective errors were about 10° and were not significantly different. Runcorn argued that, despite the far greater number of samples, "their result only confirms Irving's result but does not add to its accuracy." He now clearly acknowledged Irving's 1954 result as significant, and applauded its frugality. The facts had not changed, but Runcorn's attitude about the significance of Irving's work certainly had changed once he came to accept mobilism to explain his own results (§4.3, §4.11).

At this stage, however, Runcorn was not yet ready to follow his students and to support mobilism. Irving's Deccan results fit with Wegener. Irving had reversals, and his samples had been collected over a distance of 300 km; Creer had already argued in favor of mobilism, appealing to the discrepancy between his APW path and Graham's results. They became inclined toward mobilism fully two years before their former supervisor did.

### 3.10 Runcorn continues to favor polar wandering after his first North American survey

Runcorn's sampling plan was bold, ranging from Precambrian clear through to the Cretaceous, 102 oriented samples: sixteen from the late Pre-Cambrian Hakatai Shale in the Grand Canyon, eight from the Carboniferous Naco Sandstone near the Carizzo Creek Bridge, thirty-one from the Permian Supai Formation in the Grand

Canyon and nearby area, fifteen from the Permian Cutler Formation mostly in Monument Valley, eighteen from the Early Triassic strata of Zion National Park and elsewhere, eight from the Triassic Springdale Sandstone in Zion National Park, and six samples from the Cretaceous Dakota Sandstone, Arizona.

Back in Cambridge, J. C. Belshé, one of Runcorn's new students, and G. Turnbull, a research assistant who had replaced Collinson, made the measurements. Runcorn presented the results at a colloquium on polar wandering in Cambridge in early 1955 (Day and Runcorn, 1955) and in two other papers (Runcorn, 1955a, b).[15] At the Cambridge colloquium there were four invited speakers: Runcorn and Gold, and two paleontologists, J. W. Durham from the University of California, Berkeley, and W. J. Arkell from Cambridge. Durham, a specialist in Tertiary marine fauna and a vehement fixist, certainly knew where he stood, arguing against any sort of polar wandering or continental drift. Arkell, more diffident, emphasized the incompleteness of the paleontological and paleoclimatological data; he was sympathetic to polar wandering, as he would show in his magisterial 1956 *Jurassic Geology of the World*. He had favored continental drift in 1948, had changed his mind during its writing and no longer did so, citing data that appeared to be inconsistent with paleomagnetic pole positions (I, §8.13). Gold presented his polar wandering story and to a questioner replied that if continental drift were to occur it would bring about polar wandering, but polar wandering itself would not cause drift.[16] Runcorn argued that mantle convection could cause polar wandering, and after comparing paleopoles from Britain and North America, claimed that it offered the best explanation of them.

The palaeomagnetic record prior to Tertiary times is fragmentary; but it is clear that, apart from reversals of polarity of the field, its mean direction has changed slowly through geological time. The simplest interpretation of this is to assume a wandering of the pole of rotation of the Earth carrying the mean magnetic pole with it. From the results of a palaeomagnetic survey of sedimentary strata and lava flows undertaken by the Department of Geodesy and Geophysics at Cambridge, the motion of the Pole throughout the past six hundred million years has been traced in outline. The examination of the palaeomagnetism of rocks in other continents has not proceeded so far as in Britain, but the available data … are in approximate agreement with the British results.

(Day and Runcorn, 1955: 425)

Runcorn, in a short *Nature* note, drew up a table comparing his North American results with those from Britain and accented their similarities.

The motion of the pole as deduced from the measurements of British strata consists of a gradual movement from near Hawaii in late Pre-Cambrian times towards the Pacific coast of Asia in Palaeozoic times, reaching a high latitude in Triassic times and substantially the present pole position in Cretaceous times. The measurements on the Arizona strata give a similar picture.

(Runcorn, 1955b: 505)

But he did note discrepancies: North American poles were to the west of the roughly "contemporaneous" from Britain. Except for the Permian poles that were 40° apart, the discrepancies were about 20°, comparable to that found by Creer (1955, Ph.D.

thesis) in Graham's Rose Hill Silurian data. Following Creer's earlier interpretation, he set aside the Exeter traps, which might not have averaged out secular variation. Accenting the general similarities between results from Britain and North America, he interpreted them as a single path that had the form of "a random walk," indicating polar wandering and not drift. He noted that it is

impossible to tell at this stage whether the discrepancies between the pole positions inferred from "contemporaneous" strata in Great Britain and Arizona are due to systematic errors or to the strata not being truly contemporaneous.

*(Runcorn, 1955b: 505)*

He concluded:

Strong evidence now exists that considerable polar wandering has occurred throughout geological time. These results appear to dispose of the possibility that since pre-Cambrian times the continents of America have drifted any appreciable distance from the continents of Europe and Africa. Whether any such continental drift has occurred must await refinement of the accuracy of the measurements.

*(Runcorn, 1955b: 506)*

Runcorn sent his more extensive analysis to the GSA *Bulletin*. It was received on June 24, 1955, and published the following year (Runcorn, 1956a). He plotted his poles with circles of confidence, and constructed a meandering path (Figure 3.7). His conclusion remained unchanged.

Runcorn (1955) considered that the agreement between the European and American results is sufficiently good to conclude that no appreciable continental drift of the amount postulated by Wegener and du Toit – i.e., 60° of longitude – could have taken place, at least since Precambrian time. Only the Permian poles are displaced by this amount, and this is probably the result of inadequate sampling. Consequently if the coincidence of the mean magnetic axis with the geographical axis is accepted, these measurements must be regarded as giving considerable support to the reality of polar wandering.

*(Runcorn, 1956a: 316; the 1955 reference is to Runcorn's note in* Nature *referred to here as*
*Runcorn, 1955b)*

At the Cambridge colloquium and in all these reports, Runcorn clearly favored polar wandering only. He had ruled out the large longitudinal drift between Europe and North America invoked by Wegener and du Toit.

### 3.11 Paleomagnetism at Australian National University: Jaeger's key role

Irving sailed for Australia in November 1954, arriving just before Christmas. He was very pleased with what he found at this brand new institution. As he describes it:

Jaeger had meticulously laid out the groundwork. As he had instructed me in Cambridge, I had sent ahead detailed plans for a non-magnetic field laboratory (a larger version of the

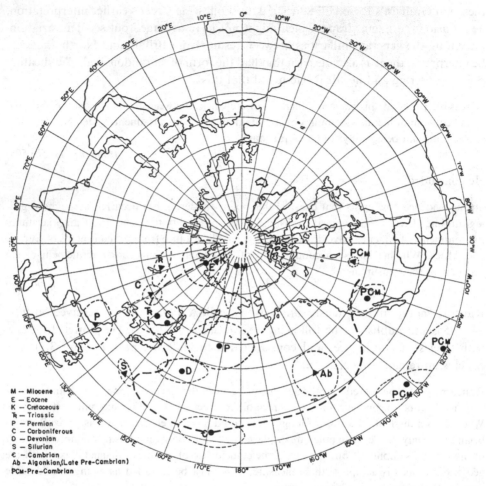

M — Miocene
E — Eocene
K — Cretaceous
Tr — Triassic
P — Permian
C — Carboniferous
D — Devonian
S — Silurian
ϵ — Cambrian
Ab - Algonkian,(Late Pre-Cambrian)
PCM-Pre-Cambrian

Figure 3.7 Runcorn's Figure 9 (1956a: 314). The dashed line is Runcorn's APW path for both North America and Europe. The North American results are represented by solid triangles; solid circles represent the European results. The North American paleopoles are as follows: PCM (Pre-Cambrian) is Creer's calculation of Graham's study of diabase dykes from Michigan; Ab (Algonkian) is Runcorn's Hakatai Shales; S (Silurian) is Creer's calculation of Graham's study of the Rose Hill Formation in Maryland; C (Carboniferous) is Runcorn's Naco Sandstone; P (Permian), the one near Taiwan is Runcorn's Supai Formation; Tr (Triassic) is Runcorn's Springdale Sandstones and Moenkopi Formation; and K (Cretaceous) is Runcorn's Dakota Sandstones. The European results are: PCM (near present-day Arizona) is Irving's Lower Torridonian; PCM (at 131° W) is Irving's Upper Torridonian; PCM (at 118° W) is Creer's Longmyndian; C (Cambrian) (horizontal short line through C) is Creer's Caerbwdy Sandstones; D (Devonian) is Creer's Old Red Sandstone; C is Belshé's from Derbyshire; P, the one in the North Pacific is Creer's Exeter Traps; Tr is Clegg's Keuper Marls; E (Eocene) is Hospers' from Northern Ireland; and M (Miocene) is Hospers' from Iceland.

Cambridge non-magnetic laboratory) and when I arrived architectural drawings were awaiting my approval: I just had to find a site. I had sent ahead plans for some instrumental components which had been made in the workshop. In Cambridge, Jaeger had authorized me to order through the ANU purchasing office many special materials not available in Australia; they were waiting for me. Small components came in my luggage. We built our laboratory and instruments from scratch.

*(Irving, February 2011 letter to author)*

The atmosphere seemed to be just right. Jaeger gave me the job, and then he came to Cambridge in the spring of 1954, and I met him first there. Now he had got interested in magnetism because of dolerite – because it was a way of studying the distribution of iron minerals and this was important for the cooling history and differentiation of sills. And he was the one who really began paleomagnetism in Australia, although his equipment was enormously primitive. In fact, he was so sheepish about it when I went there that he hardly dared show it to me. And it really was very simple, but that was just fine, and the measurements he took were perfectly right – the ones that were published in 1955.

The Geophysics Department staff was small. It consisted of Jaeger, the founding professor who had arrived in 1952, M. S. Paterson, who was an engineer and ultimately has done some enormously elegant work in high pressure studies of rocks, deformation of rocks ... and Dr. Joplin who was a classical petrologist. Jaeger, himself, was interested in heat flow.

*(Irving, 1981 interview with author; see also Paterson, 1982)*

The make-up of the department was unusual; it did not initially include researchers in well-established fields of geophysics such as seismology and gravity studies. Jaeger, as Irving later recalled, wisely recruited researchers in relatively new fields.

The choice that Jaeger made when building up the Geophysics Department was a rather interesting one. I asked him why he made those choices, which I've just explained, which was, after all rather strange because it was a geophysics department, and you would have thought, perhaps he would have wanted a gravity man, somebody doing seismology, and someone doing classical geomagnetism. And, I once asked why he didn't do this, and he said, "Well, the reason is very simple. If you want to be directing original research what you have to do are the things people aren't doing, and this is particularly important for us because we don't have an awful lot of money. We have some money, but we don't have a lot of money to really compete with the big overseas groups. So, if I got a high price person in seismology, he'd cost me a fantastic amount of money to set him up with instruments, and I'd be landed with him for life. Whereas you – you're just a temporary fellow – and if you don't find anything interesting, I can fire you after three years and if I do, you wouldn't have cost me very much. And, the chances are that your work, since it has not been looked at before, will turn up something interesting." Paterson's work on high pressure was breaking new ground and his own work in heat flow, of course; he was the first one with students to discover the difference between heat flow in orogenic zones and over old shield areas. These were all things in which new ground was comparatively inexpensive to break. This was his philosophy in planning a new department, and, of course, it made an enormous amount of sense, and it just seemed so eminently sensible for him to have done this in this way – unorthodox perhaps but enormously sensible and very wise judgment.

(Irving, 1981 interview with author; slightly altered, 2010)

### 3.12 Irving's initial work at Australian National University: his move to Stage IV

Once at ANU, thanks to Jaeger's ground work, Irving was immediately able to begin equipping the new laboratory. A magnetometer for strongly magnetized igneous rocks was quickly installed, and later with Ron Green, his first student, one for weakly magnetized sedimentary rocks.

In early 1955, Irving collected Mesozoic dolerites of Tasmania from across the eastern half of the island state. Jaeger and Joplin had already found steep inclinations indicative of high latitude for Tasmania during the Mesozoic, but could not determine the pole, their samples being from incompletely oriented bore cores. At the time, the age of the Tasmanian dolerites was only known within wide limits – post Late Triassic to pre Early Tertiary.[17] With his "igneous" magnetometer, Irving determined a north pole in the eastern North Atlantic. According to Creer's British path it ought to have been in eastern Siberia (Stage III); it could hardly have been more discordant, indicating "that there has been considerably relative movement between Tasmania and Europe," in fact over 5000 km since this time (Irving, 1956b: 164). With G. S. Watson, he also co-authored "Statistical methods in rock magnetism," a popular user's guide to Fisher's statistics (Watson and Irving, 1957).

Picking up where his thesis stopped, Irving began making global comparisons of latitudes calculated from paleomagnetism with geological evidence of paleoclimates, thereby moving to Stage IV.[18] This led to his *Geofisica* paper, which marked the next step in the development of the paleomagnetic test for mobilism (Irving, 1956a). Like his dissertation, it was rejected by the first set of reviewers.

The work on the paleoclimatological paper was begun almost as soon as I arrived in Australia. In mid-1955 I received Runcorn's paper on the polar wander path for North America in which he drew a common path through both the European and North American paths, and I replotted them and immediately drew their circles and saw that they were different, and, of course, this fitted with the data that I was then collecting from the Tasmanian dolerites which confirmed Jaeger's original work, on totally oriented samples from the Tasmanian dolerites. Jaeger's work was on incompletely oriented bore cores in which you have only the inclination not the declination. I wrote the paper, which is by far the most original paper that I've ever written. It grew out of my thesis work, as I've already explained. It was sent to the Geological Society of Australia, and they rejected it. One of the reviewers was a man called A. R. Alderman, Professor of Geology in Adelaide, another sort of anti-drifter. The other reviewer was W. R. Browne. But, Alderman just said that the paleoclimatic evidence couldn't be used in this way, and this was much too facile a way of doing it. Of course, it wasn't facile. It was based on an enormous amount of reading and study, and it was really rather a new look at the whole paleoclimatological thing. And, so what I immediately did was to go to Jaeger and he said, "Send it off to *Geofisica pura e applicata* because I'm an editor of that, and I'll send it and support it." Well, *Geophysica pura e applicata* accepted it and published it February of 1956. But, if it hadn't been previously rejected it may have been published in 1955.[19]

*(Irving, 1981 interview with author)*

He also corrected a mistake in Creer's formula for determining confidence limits for paleopoles. The correction had its greatest effect in high latitudes.[20]

Irving began by listing all paleomagnetic paleopoles and their errors. This compilation included his own, Creer's and Clegg's work in Britain, his results from India and Tasmania, Hospers' work from Northern Ireland, Runcorn's work from North America, and additional findings by Doell, Graham, Gough, Du Bois, Belshé, and Jaeger. Dismissing a few results as unreliable, either because of difficulties with the stability of their remanent magnetism or age uncertainties, he calculated the paleolatitudes derived from the remainder and compared them with paleoclimatic data. Although Irving cautioned that paleomagnetic and paleoclimatic results might not be of strictly comparable ages, sets of data for Britain yielded consilient results for every period except for the Pre-Cambrian. He (1956a: 37) found that the paleoclimatological results for North America were "in broad agreement with the magnetic considerations." He uncovered only one exception that subsequently became notorious (III, §1.2), but found (1956a: 37) a way to dismiss it. "There is one exception to this correlation, the Squantum Tillite, but this is of small extent and may be the product of a mountain glacier."[21] The paleomagnetic data from Tasmania showed that Australia was in very high latitude in the Mesozoic and paleoclimatic studies suggested a relatively cool climate there at that time. He claimed:

It may be concluded that the information at present available on the points discussed in this section, provides evidence in favour of the view that the magnetic and rotational axes of the Earth have coincided approximately since the Palaeozoic, as they are known to have done during the last 20 million years.

*(Irving, 1956a: 37)*

With this ingenious defense of the GAD model, Irving was able to extend, back through the Phanerozoic, Hospers' purely paleomagnetic support for GAD for the Late Cenozoic. He also noted Runcorn's theoretical support. He now turned to the questions of continental drift and polar wandering. Reinterpreting Runcorn's results, he compared the poles from Britain and North America and found they did not agree with one another or with his poles for India and Tasmania. He constructed the diagram shown in Figure 3.8.

Unlike Runcorn, Irving drew separate APW paths for Britain and North America, arguing that they were formerly closer together.

The positions of the magnetic pole calculated on this GAD dipole assumption, from the magnetisation of rocks of Pre-Miocene age, differ greatly from the present geographic pole, *polar wandering having occurred*. The poles given by data from four continents do not

Figure 3.8 Irving's Figure 2 (1956a: 32). The ancient pole positions given by paleomagnetic observations from North America, India, and Australia. Pole positions 1 through 7 are from North American rocks and include findings by Runcorn, Graham, and Doell. Irving drew an APW path from North America through the points. The thin-lined APW path is essentially Creer's APW path with the addition of Belshé's Carboniferous paleopole. The Tasmanian Mesozoic paleopole is 8, and that from the latest Cretaceous/earliest Tertiary of India is 9. Irving displayed the paleopoles for Britain's APW path in a previous diagram (shown here in fainter line), using ten paleopoles: four from Creer, two of his own, and one each from Clegg, Belshé, Hospers, and Roche. Roche's pole was from France (see Roche, 1951). Belshé's results from Britain were reported in Runcorn (1955a).

however agree one with another. The discordance between the poles calculated from the magnetic data from Palaeozoic rocks of North America and Europe is particularly notable, since it is shown by several determinations made by independent observers. Agreement could be achieved by supposing that North America was closer to Europe in the Palaeozoic and early Mesozoic.

*(Irving, 1956a: 39; emphasis added)*

I have italicized "*polar wandering having occurred*" in the above quotation because it is ambiguous. If by "*polar wandering*" Irving meant there had been an apparent movement of the geomagnetic axis relative to the landmass (APW), then that was obviously so and hardly needed comment. If by "*polar wandering*" he meant polar wandering in the strict sense, then he was wrong. The differences between poles from the two hemispheres are far too large for them to have been the effect of polar wandering only: the Southern Hemisphere poles are nowhere near the two Northern Hemisphere polar paths. However, if polar wandering was predominant, then poles of similar age ought to have lain in close proximity, not thousands of kilometers apart. Clearly the evidence of Figure 3.8 is that the effect is that of large relative motions among continents. There is a further argument to be made here, but first I need to explain Irving's final step in his paleomagnetic/paleoclimatic test of mobilism.

Irving developed another type of comparison, an indirect proof. Assuming the GAD hypothesis, and *no relative displacement of the continents*, Irving calculated the paleolatitudes of other continents from the APW path for Europe. He then compared them with the paleoclimatic data for these other continents. If only polar wander had occurred, they would be consilient, and they were not. He therefore argued that the assumption of no drift was incorrect.

This comparison would not have yielded greatly different results if any other set of randomly distributed points had been selected. For North China and South Eastern Australia the contrast between the palaeomagnetic latitude given from the N.W. European observations, and the local palaeoclimatology is complete; for Southern Brazil and Eastern U.S.A. some of the evidence is compatible and some conflicts: for South Africa there is no substantial dishar-mony. In the last case, it seems possible to explain the palaeomagnetic evidence from N.W. Europe by the simple hypothesis of polar wandering ... However, considering the evidence from the other four regions this hypothesis alone is inadequate, and some relative movement between N.W. Europe and these regions seems also to be indicated.

(*Irving, 1956a: 39*)

Of course, he could have rejected GAD. However, the consilience of the paleocli-matic and paleomagnetic data from the same continents strongly supported GAD; data would hardly be correct in one instance and incorrect in the other. He concluded:

Using the dipole assumption again, a comparison of the latitudes predicted for different parts of the world from the pole position deduced from the European data, with the palaeoclimates in these parts, also yields the result that relative movement of the continents in the past has occurred.

(*Irving, 1956a: 39–40*)

Irving's *Geofisica* paper is strangely titled, "Paleomagnetic and paleoclimatological aspects of polar wandering." He gives abundant evidence of continental drift but nothing substantive about polar wander in the strict sense. Why then is "polar wandering" in the title and "continental drift" is not? In neither this nor his paper on Tasmania dolerites did he use the term "continental drift"; he studiously avoided it. I can only conclude that he considered the term "continental drift" to be poten-tially toxic and best avoided, and so he simply claimed, as quoted above, "that relative movement of the continents in the past has occurred." Did he also believe that both polar wandering and continental drift had occurred or was he speaking only of apparent polar wandering and that continental drift alone explained the differences in the northern hemisphere APW paths and poles of Tasmania and India?

I think it likely that he meant strict polar wandering, and that he was claiming that polar wandering and continental drift had both occurred. If so, he was wrong. He should have concluded that continental drift had occurred and may or may not have been accompanied by strict polar wandering.

His use of paleoclimatology as an independent means for testing GAD or specific paleomagnetic results was the first among many applications of paleoclimatology to paleomagnetism, and it initiated Stage IV of the application of paleomagnetism to the mobilism controversy. Somewhat later, Runcorn and Neil Opdyke used ancient wind directions to test paleomagnetic predictions (§5.12–§5.14). Runcorn later encouraged Alan Nairn to work in paleoclimatology (III, §1.3), and at the beginning of the 1960s Blackett turned to paleoclimatology (III, §1.7). Meanwhile, paleoclimatologists themselves began to pay attention; they used the paleomagnetic data as an aid to their own work, which some offered as support for mobilism (III, §1.3; III, §1.9).

### 3.13 Imperial College moves to Stage III

In October 1954, the Imperial College group moved to Stage III by launching a general investigation of India rocks. Blackett was a friend of Nehru, who was then Prime Minister of India, and his name would have opened many doors in assisting with this project, which later involved collaboration with Indian scientists (Nye, 2004: 162). Clegg and his co-authors explained their motivation in their first paper on Indian results:

In October 1954 the Physics Department of Imperial College embarked on a three-year project of paleomagnetic research in India. The scheme arose as a result of a previous survey made in Britain when it was found that certain sediments laid down over a wide area during the late Palaeozoic and early-Mesozoic eras are consistently magnetized along a north-east south-west axis, and had a considerably lower inclination than the present earth's field ... Further evidence of past movements of the British landmass has also been obtained from palaeomagnetic measurements made by Creer, Irving and Runcorn (1954), and the data now to hand are sufficiently conclusive to encourage the belief that by carrying out paleomagnetic surveys on a global basis it may be possible to determine the positions occupied by the various land areas of the earth relative to the poles during different geological epochs.

*(Clegg et al., 1956: 419–420)*

Blackett and Clegg knew of and acknowledged Irving's Indian work (§3.4), although in the first place they may have decided to go there quite independently; India was an obvious choice because, on drift theory, it had such a large northward motion. Regardless, Irving's success in obtaining good results from the Deccan Traps helped them decide where to collect.

Its [India's] advantages as a field for palaeomagnetic work were first pointed out by Irving (1954), who obtained a few samples of the Deccan Trap lavas from different localities, and found them to have a consistent axis of magnetization, directed roughly south in azimuth, but dipping towards the south. Despite the small number of specimens examined, the consistency of the results suggested that a more extensive magnetic survey of these rocks would be of interest, and we decided, in planning the present expedition, to concentrate initially on the Deccan Trap area.

*(Clegg et al., 1956: 420; Irving (1954) refers to his dissertation; my bracketed addition)*

E. R. (Ernie) Deutsch, a new member of Imperial College's paleomagnetics group, D. H. (Don) Griffiths, from the Department of Geology at the University of Birmingham, and Clegg made their first collection in December 1954 and January 1955. Deutsch, who had just received a Ph.D. from Imperial College where he studied magnetic properties of rocks at high temperatures, was hired as a research assistant in Blackett's physics department. Griffiths had already worked on Quaternary paleo-magnetism (§3.3). The Deccan Traps are extensive basaltic lava flows covering about 200 000 square miles that were extruded over a few million years at or along the Mesozoic/Cenozoic boundary. They took over 400 samples in the Linga and Khandala areas separated by 500 miles. The samples, weighing over a ton, were shipped to Imperial College where they were cut and measured.

Because the paleomagnetic results derived from both areas differed only slightly, they averaged them, obtaining a declination 155° E and inclination +53°. They were all reversed. After noting that there were two possible interpretations of India's movement consistent with the paleomagnetic data, they opted for the following:

The results are best explained on the postulate that India lay about 34° south of the equator during the late Secondary and early Tertiary eras, and that at some subsequent time it moved into the northern hemisphere and rotated about 25° [anti-]clockwise relative to the earth's geographical axis. This shift from 34° south to the present 20° north, that is, of 54°, corresponds to a linear movement of 5000 km.

*(Clegg et al., 1956: 428; my bracketed addition: in all other passages they speak of an anticlockwise rotation, which is the correct interpretation, e.g., see passage quoted below from p. 419.)*

Although they made no comparison with Irving's result, the agreement between them is truly remarkable. Irving obtained a mean declination of 149° E with inclination of +56° from his seven localities, both normal and reversed, and suggested as one possible interpretation that India had moved through 53° of latitude from the Southern Hemisphere and rotated 28° counterclockwise (§3.4).

Comparing their Indian results with those from Britain, they argued that "India must have moved relative to" Britain and North America.[22]

It is interesting to consider our results for the Deccan Trap in relation to these hypotheses [of polar wandering by Graham, Gold and Runcorn]. The position of the north pole in Eocene-Cretaceous times (70 million years ago) as deduced from our measurement is 28° N. 85° W. This can be compared with the estimate by Runcorn of the pole position in Eocene-Cretaceous times, as deduced from measurements of British, North American and Icelandic rocks, which is 76° N. 130° E. It is evident that this large difference between the pole position calculated from our results and that calculated by Runcorn indicates that India must have moved relative to Northern Europe and North America.

*(Clegg et al., 1956: 429; my bracketed addition)*

This is simply and beautifully explained, and is the first explicit and unequivocal endorsement of continental drift by Clegg and co-workers at Imperial College. Although they did not question Runcorn's (and Graham's) rejection of drift between North America and Europe, they accepted that their results supported displacement of India relative to them both regardless of whether the two had drifted relative to one another. They even suggested drift between India and Africa based on a comparison of their data with Gough's work on the Pilansburg Dyke system. However, because Gough's rocks were much older (then thought to be between 300 to 400 million years old and now known to be Precambrian) they noted (1956: 430): "the wide difference in ages of the rocks makes this comparison of only suggestive importance."

Clegg and company clearly preferred continental drift over any other explanation of the discordance between poles. Their language is strong. Were they, however, willing to fully accept mobilism? It is difficult to tell exactly, for elsewhere in their conclusions they tempered their support for drift.

The most plausible interpretation of this result [from India] is (a) that India has drifted north from a position about 34° south of the equator when the rocks were formed some 70 million years ago and has rotated anti-clockwise through 25°, and (b) that either the earth's field was reversed when the rocks were formed or that the rocks became magnetized in the opposite direction to the field by some physico-chemical mechanism ... The pole position corresponding to these results does not agree with that found by other workers for British and American rocks of the same period. This suggests that a movement of India, relative to North America and Europe, has taken place at some time during the past 70 million years.

*(Clegg et al., 1956: 419; my bracketed addition)*

They drew back from their earlier firm language, apparently not yet ready to be fully committed to drift but were certainly strongly inclined toward it.

During this period, the Imperial College group was also involved elsewhere in mobilism-related projects. Almond, Clegg, and Jaeger (1956: 733) analyzed several of Jaeger's unoriented Tasmanian dolomite bore cores, and found values that matched those determined earlier by Jaeger and Joplin (1955), which led them "to therefore suggest that a relative movement has occurred between the land mass and geographical poles at some time since the Jurassic period."

Clegg and Stubbs made and studied several collections from France and Spain to test the rotation of the Iberian Peninsula, an idea suggested by Argand in the early 1920s (I, §8.7), seconded by du Toit in the thirties (I, §6.7) and more recently much amplified by Carey (§6.7). Blackett, who had met Carey at the University of Tasmania in 1953, decided they should focus on Carey's claim that the Pyrenees and

Bay of Biscay had been formed by an approximate 35° counterclockwise rotation of Iberia relative to France. They tentatively concluded (Clegg *et al.*, 1957: 227): "the postulated rotation of Spain has in fact occurred."[23]

## 3.14 Blackett expresses strong preference for mobilism

Blackett was invited to give the second Weizmann Memorial Lectures at the Weizmann Institute of Science in Rehovoth, Israel,[24] and used them as the basis for a book about rock magnetism published in 1956.

When invited to give the second series of Weizmann Memorial Lectures at Rehovoth in December 1954 I chose the subject of Rock Magnetism because of my great personal interest in it and because it is a relatively new and little-known field of active research which has already yielded exciting and unexpected results and will, I am convinced, continue to do so for many years. It is only in recent years that it has become clear that the detailed study of the natural magnetization of rocks is likely to allow us to trace back to the beginnings of geological time both the history of the earth's magnetic field and motion of the continental land masses relative to each other and to the geographical pole – that is, to settle the long-debated and highly controversial problems of continental drift and polar wandering.

*(Blackett, 1956: 3)*

Although he gave the lectures in 1954, his opinions about continental drift and polar wandering as described in his book reflect his views during the first half of 1956; his prefatory remarks are dated May 1956, and he finished the chapter entitled "Recent results on land movements" in spring 1956.

In the third chapter I have collected together and commented upon the main results available in the spring of 1956, when the book was at last completed, relevant to the movement of land masses and the history of the earth's field.

*(Blackett, 1956: 4)*

He began by discussing results from Great Britain by Clegg, Almond and Stubbs, Creer, Hospers and Charlesworth, and Irving, and concluded:

... the British Isles have drifted rather irregularly northwards from a latitude of 20° south, 450 million years ago to their present latitude of 53° north, that is by 75° of latitude or some 4500 nautical miles. This conclusion, if true, is of such outstanding importance, that the evidence needs careful scrutiny and further checking. Of special importance is the extension of the measurement to neighbouring parts of Europe.

*(Blackett, 1956: 71)*

Turning to the recent work of his group at Imperial College, he noted (1956: 72) that, because the results from France were widely scattered, "it is not possible yet to make any useful deduction as to the magnetic latitude of France in Triassic times." He cited similar difficulties with the results from Spain, which all predated magnetic cleaning techniques (§5.5).

Similar arguments apply to the inclination of the Spanish rocks. Unfortunately, the experimental results for the Spanish Trias are also quite inconclusive as yet, since different sites give markedly different results.

*(Blackett, 1956: 74)*

He then turned to North America. After describing Runcorn and Graham's results, he advised caution about accepting differences between them and British results as real and as strong support for continental drift, especially in view of possible inclination error in sedimentary rocks (§3.3) from which most were obtained.

This discrepancy between the pole positions as calculated from the British and American data might suggest that an appreciable separation of the two land masses has occurred during geological times. One must, however, exercise some caution here as many of the rocks studied were sandstones and, as emphasized already, the measured inclinations of the magnetization may be appreciably different from, and probably less than, the true inclination of the ambient field, thus leading to the calculated latitude being probably too low. It can easily be seen that the British and American pole positions can be brought more closely into agreement by supposing such an inclination error. However, it is unlikely that all the discrepancy can be caused, and so one concludes that, so far as the present results go, some widening of the gap between Britain and, in the last 100 million years, America is possibly indicated. One will await with interest further measurements and analysis of the American and British rocks.

*(Blackett, 1956: 76–77)*

Moving to the Southern Hemisphere, he paused first in South Africa. Apparently, unaware of Nairn's new results (§5.4), he discussed only those of Gough. Blackett believed that they, when compared with findings from the Northern Hemisphere, suggested a common movement of Africa and Europe.

The most obvious explanation of this surprising result [obtained by Gough] is not only that the earth's field was reversed when these rocks were formed but that the south magnetic pole was then near Mombasa in East Africa. This would indicate that South Africa has drifted, since the rocks were formed, right across the south geographical pole. This would mean a drift of some 4000 miles (southward initially and later after the pole was crossed northwards) in the last 400 million years. A less likely possibility is that it has rotated by nearly 180°[25] ... It will be noticed that this supposed drift is quite similar to that deduced from the British results, so that these measurements are consistent with the view that Africa and Europe have drifted several thousand miles more or less as one unit.

*(Blackett, 1956: 77–78; my bracketed addition)*

Turning to India, he remarked on its important role in the mobilist explanation of the distribution of Permo-Carboniferous glaciation. After describing the strong agreement between Irving's initial work on the Deccan Traps (§3.4) and the more extensive work by Clegg, Deutsch, and Griffiths (§3.13) indicating about a 5000 km northward movement since the beginning of the Tertiary, he noted the further agreement between the paleomagnetic and paleoclimatic data, although he was decidedly stretching a point as there were, at the time, no Late Paleozoic paleomagnetic results from India.

As far, then, as these preliminary results go, the rock magnetism measurements support strongly the deduction of many geologists from the study of the climates of the past that India must have drifted a great distance northwards. The glaciation suggests a drift of 5000 miles [8000 km] in 200 million years [now known to be ~250 million years], whereas the rock measurements indicate almost half this distance in about half the time. So it is reasonable to suppose that the drift has been at a fairly constant rate. This, of course, will be tested by measurements on Indian rocks of other ages.[26]

*(Blackett, 1956: 80; my bracketed addition)*

Blackett turned to continental drift, as indicated by relative rotations in opposite senses of Africa and India, as did the disagreement between contemporaneous poles derived from India and Britain.

It is interesting to note that India has apparently rotated 20° anti-clockwise in the last 80 million years, whereas Africa appears to have rotated 17° clockwise in the last 300 million years. The differences in age of the rocks prevent much significance being attached to the direct comparison of these figures although, as far as they go, they do indicate a relative rotation of India and Africa, as indeed has often been suggested from geological data ... The pole position corresponding to these Indian results is 28° N and 85° W, whereas that corresponding to the contemporary British rocks is about 60° N and 125° E. Clearly a large relative movement of India and Britain must have occurred.

*(Blackett, 1956: 80)*

He also found support for continental drift in the results from Tasmania. After discussing Almond, Clegg, and Jaeger's work (1956), which indicated that Tasmania had been at higher latitude during the Jurassic, and Irving's more general study of the same dolerite intrusions, he argued that Tasmania and India had probably drifted relative to one another.

Irving ... has deduced a [north] pole position 50° N and 156° W. Since this is widely different from the pole position deduced above from the Indian Cretaceous-Eocene rocks, it is probable that a considerable relative motion of Tasmania relative to India has taken place, though the difference in the ages of the two sets of rocks makes the conclusion not quite certain.

*(Blackett, 1956: 81; my bracketed addition)*

Not unexpectedly, Blackett, echoing Clegg, Deutsch, and Griffiths (1956), believed that the strongest paleomagnetic support for mobilism came from the large disagreement between the contemporaneous poles from India and Britain. In fact, Blackett and his colleagues were in full agreement about the support for mobilism afforded by the intercontinental comparisons of paleomagnetic results; there were shades of difference however. In their Deccan paper of May 1956, Clegg and company offered only a wavering acceptance of mobilism; in the same month Blackett wrote the preface to his *Lectures on Rock Magnetism* with a stronger endorsement.

There can be no doubt that the rock magnetic measurements already made very strongly support the view that there have been during geological times very large movements of land masses relative both to each other and relative to the geographical poles. Thus polar wandering and continental drift have both occurred.

*(Blackett, 1956: 81)*

If Blackett really did think that the paleomagnetic support was sufficiently strong for him to accept continental drift, he was willing to do so even though he acknowledged that the paleomagnetic case was not without difficulties. Like Clegg and company, he was cautious about the strength of the paleomagnetic support for mobilism. Here are two passages from the first chapter of his *Lectures*, more cautiously worded than the above:

The results to date are clearly only of a very tentative character and a great amount of further work is needed before any certain conclusions can be drawn. One must aim at being able to plot curves showing the azimuth and latitude of all the major land masses throughout geological history.

*(Blackett, 1956: 33)*

Holmes, in the book [*Principles of Physical Geology*] already referred to, and Gutenberg, in *The Internal Constitution of the Earth* (1951), give very balanced surveys of the evidence for and against the assumption of Continental Drift and of Polar Wandering. In the former hypothesis, one assumes large *relative* movements of the different land masses, and in the latter, one assumes a shift of the crust as a rigid whole relative to the rotational axis of the earth. I feel fairly sure that rock magnetism studies, if energetically pursued, can settle many of these highly controversial questions within the next few years.
    *(Blackett, 1956: 36; my bracketed addition; Gutenberg (1951) is my Gutenberg (1951a))*

The second passage is particularly interesting because it implies that he felt fairly sure that energetic pursuit of paleomagnetism alone would end the controversy over mobilism: that paleomagnetists were on the road to a difficulty-free solution. He himself was excited by the prospect that such work could, within a decade, settle the question of continental drift. He sought to encourage others to work in paleomagnetism by expressing his personal enthusiasm in this way.

Thus, as it has taken shape, the content of this book has become something of a mixed bag. I hope, however, that it will be a stimulus to workers in many lands to enter this exciting new field lying both in Physics and Geology. It is clear that the full exploration of the methods for studying the movements of continents will need much fieldwork all over the world. Major countries will have to study the magnetism of their own rocks just as they do their own geology. I have no doubt at all that the results of this work will, in the next decade, effectively settle the main facts of land movements, and in so doing will have a profound effect on geophysical studies of the earth's crust.

*(Blackett, 1956: 4)*

This passage, from his preface (dated May 1956) makes plain that he did not think the issue of continental drift had yet been settled.

But, there still is his remark in the first chapter about feeling "fairly sure that rock magnetism studies, if energetically pursued, can settle" the questions of continental drift and polar wandering "within the next few years" that would seem to be too sanguine.[27] The first chapter likely was written well before the preface, and Blackett may not have subsequently edited it. Probably it was not his purpose to provide the reader with a precise statement of his epistemic attitude toward mobilism. He thought drift was strongly supported by paleomagnetic work, which made sense in light of what he had learned from reading Holmes. He believed that at least India had drifted relative to Great Britain. It was not really important that his readers should know whether he accepted mobilism or was strongly inclined toward it. What was important was that they should understand the need to build more magnetometers and keep them loaded with specimens, preferably collected from faraway places.

Blackett also spoke about the paleomagnetic support for mobilism at a conference billed as the first international conference on rock magnetism and held in November 1956 at Imperial College (§1.5; §5.16). He chaired the conference and in his welcoming address he tempered his remarks, making it once again hard to determine if he had already accepted mobilism or was only strongly inclined toward it.

If it is indeed true, as many of us here including myself believe, that the study of magnetism of rocks from different parts of the globe has already made it appear probable that both Polar Wandering and Continental Drift have occurred, and that further work in the next few years will either confirm these preliminary findings in detail or refute them *in toto*, then it behooves all workers in the field to exercise an especial sense of responsibility as regards claims for the reliability of deductions from our experimental results.

*(Blackett, 1957: 147)*

### 3.15 Differing reactions of the British and Carnegie groups to the paleomagnetic results

I want to pause for a moment and, at the risk of some repetition, identify and stress the vast differences between the reaction of paleomagnetists on either side of the North Atlantic to the new paleomagnetic data supportive of mobilism and geomagnetic field reversals. Paleomagnetists working in Britain had much greater confidence than their counterparts in the United States in the methods that both had helped develop. If results had been obtained in accordance with these new procedures and techniques, the British groups were much more willing than their US counterparts to accept them as reliable indicators of the ancient geomagnetic field, and they did so *regardless of whether they indicated geomagnetic field reversals, continental drift, or polar wandering*. This is not to say that the British groups did not care about reliability; they certainly did, and they spent a lot of time caring about it. But, having developed and followed the new procedures, they did not then turn around and question or dismiss their results as unreliable or uninterpretable because of concerns as to whether reversals, polar wandering, or continental drift had or had not occurred. Nor is this to say the British workers spoke with one voice; they did not. Runcorn, for instance, was not at all sympathetic to continental drift and remained so for another two years, and Blackett, although he became sympathetic to drift as soon as he began to read about it, was reluctant for even longer to accept reversal of the geomagnetic field. Other British workers quickly accepted continental drift, and reversals. In contrast, Carnegie workers, believing that drift and reversals faced prohibitive theoretical difficulties, were much more likely to question either the general adequacy or specific use made of the very techniques they had helped develop, particularly if results supported reversals or drift. There is no better way to bring out this difference than by comparing Ken Creer and John Graham's reaction to the latter's paleomagnetic results from the Rose Hill Formation in the Appalachians of Maryland. Because it is so indicative of the wide difference between trans-Atlantic attitudes, I shall review again what happened.

When Creer moved to Stage III in the development of the paleomagnetic test for mobilism, he compared Graham's result from the Rose Hill Formation with the APW path he had constructed for Britain and noted their difference. Creer and others at Cambridge decided that Graham's positive fold test from the Appalachian Silurian Rose Hill Formation, whose directions of magnetization were strongly oblique to the current field and which Graham thought had reversed polarity, were reliable ancient field indicators, whereas Carnegie results from flat-lying rocks outside the Appalachians with no positive fold test were not: they had been remagnetized. Cambridge workers simply followed the logic of Graham's own field tests.

When Graham (1949: 151) first described his fold test, he gave special prominence to the Rose Hill Formation. He thought incorrectly (§1.8) that these rocks were reversely magnetized, and this puzzled him; nonetheless, he began by arguing that they had acquired their magnetization before folding (§1.8). However, within two years he had changed his mind (§1.9), claiming that the more recently discovered magnetizations from flat-lying platform rocks were stable, whereas the folded strata from Pinto in the Appalachians were not, and attributed their instability to unspecified physical and chemical processes associated with their deformation. With this in mind and finding the idea of field reversals unattractive, Graham abandoned his fold test. Somewhat later Néel made the idea of self-reversal theoretically respectable, and Graham reclaimed his fold test, arguing that the Rose Hill Formation had undergone self-reversal before they were folded. This relieved him of the need to consider reversals of the main geomagnetic field (§1.10). Tuve, Director of the Department of Terrestrial Magnetism at the Carnegie Institution of Washington, and Torreson, Graham's co-worker, also vacillated about the fold test. Tuve supported it in 1949, but abandoned it a year later saying that Carnegie workers need no longer worry about a reversing field or drifting continents; his dissatisfaction with mobilism was explicit. But Graham himself also must have found continental drift and polar wandering unpalatable.

Continuing into the mid-1950s, Graham sampled on the Colorado Plateau in 1955, about six months after Runcorn and Doell had sampled there, and he found results in agreement with Runcorn's results. Like Runcorn, he (1955) now argued for polar wandering but rejected continental drift (§7.4). A year later, Graham raised the possibility that stress effects could alter magnetizations, causing them to give an unreliable record of the ancient geomagnetic field. He was always shifting his ground on this vital question of data reliability. He needed no help from Tuve to disavow continental drift.

Paleomagnetists at the Carnegie Institution emphasized what I have called theoretical difficulties. Runcorn apart, the British groups took an empirical approach, and even Runcorn bowed to empiricism in 1956 (§4.3). British groups had more confidence in their procedures than did American workers, more confidence than Graham did in his own fold test. Clegg, Creer, and Irving were aware of theoretical difficulties faced by field reversals and continental drift, but they were not going to avoid these ideas by categorizing as unreliable those results that their experiments told them were reliable. Nor was Hospers willing to dismiss reversals of the geomagnetic field because some viewed them as theoretically impossible (§2.9). The British groups refused to renounce the techniques and results, even though Tuve (1950: 63) claimed, "the most straightforward hypotheses for the observed anomalous magnetization – a reversal of Earth's magnetic field or a drifting of the continents – are not satisfactory"; the British were more empirically minded than Graham and Tuve. Whether this new empirical approach would outstrip old theory was, in the mid-1950s, very uncertain. But the immediate future at least, lay with the empiricists in Britain.

There were other significant differences between the British and Carnegie groups. The former accomplished much more: all round they were more progressive, much more active; they provided strong empirical support for the GAD hypothesis, they developed and rigorously applied statistical techniques for analyzing paleomagnetic data; they recognized that fine-grained red sandstones and siltstones of many ages yielded reliable results giving focus to their sampling; they energetically carried out extensive and intensive surveys; and they set up, not without substantial internal dispute, the logical basis for testing mobilism.

Young paleomagnetists in Britain had support in high places. In the later 1940s and 1950s, P. M. S. Blackett was a towering figure in British science and public affairs (Nye, 2004), giving upbeat lectures in Britain and around the world that were very influential. Blackett pioneered the instrumentation, and in swift and timely fashion, R. A. Fisher provided the statistical methods. No comparable major figures in North America championed paleomagnetism and the means it supplied to spearhead an entirely new approach to the long-standing problem of continental drift. Tuve offered at best minimal support, and then as the problems became more complex, and required a broadened effort, interest waned and his support dwindled. Unlike Tuve, Runcorn and Blackett did not pressure their students and co-workers to agree with them, but were helpful and encouraging, whatever heterodoxical thoughts they might have. Graham felt that he would be thrown out of the Department of Terrestrial Magnetism for proposing self-reversals. Runcorn, although not initially supportive of field reversals, did not put pressure on Hospers to reject them, and he himself soon came to accept them. Even though not sympathetic to drift at the time, Runcorn found Irving's idea about using paleomagnetism to test continental drift worth telling Browne, and when Browne showed little enthusiasm, Runcorn did not tell Irving to forget it, but instead set the ball rolling by getting Irving to talk to Fisher. Fisher, a scientist of immense prestige, also encouraged young paleomagnetists, entertaining them with witty remarks, poking fun at fixism, and expressing sympathy with drift. Runcorn, favoring polar wandering, had no problem with Irving's oblique directions from the Torridonian, and Creer's APW path for Britain; in fact he adopted it as if it were his own.

In the first half of the 1950s, however, Runcorn resisted continental drift, later admitting that Gold and Vening Meinesz might have influenced him to do so. However, he did not pressure Irving or Creer to reject drift; he did not tell them they were being foolish to pursue it. In fact, I shall show in the next chapter that Creer and others in Runcorn's circle argued with Runcorn over drift, telling him that he was wrong to reject it in his initial analyses of his own North American results, telling him that his continued resistance was stubbornness. Of course, Hospers, Irving, and Creer knew about the arguments against reversals of the field and continental drift, and argued over tea with others in the Department of Geodesy and Geophysics at Cambridge. They had to stand up for their views, but they had each other, and Runcorn's, Fisher's, and Blackett's support, not necessarily for their own particular ideas or ways of doings things, but in pushing on with their projects, building magnetometers and feeding samples to them. If during this period Graham showed a brief lack of courage when he gave up his fold test, it is perhaps understandable because he was so isolated. Creer and Irving would find resistance to their work much greater once they left Cambridge, and had to defend their mobilist arguments from attacks by old and new fixists. Meanwhile, Graham became a steadfast defender of fixism, and attacked both of them.

## Notes

1 Creer and Irving (forthcoming, 2012) have themselves now written their own account of their early years working under Runcorn's supervision in the Department of Geodesy and Geophysics at Cambridge. They describe Irving's painstaking work on the Torridonian, the construction of the Cambridge magnetometer, Creer's fieldwork, and his construction of the first (apparent) polar wander path.

2 Introducing the term "apparent polar wander" at this stage is anachronistic; it did not come into vogue until much later, but for clarity it is necessary to begin using it at this stage. In his textbook, Irving introduced the expression "apparent polar wander path" to refer to paleomagnetically determined polar wander paths and distinguish them from polar wandering (Irving, 1964: 130). As Irving recalled (March 4, 1999 comment to author), "Apparent polar wander path was a term introduced by me in my 1964 book. Before that we called them simply 'pole paths'." To claim that there is an APW path does not commit one to polar wander. It commits one only to polar wander, continental drift, or their combination. Indeed, this is why Irving called them *apparent* polar wander paths. Deutsch (1963a, b) was the first paleomagnetist to explicitly propose that continental drift without polar wandering could explain differences in APW paths (III, §1.8). The usual approach at the time was to maintain that polar wandering explained the common elements of polar wander paths (APW paths), and then to argue that differences in APW paths required drifting continents.

3 On 15th December 2011, E. Irving kindly provided an update of much later research that has led to changes in Creer's original APW path. There have been three major revisions. Irving described them as follows:

(1) Paleomagnetic poles results refer only to the tectonic plate from which they were derived. Hence, connecting in sequence the Precambrian poles from the Torridonian of northwest Scotland (see endnote 12, Ch 2) and Paleozoic poles from south Wales and central England as is done in Figs. 3.3, 3.4 and 3.5, can now be seen to be incorrect, because, as we now know, they belonged to different plates. The Iapetus Ocean lay between them in the late Precambrian and Early Paleozoic. Iapetus closed in the Silurian causing the Caledonian Orogeny. It was at this time that the Torridonian (originally part of the Laurentian Shield of North America) was welded to Britain. It remains attached to Britain because when the North Atlantic opened in the early Tertiary, it did so along a line to the west of the Iapetus suture.

(2) The Longmyndian, which was formerly considered Precambrian, has now been shown to be Lower Cambrian, several hundred million years younger than the Torridon Group and so the two cannot now be compared. Later detailed sampling and extensive demagnetization by Smith and Piper (1984) yielded magnetizations steeply inclined down to the ESE considered by the authors to be an overprint acquired during the Caledonian orogeny.

(3) Compared to the Early Paleozoic and Devonian poles compiled over 40 years later by McElhinny & McFadden (2000: 258) Creer's poles are situated further north, displaced towards his Permian pole (Creer and Irving, 2012) as if affected by later geomagnetic fields. Based on what were then new results from the Russian Platform which appeared to reflect better the original field than British poles, Creer (1968) argued that rocks of Early Paleozoic and Devonian age from Britain have been partly or wholly remagnetized during the Permian. This has been confirmed by thermal demagnetization studies of Chamalaun and Creer (1964), Tarling (1985) and other workers.

By contrast, Creer's Permian and later path is in good agreement with that of McElhinny and McFadden (Creer and Irving, 2012) for that interval and it has not, to any substantial extent, been overprinted. Creer's path indicates a near equatorial position for Britain in the Permian, followed by a generally northerly drift since then, similar to that proposed by Koppen and Wegener (1924) on the basis of paleoclimatic evidence showing good agreement between latitude results obtained by independent methods. Thus, although the Devonian and older pole path from Europe as it is known today is very different from Creer's, his path from Upper Paleozoic onwards, reinforced by additions to it in following years, proved fully adequate as a basis for testing Wegenerian drift. This was because the results on which it was based showed no evidence of being strongly overprinted, and because the motions especially between the northern continents, India and Australia in the Mesozoic and Cenozoic were so large. These major revisions of Creer's original path did not affect the global paleomagnetic mobilism test.

4 Irving's notebooks contain an undated entry about his preparation of the Indian samples. But the entry is sandwiched between one about his third field trip to the Torridonian, dated by an expense account and routes driven from November 5 to 25, 1952, and a later entry of November 1952 about his preparation of Torridonian specimens (Irving, *Synopsis* (1985), p. 3).

5 Irving's entry concerning the compilation of his measurements and use of Fisher's statistics on the Indian samples appears in his "Torridonian 2" research notebook, pp. 133–137. See, Irving, *Synopsis* (1985), p. 7. Although the entry is not dated, it follows an entry of June 1953. He believes (*Synopsis*, p. 7) that the "compilation was made in 1953 with the actual measurements being taken some time before."

6 These Deccan Traps are now known from radiometric studies to have been extruded over a few million years at or along the Mesozoic/Cenozoic boundary at ~65 Ma.

7 This idea of subjecting their samples to direct and alternating magnetic fields and increasing temperatures should be distinguished from later techniques developed to remove or clean unwanted secondary components of remanent magnetization. Here the idea was to see if magnetization imparted in the laboratory was magnetically stable in the expectation that this would provide information about the stability of the NRM. Creer, who also used these stability tests, which he got from Clegg, recalled their purpose.

This rather crude kind of test was introduced by Clegg, and following his procedure, I did so too. Such tests showed that the remanent magnetization carried by our rocks was very much more resistant to ambient (laboratory) conditions (e.g. effects of elevators, other people switching electromagnets on and off etc.) than man-made magnets are known to be.

*(Creer, January 7, 1999 email to author)*

Creer also heated up samples. He described these procedures in a twelve-page report which he prepared on June 16, 1953 and entitled "Report on research carried out during the period from October 1951 to June 1953."

8 It also appears that Creer actually calculated paleopoles by the beginning of June, 1954, but did not construct an APW path. The information that Creer and Irving provided Runcorn for the Rome talk appears in an earlier version of their 1954 paper (Creer *et al.*, 1954). The document, dated June 1, 1954, has the same title as their subsequent paper, and is part of the Technical Communications of Palaeomagnetism presented to the Xth Assembly of IATME, IUGG held in Rome. Besides Creer's NRM directions, Hospers and Charlesworth's Northern Ireland results are included. Irving's Upper Torridonian NRM directions are not included, although they are mentioned elsewhere in the report. Irving's Lower Torridonian directions are not mentioned at all. Moreover, Creer calculated the paleopoles, and their positions are very similar to those he used to reconstruct the *JGG* APW path (his first version), which is consistent with his recollection that the *JGG* path was based on data available to him in June. It is also worth emphasizing that there is no APW path in the technical report, which is consistent with Creer's claim that Runcorn did not present an APW path in Rome. I add that the minutes of the Rome meeting make no mention of an APW path. The minutes include comments by Deutsch, Gold, and Lowes and a note that Irving, who did not give a paper, took part in the discussion. Irving has no recollection that Runcorn showed a path.

9 Creer wrote out his Oxford talk before presenting it, and kindly sent me a copy of his prepared talk. The talk was entitled "Rock magnetism and polar wandering." There are seven pages of text, one table, and an apparent polar wander path. I refer to this as Creer, "Rock magnetism and polar wandering," 1954, unpublished.

10 Except for Creer's on the Caerbwdy sandstones from which he derived the Cambrian paleopole, all his other collections were drawn from several locations. Moreover, even his collection of fourteen samples from Caerbwdy sandstones was obtained from two sites spanning a stratigraphical thickness of 350 feet. The rocks were purple, were shown to be stable in the laboratory, and considered reliable (Creer, 1955: 72–74).

11 Later work has shown that both Graham's Rose Hill and Irving's Deccan results accurately reflect the paleomagnetic field (French and Van der Voo, 1979; Clegg *et al.*, 1957, respectively).

12 This conference was supported by NSF. Many renowned workers in geophysics and geomagnetism attended, including David T. Griggs, Walter Elsasser, Ernest Vestine, John Verhoogen, Gustaf Arrhenius, and Arthur Buddington. Included among those working in rock magnetism were Takesi Nagata, Emile Thellier, and John Graham. Clegg, Lawrence W. Morley, and two of Runcorn's new students, Philip M. Du Bois and John C. Belshé, were present. Finally, Linus Pauling was also there for reasons unknown. See Glen (1982: 96–97) for a photograph of the participants.

13 Four years later, Runcorn, now a mobilist, returned to the fit between Africa and South America. Noting Carey's 1955 criticism of Jeffreys, he claimed:

That the coast line of much of South Africa and South America fits together is of course a fact which the exponents of continental drift have thought very significant. Jeffreys'... statement that the fit is a poor one has recently been shown to be untrue by Carey.

*(Runcorn, 1959d)*

14 Runcorn, as just seen, said that he "may have been influenced by Tommy Gold." I suspect he was. They were acquaintances, and Gold was already an influential physicist. Gold cited the paleomagnetic work in his 1955 *Nature* article. Like Runcorn, Gold thought polar wander had probably occurred.

The recent observations of fossil magnetism of rocks and sediments show such great promise that we can expect before long to have a good indication of the movements of the pole that have, in fact, occurred in geological history. It is known already that a movement of at least some regions relative to the position of the pole is indicated by the results, and this constitutes a further item of observation along with such geological items as the glaciation of now tropical zones, that force geophysicists to consider that either continental drift or polar movement must have occurred on a substantial scale. In due course the magnetic data will make the distinction between the two. The occurrence of continental drift over great distances would imply new and surprising data about the construction of the earth and in particular its crust; while the occurrence of wandering of the poles over great distances would fit in well with all that is known about the earth, and would reaffirm what can already be inferred from other data.

*(Gold, 1955: 529)*

Interestingly, however, Gold seemed more open to the possibility of continental drift than Runcorn.

15 Runcorn (1955c) also discussed his results in *Scientific American*. The article appeared in September 1955, and contained the first published account of Runcorn's APW path based on the paleomagnetic work from Britain and his study of North American rocks. This path is less complete than in Runcorn (1956a). In his *Scientific American* article, Runcorn did not even mention continental drift. He did discuss polar wandering, describing, without citing either Gold or Vening Meinesz by name, the mechanisms they had proposed.

16 When I showed Runcorn the summary from *Nature* of the colloquium on polar wandering during the 1985 interview, he fondly recalled O. T. Jones, and noted that he greatly respected Jones and had learned a lot of geology from him. Jones was a fixist.

E. R. Deutsch, who became a member of the Imperial College paleomagnetic group near the end of 1954 and immediately left for India with Clegg and Griffiths to collect samples, also attended the colloquium. Deutsch was in the process of measuring the Indian samples during the early part of 1955, but took time out to attend the meeting. Several years ago while reminiscing about the early days of paleomagnetism in a letter to Ken Creer, he remembered how votes were taken about polar wandering and continental drift at the meeting.

The Cambridge symposium mentioned was between Jan. and March 1955, as I recently had confirmed by Runcorn, who was also a speaker there. I think the third speaker was O. T. Jones and that he was the one who called for two votes: the vote on the first motion supporting polar wandering was strongly if not unanimously in favor. The second vote, to the effect that significant drift "also" occurred was resoundingly defeated, with only one vote in favour, that of Derek Blundell! Derek recently confirmed that for me. I was then in the middle of measuring my first India collection, and though the results looked promising in favour of drift, I was still not quite convinced. Of course you may have been at the symposium and perhaps recall all the fun.

*(Deutsch, May 3, 1984 letter to Creer, copy sent to author)*

Creer and Irving were not at the meeting. Creer was working as a Senior Geologist for the British Geological Survey, and was doing fieldwork. Irving was in Australia. If they had been at the meeting there would have been three votes in favor of continental drift.

The votes, strikingly in favor of polar wandering but not continental drift, are interesting, and show, as suggested by Deutsch elsewhere in his letter, the influence of Gold and the generally greater resistance to continental drift than to polar wandering. Blundell received his Ph.D. in 1957 from the University of London. His supervisor was Bruckshaw. Blundell submitted his dissertation, "Geological applications of rock magnetism," in September 1957. He was associated with Blackett and Clegg's group, and made use of their apparatus.

17 Ian McDougall (1961) later showed them to be Early Jurassic. Much later, Schmidt and McDougall (1977), on the basis of detailed demagnetization studies, showed that the main magnetization of the Tasmanian Dolerites was Late Cretaceous not Early Jurassic in age, and skillfully isolated a small, much more stable underlying magnetization whose paleopole agreed beautifully with the Ferrar pole after reconstruction of the relative positions of Tasmania and Antarctica in the Early Jurassic.

18 There is some confusion about whether Irving decided to compare paleoclimatological data with the paleomagnetic results on his own or whether he got the idea from Jaeger. For example, Le Grand (1988: 146) attributes the idea to Jaeger. But Le Grand, apparently unaware of the fact that Irving proposed the idea in his dissertation, was misled by Irving's too generous acknowledgment of Jaeger. Irving wrote:

> The general idea of comparing palaeomagnetic and palaeoclimatological observations was suggested to me by Professor J. C. Jaeger of the Australian National University, and my grateful acknowledgments are due to him for this, and for subsequent help with the writing of this paper.
>
> *(Irving, 1956a: 40)*

However, as we have seen, Irving introduced the idea in his dissertation, and later noted:

> I've often read [the above] acknowledgment, and it really is not, perhaps, a very fair statement of what happened. Jaeger had opened up huge opportunities for me. I think it is true to say that I felt almost boundless gratitude towards him, and I suspect I wrote my acknowledgement affected by that and on the spur of the moment. Jaeger encouraged me enormously, and he helped me a great deal in setting out the tests. The way of plotting the paleolatitude variations as a function of time almost certainly grew out of a discussion with him, and whether he actually suggested it to me I don't know. He helped me enormously with that paper. But, the idea of testing paleomagnetic results and paleomagnetic latitudes against paleoclimates is firmly set out about two years earlier in my thesis of 1954. So there is no doubt that my *Geofisica* paper was the outcome of at least a couple years of cogitation on this matter. Jaeger was certainly not the one to suggest that I make the comparison; the suggestion was made in my thesis. So far as I know, I was the first person to try and do this.
>
> *(Irving, 1981 interview with author)*

19 Actually, Irving submitted the paper to *Geofisica* in February 1956, it was received by the journal on February 14, and published in the April issue. Because the paper had earlier been rejected by the *Journal of the Geological Society of Australia*, he had the opportunity to include some of the more recent paleomagnetic findings, for example, he included very recent work from Doell (December, 1955b).

20 Irving discovered the mistake after analyzing his Tasmanian samples. Because of the initial rejection of his paleoclimatological paper, he had the chance to correct the mistake while revising the paper. According to Irving,

> The rocks, as my notebooks show, were collected in January and February of 1955, and I built the small astatic magnetometer in Canberra by about March of that year. I measured them very quickly. I had the results sometime in mid-winter Australia in 1955.

I immediately began to write. So the Tasmanian dolerite paper was written and sent off just before the paper "Palaeomagnetic and palaeoclimatological aspects of polar wandering" was resubmitted to *Geofisica pura e applicata*. You can date that very well because of the polar errors in the two papers: the Tasmanian dolerites are in a different form, and that was because I discovered an error in the formula for calculating areas of circles which Creer had invented in his thesis, and which I had been using. I corrected that in the second version of "Palaeomagnetic and palaeoclimatological aspects of polar wandering," but it is uncorrected in the Tasmania dolerite paper.

*(Irving, 1981 interview with author)*

21 Irving cited Dunbar as his authority. Of course, it should be no surprise that the Squantum Tillite proved to be an exception. During the classical stage of the drift controversy, fixists had raised the Squantum Tillite as an anomaly to the mobilist solution to the origin of Permo-Carboniferous glaciation, and Wegener and others had countered by arguing that the Squantum Tillite was inconsistent with the body of North American paleoclimatological data, which indicated that North America had been generally warm and dry during Permian times (I, §3.10–§3.12). I shall return to the Squantum Tillite when discussing the reception of the paleomagnetic case for mobilism (III, §1.2). Current thinking still is that it is glacial. Modern dating techniques show it to be Late Precambrian. Thus the Squantum Tillite turned out not to be anomalous with continental drift not because it was glacial but because it was not Permian.

22 It is also worth noting that they, unlike Irving, did not question or disagree with Runcorn's rejection of any relative movement between Britain and North America.

23 Van der Voo (1967, 1969) redid Clegg and company's work. Given the tentative nature of Clegg and company's conclusion, Van der Voo redid their work, demagnetizing his samples to remove any possible overprinting. He (February 26, 2012 email to author) too found an approximate "35° – declination deviation (and, hence counterclockwise rotation) with respect to Triassic results from the United Kingdom."

24 Blackett thanked Chaim Pekeris and his wife for their kindness and hospitality during his visit to the Weizmann Institute of Science. Pekeris, who worked on convection in the 1930s, set up the Department of Applied Mathematics at the Institute in December 1948, upon arriving in Israel: perhaps he played a role in the decision to invite Blackett.

25 The less likely possibility arises if the remanent magnetization had an opposite polarity. This ambiguity in latitude was also noted by Irving (1954) in his discussion of his Indian results.

26 Blackett's comment that India had moved northwards since the end of Paleozoic glaciation is qualitatively correct, but it did not move "at fairly constant rate" as he said. The paleomagnetic results showed that during the Cenozoic (since the extrusion of the Deccan Traps) India had moved ~5000 km northward, over half the more than 8000 km required to bring India from southern polar latitudes to its present latitude since the Late Paleozoic. The rate of motion in the Cenozoic therefore was, on the basis of the evidence cited, at least five times greater than that in the Mesozoic, hardly "fairly constant."

27 Irving (April 2000 comment to author) suggested that Blackett's overly sanguine estimation of the time needed to end the controversy might be attributable to the fact that Blackett did not do any of the measurements himself. "So perhaps he did not develop the experimental feel for the rightness or otherwise of results. This was very different from his earlier very hands-on involvement in physical experiments. He was an older and very busy man." But, as Irving also noted, if, as they surely must be, marine magnetics are included within the purview of rock magnetism because they are recordings of the ancient geomagnetic field, then Blackett got it about right.

# 4

# Runcorn shifts to mobilism: 1955–1956

## 4.1 Runcorn returns to North America then moves to Newcastle

In the mid-1950s rapid acquisition of paleomagnetic data resulted in the willingness of paleomagnetists working in Britain progressively to consider mobilism, to view it favorably, and, soon, to accept it. It happened individually, there was no mass acceptance. No individual's acceptance is more interesting than that of Runcorn, and I shall devote much of this chapter to describing his conversion and the reasons for it. He is the first major figure who, beginning from a fixist worldview, rethought his position and adopted continental drift, immediately becoming one of its most prominent advocates. Later there were other prominent "conversions." Strong fixists such as Hess (III, §3.12) and Wilson (IV, §1.9) also switched to mobilism and immediately became its ardent advocates; over the next decade their conversions marked important milestones in the mobilism debate. Bullard, although not an ardent fixist, and certainly not a mobilist, also switched to mobilism, and became another strong defender. Others such as Creer, Irving, Du Bois, and Parry, who had substantial prior knowledge of continental drift, and Clegg and Blackett, who did not, simply accepted the new paleomagnetic case without qualms. Opdyke, initially a rigid fixist from his American education, being now in the British milieu, immersed himself in the new evidence and switched his allegiance to mobilism (§4.2, §5.12–§5.14). Hales, one of the few ardent fixists from South Africa (I, §6.12), also changed his mind, finding mobilism's paleomagnetic support sufficiently strong to do so (§5.4).

Runcorn returned to North America in June 1955, after participating in the colloquium on polar wandering at Cambridge (§3.10) where he first presented results from his initial 1954 sampling of the Grand Canyon, and where he continued to argue in favor of polar wandering without continental drift. He expanded his critique of drift in two subsequent papers submitted soon after his arrival in the United States (1955b, 1956a), having developed the habit of writing papers in transit abroad. He divided the summer between the Museum of Northern Arizona, where he and Neil Opdyke, his hired summer help, stayed, and where he had stayed in 1954, and the Dominion Observatory in Ottawa, Canada, where he spent three months (Runcorn, 1956b: 85).

While at the Museum of Northern Arizona, Runcorn and Opdyke (who will be introduced in a moment) sampled lava flows of Pliocene-Pleistocene age in northern

Arizona in which they found reversed and normal polarities. Focusing on reversals, they opted for geomagnetic field reversals as their cause, not self-reversals, based on their discovery that one of the reversely magnetized lava flows overlaid a layer of reversely magnetized baked clay (Opdyke and Runcorn, 1956). They argued, as David (1904), Brunhes (1906), and Roché (1953) had done, that it was more reasonable to suppose that both the clay, baked by the flow, and the flow itself had become magnetized in a field of reversed polarity than to suppose that these two very different materials should both have undergone self-reversal.

Importantly in the long-run, Runcorn and Opdyke had, within a few months, figured out a new paleoclimatic check of paleomagnetism. Runcorn decided to get Opdyke back to England as a research student. Opdyke devoted most of his graduate work to developing and testing this idea, and they eventually co-authored two important papers.

Several months after returning to England, Runcorn left the Department of Geodesy and Geophysics at Cambridge to become professor of physics, King's College, Newcastle upon Tyne, taking up the position in January 1956.[1] The move enabled him to build a school of geophysics. When he learned of the opening, he asked Blackett and Chapman if he should apply.

I wanted to build up a school of geophysics, and was feeling that I wouldn't do it at Cambridge. I saw that this position had fallen vacant, and I asked Blackett and Chapman, and they were ... very keen I should do it. There weren't any jobs at Cambridge for me to build up a group. There were research students. Of course, they all just left upon graduation. And, there were very few opportunities for getting post-doctoral fellows. So the only way you built up a group in England was to become a head of a department and appoint people to your staff who were interested in the same field.

*(Runcorn, 1984 interview with author)*

Allowed four new appointments, he hired Creer, Hide, Lowes, and Jim Parry. All were former students or in the case of Lowes a post-doctoral fellow of Runcorn's. Parry, a graduate of Oxford University, although publishing little became an important member of Runcorn's paleomagnetic group, being its best internal critic. Runcorn later reminisced about Parry and his role.

In those days a new professor was told that he could appoint some new members of faculty. They gave me four new faculty positions. I talked to Hide, Lowes, Creer and Jim Parry, who had come from Oxford and was working on rock magnetism ... He is probably one of the most intelligent people we have ever had in our group, but he has never done any research, which sometimes happens. He is still with me. He has a first-class brain, but something inhibits him from doing research. I sometimes think that you can be too clever to do research. You think about a problem. He is very good working with research students ... He is a bit inclined to think about a problem, solve it to his satisfaction, and then say, "Let's turn our attention to something else."

*(Runcorn, 1984 interview with author)*

Runcorn also convinced Collinson to rejoin his group. Collinson, recall, had helped Irving and Creer in Cambridge.

I wanted someone to run the magnetometer to do the measurements. We got the position on the research grant for him, and he came in 1952 to Cambridge, and he had just taken a degree ... what they call an ordinary degree, in Manchester. He came and worked as, I suppose you would call it, a laboratory assistant. And then the year before I left to go to Newcastle, he got tuberculosis, and he went into a hospital and spent a full year there. When he came out I was just planning to go to Newcastle, and was ready to move the operation. I said, "Would you like to come and help set up the apparatus. And, Collinson came from near London, and, you know, there is a north-south difference just like the US, and he was a bit uncertain whether he had really fully recovered from TB, and in any case he hadn't decided what he wanted to do. He said to me after I had persuaded him, "Well, I'll stay three months, and help you get the magnetometer set up in Newcastle, but I don't want to live up there." He is still there! After he got to Newcastle, he got interested and got a Ph.D. He is now a reader, which is our associate professor.

*(Runcorn, 1984 interview with author)*

On arrival in Newcastle, Runcorn met Stanley Westoll, head of the Department of Geology. Westoll was one of the few vertebrate paleontologists to support mobilism during the 1940s and 1950s. He and Runcorn began having friendly discussions about continental drift. Westoll defended mobilism; Runcorn argued for polar wandering without drift. Before 1956 was half over, Runcorn became a "drifter."

### 4.2 Runcorn hires Opdyke to help collect rocks and they formulate a new paleoclimatologic test of the paleomagnetic method

It is time to introduce Neil Opdyke. As Opdyke explains, his road to Columbia University was different than most.

I was born on Feb. 7th 1933 to an unwed young woman, in the midst of the great depression at 2 Kingwood Ave, a nineteenth century house that still stands. My mother worked in the local porcelain factory putting holes in spark plugs. A mind numbing activity! I lived with my grandparents and my mother's siblings. No indoor plumbing but an outhouse at the end of the hundred foot walk. I remember as a child being washed in the kitchen sink. I loved my mother and grandparents and the father figure was my uncle Howard, who made his living as a carpenter. My mother married Walter S. Bird in 1940, the same year my Uncle Howard married and as a result I met my wife-to-be. My uncle married her aunt and she sometimes came to stay with them. She had beautiful long blond hair that she wore in pigtails and reached to the middle of her back, she was also smart and shy. In high school I took her to proms.

I can vividly remember the attack on Pearl Harbor in 1941 which caused me to strap on my toy pistols and go to war. I was growing up and joined the Boy Scouts and by the end of the war I was in the Explorer Scouts. I was at scout camp when Hiroshima and Nagasaki were bombed. Like most Americans I didn't understand the significance of this! I discovered young ladies and football about the same time and graduated to high school. I spent K through 12 in the same building with the same principal for twelve years (Mr. Light) and we became good friends but he was the chief disciplinarian and I was not without fault. The school had some good teachers but I enjoyed reading historical novels and read a lot. I started out with the cowboy books by Zane Gray. There was a small library in town, the size of one room, and I think I read every book in it. The most important being "The Seven Pillars of Wisdom" by T. E. Lawrence. The upshot of this was that in high school I knew more history than the teacher who was also the football coach. When I graduated I won the History Prize.

Uncle Howard had played semiprofessional football in Lambertville and I watched him play. I fell in love with the game. Frenchtown High School did not play football when I was a freshman. The second year they began a football program with a new coach (Rickenbacker) and of course we got hammered. In my junior year we did better and Ed Rosenmayer and I were elected co-captains for our senior year. I played guard. It was a good year and we won the county Championship and I was named All State in my division. This was good for me since it opened opportunities and I was interviewed at Columbia and Princeton and Columbia offered me some support. I had opportunities at some smaller schools as well but I thought that Columbia was the best opportunity so I took it.

*(Opdyke, November 12, 2011 email to author)*

## Coming from a small town, academics and football were a challenge for Opdyke at Columbia.

I was a small town boy in the big city and the shock was considerable. The academic shock was scary. I went from essentially writing nothing to three hour written exams. I was in remedial English. I knew why I was there but I didn't know why the others in the class were there. I was taking two heavy reading classes and Spanish. When the end of the semester arrived I took the exams and went home to recuperate. I had a meal job at the Lion's Den so when I returned to Columbia I went directly to the Lion's Den. There was another freshman football player working there as well. He told me that he had failed two courses. This was scary so after work I trudged over to the gym where all the grades were posted on the wall. No protecting the student in those days. I had a solid "C" average and I was elated and could have cried. I walked on clouds to my room.

What about football? Well I played with the freshman team, did not start, but I learned and improved. Football was a slog at Columbia since every afternoon you got on a bus and went ninety blocks to Baker Field on the north end of Manhattan. It was a good time to read something. Sophomore year was a bad year football wise and I practiced with the Rinky-Dinks. We were not exactly cast offs but we certainly were not expected to play much. I remember one day we were playing against the varsity line, which was protecting the punter. I broke through and blocked the punt. The coach was incensed and reamed the varsity line. The next play I fell down and stayed there. But life goes on. I made a momentous decision at the end of the year and decided that I wanted to be a geologist, a strange decision since I had never had a course in the subject. Most of my friends wanted to be doctors, lawyers and work on Wall Street; these options did not appeal to me. I wanted to see the world and at least spend a little time outside the office. So the summer between my sophomore and junior year I bought the book that was used in the first semester of Geology and read the book during the summer break since I knew that the autumn would be dominated by football. I also decided that if there was no chance that I would play I would leave the team and do academics. At the end of field camp I was the second string right guard playing behind the captain. Before the year was out I became the starting right guard. I was elected captain of the team for my senior year.

My academic career picked up as well and I began to get good grades in Geology as the summer study program paid off. I had some excellent teachers as an undergraduate, like John Imbrie, Carl Turekian and others. Had some girlfriends and things were good. I continued to do Geology and had a little above a B average. I wanted to go to graduate school at Columbia. So I applied but was not admitted since Imbrie had misplaced my application. I was admitted at the University of Wyoming instead, so Wyoming it was to be.

*(Opdyke, November 12, 2011 email to author)*

Runcorn liked what he saw when he met Opdyke.

Neil Opdyke. He was another person who was obviously very bright. He was from Columbia, and he was intending to go into a Ph.D. program on some stratigraphical problem at the University of Wyoming. But, after a summer as a research assistant just helping with the collection of rocks, he became interested [in what we were doing]. John Nance, who was a very good geologist at the Museum of Northern Arizona where I was working, and became head of the geology program at the NSF, was as impressed with Neil as I was. By this time I had got a research grant, which had money available for a research student, so I asked Neil if he would like to come to Cambridge. He spent one term in Cambridge with me, and then went in January to Newcastle. Because he was a geologist, I thought it might be tactful in the new university if he worked nominally under Westoll in geology, which he did. During that first summer when I was with him, we got interested in this question of whether the paleoclimatic data really fitted the paleomagnetic angles of dip. And so you know I thought from what he said about his interests and background that he might be a very good person to do this.

*(Runcorn, 1984 interview with author; my bracketed addition)*

Upon arriving in the United States in June 1955, Runcorn visited Bucher at Columbia University. He traveled to Washington, DC, where on the 27th he saw Graham at the Carnegie Institution. They discussed their work on the Colorado Plateau, and agreed that the results supported polar wandering and not continental drift (Graham, 1955: 345).

Runcorn, planning to collect more samples from the Grand Canyon, found a student from Princeton University to help, but he fell ill and canceled. Runcorn got Neil Opdyke instead, and his finding him is reminiscent of his discovery of Irving (§1.15). Opdyke, like Irving, became a major paleomagnetist, playing a key role in establishing mobilism. Runcorn found out about Opdyke from a lifeguard.

I used to swim every day when I was at Columbia ... I had to get some graduate student in geology to come and help me, and someone from Princeton had agreed – I suppose I got in touch with him through Harry Hess. [Runcorn thinks that he first met Hess in 1954.] When I arrived, I suppose, I spent the week in Walter Bucher's apartment. He [the student from Princeton] rang up and said that he had fallen ill quite seriously. So I didn't know who would be available. Of course, this was fairly late. It would have been the end of our term, which is the end of June. Most people were looking for summer jobs. I didn't really know anybody. But the lifeguard at the pool had been there the previous year. And so when I went in – I wish I could remember his name – he said, "Oh you've come back." Those were, I suppose, days when English people were a bit more rare in the States. He had been quite friendly. He said, "I suppose you have come back to do some more rock collecting." And so I explained to him all about it. Then I said that I hadn't got anyone who could serve as my field assistant, and he thought for a moment and said, "I know someone who is a geologist." He was Neil Opdyke. I said, "Well, do you think he would be interested?" He said, "I know he has finished his studies. I suppose he is not doing anything this summer." And, so I rang him up. It was very difficult to get people starting to do research in geology working in this field. It seems surprising now.

*(Runcorn, 1984 interview with author; my bracketed additions)*

Opdyke recalled why he was available, and what happened when Runcorn picked him up at his home in New Jersey.

I went to Columbia because I played football there. When I got out of Columbia I applied to graduate school at Columbia. John Imbrie lost my application form and never gave me a recommendation. So I never got into Columbia. I was accepted in the masters program at the University of Wyoming. The summer after my senior year I went home to Frenchtown, New Jersey, and started to work again as a carpenter – I had worked as a carpenter all through my college years in the summer time. I got a telephone call from this guy Runcorn. It was really amusing. I didn't know Runcorn from a bag of beans. The person on the end of the line had a very British accent. He said "this is Runcorn here." Then he spelled out his name R U N C O R N. I thought that it was a joke that one of my friends was playing on me. I said "sureee." Apparently, some student from Princeton was going to be his field assistant and backed out. He was swimming in the pool at Columbia when he met a friend of mine, Max Pirner, the lifeguard at the pool. He played halfback. Runcorn said he was looking for somebody who would be free to do geology. Pirner said there was this guy Opdyke. So, Runcorn called me from Connecticut, (it was on a Friday) and said he had to leave on Monday – which is typical of Runcorn. So I talked it over with my grandfather, my parents, my uncle. I said, "We'll I guess I better go because it is the only job I have going in geology, in the earth sciences, so I better do it." He came through on Monday morning. I was sitting outside my grandmother's house – It was on the road from New York. This ginger-haired guy with a smelly tee shirt pulled up across the street in an old Plymouth, and said, "I say do you happen to know where Neil Opdyke lives?" I said, "I am he." We lived on a little street in this little town. We went over to my house, and had lunch. He parked in back of a big stump – an old maple tree had rotted and they cut it down. So we started to drive out to Arizona, waving good bye to my parents. Runcorn starts out to drive right across the stump. My mother said that she didn't think we would ever make it to Arizona. In fact, Runcorn immediately launched into telling me about paleomagnetism, which I didn't know a thing about. And he is telling me all about it while we are driving the first ten miles. In the first five miles he hadn't yet shifted into high gear. I'm thinking to myself, "Should I tell him he isn't in high gear?" So I finally did tell him, and then we traveled out to Arizona.

*(Opdyke, 1984 interview with author)*

Before heading west, Runcorn stopped to see Graham in Washington, DC. Opdyke (Opdyke, November 12, 2011 email to author) recalled, "It rapidly became apparent that there was no love lost between the two."

Continuing to the Grand Canyon, they used the Museum of Northern Arizona in Flagstaff as their headquarters, where Runcorn had stayed the previous summer. Fieldwork with Runcorn did not always go smoothly. Opdyke remembers two particularly momentous rock-collecting trips.

Two English friends of Keith's arrived from Cambridge. She was from Girton College and he was a mathematician. Keith decided we would take them into the Grand Canyon. Off we went accompanied by a glamorous female graduate student from University of Arizona. We had not much trouble going down but when we started back up the young lady decided that she didn't want to carry her water bottle so she sent it up with a mule train. Big mistake! I shared my water with her and others did also; we managed to survive but I did learn that desert travel requires water. We made our way to the top and the trip was made more pleasant by the swaying tush in front of me. Near the top we came upon a group of geologists measuring section for some oil company. I learned later that one of them was Loyd Burckle who became a good friend and collaborator of mine at Lamont. He did not remember me but he remembered the girl.

*(Opdyke, November 12, 2011 email to author)*

The other trip occurred late in the season. Runcorn wanted to collect some Hakatai shale.

Runcorn's aim was to sample redbeds through the geologic column. The work in England had previously shown that red sediments were the best magnetic recorders. Therefore they were our targets throughout the geologic column and red formations were present in the Precambrian, Carboniferous, Permian, Triassic and Jurassic. Outcrops were readily available throughout the Colorado Plateau. Towards the end of the field season Keith decided that he wanted to collect the Hakatai shale, a Precambrian formation that outcrops in a stream bed near Phantom Ranch (a rest house and hotel). We hiked down without incident and spent a couple of days collecting hand samples. He arranged to have the samples brought out of the canyon on mules. This was a relief to me since I would not have to be the mule.

The final day's work finished at about noon and Keith had decided that we would wait until 3 o'clock on the Kaibab trail, which is shorter than the Bright Angel trail but steeper. Keith did not have a hat since he had lost it on the way down. It blew off his head into the river and he had borrowed a hat from someone at Phantom Ranch so he had to give it back before we left. We started up the trail with Keith without a hat. We went pretty well but by the time we got out of the inner gorge and onto the Tonto platform Keith began to show signs of stress and he was out of water and was drinking mine. We came to an emergency telephone on the trail and I tried to persuade him to allow me to call for help. He would not allow me to make the call. Big mistake! We continued on and were going slower and slower. Keith would walk about hundred yards and rest. At dark we were on the Supai formation and were still slowly climbing. We wandered off the trail in the dark and I told Keith that I would go no further in the dark for fear of falling off a cliff. So we just lay down where we were to wait till dawn. Later a thunderstorm came up and I thought that would be great to get some water and tried to catch some water in my hat. We did not get a drink but we did manage to get chilled to the bone. We hugged one another to try to get warm.

We started up the trail at dawn while it was still cool and collapsed every hundred yards but we finally made it to the top where the mules were pastured. Kicked the mules out of the watering trough and drank our fill! It wasn't bottled but boy was it good. We hitched a ride back to Bright Angel lodge in the back of a truck. We were dirty and hungry. Even so we managed to get served in the dining room and had a great breakfast. We returned to the Museum for R&R.

*(Opdyke, November 12, 2011 email to author)*

Runcorn, impressed with Opdyke, later asked him if he wanted to pursue a Ph.D. at Cambridge. Opdyke agreed, and Runcorn got him into Cambridge. Opdyke fondly remembered what happened.

We got to be pretty good friends. He said, "Why don't you come to Cambridge University to do your Ph.D.?" I said, "Well, I don't think I'm ready for that." He said, "That is all right." I said, "How are you going to get me into Cambridge?" He said, "I'll do it." I said, "OK, if you can get me into Cambridge, I'll go." He did. In one month's time he got me both into University and Gonville and Caius College. So we spent the summer collecting rocks, and so I had to tell the University of Wyoming that I would not be there in September, and traveled across the North Atlantic on the Queen Mary, and came into Cambridge with Runcorn. I was in Cambridge for three months. Then Runcorn got appointed Professor of Physics at Newcastle. What he wanted me to do was paleoclimatology, and relate it to polar wandering. So I decided that I could probably do that. That is what my thesis was on. The title of my thesis was "Palaeoclimatology and Palaeomagnetism in Relation to Polar Wandering and Continental Drift."

*(Opdyke, 1984 interview with author)*

Runcorn and Opdyke jointly figured out a new way to use paleoclimatology to test the GAD hypothesis and hence the paleomagnetic support for mobilism. They did so in a hotel in Chicago while returning from Arizona to the East Coast.

I didn't know anything about paleoclimatology, but the object was to try and find some independent means of checking the axis of rotation of the Earth, and we were in a hotel in Chicago, while coming back from the Grand Canyon to the east coast, just batting ideas back and forth, and I remember discussing several things with him. One of which was that the magnetic field comes in at the poles and you probably have a higher incidence of low energy particles at the poles. Then I don't know who came up with the idea that the wind system as well as climate, you know, are determined by the Coriolis force, and I came up with the idea of using the movement of aeolian sands to check directions of motion of these wind systems. And I had been stimulated by the fact that we had seen a bit of aeolian sand in the Grand Canyon area, and so that is where that idea came about from.

*(Opdyke, 1984 interview with author)*

Although Opdyke is unclear whether he or Runcorn brought up the question of studying the link between prevailing wind directions and the Coriolis force in the geological past, he believes it was he who suggested that paleowind directions determined from aeolian sandstones could be used for this purpose. Runcorn himself may well have come up with the general idea of using wind directions because one of his former Ph.D. students, Raymond Hide (1953), carried out laboratory experiments designed to study the effect of the Coriolis force on convection within Earth's core, and later its relevance to atmospheric circulation.[2] Regardless of who came up with what, the idea arose as a result of an exchange of ideas during a Chicago conversation. Left alone, neither Opdyke nor Runcorn might ever have thought of and pursued it. Their idea was original, and very different from Irving's paleomagnetic test (§3.12).

The Coriolis force plays a major role in determining general planetary circulation of the atmosphere. Because it is dictated by the Earth's rotation, they proposed that the general planetary circulation of the atmosphere throughout geological time has remained, as it is today, divisible into latitudinal zones, dominated by tropical easterly trade-wind zones, mid-latitude westerly zones, and polar zones. Consequently, if the direction of ancient prevailing wind for a particular place could be ascertained, it should conform to the latitude and direction of geographic north for that place as determined paleomagnetically on the basis of the GAD hypothesis. If Opdyke and Runcorn could come up with a way of determining the ancient wind direction for a particular region during a period for which paleomagnetic data were available, they could use it to test the GAD hypothesis and independently validate that paleomagnetic data.

Opdyke thought he knew how to do it. The orientation of modern sand dunes conforms to prevailing wind directions, and past wind directions can in principle be determined by studying fossilized sand dunes. Sand dunes have gentle windward and steep leeward slopes. The crescent-shaped leeward slope forms at the characteristic angle of about 33° to horizontal, and the tips of the crescent point downwind. Because, in well-researched instances, the orientation of ancient sand dunes is faithfully fossilized in aeolian sandstone, the direction of the prevailing winds at the time the dunes were formed can be determined. Opdyke knew of the abundance of aeolian deposits in the western United States ranging in age from Upper Paleozoic through Lower Jurassic; he remembered seeing them in the southwestern United States where he and Runcorn had just been. He reminded Runcorn of them, while batting ideas about in their Chicago hotel room. I shall describe (§5.12–§5.14) how they returned to the United States the following summer (1956), and studied four aeolian sandstone formations in Wyoming and Utah, work that formed an important part of Opdyke's Ph.D. thesis (1958b).

### 4.3 Runcorn changes his mind and supports continental drift

Runcorn reversed his position on continental drift about six months after he took up his new post at Newcastle. He signaled this in two short, pro-drift papers. The first

(Runcorn, 1956b), "Palaeomagnetic comparisons between Europe and North America," was published in November, 1956, in the *Proceedings of the Geological Association of Canada*, the second (Runcorn, 1956c), "Palaeomagnetism, polar wandering, and continental drift," in the Dutch journal *Geologie en Mijnbouw*. I shall call them the "Canadian paper" and the "Dutch paper." Runcorn ended the Canadian paper with the announcement of his conversion to mobilism.

It is therefore concluded tentatively that North America was displaced westwards relative to Europe by about 24°, in post-Triassic times, probably in late Mesozoic times. In figure 4 [reproduced here as Figure 4.1] the relative position of the continents in the present northern hemisphere is shown in pre-Mesozoic times with the path of the poles marked on, based mainly on British measurements which are extensive enough to be free of isothermal remanent magnetization.

*(Runcorn, 1956b: 83; my bracketed addition)*

Runcorn's Figure 4 (Figure 4.1 here) showed the APW curves for North America and Europe with the continents in their Mesozoic positions. It is to be compared with his Figure 1 (reproduced here as Figure 4.2).

The Canadian paper grew out of a talk at the annual meeting of the Geological Association of Canada while visiting the Dominion Observatory in June 1955, which he gave before he changed his mind. As I shall show, he changed his mind later, back in Newcastle in June 1956 when revising the Canadian paper. In the Canadian paper, which was published in November 1956, he discussed only paleomagnetic questions, and only results from North America and Europe. Previously he had rejected mobilism, attributing the longitudinal discrepancy between their APW paths to experimental error (§3.10). This he was now about to reconsider and to make the most important scientific decision of his career.

Making doubly sure that readers understood he had changed his mind, he wrote in the Dutch paper:

The palaeomagnetic directions in the geological column in Great-Britain have been examined by Creer, Irving and Runcorn (1954) who explain the results in terms of polar wandering. Runcorn (1955, 1956b) finds substantially similar pole positions from sediments collected from the southwestern states of the U.S.A. Thus he concluded that there was inappreciable continental drift between Europe and North America, at least since Pre-Cambrian times. There is however a small but systematic westward displacement of the pole inferred from the American data from the poles calculated from the British data. Runcorn (1955c) discusses this discrepancy and concludes that it cannot be explained by the errors in the palaeomagnetic determinations. He thus concludes that the separation between Europe and North America increased by 20° during the late Mesozoic times, but had remained constant between that period and the late Precambrian.

*(Runcorn, 1956c: 253) (Runcorn's references to Creer* et al., *and his own papers are, in the order he lists them, equivalent to my Creer* et al., *1954, Runcorn, 1955b, 1956a, b). His 1955c reference was mistaken and internally inconsistent because it refers to the Canadian paper (my Runcorn, 1956b) which he also referenced as 1956b.)*

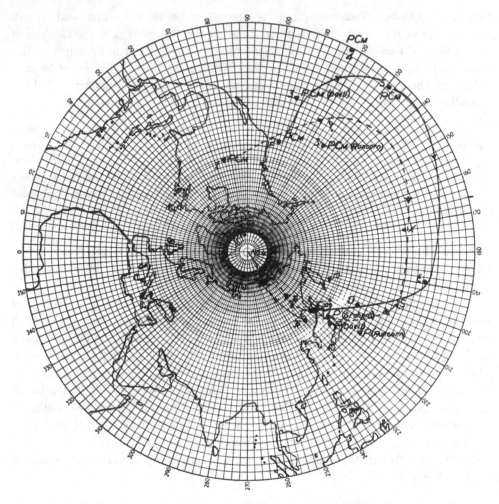

Figure 4.1 Runcorn's Figure 4 (1956b: 81), which had no caption. The European poles are connected with a solid line; the North American, with a dashed line. The poles as determined by Doell, Graham, and Runcorn from the Permian Supai sandstones are labeled. Keeping Europe fixed, Runcorn rotated the Americas approximately 24° eastward toward Europe to show the position of North America relative to Europe in late Mesozoic times. Moving the APW curve for North America with the continent, the gap between the APW paths is accordingly reduced. Notice that Runcorn showed his preference for Doell's and Graham's results rather than his own. Also notice that the three Permian poles differ in latitude but not in longitude.

In the Dutch paper, which was probably submitted in July 1956 and definitely after submission of the Canadian paper, even though it was published in August three months before the Canadian paper, he no longer restricted himself to paleomagnetism or to the paleomagnetic results from North America and Europe. He proposed a paleoclimatological argument in support of drift and began to reconsider the mechanism difficulties it faced. He discussed the Indian work of Clegg and his

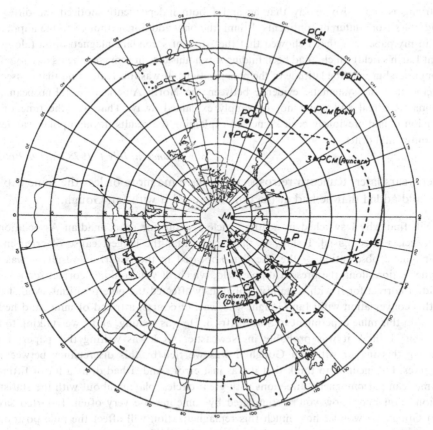

Figure 4.2 Runcorn's Figure 1 (1956b: 78). It also had no caption. It shows the continents in their present positions.

co-workers, noted that it confirmed Irving's earlier work, and argued that both supported continental drift because they indicated that India had been in the Southern Hemisphere during the Cretaceous. Runcorn referred to Irving's Indian results, now citing Irving (1956a), who in that paper earlier that year had already interpreted the paleomagnetic results from Europe and North America in terms of continental drift, and presented paleoclimatological arguments in their support (§3.12).

### 4.4 The Canadian paper

Although it is abundantly documented that Runcorn changed his mind in mid-1956, it is more difficult to determine precisely what motivated him to do so. Twenty-two years after the fact, Runcorn retrospectively told Bullard that he and Irving "independently noticed the difference among poles from different continents," and offered the following explanation as to why he began arguing for continental drift:

Ted Irving is right also to say that he and I both independently noticed the difference among poles from different continents. I think the only additional (though to me important) point in my paper was that I showed that the effect of secondary magnetisation (along the present Earth's field) which I had been finding was a nuisance in some sediments did not affect the longitude but only the latitude of the pole positions. It was this realization that caused me to decide that the systematic evidence between the North American and European pole positions were real and could not be simply explained away. That was the time of my conversion (I had previously been keener on explaining the results through polar wandering than continental drift).

*(Runcorn, July 1, 1978 letter to Bullard)*

Sixteen years later, Runcorn retrospectively gave me more or less the same explanation, and added that he had at the time discussed it with Ian Gough.

You will find why I switched in a little article I wrote for the Canadian Association of Geophysicists [Geologists]. I show in that article that the biggest cause of error in the paleomagnetic observations – the one that appeared in the American results – was this remagnetization along the present dipole field, which, if you couldn't correct for it – and I couldn't correct for it with certainty – obviously affected the pole positions. And, I then [had this conversation with] Ian Gough, who was, of course, a friend of mine, and he had tried to do this mine experiment in South Africa. He was visiting, and I was talking to him about this, I think it was probably in Newcastle. As I was writing this paper, I was discussing this matter with Ian Gough. I was worried by this discrepancy between the two curves, but inclined to think that it was just error. And, I had done a lot of fitting of the American paleomagnetic directions with great circles, playing about with the statistical question. You know how you're stimulated by someone else very often. I started saying to Ian Gough, "I wonder how much this remagnetization will affect the pole positions?" I was always interested in vectors. I knew that the dipole could be treated as a vector, and it suddenly dawned on me that these paleomagnetic points spread out on a great circle, which goes through the present dipole field, corresponded to a series of pole positions, which went through the geographic pole. You know, if you ask the question, "Where is the pole position?" Suppose you have remagnetized the rock with a component along the present dipole field direction. Then, of course, that remagnetized rock will have a component along the present dipole field direction, and the original geomagnetic field. And, it suddenly occurred to me – you know it was one of those simple results that in retrospect you think that anybody could have done – that you got the pole position by adding vectors along the present axis of rotation and the ancient magnetic axis of the Earth. So I saw that the effect of this remagnetization is to move the poles in latitude but not in longitude. When you looked at these two curves, the striking thing was, you see, that they were displaced in longitude not in latitude. And, it was at that point, that I really thought that the biggest cause of discrepancy, which made me distrust the data to a certain degree, didn't affect this issue. So that is why I wrote this little Canadian paper. I had spoken to this Canadian society while I had been in Canada. I'm almost certain that this is the moment that I became convinced that the paleomagnetic data had to be explained in terms of continental drift.

*(Runcorn, 1984 interview with author; my bracketed additions)*

Runcorn said "this is the moment," but did not say when it was, and when he had talked with Gough. Gough himself cannot recall exactly but he does remember talking about paleomagnetism and continental drift with Runcorn in Newcastle during the first half of 1956, and does not rule out the possibility that the above conversation did occur (Gough, October 1995 interview with author).

So, in an attempt to investigate further why he converted to mobilism, I shall follow Runcorn's instruction and return to his Canadian paper. He himself said he did so because of the presence of viscous remanent magnetization (VRM). VRMs are commonly acquired along the direction of the mean field since the last reversal. This magnetization, which Runcorn referred to as isothermal remanent magnetization (IRM), contaminated many of his North American samples. At the time nobody was able to physically remove this unwanted VRM without affecting the direction of the original magnetization. If VRMs were present, the estimated pole would differ from the true position; it would lie along the arc of a great circle connecting the true ancient pole and the present pole of the GAD field, namely the present geographic pole.

Runcorn, however, knew about error arising from VRM before he switched to mobilism; he knew that it affected magnetization directions (and hence the corresponding pole) and gave rise to an unreliability difficulty. Here is a passage from his last paper that he wrote as a *fixist*.

Thus it may be expected that the directions of the magnetizations of specimens from one series of rocks, neglecting scatter, will lie between these limits clustering toward the one or other depending on the relative importance of the I.R.M. [should strictly be VRM] in different samples. Thus in such cases the directions as plotted on a stereographic projection should be scattered along a great circle passing through the mean direction of all the specimens and the present axial dipole field direction at that locality. It is with this principle that we proceed to interpret the measured directions of magnetization where the calculation of a simple mean is clearly inappropriate.

*(Runcorn, 1956a: 308; my bracketed addition)*

What he did not at first seem to have realized was that such errors did not affect the longitude of pole; he did not explicitly state that VRM has no effect on the longitude of poles. However, he did discuss the effect of VRM on the latitude of poles, and he did so by comparing his results from the Permian Supai sandstones of the Grand Canyon with those obtained from the Supai by Graham (§7.4) and Doell (§3.8).

Now the [geocentric] dipoles corresponding to the different palaeomagnetic directions represented by points on the great circle lie in the same meridian [that is, longitude on the present day grid]. Thus insufficient correction for isothermal [that is, viscous] remanent magnetization along the present dipole field causes the computed north magnetic pole for a rock series to be nearer to the present geographical north pole than it should be. This fact is illustrated in figure 1 where the Supai poles are plotted derived from Graham, Doell and Runcorn's measurements ... That of Runcorn ... is clearly more affected by unstable specimens than the better grouped specimens on which Graham and Doell base their mean palaeomagnetic directions ... This consideration ... would however place the true south magnetic pole of the Supai formation still farther north, with the site (Arizona) to the geographical north of the equator at the time. However, in the absence of any physical method of cleaning the rocks of isothermal remanent magnetization, this conclusion must be viewed with reserve. Further sampling of the Supai formation may throw light on this matter.

*(Runcorn, 1956b: 81–82; my bracketed additions)*

But there is still no mention of the absence of an effect on the longitude of poles. Compare, for example, how Creer, Irving, and Runcorn in the following year clearly said that VRM affects the latitude but not the longitude of poles.

As dipole fields add vectorially, the effect of secondary magnetization is to move the computed pole along the great circle joining the pole of the original magnetic field and the pole of the field corresponding to the secondary magnetization. Thus inadequate allowance for i.r.m. [VRM] causes an error in the latitude of the computed pole *but not in its longitude*. Correction for the effect of i.r.m. consists therefore in moving the north magnetic pole, determined from the mean paleomagnetic direction, away from the present north geographical pole.

*(Creer et al., 1957: 150; my bracketed addition; emphasis added)*

Below (§4.8), I shall argue as Creer suggested that Runcorn made substantial changes in proof to his Canadian paper and illustrate how he could have done so by adding a new section entitled "Continental displacement of Europe and North America," and a new concluding paragraph in which he claimed that North America had drifted west from Europe.

Thus with our present inadequate knowledge of the properties of isothermal [viscous] remanent magnetization we are inclined to put more weight on the longitude than the latitude of the computed pole positions. With this simplification, the discrepancies between the pole positions for the same or nearly the same geological epochs, based on British and American rocks, may be examined. It will be seen from figure 1 ... that those based on the American rocks are systematically to the west of those based on British rocks ... Consequently we have here evidence of a continental displacement with a mean value of about 24° between Europe and North America in the Mesozoic. The effect is sufficiently consistent between different geological periods, except for the Permian [where the difference was 41°], to be the most unlikely to result from errors.

*(Runcorn, 1956b: 82–83; my bracketed addition)*

There still is no explicit statement about the lack of effect of VRM on longitude. Thus, although Runcorn's recollected reason for changing his mind is not inconsistent with what he wrote in the Canadian paper, it is not strongly supported by what he actually wrote in it.

## 4.5 Opinions of others as to why Runcorn changed his mind

I think that Runcorn had other reasons for switching from polar wander to a combination of polar wandering and continental drift. I shall argue that despite his retrospective explanation to Bullard and to me, Runcorn changed his mind principally, if not entirely, for these other reasons. Perhaps Runcorn simply misre-membered why he changed his mind, perhaps he eventually came to believe that he changed his mind for the reason he later gave. I do not know. Five other explan-ations as to why Runcorn changed his mind have been proposed. Walter Sullivan, former science editor for the *New York Times*, claimed that Runcorn changed his mind because Harry Hess convinced him in 1960 that continental drift had occurred. H. W. Menard, a marine geologist who later made important contribu-tions to the mobilism debate, argued that Runcorn switched to mobilism because of new data. Opdyke suggested that Westoll (Opdyke's other supervisor), who badgered Runcorn about continental drift, was the catalyst; Westoll, a longtime drifter (I, §3.5), also believed that he helped Runcorn change his mind. Creer believes that Runcorn finally changed his mind when he showed Runcorn a paper by John Bradley of New Zealand, who argued that the APW paths for Europe and North America should be interpreted in terms of continental drift, not solely as polar wandering. Creer maintains that he and others within Runcorn's circle had been telling him the same thing for some time, and that when Runcorn saw Bradley's paper, he finally changed his mind and began making appropriate changes to the proofs of his Canadian paper. Creer is also convinced that Runcorn knew about Clegg and his co-workers' Indian results before he returned the proofs of his Canadian paper to the publisher, and about Irving's (who was then in Australia) earlier mobilist analysis of the European and American APW paths (§3.12). Ron Green, Irving's first Ph.D. student at ANU, who together with Irving developed the APW path for Australia (§5.3), also believes that Runcorn changed his mind because he knew of Irving's pro-drift analysis of the European and American APW paths (Ronald Green, February 1979 letter to author). I am inclined to agree with Opdyke, and after examining Runcorn's papers at Imperial College, London, I definitely agree with Creer and Green. I say this despite the fact that Irving himself formerly believed it was unlikely that Runcorn knew of his pro-drift interpretation before changing his mind. Why am I so sure? Because of the now strong archival evidence, which became available only after Runcorn's death, that he knew of both Irving's and Clegg's work. I believe that Sullivan's and Menard's explanations are incorrect. However, because their suggestions have appeared in the literature, I begin by examining them. I shall then consider Westoll's role before considering the explanations offered by Creer and Green.

## 4.6 False accounts of why Runcorn changed his mind

According to Sullivan, Hess convinced Runcorn to adopt continental drift:

Runcorn remained loyal to the idea of polar wandering (as opposed to continental drift) until he ran into Harry Hess at an Atlantic City meeting of the American Association of Petroleum Geologists. Hess persuaded him that the magnetic findings could better be explained by changes in geography due to sea-floor spreading, and from then on Runcorn was Britain's most vociferous protagonist of the theory.

*(Sullivan, 1974: 84)*

Sullivan's suggestion is absurd. Unfortunately, more people are probably familiar with Sullivan's claim than with any other, because it appeared in a book and was later repeated by Robert Muir Wood.

Runcorn collected samples from North America in 1956 and by 1959, after encountering Harry Hess at a meeting of the American Association of Petroleum Geologists in Atlantic City, had convinced himself of the opening of the Atlantic.

*(Wood, 1985: 115)*

Sullivan sent Runcorn a copy of his book; Runcorn gently let Sullivan know that he had got it wrong.

I much appreciate your kindness in sending me a copy of your fascinating and beautifully produced book "Continents in Motion." I was very pleased indeed to receive a copy. I think it is the most comprehensive account of the development of the subject that I have seen. But there are some important misunderstandings about the sequence of the development of ideas. The last paragraph on page 84 is wrong. The meeting to which you refer was held in April 1960: a symposium under the title "Palaeontological and Mineralogical Aspects of Polar Wandering and Continental Drift" held during the annual meeting of the AAPG and the SEPM [Society of Economic Paleontologists and Mineralogists] held at Atlantic City. I can send you a copy of the abstracts. I remember Harry Hess and some of his students coming over from Princeton for the meeting and I remember him telling me at the end of the meeting – and I think this was the first time he said it "that continental drift now has to be taken seriously". But I had, however (four years previously in 1956) interpreted the discrepancy between the polar wandering curve from America and Europe as evidence for continental drift. In this paper I showed the discrepancy could not be explained by the secondary magnetization which we knew was often present and confused the palaeomagnetic story. Like many others I accepted continental drift when it seemed to be the most satisfactory explanation for the data which most interested me at the time. When I wrote this paper I was still in Cambridge and was well aware that both Blackett and R. A. Fisher considered that the rock magnetism data would prove continental drift and not only polar wandering, but I did not believe that until we had data from different continents it was the least bit plausible to explain the changes in the palaeomagnetic directions in terms of anything other than the hypothesis of polar wandering.

*(Runcorn, February 10, 1975 letter to Sullivan; my bracketed addition)*

Runcorn correctly identified the meeting in question, and it is indisputable that he began favoring drift four years before, even though, contrary to what he told Sullivan, he wrote the paper while at Newcastle after he had left Cambridge, and had other reasons for supporting drift.

Sullivan did not date the meeting. The AAPG held its 1960 annual meeting in Atlantic City from April 25 through April 28, and this was the only year from 1955 through to at least 1964 that the association met in Atlantic City. Hess gave a talk there. Moreover, the Society of Economic Paleontologists and Mineralogists (SEPM), a division of the AAPG, held its annual meeting in conjunction with the AAPG, and Runcorn spoke at a special symposium entitled "The Mineralogic and Palaeontologic Aspects of Continental Drift and Polar Wandering," which was sponsored by the SEPM and held on April 25, 1960.[3] In his talk, Runcorn argued in favor of continental drift and polar wandering. This joint meeting is probably the one that Sullivan had in mind; I agree with Runcorn. However, Hess could not have convinced Runcorn "that the magnetic findings could better be explained by changes in geography" rather than polar wandering because in 1960 Runcorn had already been advocating continental drift for four years. Moreover, Hess's famous pre-print in which he first suggested seafloor spreading and announced his support of mobilism was not completed until December 1960. Although possible, it is highly unlikely, as I shall show later (III, §3.14), that Hess already had come up with seafloor spreading by April, even though he made no mention of it in the published account of his talk. But even so he could not have convinced Runcorn to support continental drift because Runcorn already favored it, and had done so for years before Hess ever began arguing for mobilism. Moreover, Runcorn continued to support continental drift and polar wandering after 1960 and therefore Sullivan's account cannot be interpreted to mean that Hess convinced Runcorn to adopt continental drift without polar wandering. Finally, Hess told Vening Meinesz in a letter written on July 6, 1959, that he had not favored continental drift until "recently" and that paleomagnetic findings almost compelled him to accept it (III, §3.12). Hopefully, there will be no more erroneous claims that Hess convinced Runcorn of continental drift.

Menard's suggestion is incorrect but not absurd. He claimed in his book, *The Ocean of Truth*, that Runcorn changed his mind because he had more paleomagnetic data from North America in 1956 than he had in 1955.

By summer he had four more points in North America, and could show that the apparent paths of polar wandering were different for the two continents. Thus the continents had drifted apart.
*(Menard, 1986: unpaginated page within the group of pages containing figures)*

Menard (1986) provided two crude figures he adapted from Runcorn (1956a, b), showing what he believed are four new points (poles) from North America. Although Menard did not identify their origin, Runcorn identified three of them in Figure 9 in 1956a (Figure 3.5, and Figures 4 and 1 in 1956b reproduced here as Figures 4.1 and 4.2); two of the poles are from data of Doell (1955b), and one from Graham (1955). Runcorn did not identify the remaining point, which he denoted by "x" in Figures 1 and 4 (1956b), which, I believe was not a pole at all (Runcorn labeled poles with an abbreviation of the appropriate geological period; x is not such an abbreviation). Moreover, I have found no pole available in 1956 with x's coordinates, 20° N, 164° E in Figure 1, and 172° W, 20° N in Figure 4. What then is x? I think it is simply a marker that was made while drawing and not erased.

Menard does not explain why the new data could have caused Runcorn to adopt mobilism. One possibility is that having more poles gave him confidence to claim that Europe and North America had separate pole paths. A second is that the new poles increased the separation between the two continents. Neither is correct – one of the new poles decreased the difference between the two curves and the others had no effect. There is a third possibility, namely that the new data gathered by Doell and Graham from the Supai Formation helped Runcorn realize that scattering caused by variable VRM did not affect the longitude of poles, a point he illustrated by comparing his Supai data with theirs. In fact, Runcorn (1956b: 81–82) does seem to have made this point (§4.4), and Creer, Irving, and Runcorn (1957) certainly did; after describing how VRMs affect pole latitude but not longitude they compared the three paleopoles from the Supai sandstones.

Since the first palaeomagnetic collection from Arizona was made, the results of which have been described by Runcorn (1955b, 1956a), the Supai formation has been examined independently by Doell (1955) and by Graham (1955). These three sets of results are compared ... The longitudes of the pole positions inferred for the same rock series by the different authors agree, more closely than the latitudes ... thus demonstrating the result of an isothermal [viscous] remanent magnetization by the present dipole field.
    *(Creer, Irving, and Runcorn, 1957: 150–151; the references are the same as mine, except
        Doell (1955) is equivalent to my Doell (1955b); my bracketed addition)*

Menard made no mention of this critical argument.

### 4.7 Westoll's influence

Opdyke, asked about his attitude toward continental drift and why he and Runcorn changed their minds, raised this possibility:

When I first went to England I was very much anti-drift. I remember going down to visit Blackett's group at the University of London and they were pro-drift, and Runcorn was hot on polar wandering. I had a long argument with one of the research students down there about continental drift, which I was on the wrong side of. I guess ... I was trying to reconcile the Permo-Carboniferous glaciation with the polar wandering curves. You just cannot do it, and that is the long and short of it. I don't care how you manipulate it. I spent hours trying to manipulate it to see if you could do it, and I realized that you couldn't. Then I read Irving's paper [1956a], which demonstrated that polar wandering couldn't work. So I really favored continental drift, I think, before Runcorn did.

Runcorn was converted by Westoll. I had dual thesis advisors. One was Runcorn; the other was Westoll. He was in favor of drift, and he was twitting Runcorn at lunch one day about the differences in the polar wandering curve for North America and Europe. And Runcorn went back and thought about this, and three days later he produced a manuscript ... It was like light banter back and forth. And Westoll has a very keen mind. He can remember every damn thing in the world. I didn't really participate very much. Westoll was a very keen drifter and he was just sort of nudging Runcorn all the time. And Runcorn would say something about the confidence limits and things like that. "But still," he (Westoll) said, "There is a difference, you know." And so finally, Runcorn, I guess, really what he did was go back and think about it more and changed his mind.

I think his argument [in the Canadian paper] was that the viscous remanence, the overprint, wouldn't change the longitude of the pole position, it would just change the latitude, and therefore there were in fact discrete polar tracks. I cannot remember now, but I think that was the argument he used in the paper, and maybe that was the argument – he finally realized that there was a difference between them ... But it was at that lunch ... [Westoll] acted as a catalyst.

*(Opdyke, 1984 interview with author; my bracketed additions)*

Parry also believes that Westoll played a major role in getting Runcorn to change his mind. But he thinks that it happened at a meeting of the Cambridge Philosophical Society.

My impression is that it [i.e., Runcorn changed his mind or was forced to reconsider his non-drift interpretation] happened particularly at this Cambridge Philosophical Society meeting to which he went with Westoll ... Westoll had a fairly strong influence in doing this ... Westoll pushed him to rethink the question. It was a somewhat subjective thing in the sense that we had no absolute mathematical criteria as to whether these were significantly different or not. It was rather a question about probabilities over these overlaps. I think the extra argument was the viscous magnetization argument.

*(Parry, 1987 interview with author; my bracketed addition)*

This meeting, held at Cambridge on March 5, 1956, was entitled "The Wandering of the Earth's Pole," and Runcorn, Westoll, and T. Gold spoke.

Although Runcorn remembered the meeting, and recalled two amusing anecdotes, he said nothing about Westoll causing him to reconsider.

The Cambridge Philosophical Society wanted to have it, and I organized it. Westoll came down. Graham was there too. I think that was the time I asked Jeffreys. But he wouldn't talk. Opdyke came down. Neil was, of course, a geologist, and was just getting to know something about geophysics. Jeffreys made some remark that indicated that he thought that the Earth's core was a permanent magnet. Well, the dynamo theory was then being talked about a lot, and I always remember him [Opdyke] telling me either in or after the lecture, "So Harold [Jeffreys] thinks that the core is a permanent magnet." So that shows that Jeffreys wasn't really interested in the Earth's magnetism. The other thing which came out of that meeting, which I remember quite well because Tommy Gold has often mentioned it to me: There was some geologist there who made some futile objection at that time to what we had been saying, and Tommy Gold turned to me and said rather loudly, "Geology rots the mind." Tommy has often referred to that occasion, and asked if I remember that. Of course, I do.

*(Runcorn, 1984 interview with author; my bracketed additions)*

And what does Westoll think? He believes that he significantly influenced Runcorn to consider mobilism. According to Westoll:

I'll tell you one thing, and that is that Keith did not discover by his own volition the implications of paleomagnetism to continental drift. What he did was start studying polar wandering, and he wrote a paper on this in *Nature* in 1955. He had the [apparent] polar wandering curves for North America and Eurasia together, and they converge. And, he said, "Now isn't that interesting?" I said, "Yes, and do you realize that the way in which they converge, the shape of the curves, indicates that there is a polar angle that subtends both these and they are the same curve?" He hadn't seen this. And you could talk to Collinson, perhaps, Tarling, and some others that this actually happened in our senior commons room. I don't know about Tarling. Collinson was certainly there. Neil was there ... That was on a Saturday.[4]

*(Westoll, 1987 interview with author; my bracketed addition)*

Westoll also believed that this happened over lunch. Although there are differences among Westoll, Opdyke, and Parry's accounts, all agree that Westoll had a catalytic effect on Runcorn in forcing him to reconsider mobilism.

Runcorn himself, however, was less clear about Westoll's influence, although he remembered having had discussions with him, he did not recall the one mentioned by Opdyke.

When I went to Newcastle, this matter was always coming up whether the data could be explained purely as polar wandering or whether you needed continental drift. I mean this was debated a lot in my last year or so at Cambridge. By then I'd got this [apparent] polar wander curve for the western states and my first conclusion was that it was roughly in the same area in the North Pacific as the European one. The Geological Society of America were rather anxious to get this published. Of course, it was because they had given me the money. I said that the data from North America substantiated polar wandering. I was chiefly interested in asking the question, Does this very big sweeping around in the Pacific show in the American data? ... So I contended in that first paper simply to say that the [apparent] polar path of North America was much the same as for Europe, and therefore this was an argument in favor of the reliability of remanent magnetization as the record of the ancient field. Then I went on in that paper to say that I didn't think that continental drift was indicated by these results. And, I knew of course that there was a discrepancy, but I couldn't make up my mind that these differences in pole positions could not be very easily dismissed as errors. Moreover, all the work that had been done – you know the scatter of the data was quite considerable – and there were clear indications, particularly in the American work of remagnetisations along the present field which we couldn't get rid of at that time, but which, you know, I did a lot of fitting to great circles in showing that many of the points obviously laid between what I interpreted to the original direction of magnetization and the present dipole field ... I remember discussions with Westoll, but I don't particularly remember that one. But, certainly Westoll had become a bit of an advocate of continental drift. I was struck by the fact that he seemed to be more open-minded on the matter than most geologists that I knew.[5]

*(Runcorn, 1984 interview with author; my bracketed additions)*

I think there is good reason to conclude that Westoll and Runcorn extensively discussed continental drift, that Westoll pressed Runcorn claiming he was mistaken not to argue in its favor, that Westoll influenced Runcorn to consider the issue, but Westoll alone did not convince Runcorn to change his mind.

## 4.8 Creer shows Bradley's manuscript to Runcorn

I, and his [Keith's] other paleomagnetic friends had been trying to get across to him that it was highly improbable that the westward displacement of the N. American paleopoles was accidental. The sight of Bradley's m/s succeeded in penetrating his somewhat stubborn attitude where we (including Westoll) had failed.

Through the years he had dug himself into a hole by repeatedly stressing polar wander and neglecting drift, or mentioning drift unfavorably, in his many lectures given on tour. The sight of Bradley's m/s, which criticised him sharply in print, shocked him into the realization that the time had come to express in print, too, his acceptance that the two [apparent] polar wander paths were different and thus suggestive of drift between Britain and N. America.

*(Creer, January 17, 1999 letter to author; my bracketed addition)*

Ken Creer was asked to refcree John Bradley's paper "The meaning of paleogeographic pole," by the *New Zealand Journal of Science*, which received the paper on June 19, 1956. He received the manuscript several weeks later at his office at the British Geological Survey in London. He immediately read it and recommended publication, for "it expressed pretty closely what Jim Parry and I had been telling Runcorn for a year or more" (Creer, April 9, 1984 letter to author). Creer had been planning to visit Newcastle before receiving Bradley's manuscript. He had a Survey project that required fieldwork near Newcastle and, as he recalled in the same letter, "I wanted to look over the labs and meet with my new colleagues in the Department prior to taking up my post later in the year." Creer took Bradley's manuscript with him. He wanted Runcorn to change his mind, and he thought that showing him Bradley's manuscript might help. Creer thought that by continuing to neglect the importance of drift as an explanation for paleomagnetic results, Runcorn was hurting himself and his colleagues. It was not strictly proper. As Creer remarked in his letter, "In retrospect I suppose that I should not have shown him this m/s because reviewers are sent papers in confidence." But Creer already had decided to recommend publication. Bradley's paper contained no new data; he only said that Runcorn had wrongly neglected continental drift, which is what Creer and others had been saying to Runcorn "for a year or more." Forty years later, Creer recalled:

I can't remember any of the exact dates, having not kept a diary.

In 1956 I was working for the Geological Survey and visited Newcastle during the summer prior to my taking up my appointment there in September when I went to S. America direct from London to make my first rock collection. I called in at Newcastle for a few days during my program of fieldwork – Keith was there so that places the date early summer before the end of the University Term and before the end of the formal examination period.

I only saw him briefly when I brought up Bradley's paper and showed it to him. I left it with him to peruse and collected it after a few hours. When I (later the same day) put it to him, his unwillingness to accept that the palaeomagnetic evidence pointed to the occurrence of drift, he went off in a "huff" and that was the end of any exchange of views. But it was clear (from the papers on his desk visible from my side – i.e. "upside-down") that he was about to start revising the proofs of his own paper [for the *Proceedings of the Geological Association of Canada*]. He was about to leave for his summer trip to USA and I had to go off and join my fieldwork group so that was the end of our brief exchange of views and news.

His overall reaction then was typical of his behaviour when his views or conclusions were questioned. No reasoned discussion. But what had been said invariably "sunk in" and later invariably came to the surface weeks later as his own original idea.[6]

*(Creer, April 5, 2000 email to author; my bracketed addition)*

What about Runcorn? Did he remember seeing Bradley's paper? Yes he did, but he was not sure when and he did not think it affected his views. Upon being asked about the episode, Runcorn recalled:

My recollection of Bradley's paper is that there was no analysis there. He just simply said that the data could be interpreted in terms of continental drift ... But, I remember Bradley's paper well, and, without looking at it again, I think that I would have read it and said, "Oh this is quite interesting, but obviously he is a person who on geological grounds thinks that continental drift is a good idea, and he wants to press this data into his mold," if I was indeed shown it prior to the '56 paper.

*(Runcorn, 1984 interview with author)*

He remembered seeing Bradley's paper, but said nothing about the episode in question. He also said that he did not know if he saw the paper prior to "the '56 paper," his Canadian paper. However, he did not deny it and at the time that I questioned him, I knew none of the details that I might have used to help jog his memory. Nevertheless, I would think that he surely would have remembered his encounter with Creer and changing proofs so extensively. But, according to Creer, it was not uncommon for Runcorn to make extensive revisions.

Keith made extensive changes to his papers at the proof stage very frequently – most people in those days were careful not to (or were afraid to) make substantive alterations because it required expensive type re-setting at the publishers (long, long before the days of word processors). But Keith never showed any reserve: sometimes every page of the proofs would be covered with handwriting inserted between the printed lines and also filling the margins, sometimes extending to the back of the printed sheet. I was amazed at what he got away with. But though proof changing was not remarkable, it never involved such substantive changes as in this particular case: new ideas were often added, but the changes made were not so basic as to effectively reverse the conclusions of the article as originally submitted. When I saw what he was starting to do, I remember thinking, "He'll never get away with it." But he did! No one from the department who is still around remembers anything of the episode: he kept it all quiet himself because he was not a person who liked to admit that he had been persuaded to take a new view of things following the arguments of others. But largely because I don't have a written diary I have nothing new to add to what I have already said that is solid enough to warrant including in your definitive printed record.

*(Creer, March 30, 2000 email to author)*

Just how extensive were the needed changes to the proofs? And, how extensive could they have been and remained acceptable to the journal? I believe that they could have been made and probably were extensive, but also think that the changes were fairly self-contained. With Creer's help, I have analyzed the Canadian paper, and below I give two excerpts in which the parts that could have been the original text are in normal print, and the additions that Runcorn would have needed to make are in italics. The title of the paper, "Palaeomagnetic comparisons between Europe and North America," remained unchanged. The new concluding paragraph became:

Runcorn (1955, 1956) has shown that pole positions derived from the remanent directions of magnetization from American sediments of various geological ages are in rough agreement with those derived from British rocks of the same age. In this paper paleomagnetic comparison between the continents of N. America and Europe is considered further, *particular reference being made to such discrepancies which exist. It is shown that these seem to require about 20° of displacement of America west from Europe in post Triassic times.*

*(Runcorn, 1956b: 77; italics indicate needed changes. The non-italicized passages would have been appropriate as support for polar wandering without continental drift.)*

Here are three paragraphs from the new section entitled "Continental displacement of Europe and North America."

*Thus with our present inadequate knowledge of the properties of isothermal remanent magnetization we are inclined to put more weight on the longitude than the latitude of the computed pole positions. With this simplification,* the discrepancies between the pole positions for the same or nearly the same geological epochs, based on British and American rocks, may be examined. It will be seen from figure 1 or from table 2 that those based on the American rocks are systematically to the west of those based on British rocks ... Consequently we have here evidence of a continental displacement with a mean value of about 24° between Europe and North America in the Mesozoic. *The effect is sufficiently consistent between different geological periods, except for the Permian, to be most unlikely to result from errors. This conclusion is not dependent on the assumption of the coincidence between axes of rotation and the mean geomagnetic axis, on which the conclusion of wandering of the axis is based.*

*A small displacement of the globe can be represented by a rotation about a point on its surface, and so this continental displacement is a rotation of North America with respect to Europe of about 20° about a point near to the present north pole.*

It may be thought that this displacement results from experimental error: the scatter of palaeomagnetic measurements is such that a circle of 10° radius around the computed pole contains the true pole position to a probability of about 95 percent, assuming the rocks selected present a fair sample of the behavior of the field during the geological period. *Thus about a one in twenty chance exists that the displacement between poles computed for the same geological period are in disagreement by twenty degrees. To obtain an error of the same order in the same direction for five separate geological periods seems unlikely.*

*(Runcorn, 1956b: 82–83; italics are his needed changes. Again the non-italicized passages would have been appropriate as support for polar wandering without continental drift.)*

These are the main changes he would need to have made. Even the two paragraphs about the effect of VRM on the computed position of the poles based on his, Doell's, and Graham's collections from the Supai Formation could have remained unchanged because Runcorn drew no conclusions about the general effect of VRM on the latitude of poles or the absence of its influence on longitude. He could have returned the proofs with the changes above, a new title, and opening and closing paragraphs, and a new diagram. Creer's contention is therefore possible with about a 15% change.

### 4.9  Bradley and his paper[7]

John Bradley had very rough treatment here in New Zealand, which was a real shame as he inspired a significant number of students and was an unrecognized world leader in continental drift studies.

*(Graeme Stevens, April 2000 letter to author)*

John Bradley was born in Gateshead, England, in 1910, just a mile or two from where Runcorn was later correcting his proofs. He attended the University of Durham, where he received M.Sc. and Dip.Ed. degrees. During the 1930s he taught in primary schools, and later during the classes he gave at Victoria University would occasionally talk about the hardships facing slum children. Attending the University of Durham during the late 1920s and early 1930, and taking courses in geology, it is very likely that some would have been from Arthur Holmes who was there from 1925 until 1943. He may therefore have been introduced to continental drift by Holmes. Bradley served in the British Army during World War II. Even if he had not been introduced to drift at Durham, he certainly was when he took a lectureship in the Department of Geology at the University of Tasmania under Carey.[8] Just when Bradley went to Tasmania is unclear. By the time he left for Victoria University, New Zealand, in 1951, where he was appointed senior lecturer in the Department of Geology, he had become a strong advocate of continental drift. He was awarded his D.Sc. from the University of Durham in 1960, and promoted to associate professor at Victoria University the next year. He retired from there in 1975.

   New Zealand geologists were solidly opposed to mobilism. D. A. Brown was a notable exception (III, §1.18), and Bradley faced stiff opposition. When he arrived at Victoria University, he enthusiastically discussed continental drift in his classes. Graeme Stevens, who began taking courses from Bradley in 1952, recalled his support for continental drift and New Zealand geologists' opposition to it.

In 1953 I was undertaking a double B.Sc. major in zoology and geology. In the Geology Department I was getting a large dose of enthusiasm for continental drift from John Bradley. However, through contact with other geologists, mainly via a vacation job with the New Zealand Geological Survey, I was aware that these ideas were very much out on a limb in the earth science community.

*(Stevens, 1988: 31)*

Stevens (April 2000 letter to author) later said that Bradley was viewed as a joke by people in the New Zealand Survey who were outspokenly critical of him. Soon after, Bob Clark became Head of the Department of Geology at Victoria University in 1954, and told Bradley to stop lecturing on continental drift. Clark, despite being educated at the University of Edinburgh while Holmes was there, was strongly opposed to mobilism. Stevens recounted the rift between Bradley and Clark, and added an amusing anecdote involving Lester King.

> In 1955 I had a very busy year juggling the lecturing commitments with the writing up of my M.Sc. thesis. However, via tearoom talk and almost daily trips on the train with John Bradley ... I was in a very good position to see the developing friction between John Bradley and Bob Clark. Eventually things came to a head with the result that Bob gave instructions that John was to cease all talk about continental drift ... and to stick to "straight" petrology, structure and geological mapping. However, towards the end of 1955 there occurred an interesting footnote to the disagreement ... Lester King visited New Zealand in 1955 and offered to give a lecture at Victoria. I well remember the discussion that ensued between Bob and John. Bob was absolutely adamant that King would not be allowed to publicise his crackpot ideas on continental drift in HIS department!! Eventually a compromise was arrived at and Lester King gave his lecture to a capacity audience in one of Victoria's large lecture halls (in the Biology building, not Geology!). On his part King gave a masterly lecture, supposedly on geomorphology, but with a large measure of continental drift thrown in, in the guise of the landscape evolution of the Gondwana continents. As Lester King rolled out the Gondwanaland story with great aplomb Bob Clark's face was a study!!
>
> *(Stevens, 1988: 33–34)*

Apparently, Clark ran the department tyrannically.

In 1956 Bradley began writing a book on continental drift in which he gave an account of seafloor generation that involved injection of dyke swarms at mid-oceanic ridges (Stevens, 1988: 34). It was never published. Stevens also recalled that Bradley

> said in one of his lectures that if you added up the total thickness of basaltic dykes exposed on land in Iceland (the offshore story was not known at the time), the accumulated total provided a quantifiable measure of the rate of opening of the Atlantic.[9]
>
> *(Stevens, 1988: 34)*

Unfortunately, the only paper that Bradley appears to have published on mobilism was his critique of Runcorn. Bradley must have been pleased and also dismayed by Runcorn's work; pleased with his new results from North America, but dismayed at his rejection of mobilism. Perhaps Bradley was sufficiently inspired by King's talk or aggravated by Clark's prohibition on drift that he felt obligated to respond in print.

Bradley strongly disagreed with Runcorn's non-mobilist interpretation of the British and North American results. Citing Runcorn's *Scientific American* paper, which appeared in September 1955 (1955c), Bradley, in the paper Creer refereed, announced his disagreement with Runcorn by referring to a map of his that showed poles from both regions.

The present writer, adopting a different view, considers that Runcorn's map presents clear evidence wholly compatible with, and perhaps only interpretable in terms of, continental drift.
*(Bradley, 1957: 354)*

Bradley accused Runcorn of begging the question by assuming that continental drift had not occurred.

It seems to the writer that Runcorn's argument, like many arguments which have been used to discount continental drift, tends to begin with the premise rather than end with the conclusion that continental drift is unnecessary. The method of magnetic surveying and the interpretation of its facts suggest this. Despite universally accepted evidence that mountain building involves the approach of land masses over tens of miles and the evidence of transcurrent faulting, the magnetic surveyor apparently chooses to ignore the most elementary law of surveying – that he should be sure of each datum. Although he checks that the magnetism of the rocks is unchanged, he begins with the assumption that his survey stations (the rocks he uses) have not moved in relative place or orientation since their establishment perhaps a thousand million years ago. It is a common practice among surveyors to examine even the faintest rumour that base sites are unstable, and it would seem that the cautious magnetic surveyor should, in courtesy at least, accept continental drift as rumour and examine his stations before and not after his survey.
*(Bradley, 1957: 359–360)*

But Bradley did more than simply criticize Runcorn; he argued that mobilism, not polar wandering, was the better explanation. Hypothesizing a center of rotation about which one or both continents had rotated, he argued that his own mobilist solution offered a better fit than Runcorn's (1955c) reconstruction, and was consilient with those proposed by Wegener and du Toit.

The linear and consecutive arrangement of poles indicated by Runcorn is seen to be considerably enhanced by the revised plotting. Whereas Runcorn's map displays a broad zone of scattered poles, the revised map [Bradley's Figure 1; reproduced here as Figure 4.3] shows a compact, more nearly linear arrangement and a roughly circular pattern. It is extremely improbable that the new pattern is fortuitous, for the sense of rotation applied to all poles is the same, while the amount of movement given to each is closely dictated by its individual radius from the centre of rotation. The concept of rotation is, itself, not random, nor is the centre chosen with the express intention of finding a solution to the geomagnetic problem. Westward drift with rotation about a point nearer the North Pole is, in fact, agreed on by exponents of continental drift from Wegener (1920) to Du Toit (1937) and Carey (1955).
*(Bradley, 1957: 355–356; my bracketed addition; Bradley's references to Wegener and du Toit are identical to mine, but Carey (1955) is my Carey (1955a). Carey (1955a) will be discussed in §6.7)*

Bradley argued that his interpretation consistently matches the ages of poles with their spatial distribution on their APW paths. By contrast, Runcorn's single APW path placed the North American Carboniferous pole out of sequence; it was between the British Triassic and Tertiary poles instead of between the British Triassic and Carboniferous poles.

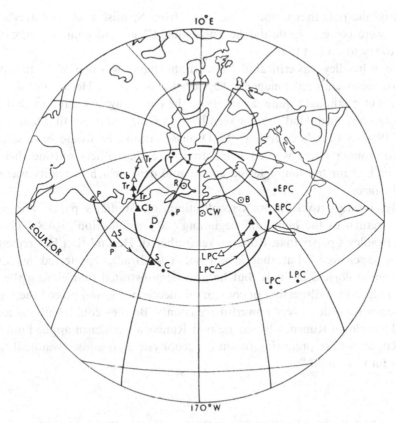

Figure 4.3 Bradley's Figure 1 (1957: 355). The effect of continental drift on the paleomagnetic poles. Solid circles are British, open triangles are American. The solid triangles are the American paleopoles as adjusted by Bradley for closure of the Atlantic. North America has been rotated about point CW, located at 50° N latitude and 170° W longitude. Bradley has rotated North America to its postulated early Tertiary position. EPC signifies early pre-Cambrian; LPC, late pre-Cambrian; C, Cambrian; S, Silurian; D, Devonian; Cb, Carboniferous; P, Permian; Tr, Triassic; T, Tertiary. Notice that the British Tr and adjusted American Tr paleopoles are almost congruent.

Apart from the improvement in linear pattern of poles, there is a distinct improvement in the time relations of poles as read in linear sequence, and anomalous successions such as Triassic-Carboniferous-Triassic which occur on Runcorn's map are rectified. This improvement greatly heightens the probability that the pattern is significant, and, irrespective of the fundamental cause of the general circular pattern of poles, it seems extremely probable that America and Britain have suffered a different amount of rotation about [a] point.

(*Bradley, 1957: 356; my bracketed addition*)

Bradley appealed to mantle convection as the mechanism for drift.

Anticipating future work, he suggested that Carey's (originally Argand's) rotation of the Iberian peninsula relative to the rest of Europe be tested by paleomagnetists, and plotted the pole that would be observed from Spanish rocks if Carey's reconstruction were correct. Little did he know that Clegg and company already had decided to try to do just that.

Although Bradley was critical of Runcorn, his paper was one of the first positive reviews of the overall paleomagnetic support for mobilism. He did not agree with Runcorn, but he thought quite favorably of the paleomagnetic support that British workers had obtained. Had Bradley seen Irving's *Geofisica* paper that had appeared in April 1956 or Creer's or Irving's Ph.D. dissertations, he would have sided with them. No wonder Creer wanted Runcorn to see that others outside their circle thought he had for too long ignored continental drift, which in reality was staring him in the face.

Finally, I return to Runcorn's recollections of Bradley's paper, and ask the reader to return to the quotation beginning "My recollection" (§4.8). There was more to Bradley's paper than Runcorn remembered. He said that he "remembered Bradley's paper well," but that is not so. Yes, Bradley preferred his mobilist solution on geological grounds, but he also demonstrated that closing the North Atlantic reduced the dispersion of contemporaneous poles, and placed them in their correct temporal order – very powerful arguments. Bradley could, with justice, have turned the table on Runcorn by saying that Runcorn's criticism applied not to him but to Runcorn who, preferring fixism on geophysical grounds, "want[ed] to press this data into his mold."

## 4.10 Further implications of Runcorn seeing Bradley's paper

If, as I have just argued, Runcorn changed his mind and corrected the final proofs to his Canadian paper in late June or July, it is very likely that he also knew of Clegg and his colleagues' new results from India, and Irving's two-APW-path, mobilist interpretation of the paleomagnetic results from North America and Europe and his results from Tasmania. Additionally, he must have quickly revised the proofs for his Dutch paper after sending off his Canadian paper because the Dutch paper was published in August (§4.11). Clegg, Deutsch, and Griffiths submitted "Rock magnetism in India" on December 15, 1955, and it was published in May 1956. Irving's paper (1956a) in which he interpreted the paleomagnetic results from North America and Europe was received by *Geofisica Pura e Applicata* on February 15, and appeared in April. There is no recorded date of reception for Irving's paper on the Tasmania dolerites (1956b), but it appeared at about the same time as Irving's *Geofisica* paper; both were published *before* Runcorn made changes to the proofs of the Canadian paper. Moreover, Irving's Tasmanian pole is discussed in his *Geofisica* paper, which also cites the volume of the journal in which the Tasmanian result was to appear. Furthermore, as I discuss in a moment, Runcorn certainly knew of both Irving's *Geofisica* paper when changing the proofs of the Dutch paper for the very good reason that he referenced it, as well as Clegg and company's paper on India's Deccan Traps, citing them as "in press" and identifying the journals. The reference to Irving's paper includes the journal and volume but not the pages, which suggests that he had been sent a preprint; the reference to Clegg *et al.* gave only the journal. Although Runcorn did not reference Irving's paper on the Tasmanian dolerites, if he had a preprint of Irving's *Geofisica* paper, he knew of the Tasmanian pole.

If Runcorn knew about Clegg and company's Indian work, he certainly would have had a very compelling reason to change his mind. They confirmed Irving's earlier results from India, showing that India in the earliest Tertiary had a pole vastly different in position from either its North American or European equivalent, and likely lay in the Southern Hemisphere. If Runcorn knew about Irving's Tasmanian pole or his mobilist interpretation of the North American and European data, he would have had even more reason to change his mind. Moreover, he also would have seen Irving's appeal to paleoclimatology that underpinned the paleomagnetic case for mobilism, and his drawing of two separate APW paths for Europe and North America.

Creer believes that Runcorn knew about both publications before they appeared.

The India results. Keith would have been aware of, even familiar with, the general tenor of the results being obtained by the Clegg group long before they were published, or even presented at a conference. He would have learnt from Blackett who was his "mentor." Blackett did not regard him as a "rival" because of the large gap in seniority. Though Clegg almost certainly did and would have attempted to delay any transfer of information. But he did not have the power to do so because Blackett kept up with every development almost with day-to-day regularity. Similarly it would be a safe assumption that Keith would have learnt of Ted's progress well before publication, through the "grapevine." So I don't think you can make a realistic historical reconstruction by checking out publication dates.

(Creer, April 5, 2000 email to author)

Unfortunately Creer (April 5, 2000 email to author) does not remember discussing either paper with Keith when he confronted him with Bradley's manuscript; he may not have known about them because he had not yet left his job in London and recommenced paleomagnetic research. Three decades earlier, Irving told Bullard and me that he and Runcorn had independently concluded that the discrepancies between the European and American APW paths required continental drift. When Irving wrote Bullard in 1978, he wanted to tell Bullard that he too had drawn two paths.

I should tell you ... that Runcorn was not the only person to notice in 1956 the difference among poles from different continents. I did too, quite independently (*Geofisica pura et applicata* (1956) *33*, 23). I was in Australia and out of touch with Runcorn, our relation being in one of their periodic times of strain. My paper would actually have been out in 1955 if it hadn't first been rejected by a more widely circulated journal for the usual reasons that drift couldn't happen ... My paper also contained the first calculations of palaeolatitudes and first attempt to make comparisons with the palaeoclimatic information.

(Irving, May 12, 1978 letter to Bullard)

Here is what Irving told me three years later:

Runcorn and I arrived at this independently. I had read his 1955 paper in which he said the poles for Europe and North America were the same. I re-plotted them, and disagreed with him, and showed that in my 1956 paper. Runcorn in his *Proceedings of Geological Association of Canada* paper in the same year changed his mind. I was having very little contact with Runcorn at this time. These were two independent conclusions. I think this is a rather important point: Keith and I came to the conclusion quite independently. But, it is true to say that I was the first person to have a go at looking at how all this worked out when you began looking at paleoclimates.

*(Irving, 1981 interview with author)*

Formerly Irving believed that he had not sent Runcorn a copy of his forthcoming *Geofisica*, the paper in which he drew two APW paths and reported his Tasmanian results, which, like his Indian results, were far removed from the equivalent poles of Europe and North America, because he does not remember doing so and because he had little contact with Runcorn at the time, their relationship being strained. I say "formerly believed" their work was independent: he now knows that Runcorn at the time had a copy of what was essentially a preprint of his *Geofisica* paper and that he had very likely sent it to him. Green and Creer are correct. Runcorn, I add, also knew of Clegg and company's India work. Creer is again correct.

I first consider the Indian work. Runcorn, just as Creer suggested, learned of Clegg and company's work through Blackett, as evidenced by the following letter that Runcorn wrote to E. D. McKee, United States Geological Survey (USGS), on February 21, 1956, telling of Opdyke's plans to look at paleowind directions in the western United States during the upcoming summer.

Dear Eddie,

We are going ahead with our plans for wind direction work and thought we don't want to do anything which our organization had already planned to do … Opdyke is thinking about the possibility this summer of looking at the Wingate and other formations. Perhaps you would consider how this would fit in with your plans …

I am certainly very convinced that we in palaeomagnetism must seriously think about palaeoclimates.

Blackett has just got some very interesting results from Deccan traps which seem at first sight to indicate a movement of India since the Cretaceous of about five thousand kilometers. This he interprets as evidence for continental drift. The magnetic results are well grouped and it is tempting to take them at face value but I feel that we must consider whether in these matters geology gives findings for or against.

With kind regards,

Yours sincerely,
S. K. Runcorn[10]

So there is absolutely no doubt that Runcorn knew of Clegg and company's results. Did Runcorn really care about determining whether or not "geology gives findings for or against" or was he just being diplomatic because he wanted McKee to help or at least not interfere with Opdyke's forthcoming fieldwork? Did he really need support from geology before committing himself to continental drift? I do not know. Nonetheless he certainly did know about Clegg and company's results and, although he did not tell McKee, he knew that they confirmed Irving's earlier Indian findings.

Irving was flabbergasted when I told him that in the Runcorn archive there is a manuscript of his *Geofisica* paper. After examining it he reconstructed what had likely happened (Irving, July 7, 2008 email to author). To understand his reassessment, it is important to recall that his paper was first rejected by the Geological Society of Australia after which Jaeger told him to send it to *Geofisica Pura e Applicata*, where it was accepted in February 1956. Recall also that he corrected a mistake in Creer's formula for calculating confidence limits for poles. I also note that on the cover of the copy in Runcorn's files, "Keith Runcorn" and "Geophysica Vol 33" are handwritten. The handwriting is Irving's, and "Geophysica" is a misspelling by him of Geofisica.

Irving first discussed the manuscript itself:

I am almost certain that this TEXT in Keith's papers is what I submitted to *Geofisica* and is close to but not EXACTLY as submitted to the Geological Society of Australia. It comprises what I submitted to the Geological Society of Australia with page 6 taken out and retyped with the formula for the lateral error in the pole corrected as in the published paper. Page 6 has at top left enigmatic marks which all other pages do not, suggesting a different history. The remainder of the TEXT was unchanged when submitted to *Geofisica*.

*(Irving, July 22, 2008 email to author)*

After pointing out minor differences between the manuscript and the *Geofisica* paper, Irving turned to the figures, noted that Figure 4 is missing from Runcorn's copy, and detailed differences with the other figures.

The FIGURES: I do not have Fig. 4. [It was not in Runcorn's copy.] Figs. 1, 2 and 3 are NOT as submitted to *Geofisica*. This confirms (especially clear in Fig. 2) again that I made correction to Ken's error formula between the two submissions. Figs. are probably close to or as submitted to Geological Society of Australia, but are not as in the printed article. Note that pole errors in Figs. 1 and 2 are not the same and this is to be expected because of change in formula; changes are small except for Tasmania because it was near the pole. In ms Fig. 2 Deccan and Tasmanian poles are crosshatched, not so in paper. In ms both Figs. 1 and 2 numbers are hand drawn, in the printed paper they have been drawn with a stencil. In both Figs. 1 and 2 the coastline breaks near poles differ from those in paper. In Fig. 3 (top third cut off) 30° S is either not there or incomplete, the numeral III is not same as in paper, similarly some error bars are not quite the same in ms as in printed paper.

*(Irving, July 22, 2008 email to author; my bracketed addition)*

Irving concluded that Runcorn's copy "was a composite of" the original and the revised paper, "a result of the paper's evolution," and that it was exceedingly unlikely that anyone else could have done this. Therefore he must have sent it to Runcorn.

I think we can say the copy I sent to Keith was a composite of what I had as a result of the paper's evolution. That is just what I could have done cobbling together pages from here and there; we had no photocopying. The changes to the TEXT were minor, only the formula change (equation 11) [Ken's error formula] being substantial. I must have redrawn ALL FIGURES before submitting to *Geofisica*. That was essential anyway because all polar errors were now changed if only by small amounts.

*(Irving, July 22, 2008 email to author)*

Irving then turned to the when and why he sent Runcorn a composite of his paper.

(A) Ron Green said that Keith changed his mind because he had pre-publication information about Australian results. It seems to me highly likely that in late 1955 or early 1956 I showed Ron the paper and told him that I had sent Keith the ms copy now in his papers. I cannot see how I would not have done that. He had by then been with me a few months and as his supervisor I would tell him what I was doing and that I was working on it. All this means that the hostile relations between Keith and me had softened much earlier than I recall. I do remember feeling very isolated in Australia with no one until Ron came along to talk to in detail about core paleomagnetism issues. Clegg never replied. Ken was off in another world never to return or so I thought. Keith I think I am correct in saying tried to keep contact by sending cards during his travels to which I have no recollection of replying. I did correspond with him minimally regarding the preparation of the Phil. Trans. papers. As the above shows, the system then being what it was, I needed Keith for publication in the key UK journal and eventually I realized this. And he came through just fine. He was getting a preview of what we were up to after all. It was mutually helpful. Presumably these sorts of thoughts came into play. (B) You report that Ken said Keith knew of the Australian data "by the grape vine." (C) (a) Reprints would come by sea from Italy. That would take 5–6 weeks. If the paper appeared in April then I would get reprints (probably) in early June and if I then sent one to Keith it would not have arrived until after he changed his mind, if he changed his mind before the end of June. I doubt very much if *Geofisica* appeared in the library before June 1956, even if Newcastle subscribed to it, so (Conclusion (a)) *Keith is very unlikely to have seen the printed article before he changed his mind.* I think we can say definitely that he did not because the reference in the Dutch paper is incomplete and misspelled as above, so he cannot have seen the printed article even when he was working on the Dutch article. When did I send the ms to Keith? There are two possibilities. (b) First let us say that I was still distrustful of Keith, and that I would not send it before I knew the paper had been accepted. It was received Feb. 15 and because publication was so swift I must have been told it was accepted not long after. Jaeger was an editor so they probably accepted it immediately, that is by late February. I would hear by air-mail which took about a week. That means I would have heard by early or mid-March. If I sent the ms by air, Keith would certainly have had it in LATE MARCH, if by sea Keith would have had it in MID-MAY. (Conclusion (b)) *So yes he could have received it before he changed his mind in late June.* (c) Next let us say that because of the happy outcomes with the statistical and the basalt papers I had either begun to trust Keith or I was so impatient to show him that he was ever so wrong about his interpretation of his North American path and he should stop holding up progress, that I sent him the ms as soon as I had completed (and had to have done this) the error corrections, perhaps even before I had redrawn the figures. This would have been before Feb. 15, 1956. (Conclusion (c)) *Then he would have received it long before his documented conversion in late June at the earliest. Of the two I think conclusion (b) above the more likely.*

*(Irving, July 22, 2008 email to author)*

Irving added that his attitude toward Runcorn must have begun to soften earlier than he had previously remembered.

I was very hostile to Keith during the last year (1953–4) at Cambridge. I did not trust him. I had softened certainly by 1957, but I did not think I had done so as early as early 1956. I thought the only contact I had with him then was about the preparation of the Phil Trans papers.

*(Irving, July 24, 2008 email to author)*

Given the letter to McKee and Runcorn's copy of Irving's manuscript, I believe that Runcorn knew of Clegg and company's Indian work, of Irving's Tasmanian pole, of his plotting separate European and American APW paths, of his paleoclimatological defense of the GAD hypothesis, and his indirect proof of continental drift (§3.12). Nonetheless, in his Canadian paper Runcorn referenced neither Clegg *et al.* (1956), which was published in May, nor Irving (1956a), which appeared in April. Even if he did not know that they had appeared, he could have referenced them as forthcoming, a common practice of his. Nor, I add, did he tell Creer that he knew of their work, or at least Creer cannot remember him doing so. It seems to me that Runcorn wanted to give the impression that he changed his mind solely because of his own realization that VRM does not affect longitude, and he did not want to acknowledge that Irving's and Clegg's work had anything to do with his conversion to continental drift. In his Dutch paper, Runcorn referenced both works; his citation of Irving's paper, however, suggests that he had not seen the published version because it is just as Irving had written it on the title page of the copy he sent to Runcorn – *Geofisica* misspelled as *Geophysica* just as Irving had done and no pages cited.

## 4.11 The Dutch paper

I said [in the Dutch paper in which Runcorn supported continental drift] that if you put the Permian equator where it is indicated by the British and American rocks, it goes through India, and I think that is right, and that is a complete contrast with the paleoclimatic data, with the glaciations. You know, there was always a hope, and that goes back a long way, that the Permo-Carboniferous glaciations of the southern continents, which were the most striking discrepancy between what one would expect about the Earth's climate and what was observed, that it was always possible that those paleoclimatic discrepancies could be explained by moving the pole around sufficiently. And, I was using in that paper, the argument, well we know now where in Permian times the pole was, let's have a look at this hypothesis that polar wandering alone could satisfy the paleomagnetic data, and in fact came out very strongly that it wasn't going to be a very good explanation alone. If the pole had moved around all over the place within the geological period, then you wouldn't, I mean even the early data wasn't very accurate, but it would not have fallen on a relatively smooth curve in a short period of time, it would have moved all over the place.

> *(Runcorn, 1984 interview with author; my bracketed addition)*

If Runcorn's retrospective comment sounds familiar, it is because in it he repeated Irving's palaeoclimatic indirect proof that paleoclimatic and paleomagnetic findings are not in agreement, if continents are fixed (§3.12). He did cite Irving's paper, but referred not to his paleoclimatic work, not to his identification of separate European and North American APW paths, not to his Tasmanian results, but only to his Deccan Trap work. He seemed unwilling to acknowledge that he was not leading, but following Irving in these matters.

The Dutch paper itself may be read as a recantation of his earlier review (Runcorn, 1955b; see §3.9), and an unacknowledged endorsement of the work of Clegg and his colleagues and of Irving's paleoclimatic arguments. After summarizing why he now thought that the 24° separation between North American and European APW paths was evidence for opening the North Atlantic and not attributable to experimental errors, he claimed that the mechanism difficulty needed reexamination.

Consequently it would appear that the geophysical objections to continental drift must be re-examined. That no forces are available to move the continents may only be held if the possibility of convection currents in the mantle is disallowed. However, the really serious objection is concerned with the strength of the basaltic floor of the ocean (see Jeffreys, 1952). It is however possible (Hills, 1947) that softening of the basaltic floor near the continental margins through heat from the radioactive continents might occur. The continents could then be slowly pushed laterally against a slowly yielding ocean floor.[11]

> *(Runcorn, 1956c: 253; both references agree with mine)*

This was the first of Runcorn's many appeals to convection as the driving force for continental drift, which he had previously considered to cause only polar wandering. For him it was a relatively easy transition.

In his discussion of the Permo-Carboniferous glaciation, he began by commenting on the variable quality of geologic and biotic evidence in support of mobilism.

The difficulty in the discussion of the geological evidence for continental drift is that it is impossible to predict the meteorological conditions of past epochs or the laws governing the migration of fauna and flora in remote geological times. It is also difficult to draw clear deductions from the stratigraphical and tectonics evidence because only a very incomplete record exists. Nevertheless the evidence cannot be disregarded. What is needed is the separation of the evidence into categories, in which different degrees of certainty attach to the inferences drawn.

> *(Runcorn, 1956c: 253)*

This should not be seen as a truism. Every solution, fixist or mobilist, during the classical stage of the mobilism debate had been saddled with severe difficulties. Following Irving, Runcorn now enlisted paleomagnetism to strengthen one of the difficulties that had been raised against fixist explanations of the Permo-Carboniferous glaciation. He first summarized the difficulty, which he noted fixists had attempted to disregard by appealing to what they considered as the undeveloped state of paleoclimatology and meteorology.

The proponents of continental drift have always pointed to the contemporaneous glaciations of India, Australia, South-Africa and South America as evidence of the existence during the late Paleozoic period of the continent of Gondwanaland. It could perhaps be argued that the possibility that a glaciation extended over what is now a whole hemisphere cannot be denied in the present state of meteorological theory and lack of knowledge of possible fluctuations in the amount of heat received at the surface of the earth during geological time (see Jeffreys, 1952). This argument is however somewhat weakened by the evidence, to which reference is made above, of hot climates in what is now the northern hemisphere. Nevertheless ignorance of climatology can still be used to avoid the conclusion of continental drift in a northern hemisphere with extensive hot regions and a southern hemisphere with extensive cold regions is a possible hypothesis.

*(Runcorn, 1956c: 254–255; the reference to Jeffreys agrees with mine)*

He constructed the paleoequators for Europe and North America during the Carboniferous and Permian, positioning them by their paleomagnetically determined poles. If continents were kept fixed, the equators cut through North America and just south of England, which dovetails with the paleoclimatological data that indicated hot climates for both regions. However, if continents are kept fixed, the equators also passed, as Irving had already noted, near formerly glaciated regions. Thus, fixists had to suppose both cold and hot near-equatorial climates.

However the determination of the pole positions by palaeomagnetic studies has made the contradiction [a hot northern but cold southern hemisphere] much sharper. The equator shown in figures 2 and 3 passes near both the hot regions [England and North America] and three of the extensively glaciated regions [southern Africa, Australia, and India]. Yet, unless the earth was not rotating, these regions must have been receiving the same heat from the sun!

*(Runcorn, 1956c: 255; my bracketed additions)*

This argument of Runcorn's is a special instance of the general indirect proof Irving had made in his *Geofisica* paper in which he calculated the latitude variation of southeastern Australia through time using paleomagnetic results from Britain. During the Late Carboniferous and Permian it remained at 10° N, in stark contrast to the occurrence of glacial deposits there throughout these periods (Irving, 1956a, 17–18).

Runcorn continued by noting that neither continental drift nor rapid polar wandering faced this difficulty, and showed his preference for the former.

One solution of this difficulty is to assume that India, Australia and S. Africa were grouped near South-America in Permo-Carboniferous times. The other solution of the difficulty is to assume that the pole of rotation made remarkably rapid movements in Permo-Carboniferous times which have so far eluded the palaeomagnetic method. This would involve the assumption that the lower Permian and the Carboniferous of Great Britain and of the U.S.A. and the glaciation of South-America are all contemporaneous. The glaciation of Australia, India and S. Africa would then involve a movement of the pole to the Indian ocean which would imply that the other pole was in the North Atlantic ... Early evidence that these glaciations were associated with the polar positions was found by Mercanton (1926a), who measured the magnetization of the Permian lavas of New South Wales finding an inclination of 87°. Even allowing for effects of secular variation, this indicates a position very near to the pole.

*(Runcorn, 1956c: 255. The reference to Mercanton is the same as mine. Mercanton (§1.18) suggested that paleomagnetism could be used to test continental drift. Because Mercanton's work was based on one sample from one lava flow, Runcorn felt obliged to mention but dismiss it as the effects of secular variation.)*

Acting alone, polar wandering would have had to have been very extensive and very rapid. But there was the possibility of continental drift. He had come to feel quite comfortable with Irving's Indian work since Clegg and his co-workers had confirmed it – work indicating that India had moved northward through 54° of latitude since the Cretaceous.

Irving (1954, 1956) determined the direction of magnetization of a number of samples of the Deccan traps of Cretaceous age from India and showed that their magnetization was as if they had been laid down in the southern hemisphere, some 90° [actually 54°] of latitude south of their present site. Since then the work has been greatly extended by Clegg, Deutsch and Griffiths (1956), who confirm Irving's result and draw a similar conclusion to his ... It would appear therefore that whereas around the North Atlantic little continental displacement has occurred since late Pre-Cambrian times, it seems probable that in the southern hemisphere it may have been very extensive, of the order of 6000 miles [actually 5000 km] since the end of the Paleozoic era.

*(Runcorn, 1956c: 255–256. The Irving references are to his Ph.D. thesis (Irving, 1954) and the Geofisica paper (Irving, 1956a); the Clegg et al. reference is to their first publication (Clegg et al., 1956) on their Indian work; my bracketed additions.)*

In this way Runcorn had moved from fixed to moving continents. He had very good reasons to switch, once he learned of the work of Clegg and colleagues and Irving; that the APW paths from Europe and North America diverged and were similarly shaped, that present field overprints VRM did not affect the longitude of poles, that the Indian and Tasmanian poles were far from those of the other continents, that the paleomagnetic and paleoclimatic evidence from the same continent were in very good agreement, and that there were flagrant contradictions between paleomagnetic evidence from one continent and the paleoclimatic evidence from another continent unless continental drift was assumed.

## 4.12 Bradley, Runcorn, and Euler's point theorem

Bradley noted that the movement of a continent could be described in terms of a rotation about a center. Such a remark is reminiscent of Euler's fixed point theorem, which is central to plate tectonics. The theorem states that any motion of a rigid body on the surface of a sphere may be represented as a rotation about a fixed point on the sphere, the latter now commonly called an *Euler pole*. Bullard *et al.* (1965) gave the first explicit application of the theorem, using it to determine the best fit between the opposing margins of the Atlantic Ocean. Although neither Bradley (1957) nor Runcorn (1956b) mentioned the theorem by name, they said things that suggest they might have had Euler's theorem in mind. I do not think they did, but what they said is worth reviewing. However, Irving thinks that Runcorn was aware of Euler's theorem, "Perhaps not initially (1956) but certainly later in the 1950s" (Irving, April 2003 note to author).

I first consider Bradley. I think he arrived at what may look like Euler's theorem because he had the idea of rotating continents and used stereographic projections to plot movements of continents. Bradley's Figure 1 (here reproduced as Figure 4.3) is an oblique stereographic projection centered on 50° N latitude and 170° W longitude, approximately the center of the small circle connecting the European poles. He then placed the pole of rotation, point R, for the movement of North America relative to Europe near to the North Pole at approximately 70° N latitude, 200° W longitude. He chose this point because mobilists from Wegener to du Toit and Carey agreed that North America had rotated westward relative to Europe about a point not far from the North Pole. Also, his point, as Bradley noted (passage quoted in §4.9), was the approximate center of the concentric arcs connecting North American poles (hollow triangles) with their "pre-drift" adjusted positions (solid triangles). Thus Bradley suggested a westward rotation of North America relative to Europe around point R, which sounds Eulerian. Moreover, the movement of North America about R relative to Europe appeared rotational in his Figure 1 because it is a stereographic projection whose point of projection, CW, is near to R. Bradley made precisely this point when explaining why he used a stereographic projection.

Figure 1 is a composite map which presents Runcorn's data and its re-interpretation in terms of westward drift of North America. In order to represent the pattern of paleogeographic poles in a geometrically correct fashion, the writer has designed the map of Figure 1 as an oblique stereographic projection (Steers, 1942, p. 134) centered on 50° north latitude and 170° west longitude. In addition to representing shapes in their true form, this [stereographic] projection also allows the rotation of land masses about its centre without involving distortion, and *permits small amounts of rotation about nearby centres without gross distortion.*[12]

> *(Bradley, 1957: 354; my bracketed addition; emphasis added)*

Moreover, Bradley (1957: 354) viewed his rotational analysis as a simplification. "In the interest of simplicity, North America has been rotated about a centre R (Fig. 1) and has been given none of the complex lateral movements that would normally be applied in a more detailed drift map." Thus, I do not think that he had Euler in mind.

Runcorn's remarks are less suggestive of Euler, but they could have been derived from Bradley's. In describing the westward displacement of North America relative to Europe, Runcorn stated:

A small displacement on the globe can be represented by a rotation about a point on its surface, and so this continental displacement is a rotation of North America with respect to Europe of about 20° about a point near to the present north pole.

*(Runcorn, 1956b: 83)*

This sounds like Bradley. Runcorn displayed the pre- and post-drift positions of North America on polar stereographic projections in his Figures 4 and 1 (reproduced here as Figures 4.1 and 4.2). North America's displacement could be represented approximately, as Bradley and Runcorn explained, by a rotation about a point on Earth's surface close to the present pole.[13] Runcorn qualified his remark by saying that "*small* displacements on a globe can be represented by a rotation about a point on its surface," which, again, echoes Bradley (1957: 354, emphasis added), who said that "*small* amounts of rotation about nearby centres" can be represented on a stereographic projection without gross distortion. Euler's point theorem, however, has no such qualification, any rotation, not just small rotations, of a rigid body on the surface of a sphere may be represented as a rotation about an appropriately chosen point on the sphere.[14] Thus I do not think that Bradley and Runcorn had Euler in mind; if either had, they certainly did nothing with it at the time. I shall return to this matter in §5.16 when discussing Creer *et al.* (1958) who do appear to have introduced Euler's point theorem into the tectonic literature.

### 4.13  Creer and Irving, Runcorn, and Graham: a study in contrasts

Of the trio Creer, Irving, and Runcorn, Runcorn was the last to accept mobilism. Indeed, he was the last of the London, Cambridge, Newcastle, and Canberra groups to do so. Moreover, if I am correct, Runcorn probably needed the work of Irving and Clegg, and Creer himself to confront him with Bradley's paper and to push him to accept continental drift. Even though it took him longer to become a mobilist than either of his first paleomagnetism students, he now began to tell anyone who would listen that he, his students, and those working with Blackett and Clegg had developed a strong paleomagnetic case for mobilism, and he began to look into the possibility of invoking mantle convection to provide the means of moving continents.

When I asked Runcorn about his unwillingness to accept mobilism, this was his response.

There were lots of people interested in the data. Some people were urging us to interpret it in terms of continental drift; some people were urging us to interpret it in other ways. I think

I mentioned in my contribution to the biography of Fisher (Box, 1978), that right at the beginning before we had really got any results, he said, "Oh, I expect that you'll get evidence for continental drift." So, you see the question for me wasn't whether other people thought that the data interpreted in terms of continental drift, it was looking at the data and seeing whether, on the basis of these very clear assumptions which were, of course, that the average field was aligned along the axis of rotation and was a dipole, and whether you could, using this basic hypothesis, which I believed was fundamentally based on the dominance of the Coriolis force in the core, whether you could then analyze what was inevitably rather rough data in a way in which the errors would not kind of vitiate the conclusion ... I think by that time Ken had become very interested in continental drift, perhaps because of Blackett or, perhaps, his own reading. For me, I wasn't at that time very concerned with the geological evidence. I still think, looking back, that the geological evidence was unfortunately, you know, could be read either way. People did; intelligent people did. So I am sure that Ken was fired up and stimulated by what he read, by Wegener's ideas, as Ted was too. I felt that they both ignored some of the geophysical difficulties and, of course, in fact then it ultimately turned out that we all had to ignore the geophysical difficulties when the first quantitative data became very difficult to explain without it. So there was a difference in outlook.

*(Runcorn, 1984 interview with author)*

Although Runcorn's retrospective comments are somewhat problematic, they capture an important difference between himself and Creer and Irving: they were empiricists who were influenced by their data and by what they read. Runcorn was more theoretically inclined and less immune than they were to the opinions of important figures whose company he sought: Runcorn was influenced by Vening Meinesz and Gold; both favored polar wandering but not drift (§3.9). Walter Bucher, a longtime foe of mobilism (I, §3.8, §5.11, §6.10, §7.3, §8.4, §8.8, §9.7), also influenced him. Runcorn cared about what some people thought; it just depended on whether they counted, and to him most geologists did not, although Bucher, who held high office in the AGU, did. Creer and Irving were, above all, influenced by their own data, which they had worked hard to get and to which they were very close. They became favorably inclined toward mobilism because of their reading of the drift literature, and because of the results of Hospers, Runcorn, Clegg and company, Gough, K. W. T. Graham, and Hales. Creer and Irving knew about the various geophysical difficulties that had been raised against mobilism, but they had more confidence in the new paleomagnetic data than in the theoretical arguments that sought to dismiss mobilism; for them the important questions were not to do with the mechanics of drift but concerned such matters as what sort of rocks gave "good" results and how could they be used to determine where South America or Australia were in the Jurassic. Likewise, Blackett, Clegg, and Deutsch and the others in the London group were more attentive to the data than to the theoretical geophysical difficulties raised against mobilism. It is, in fact, this particular contrast between Runcorn's attitude and that of the others in the Cambridge/Newcastle/Canberra and London groups that best explains his reluctance to change his mind about mobilism when he first began compiling his results from North America. True to form, once he had made the switch, he began this new phase in his thinking with theoretical speculations about the mechanics of moving continents.

If Creer and Irving are at one end of the empirical/theoretical spectrum, and Runcorn toward the other, John Graham is somewhat beyond Runcorn. Graham gave up his fold test because he did not see how otherwise to avoid accepting field reversals or continental drift. Once Néel told him that he could (in theory at least) invoke self-reversals, he reclaimed his fold test. However, he did not in the 1950s become inclined toward mobilism because of the paleomagnetic results; on the contrary, as I shall later describe, Graham spent the following years attempting to avoid mobilism by devising various ways to question the reliability of the results. This is where Runcorn and Graham differ. Runcorn took longer than the other British paleomagnetists to graduate to mobilism, but not much longer, and he had much further to go, because he took seriously the geophysical difficulties and continued to do so. Graham, however, was driven by his theoretical concerns, and conditioned by the oppressively fixist milieu in which he found himself. He was institutionally isolated from any forward-looking discussion of mobilism. Consequently, he never seems to have developed sufficient confidence in the paleomagnetic case for mobilism to accept mobilism because of it.

## Notes

1 Runcorn was appointed as Head of the Department of Physics at King's College Newcastle, which was part of the University of Durham. King's College was given university status in 1963. Runcorn later had the Department of Physics renamed "School" of Physics. The School served as an umbrella for Departments of Geophysics, Theoretical Physics, Solid State Physics and Atomic Physics.

2 Jeffreys also related Hide's work to meteorology (see Note 17, Chapter 7). Also see IV, §1.10, §3.2, and §3.9 for further discussion of Hide's work.

3 S. W. Carey also gave a talk at the symposium. According to Carey, the symposium created quite a stir.

> The campaign [that is, Carey's 1959–60 academic year lecturing tour of North America] culminated with a special session on continental drift sponsored by the Society of Economic Paleontologists and Mineralogists at the annual meeting of the AAPG at Atlantic City on April 25, 1960. I was lead speaker, and with me on the panel were Keith Runcorn, Ken Caster, and William Gussow. The hall was packed, even the aisles and the walls. After the formal papers from the panel, the questions and discussion continued until long after midnight with few if any leaving, until the chairman had to terminate the meeting. The revolution to continental dispersion had begun!
>
> *(Carey, 1988: 119; my bracketed addition)*

4 After my first interview with Opdyke, I asked Westoll whether he had influenced Runcorn. After I had told him what Opdyke had said, Westoll said the above, and added that the critical conversation had occurred on a Saturday in the Senior Common Room after the issue of *Nature* containing Runcorn's APW path had arrived. But, there is no APW path in the *Nature* paper. The APW path appeared in the papers published in *Scientific American* and the *Bulletin of the Geological Society of America*. Nevertheless, Westoll is pretty sure that it was the *Nature* paper. Although I interviewed Collinson,

I forgot to ask him about this episode. Westoll is wrong about Tarling. Tarling became a student of Irving's at ANU in 1959, and went to Newcastle after working with Irving. He was not in Newcastle in 1956.

5 It should be noted that Runcorn gave this answer after I asked him whether he remembered the meeting over lunch as described by Opdyke. At the time, I had not yet heard Westoll's account. Nor had I yet heard Parry's account about the meeting of the Cambridge Philosophical Society. Westoll had been an advocate for continental drift since the 1940s.

6 Irving, when I asked him, agreed with Creer's characterization of Runcorn's behavior when his views were questioned (Irving, December 2000 comment to author).

7 I want to thank several members and former members of various departments and schools in the Earth sciences at Victoria University of Wellington for helping me learn about John Bradley. Dr. Gillian M. Turner, a paleomagnetist at the School of Earth Sciences at Victoria University, kindly sent me materials about Bradley, and told me about Graeme Stevens and Paul Vella. Stevens took undergraduate courses from Bradley in 1952 and 1953, and wrote (Stevens, 1988) a very interesting paper on Bradley and the strong resistance to mobilism in New Zealand. Paul Vella and Bradley were colleagues in the Department of Geology at Victoria. Vella co-authored (Clark and Vella, 1985) an obituary of Bradley. Stevens has kindly provided me with additional information about him. All the bibliographic information about Bradley that is not in the above mentioned works comes from two letters from Stevens to the author sent in 2000.

8 His appointment at the University of Tasmania with Carey adds more credence to the speculation that Bradley took classes from Holmes. Carey and Holmes knew each other, Holmes having been an examiner of Carey's D.Sc. thesis (§6.5). Holmes might have recommended Bradley to Carey. However, Graeme Stevens does not remember Bradley talking about Holmes. Upon asking him about Holmes' possible influence he wrote:

As far as I can recall, John Bradley did not speak very much, if at all, about Holmes. Certainly Sam Carey was the main person he spoke about in his lectures and in conversation. He would say, "Sam has proposed ...," "But he is wrong because ..." Therefore I do feel that perhaps Carey was John's main mentor   but Holmes may have prepared the ground.

*(Graeme Stevens, June 2000 letter to author)*

9 Bradley (1965) later published a paper on the generation of dolerite sills in which he estimated the volume of injected material for various sills. Although there is one reference to Iceland, there is no mention of continental drift, generation of seafloor, or seafloor spreading as proposed by Hess or Dietz.

10 This letter is found amongst Runcorn's papers at Imperial College London. I thank Mrs Joyce K. Molyneux for permission to reproduce it and both her and E. Irving for allowing to reproduce excerpts from Irving's *Geofisica* paper that he sent Runcorn and are found amongst his papers.

11 J. Parry thinks that he lent Runcorn Hills' little book. Parry won it and Holmes' *Principles of Physical Geology* as school prizes before entering Oxford. Parry was always sympathetic toward mobilism. "I saw no reason why it should be rejected out of hand." He lent Runcorn Hills' book to examine shortly after they had arrived at Newcastle (Parry, 1987 interview with author).

12 Bradley forgot to reference Steers in his list of references. Steers was a geographer at the University of Cambridge. He wrote a very useful book, *An Introduction to the Study of Map Projections*, which helped Earth scientists understand and construct maps of different projections. First published in 1926, the fifth edition appeared in 1942. Steers included a section on stereographic projections, and explained how to construct equatorial, polar, and oblique stereographic maps. F. Debenham, professor of geography at the University of Cambridge, stressed the importance of stereographic projections to readers in his foreword to Steers' book.

Finally, I would suggest that if, confused by such a large number of projections as he will find at the page headings, he becomes disheartened and wonders where to begin, he should follow the lead of the father of all projections, Hipparchus. If the student will master the Stereographic Projection, devised over two thousand years ago, which needs but ruler and compass for construction, he will find that the others fall into line and become at least intelligible, if not always so easy to draw. Though it is rarely seen in an atlas, the Stereographic is in many respects still the most useful of all the projections, and the neatest.

*(Debenham, 1942: viii)*

13 The North Atlantic is incompletely closed by this rotation. Van der Voo (1990) has tested paleopoles from Europe and North America and found that the best fit was obtained by rotating North America 38° counterclockwise about an Euler pole at 88.5° N, 27.7° E as obtained by Bullard *et al.* (1965) which closes the North Atlantic at the edge of the continental shelf. For a more recent discussion see McElhinny and McFadden (2000), pp. 289–291 and references therein.

14 I owe this point to Dan McKenzie (February 3, 2003 email to author).

# 5

# Enlargement and refinement of the paleomagnetic support for mobilism: 1956–1960

## 5.1 Outline

During the second half of the 1950s paleomagnetists gathered rocks from regions not previously sampled, and sampled again from regions surveyed in reconnaissance fashion, determining new APW paths or filling gaps in old ones (§5.2–§5.9).[1] These regions were South America, Australia, India, Antarctica, and Africa; Soviet paleomagnetists made extensive surveys in the USSR. A new paleomagnetic laboratory was installed in India and workers were recruited.

In §5.5, I shall describe how, through development of new laboratory techniques, paleomagnetists learned to get rid of unwanted secondary viscous remanent magnetizations (VRM). This important development was magnetic cleaning, and with its advent the range of rock types that could be used was greatly expanded. Notwithstanding, selection at the outcrop still remained the most important decision a paleomagnetist made. Workers may collect all sorts of rocks but, even with these new demagnetization techniques, many still were not able to observe the ancient field because the rock may not have recorded it in the first place.

In §5.10–§5.15, I describe new work on paleowind directions and how it was used to confirm paleomagnetic results. In §5.16, I describe attempts by workers to reconstruct continents in their former geographic positions using this new paleomagnetic data. Typically they proposed several alternatives that were consistent with the paleomagnetic data, and dismissed any that were inconsistent with geological, paleoclimatological, or paleontological data. Although their results and paleogeographic maps improved, researchers still had to deal with the hemispheric and longitude uncertainties encountered before (§5.16). And here still remained substantial uncertainty concerning the relative contributions of polar wander and continental drift. But, as examples in this chapter show, it was in testing the validity of previous reconstructions rather than in creating new ones that paleomagnetic studies provided the clearest evidence of continental drift.

During the late 1950s, there was much new support for the GAD hypothesis (§5.18): poles from Late Cenozoic volcanics from four continents were shown to coincide with the present geographic pole; arguments were advanced as to why the time-averaged field should be axial; the dispersion of the field was shown to be latitude dependent and consistent with GAD models.

The paleomagnetic approach to mobilism developed by the Imperial College group differed from that of the Cambridge diaspora at Newcastle and Canberra and resulted in a philosophical schism; the latter reduced observations to poles on the basis of the GAD assumption; the former took a more phenomenological approach. The London group made no formal attempt at paleographic reconstructions.

## 5.2 Imperial College and the Tata Institute continue surveys in India

In 1957 and 1958 these groups wrote three papers. The first, authored by Clegg, Deutsch, Stubbs, and C. W. F. Everitt (a new Imperial recruit) described results from four new areas of the Deccan Traps; they (1957: 229) continued to maintain that the "India sub-continent has undergone a substantial northward movement relative to the Earth's geographical axis at some time since late Cretaceous or early Eocene times." However, systematic differences between the magnetization directions of the lower and upper Traps now became apparent, and led them (Clegg *et al.*, 1957: 230) to consider the possibility of "a northward movement of India relative to the Earth's rotational axis, during the time when the Deccan Traps were being laid down." The second paper, Deutsch *et al.* (1958), whose authorship included C. Radakrishnamurty and P. W. Sahasrabudhe of the Tata Institute of Fundamental Research in Bombay (Mumbai), where some of the measurements had been made in a new laboratory, described findings obtained between October 1955 and November 1956 from several new sites in the Deccan Traps. Confirming earlier results, they (1958: 170) maintained that their results "are consistent with the supposition that India has drifted northwards through over 50° of latitude and rotated 25° counterclockwise within the last seventy million years." For the third paper, Clegg, Radakrishnamurty, and Sahasrabudhe collected in January 1957 lavas from the Jurassic Rajmahal Traps, extending their work back to the Jurassic, or so they thought.[2] Their results showed that India was then even further south than it was in Deccan Traps times.

In November 1956, Clegg and company (1957) presented a paper at the Imperial College sponsored London meeting. They divided the poles from various Deccan Trap localities into three stratigraphic groups, defining them by their elevation above sea level. The traps have been differentially buried but have undergone little tectonic disturbance: (1) poles V–VIII were from the reversely magnetized Lower Traps; (2) poles II–IV, from the normally magnetized Upper Traps (Figure 5.1); (3) pole I, from the youngest traps, also obtained from normally magnetized rocks. Noting that the lower (older) the pole the further it was from the current geographical North Pole.

Clegg *et al.* (1958: 830–831) canvassed the three possibilities formulated by Deutsch *et al.* (1958), that the Indian pole variations could

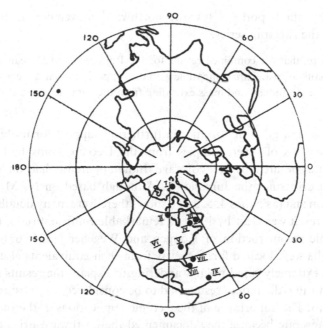

Figure 5.1 Clegg, Radakrishnamurty, and Sahasrabudhe's Figure 2 (1958: 830). Poles represented by solid circles. IX from the Rajmahal Traps then thought to be Jurassic.[3] The eight other poles were from latest Cretaceous/earliest Tertiary Deccan Traps. I is the youngest pole; IX, the oldest.

(a) ... have occurred as a result of some mechanical effect associated with different depths of burial, such as that described by Graham (§7.4).

(b) ... be indicative of partial magnetic instability which would cause a pulling of the magnetic directions towards that of the present Earth's field. This would tend to increase the magnetic dips of the reversely magnetized rocks (1) and to decrease those of normally polarized ones (2) and (3).

(c) ... represent a continuous northward movement of India relative to the equator during the time when the Deccan Traps were being erupted.

Considering their new results from the Rajmahal Traps, they raised anomaly difficulties with (a) and (b) (RS2).

The Rajmahal lavas discussed in this communication are older than the Deccan Traps. They have steeper magnetic dips [inclinations], and give a pole position (IX) farther from the present pole than any of the Deccan rocks. Considering this steeper dip in relation to postulates (a), (b) and (c) above, it appears that: (a) it is unlikely to have been caused by compressional forces associated with deep burial, since all the specimens collected came from the uppermost 250 ft. of the Rajmahal series, and it is improbable that they have ever been overlain by any substantial thickness of material; (b) it cannot have been increased by magnetic instability in the present Earth's field, as the Rajmahal rocks are normally magnetized, and this effect would tend to reduce rather than to increase their magnetic dips.[4]

(Clegg et al., 1958: 831; my bracketed addition)

This left them with hypothesis (c), the northward movement of India between depositions of the two trap series.

It seems, therefore, that the combined results for the Rajmahal and Deccan Traps conform best with the postulate (c) and are compatible with the supposition that there was a continuous northward drift of the Indian land mass extending from the Jurassic to the Eocene.

*(Clegg et al., 1958: 831)*

These results were of great importance in furthering support for mobilism. Irving's initial measurements of seven samples from the Deccan Traps had been greatly extended in time and areal overall coverage. India apparently had undergone about 60° northward drift since the Jurassic (4). Although based on NRM data without demagnetization and (as Notes 4, 5, and 6 show) there were many details still to work out, this main result was clear by 1958 – a remarkable confirmation of the mobilistic paleogeographic reconstruction of Köppen and Wegener (1924: 67) (Figure 3.1). Poles from India were located far, far away from their equivalents elsewhere, and it would require extremely rapid polar wandering to explain the results solely in that manner; realistically, the Indian results had to be considered very strong evidence for continental drift. The importance of the tectonic implications of the Indian results is difficult to exaggerate because they documented the northward drive of India that, mobilists believed, had led to the creation of the Himalayas and the Tibetan plateau, features central to the drift debate since its inception. The Indian results, together with those from Australia (§5.3), which became available at the same time, provided clear-cut evidence of continental drift; the results were so dramatically different from those from Europe and North America; they had more impact on the mobilism debate than contemporary surveys from elsewhere in Gondwana.

### 5.3  Australian National University obtains apparent polar wander path for Australia

When I asked Irving to discuss his contributions to paleomagnetism and the paleo-magnetic case for mobilism, he counted among them:

Getting the Australian group going with Jaeger's encouragement and support, and getting those [Australian] poles which were so enormously different [from the poles we already had from North America, Europe, and India] from Australia into the literature in the beginning of the late fifties with, of course, enormous support from Ron Green.

*(Irving, 1981 interview with author; my bracketed additions)*

Before discussing their work, I introduce Ron Green, Irving's first Ph.D. student who worked with him on Australia's APW path. Green, born in Brisbane in 1930, attended Queensland University where he received his B.Sc. in 1953 with honors in physics. He studied geology for the first two years before switching to maths and physics. Of the geology lecturers he recalled that none of them favored mobilism.

The chair of the department, W. H. Bryan (I, §9.3), and Dorothy Hill (I, §9.4) were outspoken fixists. "Some lecturers presented it as a speculative idea, and others that it was a heresy comparable to salting a gold mine" (Green, September 7, 2010 email to author). Nonetheless, Green came to favor drift. He (September 7, 2010 email to author) commented retrospectively, "The dedicated geology students believed in landbridges all over the place. I favoured continental drift to mind-boggling land-bridges." Although he had no objection to the idea that birds and mammals could have entered Australia by island hopping, he still did not think island hopping explained the vast differences between Australian and Asian fauna and flora. If Australia and Asia had always been as close to each other as they now are, he wanted to know why island hopping did not occur much earlier; if it had occurred much earlier, then the flora and fauna of Australia and Asia should be much more alike. He thought that a better explanation was that Australia and the Indies began far apart and developed separately, and they moved together, enabling island hopping to begin. He found the idea of landbridges to Africa and Australia "hard to believe" (September 11, 2010 email to author). He read du Toit, "thought it clever," but was not convinced by du Toit's appeal to his lithologically based geologic disjuncts:

I read Du Toit's book, I thought it clever, but one sandstone looks like another, and, even though I was aware of different granites, and tin-rich to tin-poor granites, one granite is like another. Any geologist standing on his own outcrop (or turf) usually has all the answers, but it seemed to me that the geologist on the next proprietary outcrop can find a different answer.

*(Green, September 10, 2010 email to author)*

He was not bothered by Jeffreys' mechanism objections. "Who knows what happens under extreme conditions of the Upper Mantle? It seemed to me that the 'solid' rocks of the Upper Mantle could flow" (Green, October 30, 2010 email to author). Green had escaped the regionalism of his geology teachers. Although skeptical of some evidence cited by mobilists, he preferred it to fixism.

After graduation Green went to work for the Bureau of Mineral Resources (BMR, the federal geological survey), doing magnetic and gravity surveys in Western Australia and the Northern Territory.

The magnetic results were very difficult to understand because the station density was too sparse and I soon found that the hard rocks were not magnetized by induced magnetism. A compass at some places could show you that. What could be done? I thought about using mathematical filters.

*(Green, September 10, 2010 email to author)*

So he had some experience with magnetics. Little, however, did he know what awaited him. Green reminisced.

In late 1955 while in the field I read in a week-old paper a notice that ANU wanted Scholars and Research Fellows to work in palaeomagnetism. My rocks were certainly palaeo and they were both induced & permanently magnetized, but I did not know what a *Scholar* or *Research*

*Fellow* was, or who did what. We had in Queensland arts professors who were Greek or Latin scholars and students studying for a Masters who acted as Teaching Fellows in the Labs. So I applied for both jobs. I think it was the Office who wrote back and said that they thought it was the Scholars position I wanted. I wrote back, "That suits me, OK". I finished the field [work] and had returned to Melbourne (Head Office) and was working on a filter when the Boss came and said that Prof Jaeger was there to see me. I knew who he was – I was still using his book (1951 edition) *Introduction to applied mathematics*. I think that he wanted to see what I was like. Then he wanted to see if I was worth admitting. I felt that Jaeger, with his new department, did not want a stuff-up, and he wanted to be sure. Then he, to make sure that I would be able to collect the "right" rocks and not just walk over them taking measurements, wanted to know what I knew about Palaeomagnetism. I had the previous week talking with Leo Howard (Queensland, Maths & Phys grad, and Jaeger student) about Blackett's astatic magnetometer, and fortunately, I had just been reading Creer, Irving and Runcorn, and on the mention of which, Jaeger seemed pleased and said I would officially hear from the ANU. I had the clear impression that he wanted me to be working with Irving on basalts. And so it came to pass that next week I was on the train to Canberra and ANU.

*(Green, September 9, 2010 email to author)*

Green added that "it was pure chance" that he had read the paper. He was friends with the librarian; she knew that he read a lot, and just happened to give him the issue of the *JGG* containing the proceedings of the Rome IUGG meeting with the British APW path in it. Returning to the interview, Green recalled:

He [Jaeger] gave me a short test to see if I could distinguish a basalt from a dolerite, and if I could write down the second order differential cooling equation, and what did I KNOW ABOUT Bessel Functions. Fortunately, I had just spent 3 years on magnetic ground surveys in West Aust, and I had noticed that many rocks were permanently magnetized, and I knew a basalt from a dolerite. Also I had that very week read the paper, Creer, Irving & Runcorn (1954) ... Also, I was working on digital filters for gravity measurements over the Darling Fault Western Australia. That was my first introduction to paleomagnetism, and when I arrived at ANU I was lucky to find myself working with Ted [Irving] as his first scholar.

*(Green, September 9, 2010 email to author; his emphasis; my bracketed additions)*

The Fates arranged well: Green was meant to become Irving's first scholar. "I arrived 10 AM at Canberra Rail Station, and Irving drove me in his old Riley car to University House. I dropped my bag off, and we went to the department to begin winding coils for the magnetometer" (Green, September 9, 2010 email to author).

Green and Irving set about building the high sensitivity Blackett-style magnetometer. It became operative in mid-1956 and performed trouble-free for the following seven and a half years. Its performance was essentially identical to that of Creer's Cambridge magnetometer, as sensitive as Blackett's, but with shorter period for rapid measurement. Its description formed an important part of Green's Ph.D. thesis.

Irving recalls:

Green had a happy disposition. However, when provoked he had a quick temper and used to get into fights on the rugby field. He loved playing rugby. In Australia, rocks were often

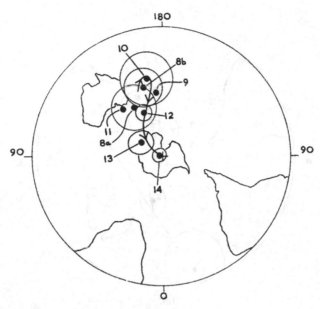

Figure 5.2 Irving and Green's Figure 2 (1958: 67). Polar movement since the Carboniferous. The older the pole, the lower the number.

weathered. Green was a strong man and quickly became expert with sledgehammers and wedges, breaking through to fresh rock. Later one of my students Bill Robertson with a technician called Snowy Pederson took to blowing off the weathered cap with dynamite. That was after Ron left.

*(Irving, December 2010 note to author)*

Irving and Green reported their findings in a series of seven single and jointly authored papers. The first three (Irving and Green, 1957a, b; Green and Irving, 1958) described their survey of basalts of Eocene and Late Cenozoic age from Victoria (13 and 14, Figure 5.2). Their last (Irving and Green, 1958), describing their APW path for Australia, was received by the new British *Geophysical Journal* at the end of October 1957, having first been rejected by the *Journal of Geophysical Research* (*JGR*) earlier that year. They summarized their results in two figures: Figure 5.2 showed the path since the Carboniferous, and Figure 5.3, during and since the Upper Proterozoic. Their path was defined by fifteen poles, including Irving's earlier Tasmanian dolerite pole (12, Figure 5.2). There were preliminary results from Upper Carboniferous glacial varves (8b, Figure 5.2) in Irving (1957a), lavas (8a, Figure 5.2) in Irving (1957b), and from Lower (9, Figure 5.2) and Upper Permian (10, Figure 5.2) in Irving (1957b). They reported reconnaissance results from Silurian and Devonian rocks of New South Wales (6 and 7, Figure 5.3), collections made under the guidance of Browne and Dr. Ida Browne, his paleontologist spouse who located accurately dated igneous rocks for Green (1961a, b). They reported results from Cambrian red sandstones (4 and 5, Figure 5.3) and Upper Proterozoic lavas and pink sandstone

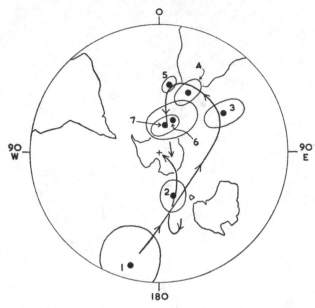

Figure 5.3 Irving and Green's Figure 3 (1958: 68). Polar movement during and since the Upper Proterozoic. The older the pole, the lower the number.

(1, 2, and 3 Figure 5.3) based on collections from the Northern Territory made by BMR under the guidance of Dr. Lynn Noakes. The results were based on conglomerate, fold and consistency tests on measurements of NRM without cleaning.

Locating dated rock formations of various ages containing rocks suitable for paleomagnetic analysis required intimate firsthand knowledge of Australian geology. Most Australian Earth scientists were against mobilism (I, §9.1–§9.6), but they were generous with their help; it was something new and they were willing to "give it a go." Irving and Green (1958: 64–65) identified a dozen geologists from various parts of Australia who did so. Irving later remarked (comment to author, June 2000), "Green and I could not have possibly got around Australia in the time" it took them to complete the work without their help. He singled out W. R. Browne (see I, §9.3 for Browne's anti-mobilism).

Most Australian geologists were very anti-drift. Notwithstanding, many Australian geologists helped us very much, most notably W. R. Browne. Browne probably knew more about geology than anyone else in Australia at that time. He was a protégé of T. W. E. David's, and he had retired from the University of Sydney, and he just completed and published, I think in about 1951 or 1952 perhaps, this monumental three volume work on the geology of the Commonwealth of Australia. [Browne edited and added much to David's work. In fact, Browne is listed as the second author of the 1950 edition, whose full title is *The Geology of the Commonwealth of Australia: edited and Much Supplemented by W. R. Browne.* (I, §9.3)] And, we struck up a sort of deal. He didn't drive but he loved to do fieldwork. So I used to take him around to all these places, and then I would collect and he would do his geologizing and he would tell me all about the rocks. And I would tell him rather carefully what I wanted and so on and so forth. He gave me only one really useless locality, and that is the real mark of a first class geologist, that is, a man for whom you can set out a description, and then he can go and find it for you. And he can describe it in its proper geological relationships and that, of course, is the hallmark of a first class geologist. Of course, I was able to help him in the sense that I took him over lots of localities where he had been as a younger person and continued to do work. He published work practically up to his death at a very late age. He was an Irishman who migrated to Australia for his health at some early stage in the 1900s, and survived to a great old age. He was rather anticontinental drift. But this never affected our relationship in any way. He was very gentlemanly about it, and he said that these paleomagnetic results were going to help this discussion. He said it was his job to see that the geological aspects were done as well as possible. Of course, he was just a fantastic help, and you can see that in many of my papers I acknowledged the great help that he did give me.

*(Irving, 1981 interview with author; my bracketed addition)*

Australians freely shared their firsthand knowledge of Australian geology with Irving and Green, and it was essential to the success of their project. Jaeger made the introductions.

Irving and Green made good use of their Australian results. Their results from the Late Cenozoic Newer Volcanics of Victoria buttressed empirical support for GAD. Because its poles coincided with the present geographic poles, they claimed, as Hospers already had shown with his Icelandic data, that Earth's magnetic field had been a geocentric dipole whose mean axis coincided with Earth's axis of rotation since the Middle Tertiary.

Our results from the Newer Volcanics of Victoria are the first confirmation from the southern hemisphere of the average dipole nature of the geomagnetic field during the later part of the Cainozoic.

*(Irving and Green, 1957a; 1065; also see Irving and Green, 1957b: 355–356)*

Results from Late Carboniferous volcanics and glacial sedimentary rocks (8b, Figure 5.2) proved further strong support for GAD. The paleoclimatological evidence indicated that Australia had been glaciated during the Late Carboniferous period – recall the centrality of the Gondwana-wide Permo-Carboniferous glaciation to the classical case for mobilism (I, §3.10–§3.12) – and the pole, obtained from glacial varves of what was then called the Kuttung Series, placed Australia at high latitude. By good luck the varves had high fine-grained volcaniclastic content and were laid down in tranquil conditions, which made them good field recorders, yielding a strongly positive tilt test. The agreement between igneous and sedimentary rocks (poles 8a and b, Figure 5.2) indicated no significant inclination error.

In his review for the Imperial College sponsored London 1956 meeting (§3.14, §5.2), Irving compared synchronous paleomagnetic and paleoclimatic data, not from the same but from different regions (§3.12). On the basis of fixed continents the two were highly inconsistent; if continental drift was assumed, the inconsistencies disappeared.

... let it be supposed that when averaged over several thousands of years the geocentric dipole has always been directed along the axis of spin. This may be referred to as the axial geocentric dipole hypothesis. The geographical latitudes of eastern Australia and western Europe now become approximately 80° and 70°, respectively. Since the pioneer work of Sir Edgeworth David, the climatic conditions under which the Upper Kuttung Series were laid down have always been regarded as cold or glacial – conditions which are most likely to have been attained in high latitudes. Western Europe, on the other hand, provides no evidence of glaciation, and geological opinion favours warm and wet conditions at this time. The agreement between the palaeoclimatic and palaeomagnetic evidence at the same place is consistent with the axial geocentric dipole hypothesis. On the other hand, the contrast between these two lines of evidence when antipodal regions are considered suggests that either western Europe and eastern Australia have moved relative to each other since the Carboniferous, or that this hypothesis is false.

*(Irving, 1957a: 281)*

Irving and Green (1958) extended this by comparing paleomagnetic and paleoclimatic data along the whole path, and found them compatible.

If this [the GAD hypothesis] is so the palaeoclimatic conditions in Australia should be broadly consistent with the geomagnetic latitude deduced from the palaeomagnetic data ... During the Upper Proterozoic, and again in the late Palaeozoic and Mesozoic, Australia was in a high geomagnetic latitude and it is at these times that the geological evidence indicates a cold climate. The tillites and varved sediments of the Adelaidian System and the outwash gravels of the Nullagine rocks provide evidence of extensive glaciation in the earlier period (David 1950, p. 79). Similar deposits in the Kuttung Series and its equivalents elsewhere in Australia indicate a glaciation in the Upper Carboniferous[5] (David 1950, p. 326). Evidence of glaciation is again found in the Permian (David 1950, pp. 341, 397), and the deposits of the Triassic, Jurassic and Cretaceous are generally regarded as having been laid down in cold to temperate conditions (David 1950, pp. 440, 476, 515). The fossil trees during these periods have pronounced growth rings suggesting the well-marked seasonal changes characteristic of higher latitudes. The great saurian reptiles are absent and the invertebrate marine faunas have a cold water aspect ... The indications of warm climate which are widespread in other parts of the world at these times, for example in the Mediterranean region, are absent, and this would seem to imply that Australia was in high or intermediate latitudes just as the palaeomagnetic results do. It may also be noted that from the Proterozoic to Devonian the geomagnetic latitudes are low, and during these times no glacial formations are known in Australia, whereas red beds, dolomites, thick limestones with algal, Archaeocyathid and coral reefs are common.

*(Irving and Green, 1958: 69; my bracketed addition)*

Although they introduced a cautionary note about the relationship between climate and latitude, they emphasized the consilience of high paleomagnetic latitude and glaciation. Like earlier mobilists, they too referred to the massive glaciation in the Southern Hemisphere during the late Paleozoic, noting that polar wandering without continental drift is unable to explain its distribution.

Because of exceptional meteorological conditions in the past the palaeoclimatic evidence may not always give a close indication of latitude. Nevertheless the broad correspondence between

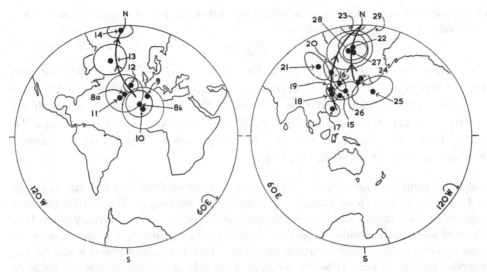

Figure 5.4 Irving and Green's Figure 5 (1958: 70). Pole positions obtained from rock formations of Carboniferous and later ages in North America, Europe, and Australia. The large discrepancy between equivalent results from Australia and those from northern continents is illustrated (1958: 68). North American poles are 15 and 16 (Carboniferous), 17 through 19 (Permian), 20 and 21 (Triassic), 22 (Cretaceous), 23 (Upper Tertiary). European poles are 24 (Lower Carboniferous), 25 (Permian), 26 (Triassic), 27 (Eocene), 28 (Oligocene), and 29 (Miocene). Australian poles are 8a and 8b (Upper Carboniferous), 9 and 10 (Permian), 11 (Triassic), 12 (Mesozoic – probably Jurassic), 13 (Lower Tertiary – probably Eocene), and 14 (Pliocene, Pleistocene and Recent).

this evidence and the palaeomagnetic latitude, in particular the high geomagnetic latitude with glaciation, suggest that on average the geomagnetic and geographic poles have always been coupled together, and that so far as Australia is concerned the great Proterozoic and late Palaeozoic glaciations may be reasonably explained by movement of the geographic poles relative to Australia. It should be noticed that this explanation for glaciations in Australia would not provide for the late Palaeozoic glaciations of South America, South and Central Africa and India, assuming that the present continental positions have always been maintained.

*(Irving and Green, 1958: 70–71)*

However, the most potent use they made of their Australia path was to compare it with paths from elsewhere. The positional differences among equivalent poles from Australia and those from Europe, North America, Africa, and India were huge. Because they were so great, Irving and Green strongly argued that Australia had drifted relative to the other continents. Polar wandering alone was insufficient.

Except in later Cenozoic times the poles determined from the Australian data do not coincide with equivalent results from Europe, North America, Africa and India. The contrast is greatest in the Upper Palaeozoic and early Mesozoic where the discrepancies between pole determinations from rock formations of similar age is between 50° and 90° arc, and are far greater than any experimental error (Figure 5 [my Figure 5.4]). It would seem very difficult to explain these discrepancies without entertaining the possibility that in the past Australia has moved as a whole relative to the Northern continents.

*(Irving and Green, 1958: 71; my bracketed addition)*

They went on to argue that there were good reasons to suppose that pre-Carboniferous as well as post-Carboniferous continental drift had occurred because the form of the pre-Carboniferous segment of the Australian APW path differed greatly from corresponding segments of the European and North American APW paths. Then continental drift was not a unique process as Wegener had envisioned, but had been operative much, much earlier.

The polar paths for Carboniferous and later times obtained from North America, Europe and Australia have an approximately similar form. They are roughly meridional in direction, begin in similar latitudes, and converge on the present pole in the later Tertiary times. Their different longitudinal positions could be explained by supposing that relative movements have occurred between these continents since the Carboniferous, movements which possibly extended into the Tertiary. However, the form of the polar path from Australia, prior to the Carboniferous is so different from that obtained from the Northern continents that the supposition of post-Carboniferous continental drift only is inadequate and it is necessary to suppose that relative movements also occurred prior to the Carboniferous. If the discrepancies between equivalent pole results from different continents are to be explained by continental drift then there seems to be no reason to regard it as a unique process occurring only at one stage of geological history; in fact there are suggestions that it was also operative in earlier epochs.

*(Irving and Green, 1958: 71)*

If Irving had not thought about pre-Wegenerian drift when first considering his results from the Precambrian Torridonian in 1954, he certainly had by 1957 as he considered the pre-Carboniferous paleomagnetic data from Australia. Theirs is, I believe, the first definite statement by paleomagnetists invoking pre-Wegenerian drift. Readers will recall that Argand (1924/1977: 139; I, §8.7) had entertained the possibility of Early Paleozoic continental drift in his concept of a proto-Atlantic Ocean.[6]

## 5.4  Surveys of Karroo System through 1959

The study by Gelletich (1937) of the magnetic anomalies associated with the Pilansberg dykes of South Africa prompted Gough to sample them.[7] He found steep downward (reversed) magnetizations giving a pole (7.5° N; 42.5° E) in the horn of Africa, which he thought indicated polar wandering, continental drift, or a combination of the two (Gough, 1956: 211).[8] Gough had joined the Bernard Institute in Johannesburg in 1947, one year after Hales (I, §5.6). Hales, an associate of Jeffreys in the 1930s, had worked on mantle convection, and was a longtime foe of mobilism (I, §6.12). He had encouraged Gough to investigate the Pilansberg dykes. Hales himself became interested in paleomagnetism while spending a six-month sabbatical in 1952 at Cambridge

University where "he spent some time with S. K. Runcorn's paleomagnetic students" (Hales, 1986: 3). Hales hired K. W. T. (Ken) Graham.

I offered a Research Assistantship to Kenneth Graham, who had just completed his BSc (Honors) in Geology. I told him that I thought that if enough measurements were made sufficiently carefully, one would find that the continents had not drifted. Graham accepted the offer but told me he was convinced that drift had occurred.

*(Hales, 1986: 3)*

It did not take long for Hales to realize that he might be wrong about continental drift. Working on Jurassic Karroo dolerites of South Africa,[9] he and Ken Graham presented a paper at the November 1956 conference on rock magnetism at Imperial College, London (§3.14), tentatively supporting continental drift and polar wandering.

It will be seen that the pole positions inferred from the measurements on the South African dolerites are not consistent with any of those found for rocks of the same age in Europe, North America and Australia. If it be assumed that the direction of magnetization, when averaged over the time of intrusion of these dolerites, is a slowly and smoothly varying function of time then the simplest way of reconciling the divergent pole positions is to postulate relative displacement of the continents, i.e. Continental drift. It is noteworthy that the displacements required are of the kind proposed by Wegener (1924), du Toit (1937), King (1953) and Carey (1955).

*(Graham and Hales, 1957: 159–160; references for Wegener, du Toit, and King are the same as mine; Carey (1955) is equivalent to my Carey (1955a))*

This was likely more Graham than Hales. However, in the next paragraph, which may have been more Hales, they noted that the imperfect reversal within the dolerites may indicate rapid polar wandering.

It should be noted, however, that there is some evidence that the assumption of a slowly varying direction of magnetization is an over-simplification ... It is interesting to note that the difference in the direction of magnetization of the reversed and normal South African dolerites is indicative of a fairly rapid movement of the pole or of the continent.

*(Graham and Hales, 1957: 160)*

They did not suggest the possibility, as the Imperial College group had done to explain the same effect in the Deccan Traps (§5.2), that the asymmetrical reversal reflected the presence of recent field (VRM) overprints; all were measurements of NRM without the now standard cleaning. Hales, speaking from the perspective of thirty years, recalled his conversion to mobilism.

To Kenneth Graham's great pleasure, the paper we presented at the symposium organized by P. M. Blackett in 1956 contained the statement that the "simplest way of reconciling the divergent pole positions [from Europe, North America, Australia, and South Africa] is to postulate relative displacement of the continents, i.e., Continental Drift" (Graham & Hales, 1957). From that time on I called myself a "reluctant drifter" ...

*(Hales, 1986: 4; Hales' bracketed addition)*

This early work at the Bernard Price Institute was followed by more extensive studies, areally and stratigraphically, of the Karroo in Central Africa by Nairn. Runcorn recruited Nairn in 1954. He gave an evening talk on paleomagnetism at the University of Glasgow, where Nairn had obtained his Ph.D. in geology, in stratigraphy and sedimentology. Nairn found it interesting. The next morning Runcorn invited him to join his Cambridge group. Nairn said he would if Shell Oil would release him from the contract he had signed with them. Shell did, and Nairn went to Cambridge in 1954 (Nairn, March 1993 telephone conversation with author).

Nairn knew about continental drift. He had attended the same high school as Arthur Holmes, whose views about continental drift were well known there, being such an illustrious alumnus. The sixth form geology course he took at age eighteen was even taught by one of Holmes' graduate students. Nairn noted, "So I heard quite favorably about continental drift" (March 1993 telephone conversation with author).

Runcorn encouraged Nairn to work in Africa. They obtained support in 1954 from the Directorate of Overseas Geological Surveys (United Kingdom) for a paleomagnetic survey of the Upper Carboniferous to the Lower Jurassic Karroo System (Nairn, 1959). Nairn sailed for Africa, collected samples, and arranged with local geologists for further samples to be sent to him in Cambridge. Returning to England, he discovered that Runcorn had left for Newcastle. Nairn recalled, "I did not even know that Runcorn had moved to Newcastle until I got off the boat from South Africa with my collection of samples" (March 1993 conversation with author).

Nairn's (1956) first African paper was published a month before Gough's paper, which had been submitted about nine months before Nairn's. It described results from Jurassic Karroo lava flows from the Bulawayo and Victoria Falls areas of what was then Rhodesia, now Zimbabwe. They gave a North Pole position in Hudson Bay. Comparing this with the very different Jurassic pole from Tasmania, he argued for extensive movement of Africa and Tasmania relative to each other and to the present pole.

> Thus it is clear that, while Britain in Jurassic times must be regarded as having approximately the same position as at present relative to the pole (the present pole lying within the circle of confidence of the Jurassic results), and the movement of America is not significant, it is equally clear that this is not the case in the southern hemisphere. The Rhodesian lavas indicate a movement of about 2000 miles, and the Tasmanian results a movement of about 2500 miles ... The inevitable conclusion is that continental drift must be accepted if the assumption of the coincidence of the axial dipole field with the rotational axis is accepted! More exact dating of the time of drift and the relative positions of the continental blocks must depend on the more detailed work at present in hand in Great Britain, South Africa and Australia.
>
> *(Nairn, 1956: 936)*

Nairn spoke at the 1956 meeting of the BAAS in Sheffield on September 4, 1956, as did Blackett and Clegg. At the meeting, as summarized by Clegg (1956), Nairn repeated his claim that Africa had shifted 2000 miles relative to the geographical pole since the Jurassic, and spoke of other work at Newcastle including Opdyke's paleoclimatological investigation of aeolian sandstones (§4.2, §5.12–§5.14). Clegg described the general tenor of the meeting:

All the speakers emphasized at the meeting their belief that while information of great importance has already been obtained from palaeomagnetic studies, the present conclusions must be regarded as purely tentative, and will be subject to continuous modification as work develops.[10]

*(Clegg, 1956: 1087)*

Nairn discussed his African results two months later at the international conference on rock magnetism at Imperial College, London, where Ken Graham and Hales also presented their Karroo results. Nairn reported (1957a: 167) a new pole from Upper Triassic Bechuanaland sandstones in Hudson Bay, close to his Jurassic pole.

Nairn expressed concern about the difficulty in Africa of getting the fresh samples needed for reliable results. Unlike outcrops in higher latitudes, they had been exposed for much longer and had been deeply and extensively weathered, and they had not been eroded clean by the Quaternary glaciation or (except in rare instances) deeply incised by rivers as in the Grand Canyon. Nairn also remarked that the "extension of the work to continents other than Europe and North America is generally handicapped by the imperfectly known geology" (Clegg, 1956: 1086). A third major difficulty was the high incidence of lightning strikes in tropical regions, strikes that remagnetized long-exposed outcrops (K. W. T. Graham, 1961). Also important was the relative infrequency, compared to Europe and North America, of deep road-cuts and quarries that gave paleomagnetists direct access to fresh unweathered rock. Irving and Green were fortunate that in southeast Australia they had abundant road cuts and quarries in Tasmanian dolerites and Cenozoic volcanics of Victoria.

Two years later, Nairn (1959) summarized his survey of the Karroo System by an APW path for southern Africa based on six poles from Upper Carboniferous to the Lower Jurassic rocks. They spanned the Karroo System (Late Carboniferous through the Late Triassic), and included the overlying Karroo lavas and associated dolerite intrusions. Instead of plotting the APW path, he plotted Africa relative to the paleogeographic grid (Figure 5.5). He called this a "drift diagram," and it showed changes in latitude and rotations relative to the meridian. He noted:

The African Karroo results seem to indicate a drift from a near-polar to an equatorial position in the period from the late Carboniferous to late Triassic, and that this was associated with an anti-clockwise rotation of approximately 90°. The movement in post-Triassic times would appear to have been small compared with that of the earlier period and especially of the Permian. Even if the minimum movement is assumed, the total shift is still enormous.

*(Nairn, 1959: 409–410)*

There was a difficulty correlating poles: poles from Triassic rocks of Rhodesia (Forest Sandstone) and Bechuanaland (Cave Sandstone) differed by 31° of latitude, and 36° of longitude; the latter was closer to the younger Jurassic Karroo basalt pole than to the former. The paleomagnetic results seemed equally reliable, and he could

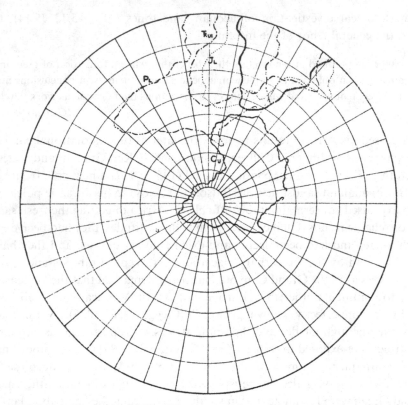

Figure 5.5 Nairn's Figure 11 (1959: 409) showing changes in the latitude and orientation of Africa during "Karroo" times. $J_L$, Lower Jurassic Karroo Lavas; $R_U1$, "Upper Triassic" Cave Sandstones; $P_L$, Lower Permian Taru Grit; $C_U$, Upper Carboniferous Dwyka varved clay.

not reject one without rejecting the other, so he suggested instead that the Cave Sandstone might be incorrectly dated; equally possible was that it had been remagnetized at the time of vast outpouring of basalts that originally covered the Karroo System.

At the end of the 1950s an all-out effort to obtain an APW path for Africa was initiated by Ian Gough at what was then the University College of Rhodesia (now Zimbabwe), and this, together with continuing work at the Bernard Rice Institute at Johannesburg, answered many of these difficulties (III, §1.17).

Nairn's work, although based only on reconnaissance, indicated very large latitudinal changes for Africa, and was put to good use by Creer once he obtained data from South America (§5.6). However, there was much that Nairn could have done at the time to strengthen his argument for mobilism. For example, surprisingly he did not appeal to paleoclimatology; his results indicated that southern Africa was near the south geographic pole during the Late Carboniferous (Figure 5.5), which is what du Toit and others had maintained on the basis of extensive glaciation of this age.

He did not compare results from India (§5.2), from Australia (§5.3), and from South America (§5.6), which were all published at least one year before his own (1959). Nairn, however, did co-author a paper with Collinson in which they referred to the Australian APW path, and mentioned that "a few points are known for South America and South Africa" (Collinson and Nairn, 1959: 390); although they had expressed interest in unraveling its history, they made no serious comparisons with results from elsewhere in Gondwana.

## 5.5 Magnetic cleaning boosts the record

Up to this point paleomagnetists had selected lithologies that they thought from experience would likely have stable NRM, relying on field stability tests to tell them if it were so; it was on this that they based their arguments. In the absence of cleaning techniques, it was all that they could do; it yielded quick results, but it would take them only so far because secondary magnetizations were, to some degree, commonly present. As they extended their collections, paleomagnetists increasingly found that the primary remanence is often masked, leading to errors in estimating pole positions, even making them indeterminable. The secondary magnetization (overprint) of immediate concern for these early workers was VRM acquired in the recent geomagnetic field; witness Runcorn's worries about its effect on the results from the Supai Formation of the Colorado Plateau (§4.4), or Khramov's concerns about his results from sedimentary rocks of Western Turkmenia (now Turkmenistan) (§5.9), or Creer's concerns about his results from the British Triassic redbeds (§2.12, §3.6). There was an increasing need to rid rocks of these unwanted VRMs, and to develop the procedures for doing so without destroying the primary remanent magnetization.

Because VRMs have lower coercive force and unblocking temperature than primary magnetization, they can be removed by applying alternating magnetic fields (AF) or thermal cleaning (TH) respectively. AF cleaning is accomplished by applying alternating magnetic fields in steps of increasing strength that are then reduced to zero in the absence of an ambient magnetic field until VRM is removed. The magnetization is measured between each step. With TH cleaning, the sample is heated, and then cooled in a zero ambient field. This is repeated at increasing temperatures, magnetizations being measured between steps. Various procedures were developed to determine when VRM had been removed. Demagnetization apparatus had to be designed and built from scratch, and during the later 1950s and early 1960s demagnetization procedures gradually came into general use (Collinson, 1983).[11]

Doell (1956) experimented with both AF and TH cleaning of sedimentary rocks, and Cox (1957) used thermally cleaned volcanic rocks (§7.6). At the November 1956 Imperial College meeting on rock magnetism (§3.14), Brynjolfsson (1957) described in some detail how he had removed the recent field VRM from several samples of Cenozoic Icelandic lavas without destroying their stable remanent magnetization, thereby reducing dispersion and revealing the underlying stable remanent magnetization. However, it was Creer (1958a, b) in Newcastle, and J. A. As and J. D. A. Zijderveld (1958) in Amsterdam who deserve most credit for the development and formalization of magnetic cleaning. At the IUGG meeting in Toronto, Creer (1958a) described experiments in which a soft unstable component of magnetization in basaltic lava flows of both Jurassic and Quaternary age from Uruguay and Argentina respectively was removed by treatment in an alternating magnetic field of about 300 oersted peak value. After "cleaning" the precision of the measurements was greatly improved and the directions fell into well-defined normal and reversed groups.

Necessity is the mother of invention and a prepared mind is conducive to discovery, and it is easy to understand how Creer came up with the idea of magnetic cleaning. Already, when working on his APW path for Great Britain, he had encountered scattering of directions because of VRM overprints (§2.12), and with his background in physics, he became interested in isolating the various magnetizations present. As he later recalled:

The result from the Keuper red sediments from Sidmouth was certainly the first which demonstrated to me that some rocks contained more than one component of magnetization. In this case I was able to demonstrate that there were three. Most important it seemed to me that something could be done about isolating the components, and as a physicist, this posed problems as interesting as the geological ones like drift.

I was very conscious of the existence of magnetic overprinting, particularly so after seeing the effects of the Serra Geral Basalts on the underlying sediments. They were all stained red ... Thus, I developed the AF equipment at Newcastle. I also had a very primitive gas furnace for thermal demagnetization while I was a student at Cambridge but could only use it to measure Curie Points because there was no time to build a zero field environment. Later with my student François Chamalaun ... we built a thermal demagnetizer in Newcastle (about 1960) [Chamalaun and Creer (1963)].

*(Creer, April 9, 1984 letter to author; my bracketed addition)*

Putting the need for cleaning into its larger context, in his report of the Toronto meeting, Creer cited the APW paths for Great Britain, North America, and Australia, and noted:

At the present time, work is proceeding in these and in other continents and in order that the possibly complex pattern of continental displacement and polar shift can be worked out in detail it is becoming increasingly important to refine the accuracy of the measurements and calculations. One great difficulty is that the natural remanent magnetizations of a great many rock formations are unstable. Though this is well known, most workers have so far concentrated on rocks of proven stability and consequently gaps remain in the record. All such gaps cannot be filled because many formations are too weakly magnetized ... but a great many formations which possess partially unstable magnetizations can profitably be studied by using appropriate techniques.

*(Creer, 1958b: 374)*

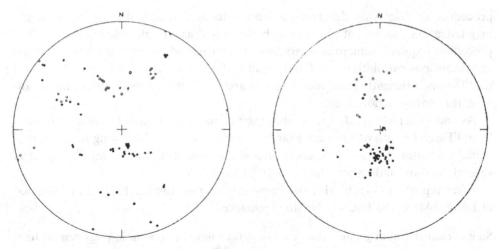

Figure 5.6 Creer's Figures 8 and 9 (1958b: 382) showing the directions of the magnetization of the Quaternary basalts from Argentina. The north-seeking poles are plotted on a polar stereographic projection. Solid circles represent downward dips; upward dips are shown as open circles. The scattering is greatly reduced after cleaning (right).

In his Ph.D. thesis, Creer had already attempted to separate graphically the primary and secondary magnetizations, but the use of this method was limited. He noted another possibility:

Alternatively the "soft" [secondary] component must be destroyed, and a procedure for doing this may be called "magnetic cleaning." In this paper a method using alternating magnetic fields is described.

*(Creer, 1958b: 374; my bracketed addition)*

Through magnetic cleaning, gaps in the then existing APW paths from Britain, North America, and Australia could be filled without sacrificing reliability.

Using alternating fields of 250 oersteds peak amplitude, Creer cleaned all his samples from Quaternary Neuquén basalts and from the Jurassic Serra Geral basalts from Brazil and Uruguay, and the adjacent Botucatu Sandstone baked by them. The reduction in dispersion of directions was spectacular, as illustrated by his "before" and "after" plots (Figure 5.6). He noted (1958b: 389), "The procedure of cleaning has been shown to reduce the scatter appreciably and thus increase the precision of palaeomagnetic data."

Although As and Zijderveld, members of the Dutch paleomagnetic group led by Rutten and Veldkamp, submitted their paper three months after Creer, they have generally been given greater credit for developing magnetic cleaning. Unlike Creer, they titled their paper "Magnetic cleaning of rocks in palaeomagnetic research," making clear their major topic; it was not hidden, as it was in Creer's paper entitled rather vaguely, "Preliminary palaeomagnetic measurements from South America." As and Zijderveld have been given the greater credit because they introduced the procedure of displaying demagnetization results as orthogonal plots in which the magnetization component present can be identified as straight line segments. Their procedure required numerous measurements on each specimen and what were then very laborious calculations, and as a result it did not really come into its own until the1970s as sufficiently rapid measurement and computation methods became available; it is now generally used.

As and Zijderveld carefully described their demagnetization of Permian specimens from l'Estérel region of southern France by heat as well as alternating fields. Partial demagnetization decreased dispersion in a striking manner, which they illustrated in several "before" and "after" diagrams (1958: 311–317).

After reporting (1958b: 310) the Permian pole position for Europe as "situated at Long. 144° E and Lat. 47° N," they remarked:

So it is clear that even greater errors than have been found can occur in poles derived from the observed magnetic directions in uncleaned rocks. Even when a great concentration of the directions is obtained, it is not certain that a systematic deviation is absent. On account of the great concentration, one gets a small cone of confidence but cones of confidence do not in fact reveal systematic error.

*(As and Zijderveld, 1958: 318)*

This rather dire statement implied that any pole determined from uncleaned rocks was suspect (RS2).

Finally, from these demagnetization experiments it follows that all pole determinations that are based on measurements of remanent magnetizations in rocks from which the disturbing magnetization of the present Earth field has not been removed should be treated with reserve.

*(As and Zijderveld, 1958: 318)*

By hinting that all previous work was unreliable, it was a serious attack on the then current paleomagnetic case for mobilism.

Such gloominess was too much for Irving, who responded with buoyant optimism. According to him, the Dutch paleomagnetists had failed to point out that their results vindicated the previous studies from l'Estérel published the previous year by A. Roche (1957). It was not that the previous work was unreliable, it was reliable by the standards of the day, and As and Zijderveld had confirmed it. Also their results agreed, within errors, with previous studies of similarly aged rocks from elsewhere in Europe; that is, they had shown that those paleomagnetists who had carefully selected rocks and relied on their field tests had not produced prohibitively "dirty" rocks, they had simply chosen them well in the first place.

On the basis of demagnetization experiments with 14 specimens from one rock formation in southern France, As & Zijderveld (1958) conclude (p. 318) that "from these demagnetization experiments it follows that the pole determinations that are based on measurements of remanent magnetization in rock from which the disturbing magnetization of the present Earth's field has not been removed should be treated with reserve."

Now the mean direction (D = 216, I = −24) of the specimens studied by As and Zijderveld after removal of the secondary component (that produced by the present Earth's field) is in broad agreement with those published earlier by Roche (1957) who gives D = 210, I = −16) ... No reference is made to this earlier work. Moreover, the pole positions they give (47° N 144° E) agree very well with that obtained from work on other Permian formations of Europe without magnetic "cleaning" ...

*(Irving, 1959b: 140)*

After citing the positions of nine similarly aged poles, whose agreement he characterized as "very good," Irving turned As and Zijderveld's conclusion on its head.

These new results given by As and Zijderveld do not cast suspicion on what has been done previously; on the contrary, they confirm it. In fact this agreement between many determinations from Permian rocks in Europe would seem to be a most convincing confirmation of procedures in palaeomagnetism.

*(Irving, 1959b: 140)*

Having shown that their new results from cleaned rocks provided increased confidence in previous work, Irving looked cheerfully to the future.

The techniques of magnetic "cleaning" of rocks are providing the means of dissecting out the unwanted secondary components which often mask the primary components in the natural state. Up till now only a few per cent of rocks have been of use, but now our range of vision may be increased many times. The mood is one of optimism for the future, not reserve about the past.

*(Irving, 1959b: 140)*

This initial work on magnetic cleaning had been applied generally only to igneous rocks. Creer (1959) now developed the means to produce stronger alternating magnetic fields and attempted, but with only partial success, to remove unwanted VRMs from his 1953 collections from the red sedimentary Keuper sediments. He had already identified graphically primary, secondary, and temporary magnetizations (§2.12). The temporary magnetization decayed after storage in field-free space for a few years. By subjecting samples to fields up to 850 oersteds peak amplitude, Creer found that present field VRM was removed or considerably reduced; fields much stronger were needed to remove them from igneous rocks. AF magnetic cleaning was not satisfactory for routine processing of sedimentary rocks, and within two years thermal demagnetization became the standard (Irving *et al.*, 1961; Chamalaun and Creer, 1964).

Ridding rocks of present field VRM enhanced the paleomagnetic case for mobilism and became the way of the future; within a few years paleomagnetists included removal of recent VRM as a standard procedure, adding it to their list of requirements before results could be deemed reliable. This list, albeit not canonical, included: the presence of normal and reversed magnetizations aligned at 180° to each other; obtaining results, where possible, from coeval igneous and sedimentary rocks; requiring a good number of samples spanning substantial stratigraphic thickness and lateral spread; studying igneous contacts; carrying out tilt, conglomerate, and slump tests; and, of course, analyzing dispersion and packaging the data using Fisher's statistics. Recently added to this list, was the consilience (§3.12) of paleomagnetically determined latitudes with paleoclimatology, and in §5.10 I shall describe the introduction of an important new way of making such comparisons.

## 5.6 Survey of South America

Creer (1962d: 154) made two trips to South America, in October through February, 1956–7, and in 1958. He aimed to repeat what he had done in Britain: sample through the Phanerozoic column and construct an APW path. Why South America? It was a major part of Gondwana, it was unsampled paleomagnetically,[12] one could expect to be able to collect from many parts of the geological column, and the results would be of importance in unraveling the history of Gondwana. Irving had staked out Australia. Hales and Gough had made a start in South Africa, and Nairn already had returned to Newcastle with his first boxes of rocks from further north in Africa. In a letter, Creer recalled why he targeted South America.

The critical question at that time was: will polar wander curves for the different continents (when obtained) superimpose when plotted on the present globe? The first indication that they did not came from Irving's result from the Deccan Traps ... and then from his Australian results. By that time, I was on my way out to South America realizing that the key lay in results from Africa and South America because they lay in the heart of Gondwanaland. Without data from these two continents it would be impossible to reconstruct the past relative positions of Australia with Europe and North America.

*(Creer, April 9, 1984 letter to author)*

In the same letter, he added that he "realized that work in Africa was being planned by Nairn (and by other groups in the longer term)."

In South America, Creer received extensive help from newly made friends in Brazil and Argentina, some of whom became colleagues. Many were pro-drift, and several, including Reinhardt Maack, who had already sent Irving samples, had personally known Wegener.[13]

In Brazil, I was helped by and had interesting discussions with Reinhardt Maack, Wilhelm Kegal and Victor Leinz, all German émigré geologists who had known Wegener. I received extensive help – field geologists to accompany me, and vehicles from geological surveys and universities. On explaining my objectives, the response often was "Well why, isn't it clear that drift has happened?" The outlook was quite different from that generally held at the time by UK and N. American field geologists. Jao Jose Bigarella, a student of Maack in Curitiba has since become a lifelong friend. Subsequently he sent me a student who obtained his M.Sc. at Newcastle, but on his return funds were not available to set up a palaeomagnetic laboratory at Curitiba and the equipment I sent out was later transferred to Sao Paulo where a world-class laboratory was developed initially by Igor Pacca.

In Argentina, I received extensive help of both a professional and logistic nature from the National Geological Survey, and from the relevant Provincial Survey Offices. At the University of Buenos Aires my initial contact was with the Head of the Engineering

Department, Ing. Baglietto, who had interests in the global gravity network. He provided me with further introductions, and helped push things along with the University geologists. There is now an excellent palaeomagnetic laboratory in Buenos Aires run by Daniel Valencio and Juan Vilas, started up largely with equipment sent out from the UK but subsequently developed through their own initiatives. Until the Falklands War, we maintained a collaborative field programme in Argentina, most recently on lake sediments.

*(Creer, April 9, 1984 letter to author)*

Creer recalled two problems, uncertainties in stratigraphic assignments (examples later) and the pace of progress.

In retrospect, a major problem was that the stratigraphy in many places had not been thoroughly evaluated in the fifties and early sixties. Many of the rock formations I worked on then have since been redated. Another problem I had, but which I enjoyed to some extent, was that they did absolutely nothing to prepare for my visit until I arrived. Then, when I did arrive, they would invariably pull my letter out and start making excellent arrangements. This usually took about a week, which gave me time for sightseeing!

*(Creer, April 9, 1984 letter to author)*

On his first expedition, he collected, as noted above, basalts from the Quaternary Neuquén lavas of southern Argentina and from the Serra Geral Formation of southern Brazil and Uruguay, part of the Paraná Large Igneous Province.[14] He wanted the Neuquén basalts to check the GAD hypothesis, and they did: "The close agreement between the axis of magnetization of the Quaternary basalts of Argentina and the direction of a geomagnetic axial dipole confirms this view" (Creer, 1958b: 383). And he wanted the Serra Geral basalts because at the time they were considered to be Late Triassic to Jurassic and correlative with the Karroo basalts of southern Africa. His new pole and that for the Karroo basalt of Nairn (1956) were not far from the present geographic pole, indicating that the latitude of Africa and South America was not very different from at present, as roughly expected on drift theory because the South Atlantic opened in an east–west direction. Creer showed (Figure 5.7) that the amount of rotation required to bring the azimuths of the two results into alignment is 22°, very much less than the 45° to 50° required by du Toit (1927) to close the South Atlantic. Relative longitudes were indeterminate and Creer placed Africa and South America an arbitrary distance apart. He accepted the likely correctness of du Toit's morphological/geological arguments that they were once side-by-side and interpreted this result as evidence of "relative continental displacement rather than polar wandering during the interval between the [eruption] of the two formations" (Creer, 1958b: 388).

Here and elsewhere in his discussion Creer was wise enough to recognize that Serra Geral and Karroo volcanics "may not be contemporaneous" (Creer, 1958b: 386), as was later shown to be so. It is difficult to ascertain exactly what he meant by "relative continental displacement." If he meant relative motion of Africa and South America between deposition of the two units, then, on the evidence available at that time, his argument was unlikely to have been correct because both Serra Geral

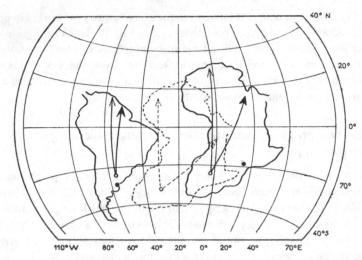

Figure 5.7 Creer's Figure 11 (1958b: 387). Equal area projection showing the suggested relative positions of South America and Africa during the Early Jurassic. Positions of collecting areas show thus: open circles, in the early Jurassic; solid circles, at present time. Darker arrow is the direction of the North Pole in ancient times; lighter arrow is its present position. Position of Africa relative to South America suggested by du Toit in his 1927 *Comparison of the Geology of South America with South Africa* is shown by dashed lines. The paleomagnetic data from the early Jurassic suggested clockwise rotations of Africa and South America of 28° and 6° respectively, which implied a relative rotation of 22° since the early Jurassic.

and Karroo lavas were then considered Jurassic, whereas the opening to the South Atlantic did not become appreciable until thick marine deposition began in the Cretaceous in the rift between them (du Toit, 1937: 101). If he meant displacement of South America and Africa relative to their geographic pole prior to their rupture, then he would have been correct (see §5.7). In either case mobilism was involved.

Creer returned to South America in 1958 and sampled seventeen new localities from which, along with his earlier work, he constructed an APW path extending back to the Pre-Cambrian. The path differed from that of other continents. It gave latitudes for South America that were broadly consilient with paleoclimatic evidence, although there were data inadequacies as noted below. This, as Creer was well aware, was important for the mobilist solution of the distribution of the Permo-Carboniferous glaciation, and he was especially interested in obtaining a pole of that age. He presented his new results at a symposium on rock magnetism at Kyoto in September 1961. Nagata organized the symposium, and from the English-speaking world Creer, Cox, Blackett, and the ever-present Runcorn gave papers.

... there is strong geological evidence of glaciation in the Permo-Carboniferous, particularly in Southern Brazil. Similar evidence of glacial conditions in these times exists in Africa and Australia and for this, among many other reasons, several geologists, of whom Wegener and Du Toit are perhaps the most famous, have suggested the existence of the gigantic primeval

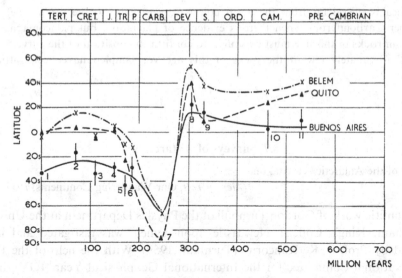

Figure 5.8 Creer's Figure 15 (Creer, 1962d: 164; my bracketed additions). Palaeolatitudes calculated for Belem [Brazil], Quito, and Buenos Aires from palaeomagnetic [data] of the following rock formations: 1, Present day; 2, Huitrinian Purple Sandstones, Neuquén, Argentina; 3, Kimmeridgian Purple Sandstones, Neuquén, Argentina; 4, Serra Geral formation, S. Brazil and Uruguay; 5, Triassic red beds, La Rioja, Argentina; 6, Mitu Formation, Peru; 7, Mississippian tuffs, Peru; 8, Devonian Purple Sandstones, Salta and Juyuy, Argentina; 9, Corrego das Pedras Formation, Corumba, Brazil; 10, Cambrian Purple Sandstones, Salta and Juyuy, Argentina; 11, Precambrian Purple Sandstones, Salta and Juyuy, Argentina.

continent Gondwanaland before the commencement of the Mesozoic. It is probable that the remanent magnetism of rocks might reveal independent evidence relevant to this hypothesis and this was one of the objects of the two palaeomagnetic surveys undertaken.

*(Creer, 1962d: 154)*

Although Creer's results showed little change of latitude of South America since the Jurassic, his Triassic and Permo-Carboniferous results suggested that South America was then at a higher latitude, as expected given the glaciation.

The youngest formation to give a pole significantly different from the present geographic pole is the Triassic red beds from Argentina, and this movement is confirmed by the results for the Upper Carboniferous or Permian Mitu Formation from Peru. Both of these formations contain normally and reversely magnetized samples and the information obtained from them is considered reliable.

*(Creer, 1962d. 163)*

Creer gave the paleomagnetically determined paleolatitudes of certain cities, which indicated that South America had been positioned at higher than present latitudes during the Early Carboniferous (Figure 5.8).

However, the results were not clear-cut; he had no firm evidence from the Late Carboniferous, that being the age of glaciation; his Early Carboniferous data were too old.

S. America is placed far from the north pole (i.e., very close to the south pole) during the Upper Carboniferous when there is evidence of glaciation, but palaeomagnetic evidence from rocks of this age must be sought to confirm the position of the curve for these times. Unfortunately few of the rocks of this age yet sampled have been sufficiently magnetic.

*(Creer, 1962d: 164)*

## 5.7 Surveys of Antarctica

The role of the Antarctic is a vital one.

*(Alex du Toit,* Our Wandering Continents, *1937: 128)*[15]

The Antarctic work of Gordon Turnbull of the Physics Department at the University of Durham, King's College, Newcastle upon Tyne, "was instigated and largely arranged by Prof. S. K. Runcorn" (Turnbull, 1959). With the help of the United States National Committee for the International Geophysical Year (IGY), he collected samples of three ages. Late, mainly Quaternary, volcanics of Hallett Cove on the west coast of the Ross Sea gave a pole very close to the present geographic pole, and reversals were present. Some samples were found to be highly stable when treated in alternating fields; demagnetization was becoming required practice. He (1959: 153) argued that his result and earlier evidence of Hospers (1955), Campbell and Runcorn (1956), Irving and Green (1957a), and Creer (1958b) showed that the time-averaged geomagnetic poles had been dipolar and aligned along the present axis of rotation for the late Cenozoic. This fundamental feature of the time-averaged geomagnetic field and basic tenet of the paleomagnetic case for continental drift (the GAD hypothesis) had now been confirmed from four continents through the efforts of the original Cambridge group and its diaspora.

Turnbull also sampled dolerite sills, of probable Mesozoic age, and sandstones from the Late Paleozoic or early Mesozoic Beacon Formation from near the Ferrar Glacier. The dolerites (now called the Ferrar Dolerites) intruded the Beacon Sandstones. The dolerite magnetizations were well grouped and little changed after treatment in high alternating fields. The sandstones were magnetized in the same directions as the dolerites, and he suggested that they had been heated by the intrusions and magnetized when they cooled. The dolerite and sandstone poles deviated ~30° from the current geographical pole implying apparent polar wander. Because his pioneering results were based on small collections, he was unwilling to draw conclusions about continental drift.

Older rocks are magnetized in a direction significantly different from that of the present mean magnetic field. This may be interpreted as evidence of polar wandering, implying a polar movement of some 30 degrees since the later part of the Mesozoic era. The data are perhaps too sparse to admit of a discussion of continental drift.

*(Turnbull, 1959: 157)*

Within the next year three more Antarctic studies were published. Nagata and Shimizu (1959) obtained some Pre-Cambrian gneisses from the Ongul Islands, where Syowa, the Japanese Antarctic Research Base, was located. They reported coherent magnetizations but gave no information about possible tilting since their magnetization, and for this reason as well as being a single Precambrian result, had no direct relevance to the mobilism debate.

The next study, published a month after the above, was by D. J. Blundell and P. J. Stephenson. Stephenson, a member of the geology department at the University of the Punjab, India, collected eight dolerite samples from mountains south and east of the Filchner ice sheet, and sent them to Blundell, then at the University of Birmingham. Blundell, who received his Ph.D. in 1957 from Imperial College, working under Bruckshaw, already had shown his support for continental drift at the 1955 Cambridge meeting, which Irving and Creer did not attend, and where only he voted in its favor (Note 14, §3.9). The dolerites were thought to be Jurassic. They were unaware of Turnbull's result. Their poles were very close, 54° S, 136° W, and 58° S, 142° W respectively. They compared their finding with K. W. T. Graham and Hales' roughly contemporaneous poles from the Karroo Dolerites of South Africa, with Irving's Tasmanian Dolerites, with Creer's Serra Geral lavas of South America (all at the time thought to be Jurassic), and with Clegg and company's younger pole from the Cretaceous/Tertiary Deccan Traps of India. They noted the huge differences between them, firmly stated their preference for continental drift rather than polar wandering, and noted that the results were "not inconsistent" with earlier mobilist reconstructions.

The age correlation between these rocks [used to derive the paleopoles for Antarctica, southern Africa, Tasmania, India, and South America] is not good, but it may be stated with some confidence that they are all Jurassic, and are the same age to within 30 million years. The difference between the Antarctic and Indian pole positions is more than 100°, and though this could be explained in terms of Gold's mechanism for polar wandering, it seems unlikely in view of the relatively small variation in pole positions from the Jurassic to the present day relative to a single continent. The scatter in pole positions [for the five regions] is more indicative of relative movement of the continents, and it is interesting to note that these results are not inconsistent with earlier reconstructions put forward on independent evidence by proponents of the continental drift hypothesis.

*(Blundell and Stephenson, 1959: 1860; my bracketed additions)*

Blundell had not changed his mind about continental drift.

Colin Bull from the physics department at Victoria University of Wellington, New Zealand, and Irving were responsible for the next survey. Their paths had already crossed. Bull recalled:

I'd met Ted Irving in September 1951, after my return from Spitsbergen, when I accepted Keith Runcorn's offer of a position in Geophysics at Cambridge. Ted and others warned me not to take Keith's words too literally. There was no money attached to Keith's offer. Rather soon

I despaired of him and took a position with the British North Greenland Expedition and did not return to Cambridge till September 1954. Ted left Cambridge shortly after that, cheated out of a Ph.D. I kept vaguely in touch with him and followed his work in Canberra fairly closely – and even tried to follow his plans for building an astatic magnetometer, in Wellington, New Zealand.

*(Bull, January 10, 2008 email to author)*

While leading the 1958–9 Victoria University of Wellington Antarctic Expedition, Bull collected from ten sites in two huge dolerite sills each over 800 m thick, and prepared them for measurement at ANU.

On that first ever University Expedition to Antarctica, 1958–59, from the Victoria University of Wellington, New Zealand, I collected two suites of oriented specimens from the dolerite sheets, one from the sheet that is exposed on the north side of Wright Valley, and the other from Dais, the mesa-like feature in the center of the western end of Wright Valley.[16] After our return from Antarctica, February 1959, I prepared rock disks from the dolerite and dyke rocks that I had collected and went to Canberra at John Jaeger's invitation in June or July 1959, to work them up on Ted's magnetometer. We wrote those first two papers while I was still in Canberra, or very shortly afterwards.

*(Bull, January 10, 2008 email to author)*

They characterized the dolerites as "very likely to be of Mesozoic age." They carried out alternating field cleaning on test samples, and obtained a pole at 51° S, 132° W, close to the two just described; three independent studies in excellent agreement (Bull and Irving, 1960a: 835). They also studied Beacon Sandstone and basement dykes (likely Paleozoic) from below the Beacon Sandstone, which in a later detailed study they showed were magnetized in the same direction as the dolerites and like Turnbull's Beacon Sandstone were likely remagnetized when these huge sills were intruded (Bull, Irving, and Willis, 1962, Table 2: 320). As the last contributors to this phase of Antarctic work done at the time of the IGY, Bull and Irving (1960b) were in a position to summarize and compare results from Mesozoic basalts and dolerites of Antarctica with those from others parts of Gondwana (Figure 5.9).

When plotted with respect to present positions of continents the poles were very widely scattered, grossly inconsistent with fixism and strong evidence against it. They listed other possible explanations, continental drift, polar wandering, a non-dipole geomagnetic field, or inaccuracies of alignment of the remanent magnetism with that of contemporaneous geomagnetic field. Noting that, given the evidence presented, they could not "decide with certainty between these," they continued:

... in connection with the first possibility [continental drift] it is of interest to test the extent to which these pole positions are consistent with reconstructions of the past distribution of the continents derived from other evidence. Clearly, if our procedures and the reconstruction are correct then pole positions, previously divergent, should unify.

*(Bull and Irving, 1960b: 223; my bracketed addition)*

Restoring poles to du Toit's reconstruction of Gondwana produced much better agreement, ordering them not into one, as would be expected if they were of similar

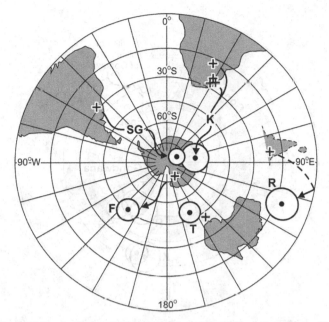

Figure 5.9 Bull and Irving's Figure 6 (1960b) with Ferrar pole updated from Bull *et al.* (1962). Paleomagnetic poles from Mesozoic basalts and dolerites of Gondwana relative to present position of continents. F, Ferrar dolerites; K, Karroo of South Africa (Graham and Hales, 1957); R, Rajmahal Traps, India (Clegg *et al.*, 1958); SG, Serra Geral lavas and baked sediments (Creer, 1958); T, Tasmanian dolerites (Irving, 1956); +, location of samples. In the original figure, the Ferrar dolerite pole was based on Turnbull (1959), Blundell and Stephenson (1959), and Bull, Irving, and Willis (1962).

age, but into two groups (Figure 5.10). Poles for Africa (K) and Antarctica (F) are situated in the southern ocean between South America and Australia, and essentially identical. Poles from India (R) and Australia (T) are also in excellent agreement, and that from South America (SG) is close by, but somewhat north and west; all these are about 30° further south than the Karroo and Ferrar poles.

Bull and Irving sought explanations for these two groups, noting:

The age relationships of these dolerites and basalts are not well known. In most cases they intrude or overlie Triassic beds, which fix their lower limit. Their upper limit, however, is less satisfactorily determined ... However, it seems probable that all of these rock formations date from the Jurassic or later Mesozoic.

(Bull and Irving, 1960b: 222–223)

They also noted:

Du Toit's reconstruction refers to the Palaeozoic while the formations here discussed are Mesozoic and it is possible that the break-up of Gondwanaland had already begun at the time of their formation.

(Bull and Irving, 1960b: 224)

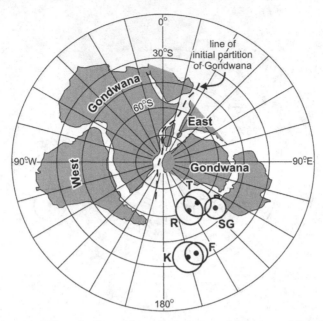

Figure 5.10 Paleomagnetic poles from Mesozoic basalts and dolerites of Gondwana labeled as in Figure 5.9 and plotted relative to du Toit's (1937) reconstruction of Gondwana.

The two groups of poles could fall within the Jurassic and Cretaceous and reflect different stages either before or during breakup of Gondwana. Because of these uncertainties at the time in the ages and intercontinental correlation of the Mesozoic dolerites and basalts of Gondwana, Bull and Irving did not have a fully satisfactory explanation of the two markedly different groups of poles after restoring Gondwana (Figure 5.10), although the very much better overall grouping after restoration implied that continental drift was involved: the spectacular agreement of the poles of such similar bodies as the Karroo and Ferrar alone would seem to essentially require drift, but at the time geological evidence of age was not precise and radiometric ages of these massive bodies had yet to be made.[17]

## 5.8 Surveys in Japan and China

Nagata and colleagues (1959) soon produced the first APW path for Japan (Figure 5.11). They based it on magnetically stable igneous and sedimentary rocks ranging in age from Lower or Middle Cretaceous (m.l.Cr) through Lower (l.M), Upper or Middle Miocene (u.m.M), to Lower (l.P) and Upper Pliocene (u.P). All were obtained within about 400 km on the main island of Honshu. Japan being geologically relatively young and tectonically active, older rocks are generally too altered or deformed to be studied. Nonetheless, by constructing Figure 5.11, they compared Japan's APW path with those from Australia, India, North America, and Europe. They (1959: 382) claimed that the APW path for Japan "gives a support to

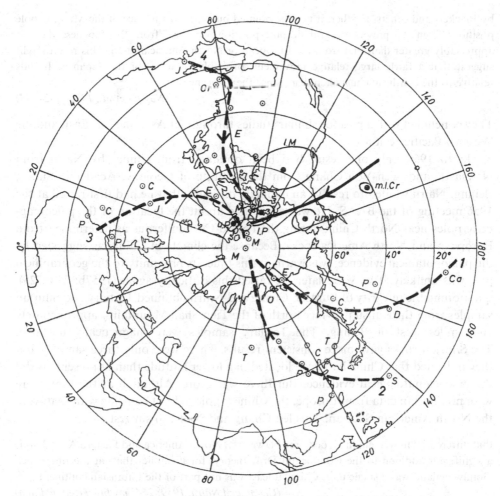

Figure 5.11 Figure from Nagata *et al.* (1959: 382). Poles with error circles are from Honshu and the solid line is the proposed APW path for Japan; 1, 2, 3, and 4 respectively represent the APW paths for Europe, North America, Australia, and India with poles labeled by geological systems. P: Pliocene; u.P: Upper Pliocene; l.P: Lower Pliocene; M: Miocene; u.m.M: Upper or Middle Miocene; l.M: Lower Miocene; E: Eocene; O: Oligocene; Cr: Cretaceous; m.l.Cr: Middle or Lower Cretaceous; J: Jurassic; T: Triassic; P: Permian; C: Carboniferous; D: Devonian; S: Silurian; Ca: Cambrian.

the hypothesis of polar wandering of the earth, proposed by Runcorn." Astonishingly, even though they were in the midst of one of the most active tectonics regions on Earth, they showed no enthusiasm for continental drift even though the APW path they gave for Japan could hardly have been more different from the others.

The mutual discrepancy in the paths of polar wandering derived from the Japanese and foreign data might be attributed to continental drift, accompanied by rotation also, as was suggested by Blackett and others. Further, it may be pointed out that the distance of the Miocene pole positions from the present geographic pole position, obtained from the Japanese data, is appreciably greater than that from the European or North American data. This result might suggest that a fairly large relative movement, probably rotation, of the Japanese Islands relative to the continents has taken place since the Miocene.

*(Nagata et al., 1959: 382–383)*

They concluded with a plea for similar studies throughout Asia, the Far East, and the Western Pacific region.

Up to 1959 only one result had been obtained from China, by Nairn from Newcastle and Wen-You Chang from the Institute of Geology, Academia Sinica, Beijing. Nairn wanted to resolve an anomaly difficulty that he had first raised at the 1956 meeting of the BAAS. He was concerned about the location of the paleomagnetic poles near North China during the Permo-Carboniferous as determined from European and North America rocks. Because the climate in China, as suggested by the paleobotanical evidence, was warm at this time, its proximity to the geographical pole was unlikely. Two years later, he raised this difficulty again (1957b: 724–725), questioning the integrity of Eurasia. Chang and Nairn obtained Tertiary and Silurian samples from the Kansu province north of the Tien Shan Mountains, approximately 1000 miles west of Beijing. The Tertiary samples were magnetically unstable. The Silurian rocks gave more consistent results. There were only three samples, but they indicated that China had been located in a lower latitude than at present, as did the paleoclimatological evidence. But there was more. Although at the time there were no Silurian data from Europe, the Chinese pole did not coincide with that from the North American Silurian. This led Chang and Nairn to suggest

that differential movement [had] occurred between ... North America and Eastern Asia. This is a significant addition to the Continental Drift theory, for it implies that, at the time when Gondwanaland was a single unit, China at least was not part of the Laurasian continent.

*(Chang and Nairn, 1959: 254; my bracketed addition)*

Their conjecture has turned out to be correct; later work has shown that the North China Block was separate and became attached to Euro-Asia in the later Phanerozoic.

### 5.9 Surveys in the USSR

Soviet paleomagnetists began surveys during the mid-1950s and soon became very active. By the late 1950s, Aleksei N. Khramov and other Soviet paleomagnetists had obtained poles for the USSR ranging in age from Pre-Cambrian to present. In his 1964 textbook, Irving cites thirty-seven works by Soviet paleomagnetists; one dated as 1954, two dated 1956, three 1958, six 1959, four 1960, thirteen 1961, three 1962, and five 1963.[18] Khramov's book, *Palaeomagnetism and Stratigraphic Correlation*, was published in Russian in 1958. Irving recognized the book's importance, and J. C. Jaeger, head of the department, agreed to have it translated and published by the geophysics department at ANU in 1960. Approximately a hundred copies were produced. About a decade later, Khramov and L. Ye. Sholpo (1967) compiled huge tables of paleomagnetic results from the USSR. The volume was translated and published at the Defense Research Board in Ottawa, Canada (1970), and edited by Irving. Summaries were given by McElhinny (1973: 292–301). Khramov and other Soviet paleomagnetists argued in favor of mobilism, and found themselves at odds with the anti-mobilist attitude of leading Soviet Earth scientists (I, §8.5). Khramov's work is memorable because it represents an independent appraisal of the paleomagnetic case for mobilism, made despite the disbelief in mobilism then prevalent in the USSR.

Khramov's mobilism arguments were in two stages. The first summarized in his 1958 book was based mainly on the Carboniferous through Triassic results from the Russian Platform and the part they played in the global test of continental drift, and it is with this that I am here concerned. The second stage was marked by the massive succession of results summarized in the Khramov–Sholpo (1967) compilations. As Soviet paleomagnetic surveys intensified, differences were found between Paleozoic poles from the Siberian and Russian Platforms indicative of their former separation and collision at the end of the Paleozoic. This was the topic of his acceptance speech to the European Union of Geosciences in Strasbourg in 1995 when he received the Wegener Medal. His work on Paleozoic mobilism is beyond the time limits of my book.

While a student at Leningrad State University (now St. Petersburg State University) and before joining in 1951 the All Union Oil Research and Geological Prospecting Institute of the Ministry of Geology (VINIGRI) in Leningrad, Khramov wrote a review favoring field reversals.[19] Learning that oil geologists were having trouble correlating poorly-fossiliferous oil-rich strata in Western Turkmenia (now Turkmenistan), he suggested that they could be correlated paleomagnetically in terms of alternating normal and reversed zones, a very early demonstration of the stratigraphic importance of reversals. From September 1953 through 1955 he sampled mainly red sediments on or near the southeastern shore of the Caspian Sea, and traced a sequence of thirteen alternating polarity zones spanning what he estimated to be the last five million years. He considered the duration of each zone to be roughly equal – about 400 000 years. This was a very important result, being the first demonstration of reversals occurring sequentially in stratigraphic sequences in late Cenozoic sedimentary rocks, echoing the demonstration in igneous rocks of similar age in Iceland by Hospers (§2.9).[20] It followed quickly the demonstration by Irving (1954) of the sequential reversals in the red Torridonian sedimentary rocks showing that they occurred in the Precambrian much earlier in Earth history (§2.5, §2.6).

Khramov sampled extensively, usually red sedimentary rocks, from the Russian Platform and western slopes of the Urals, selecting Ordovician, Permian, and Triassic deposits. He found (1958: 143) normally and reversely magnetized Ordovician deposits. He found only reversely magnetized Permian deposits, and normally and reversely magnetized Triassic deposits, and noted their poles were generally in good agreement with results from Western Europe. This was an important result; not only did it testify to the reliability of the paleomagnetic method as applied over a distance of ~3000 km, it implied that western Europe and the Russian Platform had been an unbroken block (plate) since the Late Paleozoic.

Khramov's work and that of other Soviet paleomagnetists was characterized by energetic sampling closely tied to stratigraphic sequences, and by clever graphical techniques for testing magnetic stability. They found many examples of strongly oblique magnetizations "streaked" along great circles toward the present geomagnetic field, which, on the basis of Néel's theory, they attributed to VRMs. Constructing great circles, called by Khramov "circles of remagnetization," through these distributions, he developed graphical methods of correcting for VRMs. The principle is that in strata that are tilted in different directions and by different amounts the demagnetization circles after correction for tilt will have a common intersection point corresponding to the direction of primary magnetization. This is a very important principle that has been used widely by later workers in this and other contexts. Khramov applied it by migrating the individual points along the circles of remagnetization until he obtained a common distribution (Khramov, 1958/1960: 90). In this way, he side-stepped the complicated statistical problems that later workers with improved computational techniques have addressed (Halls, 1976).

From the references he gives in his book, it is possible to reconstruct how Khramov came to favor continental drift. He references the British APW path (Creer *et al.*, 1954) but nothing else of early Cambridge and ANU work. He knew that Clegg and company favored continental drift because he referenced Clegg, Deutsch, and Griffiths (1956), in which they suggested that India had moved relative to North America and Europe. He also had Irving's results from India as reported by them, but did not know of Irving's early (1956a, b) Australian results, which appeared in obscure journals (§3.12). He had available to him the review of early British work from the Birmingham meeting (§3.3) (Griffiths and King, 1954), he had by then Runcorn's results (1956a) from North America, Doell's (1955b) and Graham's (1955) results from the Grand Canyon region (§3.7, §7.5), Du Bois' (1955) Precambrian results from Michigan's *Keweenawan* Peninsula (§3.2), Gough's (1956) results from Africa (§3.13, §5.4), and he knew about Mercanton's (1926a) work on Australian Tertiary and Permian lavas.

Khramov considered strict polar wandering before turning to continental drift. He (1958/1960: 55) presented a summary table (his Table 9) of the numerous Pliocene Proterozoic paleopoles from Europe and North America available to him. He also constructed an APW path for Europe (his Figure 5, p. 180) in which he incorporated his own Soviet data. This led him initially to favor polar wandering.

The paleomagnetic data, as we see from Table 9, suggest a very marked shift of the poles since the Proterozoic, when, according to these data, the north pole was in the central part of North America. It then started to move towards the west, and in the course of the Upper Proterozoic, Cambrian, and Ordovician, it passed across the Pacific Ocean south of the Hawaiian Islands.

In the Silurian it was near the coast of Japan, after which it turned northwards. In the Upper Paleozoic the north pole was in the eastern U.S.S.R. In the Jurassic it was near the shores of the Arctic Ocean, within the limits of which it has since remained. This movement is shown in Fig. 5.

*(Khramov, 1958/1960: 54–55)*

He acknowledged the differences between poles from North America and Europe (as known in the mid-1950s), but they did not provide "an undoubted confirmation of continental drift" because there were alternative explanations.

The scatter of results within one system and the appreciable disagreement of the pole positions obtained from Eurasia and North America make difficult any simple interpretation. It is possible that the first point indicates that the polar movements were complicated in character and that they correspond only to the first approximation to the simple picture outlined above. However, this scatter could be due to partial magnetic instability of rocks. It could also be caused by distortions of the directions of magnetization during deposition and compaction of the sediment due to non-isometric ferromagnetic particles [inclination error]. Finally, it can, to a large degree, be explained simply by sampling inadequacies.

The fact that data from North America gives a more westerly position of the pole could be regarded as a new confirmation of the hypothesis of the continental drift. Let us note, however, firstly that this divergence never goes beyond the limits of the oscillations of the poles according to data obtained from the same continent, and secondly that it is easy to show that a comparatively small decrease of the values of the inclination of the geomagnetic field in both Europe and north America could explain the observed disagreements in the pole positions. Compaction of the sediments could explain this decrease, since the data from the Palaeozoic were obtained chiefly from the study of sedimentary beds. Thus the palaeomagnetic data from Eurasia and North America are not an undoubted confirmation of continental drift, although they are evidence in its favour.

*(Khramov, 1958/1960: 56; my bracketed addition)*

Before turning to continental drift, he proposed an additional argument for polar wandering by comparing paleomagnetic and paleoclimatic data. Although generally the same, he developed his arguments independently of Irving (§3.12), and Opdyke and Runcorn (§4.2, §5.10–§5.14). It is his independence of thought that is so interesting. Like them, Khramov stressed the consilience of the different data-sets (RS1).

Let us compare palaeomagnetic and palaeoclimatic data regarding the positions of the poles (Table 10). The palaeomagnetic co-ordinates of the pole have been determined for certain periods, and for these intervals of time the mean co-ordinates of the north pole are also given, on the basis of data as to the position of equatorial zone of reefs and salt deposits in the northern hemisphere (Schwarzbach 1955) and also on the mean position of the arid belt in Eurasia for various periods ... Any drifting apart of continents was not taken into account. Data for the Pliocene and the Upper Proterozoic was borrowed from Kreichgauer (see Jardetsky 1949). The agreement between palaeomagnetic and palaeoclimatic data is very good. This agreement between data obtained from completely different areas of knowledge and completely independently from each other cannot be fortuitous. Undoubtedly the palaeomagnetic data provide a new and very powerful confirmation of the movement of the poles during geological history.

*(Khramov, 1958/1960: 58–59)*

Turning to continental drift, Khramov compared the scatter among paleomagnetically determined poles based on the current position of the continents and based on Wegener's positioning of Europe, North America, and India during the Cretaceous; Europe, North America, and Australia during the Permian; and Europe, Africa, and North America during the Upper Proterozoic. Their scatter was reduced by a factor of four for the Cretaceous, by two for the Permian, and by four for the Upper Proterozoic, thus favoring continental drift.

In conclusion, let us return to the question of continental drift. Very important data were obtained recently by Gough (1956), who studied the palaeomagnetism of the Upper Proterozoic Pilansberg dykes [§5.4], and by Clegg, Deutsch, and Griffiths (1956), who studied the Deccan traps of Central India [§5.2]. In the light of new palaeomagnetic data on Permian deposits of Europe and North America, the thirty-year-old work of Mercanton, who studied among other formations the direction of NRM of Permian lavas in New South Wales, Australia, acquires a great importance [Volume II, Introduction]. Table 11 gives a comparison of the co-ordinates of the north pole during the Cretaceous, calculated from palaeomagnetic data from various continents. These co-ordinates were calculated first from the present position of continents and then from the position of the continents corresponding to the reconstruction of Wegener for the Cretaceous. The co-ordinates in the latter case are given relative to Europe and Africa. The same calculations were made for the Permian period and the Upper Proterozoic (Tables 12 and 13). As we see from Tables 11–13, correction of the palaeomagnetic data for translation and relative rotations of continents decreases the scatter of pole positions by factors of between two and five. We can conclude, therefore, that the data at present available favours the hypothesis of continental drift.

*(Khramov, 1958/1960: 60; my bracketed additions)*

Mercanton's Australian Permian result certainly was of "great importance" for Khramov because it was all he had; he did not yet have Irving's early Australian Permian results. His reliance on Mercanton's result might also be questioned, based as it was on a single sample. However, Khramov's method of analysis was bold, his judgment sound, and his conclusions correct.

Khramov's early support of mobilism was particularly impressive because of the strong resistance to it among academic Soviet tectonicists (I, §8.5), notably by V. V. Beloussov at the Institute of Physics of the Earth, Academy of Sciences, a powerful figure in Soviet Earth science, and an adamant fixist. Khramov was not employed by one of the large better known Soviet research institutes; he may, at first at least, have been something of an outsider, not among the higher echelons of Soviet science. He worked in the petroleum industry for VINIGRI, where his work was supported and its importance recognized, and where he was presumably protected from Beloussov's influence.

Khramov was aware that Wegener's views faced "well-founded objections" among Soviet Earth scientists. But, at the same time, he reminded Soviet geologists that there was support for mobilism among geologists elsewhere and among paleoclimatologists.

On the question of polar migration and continental drift stands have been made by such authorities as the geophysicists Jeffreys ... Urey (1953), Jardetzky (1949), and in palaeoclimatology, by Schwarzbach (1955) and Koppen and by a whole series of geologists studying the geology of South Africa, South America, India, and Australia (see, for instance ... Du Toit ...). In spite of this, ideas concerning the mobility of either the poles or continents are extremely unpopular among most Soviet geologists at the present time. The hypothesis of continental drift which was propounded by Wegener is considered to be completely inconsistent (Beloussov 1954). To some degree these objections were greatly furthered by the geotectonic ideas of Wegener, which encountered well-founded objections, with the result that the whole idea of a continental drift was compromised.

*(Khramov, 1958/1960: 54)*

In contrast to the North American oil industry, the state-owned Soviet oil industry consistently supported Khramov's paleomagnetic work over many years even though he advocated mobilism. He was willing and found himself able to disagree with the fixist minded mandarins of Soviet Earth science, and in this regard he differed markedly from the contemporary North American paleomagnetists Graham, Cox, and Doell, who likewise in the 1950s were immersed in a fixist milieu. Khramov's pioneering work has been well recognized in the West by the awarding of the Wegener Medal of the European Geosciences Union (1995) and the Bucher Medal of the AGU (2008).

The "Western" paleomagnetists, Deutsch, who had recently left Imperial College, London, and begun working in Canada, and N. D. Watkins (Deutsch and Watkins, 1961), also made an early paleomagnetic survey of Russian rocks. The Soviet geologist V. V. Fedinsky, whom they met in Toronto at the IUGG meeting in 1957, sent them some oriented samples from the Siberian Traps. They were then considered Triassic (now known to be at the Triassic/Permian boundary) and furnished a pole that could be placed on the European APW path. Deutsch and Watkins (1961: 545) found this result "not surprising, for it seems likely that in executing any movements inferred from this path, Europe and northern Asia would have acted as one unit." Their results also agreed with those of Khramov and other Soviet paleomagnetists.[21]

## 5.10 Paleowind studies, previous work

On another occasion we were camped in the shelter of a big dune. Around midnight we were startled by a tremendous booming sound coming from the ground a few feet away; we could feel the sand vibrating beneath us. We had to shout to make ourselves heard. The dune had begun a spontaneous song. In the moonlight we saw an avalanche creeping slowly down the dune's slip face. The sound came from below, where the avalanche was slowing and the sand accumulating. I had often heard this "singing" in previous years, but now, after about a minute there came an answering boom, and then another, from dunes half a mile away. The vibrations from our dune must have started avalanches elsewhere. We had the eerie notion that these great beings were talking to one another in the stillness of the night.

*(Bagnold, 1990: 118)*

Brigadier General Ralph A. Bagnold (1941) wrote the book *The Physics of Blown Sand and Desert Dunes*. As noted earlier, Opdyke and Runcorn devised a plan for using paleowinds as a means for testing the GAD hypothesis and the reliability of paleomagnetic data, a plan they developed while in Chicago during the summer of 1955 on their way back to New York after collecting in the western United States (§4.2). Runcorn got Opdyke admitted to Cambridge for graduate work, but Opdyke soon moved to Newcastle upon Tyne following Runcorn. Opdyke and Nairn spent about a month transferring paleomagnetic equipment from Cambridge to Newcastle. Opdyke is a geologist, and, wanting to be around geologists, a desk was arranged for him in the Geology Department and its head, Professor Stanley Westoll, became his co-supervisor. Opdyke recalled this hectic time, and Westoll's positive attitude toward continental drift.

When we went to Cambridge, and I had only been in England from October to December, Runcorn was appointed Professor of Physics at Newcastle, and this caused a tremendous blow-up for us, and he attempted to take everyone with him who was at Cambridge. So I asked whether I could stay on at Cambridge for the rest of that year, and he was adamant about not doing that because he was trying to persuade everyone not to do that. Phil Du Bois and John Belshé refused to leave Cambridge and go to Newcastle. Alan Nairn who was at the time a post-doctoral fellow and myself had to go to Newcastle. We spent about a month driving back and forth from Cambridge to Newcastle moving the paleomagnetics Lab to Newcastle. So, by the time that spring arrived, we were ensconced in Newcastle. When I went up there, I didn't feel too badly out of place, because people were talking about the Earth (even though I was in a geophysics department at Cambridge). In Newcastle it was a classical physics department, and, as part of the agreement of my going to Newcastle, I had an agreement worked out with Runcorn that I would be half in the department of geology, and half in the department of physics, and that I would have two supervisors, Runcorn being one of them, and Westoll being the other. And Westoll, of course, was a continental drifter; he was one from the word "Go." I had gone up to Newcastle and met him, and he agreed that he would do this, although he and Runcorn had some disagreements. But still, it worked out all right. It was much more comfortable for me sitting in the geology department. I could talk to people, and the thesis I was doing was not a physics thesis.

*(Opdyke, 1984 interview with author)*

Opdyke's supervisors were not sedimentologists, and he had to teach himself how to determine paleowind directions from aeolian sandstones. Westoll was not a sedimentologist but helped him get started with the literature. Studies of aeolian sandstones have a long history.[22] Opdyke talked with other workers, and figured out how to make the needed field observations that he was to take in Great Britain and the western United States.

Direction of Wind

Figure 5.12 Holmes' Figure 560 (1965: 763). Barchans shaped by prevailing winds.

Westoll knew more about all these subjects than anyone else I knew. He was a paleontologist, and all the problems about aeolian sands I researched out all myself. The fieldwork I more or less learned how to do myself, and did it myself. Bagnold's book, I read his book, and I read everything that had been published – Shotton had published some things in England on the subject that I found and read. I talked to Shotton and a gentleman by the name of Reiche who had studied aeolian sands in southwest North America, and I read all of that – that was what was known up to that time. I didn't know at that time that the US Geological Survey was doing a lot of work on cross-stratification measurements on the Colorado Plateau. But, I subsequently learned that when I went back the following summer. So I didn't do any work on the Navajo Formation on the Colorado Plateau because the US Geological Survey did them all. I worked on the Permian aged Wyoming aeolian sands, and extended Shotton's work in England. I never actually ended up publishing the England work. No, part of it was published in that paper in Alan Nairn's book. That is more or less how it came about.

*(Opdyke, 1984 interview with author)*

Multidirectional winds, typically prevailing winds interrupted by seasonally strong storm winds, produce longitudinal or seif dunes. Steady unidirectional winds produce barchan dunes, which are crescent-shaped with their rounded noses pointing downwind; they have gentle windward and steeper leeward faces, and their width is about a dozen times their height, which can reach over thirty meters (Figure 5.12). If these features can be observed in outcrop, prevailing wind directions can be determined.

Barchan dunes migrate downwind (Figure 5.13), larger ones can migrate seven, smaller ones twenty meters a year. As Bagnold (1990: 103–104) explained:

In 1929 and 1930, during my weeks of travel over the lifeless sand sea in North Africa, I became fascinated by the vast scale of the organization of the dunes and how a strong wind would cause the whole dune surface to flow, scouring sand from under one's feet. Here, where there existed no animals, vegetation, or rain to interfere with sand movements, the dunes seemed to behave like living things. How was it that they kept their precise shape while marching interminably downwind? How was it that they insisted on repairing any damage to their individual shapes? How, in other regions of the same desert, were they able to breed "babies just like themselves that proceeded to run on ahead of their parents?" Why did they absorb nourishment and continue to grow instead of allowing the sand to spread out evenly over the desert as finer dust grains do?

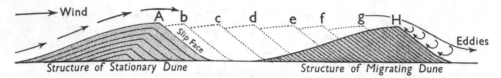

Figure 5.13 Holmes' Figure 552 (1965: 757). Longitudinal cross-section of a migrating barchan sand dune. A grows in height as its crest advances upward and forward. Prevailing winds remove sand from the windward side and add it to the advancing leeward side. The leeward side becomes a slip face with angles approaching 33°, the angle of repose for dry sand. A migrates to H. Succeeding leeward faces may become fossilized. The resulting cross-bedded planes indicate dune migration and prevailing wind conditions.

Winds remove sand from the windward side of advancing dunes and deposit it on the leeward side. As the leeward side steepens, it collapses forming a slip face angle of up to 33°, the angle of repose for dry sand. In this way the leeward face receives new sand. By this process of removal and deposition of sand, dunes migrate, but they keep their original shape and orientation relative to the prevailing wind.

Because sand varies in grain size and the way it compacts, vertical sections through fossilized dunes exhibit subparallel layering. These layers form on the leeward or advancing faces of drifting dunes, and are called "false" or cross-bedding planes because they are not parallel to the horizontal at the time of deposition.

Once the successive leeward faces of a fossilized dune are identified, the direction of advancement can be determined by measuring the azimuths (down-dip directions) of their greatest dip. Averaging azimuths over several dunes yields their mean direction of motion and thus the wind direction prevailing during formation.

Success required solving several problems. Cross-bedding in fluvial strata had to be identified and avoided. If the beds are truly aeolian, researchers then had to determine if they were barchan dunes. If so, then techniques had to be developed for determining the ancient dune's direction of migration from mapping their lee-ward slopes. Opdyke tackled these problems in his fieldwork and from the literature.

He read the work of F. W. Shotton and R. A. Bagnold from Great Britain. Bagnold developed a general quantitative account of the formation and evolution of sand dunes in his book.[23] Shotton, from the University of Birmingham, wrote a seminal paper in 1937 on the Permian of central England. He established that the outcrops were truly aeolian not fluvial, and determined that the prevailing wind was from what is now the east (Figure 5.14(e)).

Warren O. Thompson (1937), Edwin D. McKee (1933, 1938, and 1940), and Parry Reiche (1938) had investigated wind-blown deposits in the western United States, and developed and refined criteria for distinguishing fluvial from aeolian deposits. McKee and Reiche studied the Permian Coconino sandstones in northern Arizona, and their extension, the De Chelly sandstone, in northeastern Arizona (Figure 5.14(b)). Both eventually agreed that dunes in the Coconino sandstones had migrated

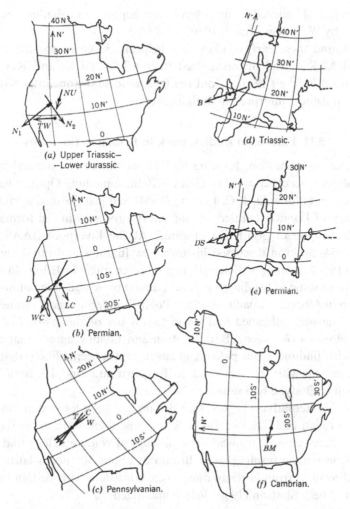

Figure 5.14 Irving's Figure 9.71 (1964: 235). Summary of paleowind directions as they were known in 1960 from North America and Europe compared with paleomagnetically determined latitudes and meridians. Directions are indicated by bold arrows, the dot on each being roughly central to the region of observation. The arrows N' point to paleonorth. The wind directions are labeled as follows: (a) *NU* Nugget Formation, $N_1$ Upper Navajo (Kiersch), $N_2$ Middle Navajo (Kiersch), *TW* type Wingate; (b) *D* De Chelly, *WC* Western Coconino, *LC* Coconino of the Little Colorado; (c) *C* Casper, *T* Tensleep, *W* Weber; (d) *B* Bunter Sandstone; (e) *DS* Dumfries Sandstone; (f) *BM* Baraboo Monadnock.

southward, under the influence of prevailing winds from the north.[24] Opdyke also read S. H. Knight's analysis of the Permian Casper sandstone in Wyoming. Although Knight (1929) developed a way of graphically displaying and statistically analyzing his results that was followed by Reiche and others, and although he surmised that the sandstone had been transported from the northeast, he attributed the movement to water currents, and believed them to have been deposited in shallow seas; his claim was disputed by W. H. Twenhofel (1950).

Opdyke found these early studies very helpful. He corresponded with Reiche, Knight, and McKee, and acknowledged their help (Opdyke and Runcorn, 1960: 960). They used Reiche's results from the Coconino sandstone, and Knight's from the Casper sandstone, interpreting both as aeolian.

### 5.11 Paleowind studies, work in Britain in the 1950s

Opdyke spoke with Shotton, learning that he had recently extended his study of aeolian sandstones to other areas of Great Britain, something Opdyke had begun to do. They were not alone. D. J. C. Laming (1954) had just finished a Ph.D. thesis at the University of London entitled "Sedimentary processes in the formation of the New Red Sandstone of South Devonshire." At the Liverpool BAAS meeting in December 1954, Shotton discussed his own new findings and published them two years later (1956). Laming's dissertation was never published, but Shotton (1956: 463) reproduced some of his diagrams, and Laming presented some of his findings at a symposium in Alberta, Canada, entitled "Polar Wandering and Continental Drift." The symposium was published in 1958 in two issues of the *Journal of the Alberta Society of Petroleum Geologists*. Both Shotton and Laming agreed that their results concurred with findings from paleomagnetism, and it seems likely that they independently came up with the same idea as Runcorn and Opdyke; both began their fieldwork before Opdyke had settled at Newcastle.

Shotton (1956) tentatively argued that Britain had been located in lower latitudes during the Permian and Triassic for three reasons: there were extensive evaporites indicative of deposition in tropical or semi-tropical climate; Clegg and company's Triassic paleomagnetic results placed Britain at about 30° north latitude; and the paleowind directions from barchan dunes were consistent with Britain then being in the trade-wind belt. Shotton (1956: 463) summarized:

In all cases there can be seen a distribution of false-bedding which indicates a remarkably constant wind direction (as Bagnold maintained was necessary to produce barchans). When these directions are inserted on a map, as has been done in Fig. 2 [reproduced as Figure 5.15], they obviously form a regional pattern which veers from north-east in the north to south-east in the south. The Permian sand sea exhibits, throughout the duration of its formation, the same sort of curving pattern in the direction of its prevalent wind as affects the large deserts of the present day. It is a system which looks anomalous for Britain in its present day latitude, but might appear more reasonable if, as has been suggested earlier [from paleoclimatic and paleomagnetic studies], New Red Britain were in a more southern latitude with the north pole somewhere in north-east Asia.

*(Shotton, 1956: 463; my bracketed addition)*

Laming's arguments were equally forceful. The wind directions he obtained from New Red Sandstone (Permo-Triassic) in Devon agreed with Shotton's from the

Permian of central England. After discussing the Coriolis force and its effect on planetary wind patterns, Laming went on to consider the relation between wind direction and orientation of barchan sand dunes, and the orientation and latitude of Britain during the period of time.

A consequence of this argument is that Britain, and all other areas of extensive barchan development must have lain between about 15° and 30° latitude at the time of their formation. It may also be deduced that rotation relative to the position of the pole has taken place, since the dominant easterly direction evident over most of Britain obviously should be northeasterly to conform to the trade wind pattern, assuming an original location north of the Equator. A rotation of about 40° to 45° is required to bring the wind to a northeast direction; this shift would be in a clockwise sense, effective since the Permo-Triassic. Indeed, the south-south-easterly wind measured in beds of the same age in Devon would be an impossibility without it. Anomalous wind directions are accounted for by local deflections due to mountains known to have been important topographic features in the Permo-Triassic landscape.

*(Laming, 1958: 183)*

Laming (1958: 183) also characterized the paleomagnetic work of Clegg and his co-workers "to be remarkably well in accord with the present author's deductions, made quite independently, especially in view of the nature of the methods used." The consilience he described between results obtained by two entirely independent methods was indeed remarkable – very strong evidence favoring the GAD hypothesis.

## 5.12 Newcastle begins paleowind studies in North America

Armed with their knowledge of previous work in North America, and Opdyke's increasing expertise in fieldwork, he and Runcorn returned to the western United States to study the Pennsylvanian Tensleep, Casper, and Weber formations of Wyoming, and Utah. In agreement with Reiche and Knight, their fieldwork indicated that the sandstones had formed under prevailing northeast winds, but unlike them, they identified these as trade winds, just as Shotton and Laming had their British results. They went on to argue that such wind directions were consistent with Runcorn's paleomagnetic results from the western United States; in Late Paleozoic times, North America was much closer to the equator than at present, and had since rotated clockwise relative to the meridian (Figure 5.14(c)). Back in the summer of 1955, when Opdyke and he first came up with their plan to study paleowinds, Runcorn could not have hoped for more.

Opdyke (1958a) spoke at the 1956 International Geological Congress in Mexico City. He summarized paleomagnetic results from Great Britain, North America, India, and Australia, making it clear that since his arrival in the United Kingdom he had become a mobilist because of the excellent agreement of paleomagnetic and paleoclimatological findings. To illustrate, he appealed to the paleowind direction work from the United States and England, and F. F. M. Almeida's (1953) study of Triassic aeolian deposits in South America. He noted that the existence of Silurian coral reefs in the northern regions of North America and Europe could be explained if the two continents were in low latitudes, as indicated by paleomagnetic results (§5.19). Turning to the Southern Hemisphere, he cited Irving's (1956a) seminal study, and the ease with which the paleomagnetic results could be used to explain the Gondwana Permo-Carboniferous glaciation. Finally, he reconstructed the positions of South America and Africa by arranging them on either side of the Mid-Atlantic Ridge, and deduced that the paleowind direction in southern Brazil as determined by Almeida's study would then have been westerly.

### 5.13 Paleowinds, the 1957 Royal Astronomical Society meeting

Opdyke and Runcorn gave an early report at a meeting of the Royal Astronomical Society (RAS). This was held on November 22, 1957, in London, and chaired by Blackett. Bagnold spoke first, beginning with a description of dune form and wind direction. He described barchan and longitudinal dunes, outlined ways to identify them correctly, emphasized the need to do so, and the serious consequences of not doing so:

In real desert areas where there was no vegetation, the most common type of dune, the longitudinal dune, was due to multi-directional winds. The resultant wind direction was due to normal winds from one direction plus occasional storm winds from a somewhat different direction ... If either of the component winds changed markedly the dunes would break up into barchan dunes, in which the tip of one dune was the source of the next dune formed. Barchans were very rare and only found in remote places. A concave slip face was not sufficient indication of a barchan dune unless the echelon pattern, tip formation, etc., were present. The indicated wind direction might be out by 90 degrees if a previous dune slip and a barchan were confused.

*(Anonymous, 1958: 65–66)*

Shotton spoke next. He outlined the characteristics of fossil dunes, discussed the relation between wind and movement of barchan dunes, and identified the features that could be used to determine their original orientation; these included leeward slip faces, curving fronts, and directions of dune tips or horns. He summarized his work on various Permian and Triassic aeolian sandstones.

A diagram [of Shotton's; Figure 5.15] showed that aeolian sands of Permian age all had distinctly preferred directions. From this it was possible to interpret both the direction of the sand cascading down the barchan and the direction of the wind driving the sand. The regional picture of wind directions in Worcestershire and in different parts of the country showed a north-east wind in the north, veering through an east wind in the midlands to a south-east wind in the southwest country. From this wind picture the Permo-Triassic north was seen to lie to the east of the present geographical north.

*(Anonymous, 1958: 67; my bracketed addition)*

In the last sentence he was proposing clockwise rotation of Britain relative to the geographic meridian.

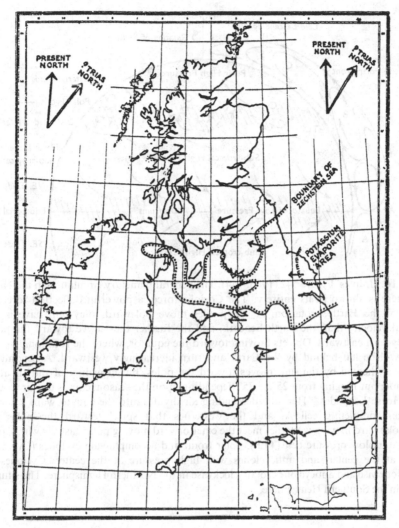

Figure 5.15 This diagram is a simplified version of Shotton's (1956: 459) Figure 2. It shows his estimation of the prevalent winds in the Permian desert deduced from his study of ancient barchan sand dunes.

Opdyke followed. He talked about analysis. He described procedures for sampling an outcrop and the statistical analysis of fossil dune directions. He compared Shotton's method of preferred orientation with that of summing individual outcrop measurements. Runcorn then linked paleomagnetic and paleoclimatic work through the GAD hypothesis, through the effect of the Coriolis force on the overall planetary wind patterns (see Figure 5.16). He presented their new data, suggested that the Permian aeolian deposits of North America and Britain had formed under the

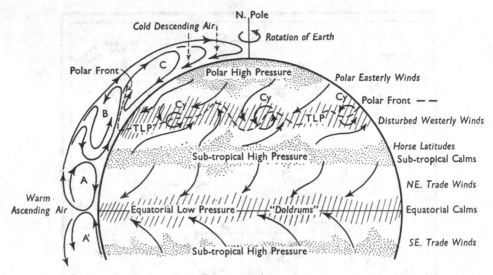

Figure 5.16 Holmes' Figure 542 (1965: 744). The general planetary circulation in the Northern Hemisphere is shown in its entirety. The Coriolis force, a powerful deflection force, arises because of the Earth's rotation. As projectiles move poleward, they will have a greater rotational velocity than the Earth beneath, and the further they move toward the pole, the more they turn eastward. Objects moving toward the equator, where the rotational velocity is greater, will be left behind by the Earth and turn increasingly westward. Ascending air at the equator, heated by the sun, moves toward the poles, but crowding of the air at about 30° latitude, but ranging from 25° to 35° depending upon the season, causes the air to descend at the "Horse Latitudes." Here the descending air divides into the equatorward trade winds (easterlies) completing cell A, and the westerlies that spiral toward the poles in cell B. The polarward moving westerlies meet the equatorward moving polar easterly winds, creating a temperature low pressure belt (TLP), polar front, and accompanying cyclones (Cy) with low pressure at the center, and anticyclones with high pressure at the center. Cyclones rotate counterclockwise and anticyclones rotate clockwise in the Northern Hemisphere. The situation is reversed in the Southern Hemisphere.

influence of Northern Hemisphere trade winds, and placed both regions in lower latitudes as the paleomagnetic evidence indicated.

Blackett was interested, and asked whether geologists could confidently distinguish ancient barchans from other dunes. Why did geologists commonly find barchan-type fossil dunes while modern barchans are rare? Bagnold replied (Anonymous, 1958: 68) that Shotton's dunes might be misidentified with their slip faces "90° wrong but climatic conditions might have been different." Shotton replied that geologists looked only where sand deposits had accumulated in great thicknesses; he asked if modern barchans formed on rock or sand. Bagnold did not know. Blackett asked if the orientation of modern dunes with steep slopes agreed with the wind directions over large areas. Bagnold again noted that longitudinal dunes dominate among modern dunes, adding that they do not have steep slopes. Opdyke asked if longitudinal dunes have slip faces on them. Bagnold said that they did not, but again noted that as longitudinal dunes faded away they broke into barchans: at critical wind speeds barchans formed. Blackett did not say how satisfied he was with these answers.

## 5.14 The Newcastle contribution to paleowind work

Opdyke completed his doctoral thesis "Palaeoclimatology and Palaeomagnetism in relation to continental drift" in 1958, and published (Opdyke, 1959) a comparison between paleomagnetic and paleoclimatological data, noting their concordance (RS1), and argued that paleomagnetism offered paleoclimatologists an independent way to determine ancient latitudes. He and Runcorn co-authored two papers (Opdyke and Runcorn, 1959, 1960). The second, more technical, was received in 1958. They emphasized that the study of paleowind directions in particular, and paleoclimatology in general – how temperature-sensitive sedimentary rocks especially – could be used to buttress the GAD hypothesis and to test the general reliability of the paleomagnetic method.

They noted the decisive role of the Coriolis force in determining the general planetary circulation of the atmosphere.

In the hydrodynamics of the atmosphere, as in the earth's core, the Coriolis force plays a decisive role because of the large length scales. It seems likely therefore that the general planetary circulation of the atmosphere throughout geological time will consist of an equatorial easterly trade-wind zone, a mid-latitude westerly zone and a polar zone. The wind pattern will therefore be symmetrical about the axis of rotation, wherever this happens to be on the earth's surface, with only minor asymmetries produced by the topography of the globe and world-wide climatic changes.

(*Opdyke and Runcorn, 1960: 960*)

They noted that when Britain and the United States were positioned closer together as their APW paths indicated, their paleowind directions fell into agreement. There was a remarkable concordance between three lines of evidence: the paleomagnetic evidence that defines the distance to the pole (the complement of paleolatitude $\lambda$) and the direction of the pole (paleomeridian or paleonorth $N'$ of Figure 5.14), the paleoclimatic evidence that can be compared with $\lambda$, and paleowind directions that are understandable in terms of planetary circulation (Figure 5.16). They claimed that these are generally consilient, and hence that the cornerstone of the paleomagnetic case for continental drift, the GAD hypothesis, is likely correct.

The possibility of tracing the direction of the wind through geological time by means of aeolian sandstones is an interesting project in itself, but it is at present important because it represents one of the few ways in which the determinations of the position of the pole in relation to the land masses through palaeomagnetic measurements can be independently checked. The palaeomagnetic evidence suggests that the western United States and Great Britain were much closer to the equator in Upper Palaeozoic times than they are at present ... It will be seen that both these areas would have been within the wind belt of the northern trade winds if the present width of this belt is typical. It is interesting to see that the evidence from the aeolian sandstones described above supports this prediction. The easterly wind deduced from the Permian of Great Britain and from the late Palaeozoic of the western United States both become north-east winds relative to the position of the equator at that time.

*(Opdyke and Runcorn, 1959: 34)*

Opdyke and Runcorn first tried to establish that the cross-stratified sand deposits they studied were aeolian and not fluvial, and divided them into those formed as transverse and those as barchan dunes. They showed that the prevailing winds during the Permian and Pennsylvanian in the western United States as determined from the orientation of the barchan dunes were northeasterly. Next they identified them as trade winds, and noted the similarity between their own and Shotton's study. Lastly, they argued that moving North America and Britain closer to the equator offered a better explanation than the alternative of greatly widening the Permian trade-wind belt and retaining the United States and Great Britain in their present positions.

In their 1960 paper, they gave a diagram that showed the paleomagnetically determined equators of North America and Europe during Carboniferous, Permian, and Triassic times, and showed the agreement between the paleowind directions and the paleomagnetic results (RS1). They also noted that the additional relative movement between North America and Europe as shown by paleomagnetic measurements was immaterial to their argument, and claimed:

The winds determined by Shotton for the British Permian and those for the North American Pennsylvanian both become northeast winds within a belt 0°–20° N. of the equator at that time. A possible explanation of these paleowind results is that they represent the trade-winds of late Paleozoic time in Great Britain and in the western United States.

*(Opdyke and Runcorn, 1960: 968)*

They then considered two fixist solutions. Both, they argued, faced anomaly difficulties (RS2 and RS3).

If, notwithstanding the paleomagnetic data, these continents and the poles were in the present relationship in the late Paleozoic time there seem to be two alternative explanations of the data: (1) The paleowinds were not planetary winds but resulted from local geography. (2) In late Paleozoic time the trade-wind belt extended over a greater range of latitude to include the Western United States and Great Britain.

*(Opdyke and Runcorn, 1960: 968)*

If (1) were correct, the paleowinds in the western United States should not be "substantially constant" "over nearly 1000 miles."

The wind directions over nearly 1000 miles in Permo-Pennsylvanian times are substantially constant. The minor variations may plausibly be the result of local geography, but a wind pattern constant over an area of this order is most easily explained as arising from planetary causes.

*(Opdyke and Runcorn, 1960: 968)*

They also appealed to a study by Forrest G. Poole of the USGS, who argued that "Northeast Trades" had prevailed during the early Mesozoic, making it even more unlikely that such constant winds were the result of local conditions. Poole argued for widening the tropical climatic belt ((2) above).

During the past 8 years data have been collected on cross-stratified sandstone units of probable eolian origin on the Colorado Plateau ... Dip orientation of cross-strata indicates that late Paleozoic and early Mesozoic winds on the Colorado Plateau were relatively constant and blew in a southerly direction. The high consistency in dip directions of cross-strata in eolian deposits is probably the result of winds with great regularity and strength similar to our present-day "Northeast Trades." It is thought that in late Paleozoic and early Mesozoic time the northern hemisphere trade-wind belt in this area was 10° to 20° farther north than at present. Tropical-type fossil plants found in continental sediments of the same age are similarly transposed from the present equatorial zone.

*(Poole, 1957: 1870)*

Opdyke and Runcorn (1960: 968) agreed that the easterly Permian paleo-wind directions that prevailed in Permian Britain were inconsistent with their being trade winds at all, if Britain had not rotated since that time. But they argued that paleomagnetic declinations indicated Britain had rotated since the Permian.

The second possibility that the trade-wind belt spread over latitudes up to 55° cannot be excluded. However, the Permian winds in Great Britain are east winds relative to the present geographical meridian with very little suggestion of a north component. This would appear to be incompatible with a trade wind and implies that the orientation of Great Britain relative to the earth's axis of rotation was not the same as at present. This of course is what the paleomagnetic angle of declination for Permian time indicates.

If, on these grounds, Britain were rotated and placed where they thought it belonged, it would be consistent with the paleomagnetic data and trade-winds latitudes would be much as at present.

They did admit, however, that their own solution faced a serious anomaly difficulty (Figure 5.14(b)), which they attempted to remove by arguing that the difficulty was itself based on results biased by secondary overprinting (RS2).

[The paleomagnetic data from] the Supai formation of Permian age gives a direction of magnetization which places Arizona just south of the equator. Unfortunately results from this formation may possess a secondary component in the direction of and induced by the present magnetic field of the earth. This component would make it appear that the Supai formation was deposited south of the equator even if it were originally deposited north of the equator. This difficulty cannot yet be resolved in the Supai, but in the case of the European Triassic and many of the Triassic results from North America this ambiguity does not arise because reversals of polarity occur in these rocks, and the effect of a secondary component if present is opposite in the two opposed groups, thus canceling itself out.

*(Opdyke and Runcorn, 1960: 970; my bracketed addition)*

The difficulties researchers find easiest to raise against their own solutions are, after all, those they think they already know how to remove from work at hand. They summarized their paleowind studies, which had been inspired by the new findings in paleomagnetism, in this way.

The positions of the equator based on the paleomagnetic data appear therefore to offer the best explanation of the wind-direction results, and it is concluded that it was only in the late Mesozoic that Europe and North America moved out of the trade-wind belt into the westerly-wind belt.

*(Opdyke and Runcorn, 1960: 970)*

Subsequent work has confirmed their conclusion.

### 5.15 A spin-off paleowind study

In his editorial foreword to the 1958 symposium on polar wandering and continental drift sponsored by the Alberta Society of Petroleum Geologists, the meeting where Laming presented his paper on paleowinds, G. O. Raasch recounted how he, weaned away from mobilism when a student, felt after hearing Runcorn talk about paleo-magnetism's case for mobilism.[25,26]

It was more decades back than I care to acknowledge when my professors set my mind at rest on the twin subjects, polar wandering and continental drift. The geologist could spare himself further speculative efforts in this direction. The physicists had shown the alleged operation of these phenomena to be out of the question. The earth's crust could not withstand the ensuing stress. Those who took Wegener seriously were either fools or foreigners.

Thus it was on a certain evening last winter, I felt "like some watcher of the skies when a new planet swims into his ken" to hear, of all people, a physicist present significant evidence in support of the wandering of the poles and shifting of continents. Granted, the case is not closed. It is, in fact wide open – open enough so that a substantial number among the geophysicists no longer maintain that phenomena such as these are physically impossible. It behooves the geologist therefore to preserve an open mind and consider the evidence.

*(Raasch, 1958a)*

Inspired by Laming, Raasch (1958b: 183) thought of another indicator of paleo-winds, "the lee and windward shores of paleo-islands on evidence supplied by shore features and detrital trains." Treating his study as a work in progress, Raasch turned to the Paleozoic Baraboo monadnock in Wisconsin, which he had and others had studied in the 1930s (Wanenmacher *et al.*, 1934). This erosional remnant had, in the mid-Upper Cambrian, been one of an island group. After describing differences between opposing faces of the monadnock, he argued that the northern shores were windward and the southern shores leeward (see Figure 5.14(f)). He proposed that southern Wisconsin

lay within a trade wind belt in which the prevailing wind direction was a few degrees east of north. When it is considered that current palaeomagnetic evidence would place the Cambrian

North Pole in mid-Pacific north of the Equator, and when to this is added the factor of possible continental drift, it becomes apparent that the trade wind belt in Cambrian time at Baraboo might prove to have lain in the Southern Hemisphere.

*(Raasch, 1958b: 187)*

Although Raasch's brilliant idea was only rarely noted (Irving, 1964: 235–237), I believe the paleomagnetic/paleowind support for mobilism awoke him from fixist slumber.

### 5.16 Attempts at paleogeographies by Newcastle and Canberra groups

Fixism (mobilism) had failed (passed) its first physical test because pre-Late Tertiary paleomagnetic poles from different continents well supported by paleoclimatic evidence most decidedly did not agree. Irving in Canberra decided to see if mobilism could explain these disagreements. Khramov in Leningrad was asking the same question (§5.9). Independently and ten thousand miles apart they had moved from testing fixism to testing mobilist paleogeographies. Irving explained.

This test may be achieved by recalculating the pole positions not, as hitherto, with respect to the present positions of the sampling area, but relative to the past positions postulated by "drift" reconstructions. The "ancient" co-ordinates of the sampling area can be read from these maps, and the ancient pole positions calculated ... If the reconstructions are valid, the discrepancies in the pole positions obtained from rocks of the same period from different continents should disappear.

*(Irving, 1958a: 47; see also Irving, 1957b: 212)*

Irving's presentations at the Hobart (March 1956) and London (Imperial College, November 1956) symposia were the first full-scale attempts to present the paleomagnetic case for continental drift. Although he had used the paleogeographic maps of Köppen and Wegener (1924) to formulate his India drift test (§1.18), he now switched to the more precise reconstructions of du Toit (1937) (I, §6.7) and those that King and Carey had presented at Hobart (§6.16). King, like du Toit, postulated two former supercontinents, Gondwana and Laurasia; Carey adopted Wegener's single supercontinent, Pangea.

Irving found that Carey's reconstruction (the one he presented at Hobart (§6.15)) provided the best fit for Laurasia; a few poles from opposite sides of the Atlantic failed to coincide exactly, but they were always much closer than with continents fixed. Laurasia reconstructions by King and du Toit gave less good fits. He cautioned:

Before any judgment can be passed on Carey's "Pangaea" from the palaeomagnetic view point very much more data from the southern continents is required, but although there are certain inconsistencies [discussed in §6.15] the palaeomagnetic data is in broad agreement with the relative positions he gives for North America, Europe and Australia.

*(Irving, 1958a: 50; my bracketed addition)*

Irving soon embarked on a more ambitious project; he began to develop paleogeo-graphies based primarily on paleomagnetic data. He even began to do this while still working on the earlier approach, which he thought of as an interim measure useful only until the data were more complete.

When complete results are forthcoming from many parts of the world it ought to be possible to produce palaeogeographic reconstructions from them, but it will be many years before there is sufficient information to do this satisfactorily. However, as an interim measure, it is of interest to see to what extent the present data are consistent with the palaeographic reconstructions provided by the various exponents of continental drift.

*(Irving, 1957b: 210–211)*

In retrospect, this might seem rather sanguine, even reckless. A decade later it would, for post-Triassic time, be superseded by plate tectonics. What Irving did was tecton-ics on a sphere based on paleomagnetic pole positions; to develop plate tectonics many other sorts of data would be required that were not available in the late 1950s. No one knew this, and Irving was not deterred. Paleomagnetism gave latitudes directly and there were, he argued, certain geometries for which it ought to be possible to fix relative longitudes.

In the publications of the March 1956 Hobart (1958a) (§6.16) and November 1956 London (1957b) symposia, Irving described the new approach to paleogeography afforded by paleomagnetism. He first noted two important uncertainties – longitu-dinal and hemispheric. He wrote:

For any single geological period – regard the pole as fixed and the continents as moveable. The magnetic dip $I$ gives the co-latitude $p$ of the sampling areas by the following equation, cot $p =$ ½ tan $I$. If the possibility of drift is admitted, both rotations and translations of the continents are equally likely, so that the declination does not fix the longitude of the sampling area, and any position on the small circle of radius p around the pole will satisfy the palaeomagnetic data provided the land mass is rotated suitably. Similarly, the $n$th sampling region with magnetic dip $I_n$ may have been situated anywhere on a small circle of radius $p_n$, so, for given sets of data from several continents an infinite number of continental arrangements are possible.

*(Irving, 1957b: 213–214)*

To illustrate, he constructed the diagrams shown in Figure 5.17 (his Figure 14). Placing paleomagnetic poles at the south geographic pole, he offered four solutions for the Jurassic Period based on his results from the Tasmanian dolerites (Australia) and the Parana dykes (South America), and on Maack's samples (§5.6) and Nairn's result from the Karroo basalts (Africa); all solutions were consistent with the paleo-magnetic data. He ruled out configurations in which continents overlapped. But for a single time interval such that these results were then thought to represent, many possibilities were permitted.

The co-latitudes given by the magnetic results are 10° for Tasmania, 49° for southern Brazil and 60° for Central Africa, and the land masses are rotated through the declination angle so

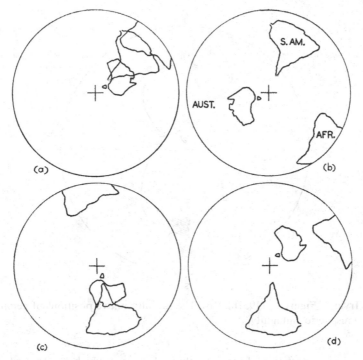

Figure 5.17 The uncertainty in longitude occurs when results are available for only one time interval. Irving (1957b: 215).

that the horizontal component of magnetization lies along a meridian. These rotations are 35° clockwise, 22° anti-clockwise, and 30° clockwise respectively. In fig. 14(b) the relative longitudes are those of the present day. The remaining diagrams give other solutions compatible with the magnetic data. Clearly, the situations in figs. 14(a) and (c) where the continents are superposed, are absurd. Nevertheless, there remains a very large number of solutions of the type 14(b) and (d) between which the paleomagnetic data cannot discriminate.

*(Irving, 1957b: 214–215; also see Irving, 1958a: 52)*

The number of viable solutions may be reduced if sequential results are available from constituent fragments of a former supercontinent yielding contemporaneous APW segments. The fact that North America and Europe had separate and matching APW paths for Paleozoic and Triassic (Figure 3.6) implies that during this interval these continents were longitudinally much closer together than at present, the Grand Banks of Newfoundland and western Ireland then being side-by-side. In this particular tectonic circumstance (supercontinent followed by fragmentation and dispersal), estimates of relative longitude can be made, of which more below.[27] Solutions would not come easily, and there would be a need to apply geological constraints.

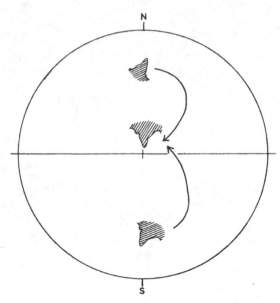

Figure 5.18 Irving's Figure 15 (1957b: 216). The two alternative positions of peninsular India because of hemispheric ambiguity.

In general, in the interpretation of those geophysical surveys, which are directed at problems concerning the nature and properties of the crust of the Earth, it is necessary to incorporate geological facts or inferences from geological facts in order to decide between alternative interpretations of the geophysical data. In all probability rock magnetism, in its relation to the problem of continental drift, is no exception, and the ambiguities which are likely to arise in the solution referred to above, may have to be finally settled by reference to geology. It should be possible to eliminate many solutions, on the grounds that they are contrary to the facts of geological structure and rock facies.

*(Irving, 1958a: 53)*

The hemispheric ambiguity arose from the uncertainty in the polarity of the geomagnetic field when rocks become magnetized.[28]

Because of the uncertainty surrounding the interpretation of reversals we cannot say to which of the geomagnetic poles the north-seeking polarizations of a rock formation points; we cannot say whether the sampling area was situated in what was at the time of deposition the Northern or Southern Hemisphere.

*(Irving, 1957b: 215, also see Irving, 1958a: 50–51)*

Using his own data from India as an example, he presented the above figure (Figure 5.18), and commented:

The results from India provide a good example of this dilemma. The author's determinations in the Deccan Traps give a mean direction of 149°, +56° and reversals occur. Even if there were no reversals the doubt about the sign of the magnetic field at deposition still remains, since there is a possibility of a whole rock formation developing self-reversed magnetization. Thus in explaining the results from India two alternatives arise (Fig. 15).

(i)  that peninsular India has moved from a latitude of 54° N rotating clockwise through 149°.
(ii)  That India has moved from a latitude of 54° S and rotated 28° anticlockwise.

*(Irving, 1957b: 215–216, also see Irving, 1958a: 51)*

From paleomagnetism alone, the hemisphere is indeterminate, but geology helps.

The palaeomagnetic data cannot distinguish between these two alternatives and at this stage geological arguments have to be introduced. For instance, the first of these alternatives is grossly out of accord with the requirements of the Upper Tertiary Himalayan compressions which supposes that the Deccan and Asia have moved together, not apart, whereas the second situation is, qualitatively speaking, that required by the drift reconstructions of Gondwanaland which are based upon geological data.

*(Irving, 1958a: 52; also see 1957b: 216)*

Thus he preferred the first alternative, now known to be correct.

These two uncertainties, especially that in longitude, hampered attempts to reconstruct continents, but as they got more data, paleomagnetists tried to reduce the number of possible reconstructions by considering results not from one but from a sequence of time intervals, by working backwards from the present and by minimizing intercontinental motions. They attempted to reconstruct Gondwana by unifying poles from India and the southern continents, to reconstruct Pangea by bringing poles together globally, and, as noted above, to reconstruct Laurasia. I now want to jump forward and show a highly idealized diagram (Figure 5.19) drawn by Ken Graham and colleagues in the mid-1960s, which helps focus discussion.

Two continents, A and B, remain together for times 1–8, during which they move in concert relative to the geographic pole tracing out a common APW path; they are imagined to have been constituents of a former supercontinent. This is just what Wegener envisaged with his Pangea, and du Toit with his Gondwana and Laurasia. Irving later explained:

If you have a supercontinent which moved relative to the pole, and then that supercontinent traces out a polar wander path and then the supercontinent splits, then those continental fragments will now have a record of that earlier polar wander path – one should be able to match them, and therefore to fix the continents unambiguously.

*(Irving, 1981 interview with author)*

A and B then split apart taking their APW paths with them; to reconstruct for interval 1–8 their paths are brought together dragging A and B with them. The divergence of paths marks the time they separated. APW paths for interval 1–8 will only match if A and B remain fixed relative to one another during that interval. If they do not, then paths will not match and no unambiguous reconstruction can be made. For reconstruction purposes it is necessary that coeval paths have matching

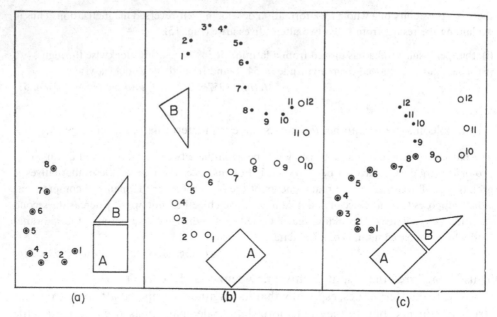

Figure 5.19 (a) Continents A and B and their common APW paths for times 1–8 when they were together. (b) APW paths after they move apart. (c) Reconstruction of their starting positions. Poles from A (B) are open circles (solid dots). A and B begin to split apart between times 8 and 9 (Graham *et al.*, 1964: 3896).

shapes and that errors of successive pairs of poles overlap one another. In the 1950s and 1960s workers were very far from having eight perfectly observed, consecutive, matching pairs for all continental blocks; the idealized situation in Figure 5.19 was not realized; at best only two or three were then available. Minimally, at best two pairs are sufficient provided they are substantially separated in time.

Returning now to my story,[29] Irving, following the 1957 Toronto IUGG meeting, visited Newcastle for three months. He and Creer, Nairn, and Runcorn pooled their work from South America, Australia, Africa, Europe, and North America and submitted it to *Annales de Géophysique* on April 17, 1958 (Creer *et al.*, 1958). They began by postulating continental drift (because of the differences among APW paths from different continents) and polar wandering, and addressed the question of how to construct paleogeographies, beginning with this remarkable statement, which *is of historical interest because it put tectonics in the context of Euler's Point theorem, and they seem the first to have done so.* As usual, they assume the GAD hypothesis:

The problem of continental reconstructions is clarified by considering the geometry of a spherical surface. A displacement of a continent to any other position on the Earth's surface may be obtained by a finite rotation about a fixed pole on the Earth's surface, which is convenient to call a pivot point. It is required to determine what relative continental displacements have to be assumed to satisfy the palaeomagnetic data. Consider two continents A and

B from which palaeomagnetic directions have been determined for a certain geological time. Let the pole positions on the present globe corresponding to these directions be PA and PB respectively. It is required to determine what displacement of the present position of B, supposing A and the spherical surface it defines are fixed, will bring B into its ancient position. If B and PB are imagined to be fixed to a spherical surface, then rotations of this surface about an infinite number of different pivot points can bring PB into coincidence with PA. These rotations correspond to moving PB along the great circle and infinite set of small circles joining PA and PB. The different pivot points all lie on the great circle equidistant from PA and PB. This will be termed the locus of the pivot points ... Because only one of the positions into which continent B can theoretically be brought by the various rotations is its ancient position, there is an essential indeterminacy in reconstructing the ancient position of the continents from palaeomagnetic data ...[30]

*(Creer et al., 1958: 496)*

For a single interval there are an infinite number of possible pivot points and rotations that would bring B into its original configuration relative to A. Acknowledging that continents cannot overlap, they went on to argue:

If pole positions for the two continents are available for more than one geological epoch, then a set of loci of pivot points can be found; one for each pair of poles of the same age. In general they will not coincide. If no relative movement between the continents has taken place between the earliest and latest geological periods under discussion, then the loci of the pivot points for the separate periods should all intersect in one point. In this case continental drift took place since the latest geological period, by rotation of one continent with respect to the other about this point.

*(Creer et al., 1958: 496–497)*

Creer *et al.* (1958) did not pursue this important idea. Instead they proceeded to plot separately the former positions of Australia, Europe, North America, and Africa relative to the rotational pole, but not to each other, essentially a sequence of what Nairn had called "drift maps" (§5.4).

Then from the paleomagnetic directions, declination D and inclination I, for a point in geological time the position of the site may be plotted ... The indeterminacy in this case will be that for any geological period the longitude may have any value. Some limit to this indeterminacy may be derived by working back from the present site coordinates assuming in each case a minimum movement.

*(Creer et al., 1958: 497)*

Much interpretation was still involved; for example, they had to decide how much rotation of a continent one should allow to minimize changes in longitude.

They offered alternative plots of Australia's motions relative to Europe, which I shall refer to as C1 and C2. With C1, they assumed that relative displacement had not begun until very late, the Late Tertiary; they superimposed their APW paths and calculated that Australia had moved at the very rapid rate of 40cm/year relative to Europe.

The polar wandering paths for Great Britain and Australia come together in a region near the present N. Pole. Consequently if the relative movement between Australia and Great Britain only took place during the upper Tertiary times the pole of displacement would have been close to the present geographical pole and about 120° of relative rotation would have taken place. This would represent a movement of about 8000 km in say 20 million years since the middle or late Tertiary (i.e., 40cm/year).

*(Creer* et al., *1958: 497)*

Implicitly, they assumed that the forms of the European and Australian APW paths were identical as are poles 1–8 in the hypothetical case of Figure 5.19. Inspection retrospectively of the paths shows that this is not so; the succession of poles in the two paths cannot be overlaid in this way, and consequently the rate of motion they obtained is far too high.

In their second approach (C2) Creer *et al.* assumed that Australia had moved relative to Europe throughout the Mesozoic and Tertiary, and estimated that it had done so more or less steadily at a more reasonable rate of about 4 cm/year. They noted that they needed a way to separate polar wandering from continental drift, in determining what part each had played. Like Wegener, they also wondered about the possibility of using geodetic techniques to determine current relative movements among continents.

In [the figure showing the movement of Australia relative to Europe] using the assumption of minimum movement it will be seen that the motion of Australia in Mesozoic and Tertiary times represents a movement of about 4 cm. a year and this interpretation appears to be more reasonable than the value inferred previously on the basis of the polar wandering curves. This discrepancy arises from the fact that it is not possible yet to separate polar wandering, i.e., movements of the whole crust with respect to the axis of rotation, from continental displacements, which are movements of the continents relative to the deeper parts of the crust. Movements of this order of magnitude occurring at the present time are not detectable by present methods of determining latitude and longitude …

*(Creer* et al., *1958: 497; my bracketed addition)*

Upon returning to Australia in early 1958, Irving (1958b) began his own attempt to deal with longitude uncertainty using a procedure that he later called "the method of paleocolatitude circles" (Irving, 1964: 268).[31]

First assume continent A with successive pole positions $a_1$, $a_2$, $a_3$ … for the time interval 1, 2, 3 … fixed relative to the crust of the Earth. Now allow continent B, for which there are pole determinations $b_1$, $b_2$, $b_3$ … for the same time intervals to move so that $b_1$ coincides with $a_1$, $b_2$ with $a_2$ and so on. The co-latitudes $p_1$, $p_2$, $p_3$ … for B during geological periods 1, 2, 3 … can be calculated from the magnetic inclinations $I_1$, $I_2$, $I_3$ … measured from rock formation laid down during these periods by the equation $\cot p = \frac{1}{2} \tan I$. The measured declination does not define the longitude so that the possible positions of B are represented by a family of small circles of radius $p_1$, $p_2$, $p_3$ … and centres $a_1$, $a_2$, $a_3$ … Since the angular distances between successive pole positions $a_1$, $a_2$, $a_3$ … is usually much smaller than the co-latitudes $p_1$, $p_2$, $p_3$ … these small circles will intersect.

*(Irving, 1958b: 227–228)*

Although in a mobilist world both continents may be expected to move relative to the pole, for the purposes of construction, continent A is stipulated as fixed and B is movable. Poles $a_1$ and $b_1$, $a_2$ and $b_2$, and $a_3$ and $b_3$ are equivalent poles, which Irving (1958b: 227) defined as "those calculated from rock formations which are contemporaneous in a geological sense." At the very least they had to be from the same geological system. If A and B initially moved in concert relative to the pole and subsequently moved relative to one another, then the paleocolatitude circles for B will have a common intersect which defines the initial position of B relative to A. As Irving remarked:

It is worth noting that *if* Wegener-type drift has occurred, and the primeval continent (or continents) remained as an entity during the early Mesozoic, Paleozoic, and earlier time, then it is a geometrical necessity that common intersections of paleocolatitude circles should occur and thus provide the means of reconstructing the present fragments.

*(Irving, 1964: 268)*

The examples of the paleolatitude circle given originally by Irving (1958b) were based almost entirely on NRM data without demagnetization – all that were available at the time. The example in Figure 5.20 six years later is based largely on cleaned results and many radiometric ages. To include it here extends my account beyond the interval covered in this chapter, but it allows me conveniently to complete his discussion of paleogeographic reconstructions prior to the introduction of plate tectonics. In Figure 5.20, Europe, which is assumed fixed, and the paleolatitude circles from Permian to Cretaceous have a primary intersection (1) at 12° E, 43° S, and Australia is reconstructed accordingly. The Early Tertiary paleocolatitude circle for Australia does not pass through this intersection, but is close and within experimental error. There is a secondary intersection (5) for Permian, Jurassic, Cretaceous but the Triassic circle is far away.[32]

Irving wrote:

The simplest explanation [I1 analogous to C1 of Creer *et al.*, 1958] of this result is that Australia remained fixed relative to Europe in position (1) during the Permian through into the Jurassic ... in the Cretaceous and Tertiary it occupied positions (2) and (3) moving with small clockwise rotation to its present position (4). The rate of movement relative to Europe may be calculated as approximately 92° of arc since the Jurassic or an average of 6 cm/y. Alternatively [I2 analogous to C2] the position (1) may be maintained from Permian to Early Tertiary ... thereafter Australia moved relative to Europe to position (4) at an average rate of 20cm/y. The first alternative seems far more reasonable.

*(Irving, 1964: 269; my bracketed additions)*

Irving can be faulted for not following through more rigorously. It would have helped the reader if Figure 5.20 had been constructed as an equatorial projection about the Greenwich meridian – as Wegener had done (I, Figures 3.1, 3.2). It would have been a much more readily understood demonstration of the broad consilience between Wegener and a reconstruction based solely on paleomagnetic results.

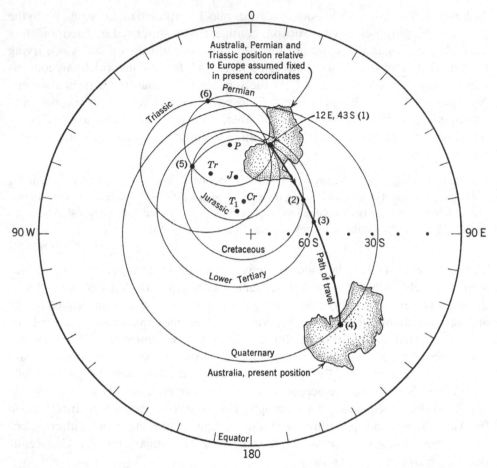

Figure 5.20 Irving's Figure 10.16 (1964: 269). Paleocolatitude circles for Australia drawn about the south poles for Europe; the latter and its paleopoles are assumed fixed. The radii of the circles are calculated for a reference point (1) in South Australia. Europe is fixed in the Northern Hemisphere, the underside of Figure 5.20 is not visible; the reader had to imagine Europe behind and aside the 0° meridian at about 50° N.

What interested Irving was that the co-latitude circles for Australia had (approximately) a common intersection over a considerable interval, which suggested that Australia had remained fixed relative to Europe for that time, that is, for the time it remained a part of a fully assembled Pangea. He argued:

Because a single palaeomagnetic result does not fix the ancient longitude of a continent it was thought at first (Irving, 1957a) that many widely differing reconstructions would be possible. However, if continental drift has happened as periods of movement separated by defined standstills during which [apparent] polar wandering occurred (Figure 2a), then other coincidences like those for Australia (Figure 5) will arise for all continents. If such coincidences should happen frequently the probability of their being significant becomes very great. It is not unlikely that relative longitudes will be more closely defined in this way, and a very limited number of solutions, perhaps even a unique solution, may arise.

> *(Irving, 1958b: 233; my bracketed addition. Irving's 1957a reference is to what*
> *I have referred to as Irving, 1957b. Irving's Figure 5 showed Australia at the*
> *common intersection of the co-latitude circles as shown on Figure 5.20.)*

His suggestion may have seemed optimistic and premature in 1958 when the paleomagnetic results were few and based almost entirely on NRM data without magnetic cleaning and radiometric dating. It was perhaps more believable in the mid-1960s when cleaning and radiometric dating provided greater accuracy. Nonetheless, it would have required decades of work to accumulate sufficient data to satisfactorily reconstruct Pangea by these methods alone, and it would never have been possible to reconstruct unambiguously the subsequent movements of its fragments in this way: only reconstruction of supercontinents is possible. Nevertheless, the APW paths of each fragment converged on the present geographic poles, and it should have been possible to trace the relative movements of each fragment back from its current position to its position within the supercontinent by minimizing motions, applying geological constraints, and disallowing superimposing of continents. If plate tectonics had not intervened a decade or so later, this method might have stood some chance of unraveling Wegenerian drift in broad outline, but far less satisfactorily than plate tectonics subsequently did.

When considering these two interpretations, first, slow Mesozoic/Cenozoic drift (C2, I2), and second, fast Late Cenozoic drift (C1, I1), both accounts preferred the second but, in retrospect, not strongly enough. The first simply would not have explained the timing of many intercontinental disjuncts; perhaps if they had had more confidence that such arguments would, at the time, have had a fair hearing, even been accepted, then perhaps they would have dismissed such fast Late Cenozoic drift outright. Retaining it as a possibility allowed Irving (1958b: 236), in retrospect rather foolishly, to write:

In short, with the exception of India, the variations of geomagnetic latitude observed in different continents can be explained by polar wandering [alone], and continental drift has, for the most part, succeeded only in changing their relative longitudes.

> *(Irving, 1958b: 236; my bracketed addition)*

This fast Late Cenozoic interpretation (C1, I1) derived from the image of APW paths, converging like bent spokes of a wheel on the present rotation pole that could be folded back upon themselves (Figure 5.11), was based on the erroneous assumption that all paths (like those of Europe and North America) could be matched. It was an interpretation that was on the road to nowhere, certainly not to plate tectonics. Well into the 1960s, the Cambridge diaspora in Newcastle and Canberra believed that their results could be reasonably explained by continental drift but conservatively still retained a belief that polar wandering may also have occurred.

### 5.17 The increasing necessity for continental drift

When oblique directions were first discovered in Britain, they were interpreted by those who made the discoveries as being caused by polar wandering, by continental drift, or by the action of both (Clegg *et al.*, 1954b: 198; Creer, as reported at the 1954 Oxford meeting (§3.6) and Creer, 1955; Irving, 1954: 109). The exception was Runcorn who considered that "movement of Earth's surface features relative to the axis of rotation" was the simplest explanation (Griffiths and King, 1954); he was at first unwilling to include continental drift as a possibility. Once he accepted continental drift, he did not follow his previous advice about choosing the simpler of two hypotheses (§3.3), but retained polar wandering and continental drift (§4.3). So far as the members of the Newcastle and Canberra groups were concerned, once they saw that the APW paths for Europe and North America were different and that there were much larger differences between poles for Australia, India, Africa, South America, and Antarctica, they accepted continental drift. Nonetheless, their belief that polar wandering could have played a role persisted, and this became a conservative theme in their thinking. Creer, Irving and Runcorn, their co-workers and students, retained the possibility that polar wandering might account for the common movements among continents, while continental drift explained the large, sometimes very large, differences among APW paths. Clegg and his co-workers at Imperial College developed their own mobilistic interpretation of oblique directions, which I shall deal with in §5.19. The Newcastle and Canberra groups proceeded to try to construct paleogeographies and to develope methods by which, under certain circumstances, the relative longitude uncertainty might be overcome. Even when minimizing continental movement, they still needed to postulate continental drift to explain the paleomagnetic data; by 1959 there seemed no way of getting around it.

Accepting the need to postulate drift, they found ways to augment its paleomagnetic support. They also made attempts to distinguish drift from polar wandering, all, as it turned out later, unsuccessful. Runcorn (1957) and Green (1958) suggested that polar wandering might not be steady and orderly but resemble a random walk. This gambit allowed for the joint or separate occurrence of polar wandering and continental drift during particular periods, and Green linked explanations for random polar wandering (such as Gold's) with irregularly occurring tectonic events, causing redistribution of mass within the Earth.

But since the geological events which may have been responsible for polar wandering have been irregular, both in their type and in their distribution and timing, they are likely to have produced random polar movements.

*(Green, 1958: 382)*

Meanwhile, Irving developed another way to marshal paleomagnetic data to support continental drift. He also sought evidence for polar wandering. Comparing results within and between continents, he reasoned that if continental drift had occurred, the agreement of contemporaneous poles from each geological period from single continents should be much better than that among several continents. He compiled 112 poles that were available to him from every continent but Antarctica. They ranged in age from present to Pre-Cambrian; 75% were from Carboniferous and younger rocks. Poles from the Neogene and Quaternary from different continents were in good agreement with each other and the present geographic pole, confirming Hospers' GAD hypothesis (§2.9). With few exceptions, poles from older rocks diverged considerably from the present geographical pole.

Polar results from Carboniferous and Permian rock formations are generally found to be in intermediate latitudes. Thereafter, in the Mesozoic and Palaeogene, they occupy higher latitudes.

*(Irving, 1959a: 69)*

Irving continued, erroneously as it later turned out, to attribute this common northward latitudinal movement to polar wandering.

Irving (1959a: 69) then presented a new argument in favor of continental drift. First, he noted that in most cases there was "good but not always exact agreement between results from rock formations laid down in the *same* continent during the *same* geological period." He noted the agreement between igneous and sedimentary rocks in various geological environments. Second, he found:

Polar results derived from rock formations laid down prior to the Neogene and in the same geological period but situated in different continents in general do not agree with one another … the contrast in Permian to Palaeogene times of agreement between results from the same continent and disagreement between results from different continents is of first importance.

*(Irving, 1959a: 69)*

Irving displayed the data as histograms, plotting the angular separation of poles taken two at a time from the same and different continents, where each pair of poles was determined from rock formations of the same time interval. If the degree of separation in the internal comparison proved to be as great as the external comparison, polar wandering would be sufficient to explain the data. But, if the external comparison showed less agreement than internal comparison, continental drift would be needed. Prior to the Neogene, external differences were appreciably greater than internal ones. Continental drift was needed to account for this. There were a few divergences that he attributed to sampling inadequacies or, more interestingly and hinting at things to come, to local rotations in tectonically active regions such as Japan and western United States.

## 5.18 Clarification and further support for the GAD model

In the late 1950s and early 1960s paleomagnetists expanded support for the GAD hypothesis, which was central to their case for continental drift, and reexamined its role in the paleomagnetic defense of mobilism. They remarked that Neogene and

Quaternary poles obtained from newly surveyed regions coincided with the present geographical poles, confirming work of the early 1950s. They discovered further paleoclimatological support for the GAD hypothesis, and Opdyke and Runcorn presented a new approach through their work on paleowinds (§5.10–§5.14). But, there was more. For example, they noted that the paleomagnetic case for continental drift did not necessarily require that the average field be axial.

The interpretation of the data [in terms of continental drift] is based on the hypothesis that the mean geomagnetic field has been that of a geocentric dipole throughout geological time. The further hypothesis that the axis of this dipole and the axis of the earth's rotation coincide is strictly irrelevant to this problem, but must be assumed if the palaeomagnetic and palaeoclimatic data are to be compared.

*(Creer* et al.*, 1958: 492; my bracketed addition)*

Paleomagnetists also assessed the possibility of long-term non-dipole components in the ancient geomagnetic field, and considered how comparisons of analyses of dispersion of present and past fields might be used to test the geocentric dipole hypothesis. I am chiefly concerned here with these two endeavors.

### Axial non-dipole fields

Runcorn was responsible for some of the most interesting new work. He (1959a) argued that if the geomagnetic field had a long-term non-dipolar component, the mean geomagnetic field over time could still be axially symmetrical. In a second paper (1959b), he considered multipolar geomagnetic fields, and argued that postulating continental drift and a geocentric axial dipolar field fitted better with paleomagnetic and paleoclimatological data than did a multipolar field and no continental drift.

Runcorn (1959a) took as his starting point the idea that secular variation originated in near-surface eddies in the fluid core, and the core rotates about the same axis as the mantle, but at a slightly slower rate causing the westward drift of the non-axial parts of geomagnetic field; he regarded the minute variations in the length of the day as evidence for differential rotation, which he attributed to slight changes in electromagnetic torques which vary as the field changes. Thus he claimed that over time the non-axial components of the geomagnetic field would be averaged out. He concluded:

Consequently in the case of the actual geomagnetic field over times long compared to the period of relative rotation of the core and mantle, the mean direction of the geomagnetic field will have a declination of $0°$ and an inclination of $\tan^{-1}(2\cot \theta)$ with respect to the axis of the earth's rotation at that time [$\theta$ is co-latitude] Thus one of the most important assumptions used in rock magnetism research appears to be well founded.

*(Runcorn, 1959a: 91; my bracketed addition)*

Runcorn then turned to the possibility that the geomagnetic field, albeit axially symmetrical, might be multipolar. He did not attack this idea on theoretical grounds, but evaluated it against observations by seeing which of the hypotheses of an axial dipole, axial quadrupole, or axial octupole fields, with or without continental drift, explained both paleomagnetic and paleoclimatological data. Restricting himself primarily to North American, European, and Australian data, he plotted the co-latitudes of each continent during the Cambrian, Carboniferous, Permian, and Triassic Periods in terms of the three hypotheses. If the co-latitude on a given hypothesis did not give concurrent poles for a particular time period, he gave it a failing mark. If it gave at least roughly concurrent positions, he then applied the further test of whether it positioned the continents at co-latitudes consistent with paleoclimatologic data for the time and place; consistency would be expected because each multipole field is axial and symmetrical about Earth's axis of rotation.

Without drift, the GAD hypothesis failed because equivalent poles from different continents did not coincide. The Cambrian results (Runcorn, 1959b: 171) were "not well represented by an axial quadrupole field," and were "very poorly represented by an axial octupole field." At best, some poles for the other periods were roughly in agreement with the quadrupole and octupole fields. In particular, Runcorn (1959b: 172) noted that the "Triassic and Carboniferous results come nearest to being satisfied by the octupole field, and the Permian by the quadrupole field, but the discrepancy is still very great." Both hypotheses, however, required large and rapid shifts in field directions. Moreover, the co-latitudes of the continents dictated by multipolar hypotheses were inconsistent with well-established paleoclimatic data.

The positions of Australia in Carboniferous and Triassic times are within 10° of the equator, and the positions of South Africa in Carboniferous times and of Australia in Permian times are within 30° of the equator. These positions are hard to reconcile with the evidence of a cool climate in Australia in Triassic times and widespread glaciations in Permo-Carboniferous times in Australia and South Africa (David, 1950; du Toit, 1937).

*(Runcorn, 1959b: 172)*

Runcorn summarized the findings in this way:

It is of course clear that the discrepancies between paleomagnetic data from different continents, which have been taken by many writers to imply that continental drift has occurred, can be explained away by postulating a sufficiently complex geomagnetic field for each epoch in geological time. But the only physically plausible fields are axial multipoles and these seem to imply positions of the continents in flat contradiction to the palaeoclimatic data for the late Palaeozoic and early Mesozoic times. Until results from other continents are available the possibility cannot be entirely excluded that the mean geomagnetic field in Cambrian times was not dipolar.

*(Runcorn, 1959b: 172–173)*

These were powerful, pro-mobilist arguments. His attitude toward continental drift certainly had changed since he first looked at his data from North America (§3.9).

## Dispersion of the geomagnetic field

Runcorn had kick-started the paleomagnetic study of old sedimentary rocks at Cambridge by attempting, with Irving, to observe the secular variation of the Precambrian geomagnetic field in the Torridonian (§1.17). They found no sign of bed-by-bed changes that could reflect paleosecular variation. Instead they discovered a strongly oblique magnetization which soon monopolized their interest. Irving abandoned bed-by-bed studies for broad-scale stratigraphic work. There soon arose, however, a new method of studying paleosecular variation from estimates of the dispersion of the ancient field made by using Fisher's statistics. These measures of dispersion were a by-product of the more detailed paleomagnetic surveys, and their use in studying paleosecular variation was initiated by Creer (1955) in his Ph.D. thesis, but it was not until seven years later that accounts were published in the scientific literature (Creer, 1962b, d; Cox, 1962). The measure of dispersion used was the half-angle of the cone around the mean direction that contains about two-thirds (actually 63% in the case of dispersion on a sphere) of the individual observations, called the spherical standard deviation, analogous to standard deviation in a linear Gaussian distribution (§1.13). Theoretical statistical models of the geomagnetic field were also developed for comparison with paleomagnetic estimates of dispersion (Creer et al., 1959; Irving and Ward, 1964).[33]

The best rocks for observing paleosecular variation statistically are sequences composed of many lava flows or very thick sills that erupted over thousands of years and took thousands of years to cool. Somewhat less good are thick sedimentary sequences or a single sedimentary bed. The magnetization of an individual flow or a single sill exposure would have been acquired over a period of months or years and constitutes a "spot reading" of the ancient field. The standard deviation of many "spot readings" taken through a thick igneous or sedimentary sequence provides an estimate of the magnitude (expressed as dispersion) of the secular variation of the field during the time it took to cool, or, in the case of sedimentary rocks, to be deposited.

Gough et al. (1964) compared results from four Early Jurassic basalt (lavas) and dolerite (sills) sequences from the Southern Hemisphere with the angular standard deviation of the present field measured around a line of latitude. The latter are plotted as a function of latitude, and they vary by a factor of two, which reflects the overall axial dipolar nature of the present field (Figure 5.21). The four experimental points show the same trend, higher (lower) in higher (lower) paleolatitudes, and are consistent with a geocentric dipole field. Irving and Ward (1964) and Irving (1964) made similar comparisons (Figure 5.22) based this time not on results from one time interval but ranging from Precambrian to Recent, and including igneous and sedimentary rocks. There was much scatter but the same effect was apparent, the dispersion decreases as latitude increases. There were substantial uncertainties in estimates of dispersion and considerable scatter in the determinations themselves (Figure 5.22), but there was no evidence of systematic departures from a dipolar field.

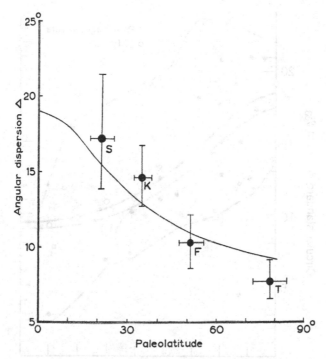

Figure 5.21 Gough, Opdyke, and McElhinny's Figure 6 (1964: 2514). Dispersion of site directions in Jurassic rocks of Gondwana continents. S, Sierra Geral Volcanics, South America (Creer, 1962c); K, Karroo Lavas, Africa (Zijl *et al.*, 1962); F, Ferrar dolerites, Antarctica (Bull *et al.*, 1962); T, Tasmanian dolerites, Australia (Irving, 1963). The error bars are at 95%. The line is the estimate of dispersion of the present field along a line of latitude for the 1945 field of Creer (1962e).[34]

The internal dispersion of paleomagnetic results as known in the early 1960s indicated a predominately dipole geomagnetic field over the Phanerozoic and at least part of the Precambrian.

## 5.19 Alternative approach of Imperial College group

In 1975 John Clegg provided an interesting retrospective on attitudes at Imperial College, London:

It is clearly possible in principle to distinguish between polar wandering and continental drift by palaeomagnetic measurements. For if a sufficiently large number of results are available for all the continents over a sufficiently long period of geological time, the pole position can be plotted as a function of time for each continent. If nothing but polar wandering has occurred, the curves for the different continents would be expected to coincide. Any deviation between the pole positions obtained for rocks of the same geological age from different continents strongly suggests that continental drift has happened.

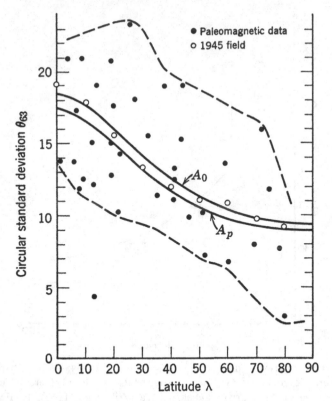

Figure 5.22 Irving's Figure 6.33 (1964: 146). Curve $A_0$ is the dispersion of the present (1945) field as in Creer (1962e). The solid dots are a selection of site dispersions available in the early 1960s. Twenty-nine are from igneous and six from sedimentary rocks. Irving excluded the outlier data-point outside the perimeter (dashed line) marked. Curve $A_p$ is the best fit of the paleomagnetic dispersions to a very simple statistical model of the geomagnetic field.

In 1959 Irving (see Blackett *et al.* 1960) carried out an analysis of this kind. It had already been known for some years that the poles derived for the different continents, for recent geological times, coincide closely with the present poles, but he showed that during earlier geological ages the polar wandering curves deviate, suggesting strongly that relative movement has occurred between the continents.

In 1960 Blackett, with other members of the Imperial College team, undertook a new systematic analysis of the direction of magnetisation of rock samples in a somewhat different and supplementary way to that used by Irving. Blackett's basic idea in embarking on this survey was to investigate all conceivable explanations of the results as searchingly as possible.
>           (Clegg, 1975: 16–17. The reference to Blackett is to Blackett et al., 1960; the Irving
>                                                       reference is to Irving, 1959a.)

During the later 1950s, while paleomagnetists at Newcastle and Canberra were honing their paleomagnetic defense of mobilism and constructing paleogeographic maps, those at Imperial College at first concentrated mainly on obtaining more data from India and Europe. However, in December 1959, Blackett, Clegg, and Stubbs submitted a paper containing a detailed account of their own different defense of mobilism and their views about the application of paleomagnetism to paleogeography. Instead of pushing data to their limit and deriving paleogeographies as the Cambridge diaspora did, they took what they described as a more phenomenological approach, attempting to present their paleomagnetic case for mobilism invoking the fewest assumptions. As they put it:

The main analysis of the observations will be based on a minimum of theoretical assumptions, both about the various ways in which the rocks acquire their magnetization and about the still very hypothetical mechanisms in the earth which give rise to the terrestrial magnetic field.

*(Blackett et al., 1960: 292)*

They wanted to see if continental drift provided the most reasonable account of the paleomagnetic data without recourse to any particular theory about the origin of the geomagnetic field or the manner in which rocks acquired their remanent magnetism. Blackett, after putting his distributive rotational theory to rest, perhaps had had enough of theories about the origin of the geomagnetic field (§1.3–§1.5).

They had a few more data than Irving (1959a), and they packaged them differently. Rather than making the calculation of poles their main analytical tool, they compared paleomagnetic directions directly with the GAD field at the collecting locality.

It is desirable to display these data with the minimum of assumptions about the eventual explanation of the results. To this end it is useful, in making a preliminary analysis, to compare the direction of the remanent magnetic vector of the rocks with that of the recent earth's field at the locality concerned. It is generally accepted that the recent geomagnetic field, averaged over periods of the order of a few thousand years, has coincided closely with that which would be expected of a magnetic dipole situated at the centre of the earth and directed along the rotational axis.

*(Blackett et al., 1960: 294)*

They found that deviations of paleomagnetic directions from the present GAD field increase with age. As Clegg later put it:

The complete data available for four land masses of continental size, viz., Europe, North America, India and Australia were examined. Instead of using the concept of ancient pole positions, in the usual way, to express magnetic directions, it was found useful for the purpose at hand to introduce a quantity which was called the "deviation," $\phi$. This quantity was defined as the angle between the remanent magnetic vector of the rocks at the site and the axial dipole field at the same place. Thus for rocks laid down during the present epoch, $\phi$ is zero, and it would always have been zero if (a) the geomagnetic field had always been that of an axial dipole, (b) the directions of magnetisation of the rocks faithfully represented the direction of the magnetising field, and (c) the continental land masses had always occupied the same positions relative to the earth's rotational axis. In actual fact $\phi$ increases fairly steadily with geological age T.

The use of the single variable φ (rather than the two coordinates required to define a pole position) enabled a statistical analysis to be carried out more easily. When this was done a high positive correlation was found between φ and T. On drawing the lines of regression of φ on T for all four continents average values of (dφ/dT) were found to vary between about 12° per 100 million years for Australia to 80° per 100 million years for India. After considering all known factors that might prejudice the measurements, it was concluded that this phenomenon could not be due to random causes but must represent some systematic change in conditions that has progressed continuously throughout geological time.

(Clegg, 1975: 17)

Blackett, Clegg, and Stubbs offered five possible explanations for the increase of φ over time.

(a) The direction of magnetization of the rock today does not represent the direction of the local earth's field at the time of magnetization.
(b) The main geomagnetic field has always been that corresponding to a dipole parallel to rotational axis, but the earth's crust as a whole has shifted relative to this axis; this is the hypothesis of geographical polar wandering.
(c) The earth's crust has remained rigid and fixed relative to the present geographical poles, but the main earth's field has in the distant past corresponded to a dipole which was not directed along the rotational axis: this is the hypothesis of magnetic polar wandering.
(d) The earth's field has had strong non-dipole components in remote geological times.
(e) Continental drift has occurred.

(Blackett et al., 1960: 305)

The first three hypotheses, they argued, made no sense. They began (a) by denying that remanent magnetization is a record of the paleofield at the time the rock was formed. What could the remanent magnetization represent if not the orientation of the field? Blackett and company cited the possibility raised by John Graham that widespread mechanical stress might be responsible for the oblique remanent magnetization directions (§7.4). As Du Bois, Irving, Opdyke, Runcorn, and Banks (1957) had already done, they noted that similar results were obtained for rocks of different types of the same age from the same general location, which would be highly unlikely if Graham's idea were correct. Later, I shall examine Graham's claim in more detail (§7.4). Although Blackett and company did not assume a particular process of magnetization acquisition, they were here attacking one proposed process.

Turning to polar wandering (b), they noted the mechanism recently suggested by Gold and Munk, and agreed with Irving, Runcorn, and others that differences in APW paths means that polar wandering alone cannot explain the paleomagnetic data. Nor could the hypothesis of a non-axial but dipolar field (c). However, they did not rule out polar wandering as a partial explanation. In this they generally agreed with the workers at Newcastle and ANU, and even plotted paleopoles.

Irving (1959) shows, as many other workers have done, that there are appreciable differences between the [apparent] polar wandering curves calculated from data for the different continents; for geological epochs dating back to more than 50 My the discrepancies are large. It appears therefore that while polar wandering may have occurred in the remote past, we cannot account for the palaeomagnetic results by this hypothesis alone.

*(Blackett et al., 1960: 307; My, million years. Irving reference is to Irving, 1959a.)*

Because (b) and (c) were paleomagnetically indistinguishable, they treated them similarly; geographic or polar wandering could partially but not entirely explain the paleomagnetic data.

The fourth explanation, fixity of the continents and a non-dipolar field (d), could not be dismissed on phenomenological grounds; a non-dipole geomagnetic field of complex configuration could always be constructed to fit the paleomagnetic data.

In general then, one can conclude that it is not at present possible to exclude, on phenomenological grounds alone, the possibility that the rock magnetic data now available might be explained by the two assumptions of a time-dependent non-dipole field together with the fixity of the continents.

*(Blackett et al., 1960: 308)*

They went on to consider the already well-advanced continental drift test.

It has become customary in discussing palaeomagnetic results to determine $I$ and $\psi$ the latitude and longitude of the ancient magnetic pole for each continent and each geological epoch. If the pole positions calculated for all continents at each age in the past had coincided within the limits of experimental error, with one another but not with the present geographical pole, then one would have concluded that the data could be explained by magnetic polar wandering alone, and that no other mechanism need be involved. We have seen that there is in fact no such coincidence, and that this has led many workers to suggest that continental drift has occurred. The now widespread use of pole positions in discussing quantitatively the problem of continental drift, however, leads to needless complication.

*(Blackett et al., 1960: 309)*

Although they were willing to use Cambridge's paleopole analysis as a way of showing that polar wandering alone was insufficient, they did not do so when presenting the paleomagnetic implications for continental drift. Instead they did so in terms of paleogeographical latitudes and azimuthal rotations, which, they suggested, had a further advantage:

Whatever changes may have occurred during remote ages the rotational axis has always been the one constant unchanged feature by which the position of an object on the earth's surface can be defined.

*(Blackett et al., 1960: 310)*

Clegg put it this way:

Again he [Blackett] discarded the custom of using ancient pole positions which, as he showed, introduces needless complications. Instead he chose to express the palaeomagnetic data in terms of ancient geographical positions of the continents. This was done in the following way:

(i) from measurements of rocks from a given site, X, the ancient latitude $\lambda$ of the site was measured in the usual way and the rotation $\psi$ (i.e. the ancient direction of true North, which is numerically equal to the measured declination D) was noted.

(ii) The assumption was made that the continents remained rigid during geological time, and that each one has drifted over the surface of the globe unchanged in shape.

(iii) On this assumption the rotation $\psi$ will be the same for all points on a single continent, and from the measured value of the ancient latitude $\lambda$ of the site X, the ancient latitude $\lambda_m$ of any other point can readily be calculated. Thus we can select an easily identifiable place M close to the middle of a continent, and express the results of palaeomagnetic measurements, made at any other place on the same continent, in terms of the ancient latitude $\lambda_m$ and rotation $\psi$ of M.

*(Clegg, 1975: 18–19; my bracketed addition)*

Taking as reference points Paris for Europe, Denver for North America, Nagpur for India, and Alice Springs for Australia, they presented their analysis in the diagram shown in Figure 5.23.

Strikingly demonstrated was the general northwards movement of continents during the past 200 million years. They also noted the very different rotations, the prior southward movement of Australia from 500 to 200 million years ago, and the greater and faster northward movement of India. They noted that their representation provided no information on relative longitude; unlike the Newcastle and Canberra groups, they thought it unwise to make even any attempt to pin them down.

It is obviously impossible by purely physical measurements alone, to determine the relative longitudes of two land masses in the past. To take a specific example, we have no way of ascertaining by rock magnetic evidence whether America has moved further from or nearer to Europe during the past few hundred million years.

*(Blackett et al., 1960: 313)*

This is in stark contrast to the Cambridge diaspora and Ken Graham and colleagues who thought that by bringing together matched segments of APW paths and moving continents along with them, the relative longitude of continents in reassembled supercontinents could be estimated (§5.16). Blackett and colleagues continued:

The palaeomagnetist is in exactly the same situation as a seaman only able to observe the altitude of the pole star. He can deduce the orientation of his ship and its latitude but can say nothing about its longitude of a site and its orientation relative to the ancient meridian, but the ancient longitude is indeterminate.

*(Blackett et al., 1960: 313)*

They ended by discussing secondary processes that might change the directions of magnetization of rocks over geological time, and the past existence of a long-term non-dipole field. They considered the former improbable because the increase in the deviation angle $\phi$ over time was steady, not random, as would be expected from such processes. But, they could not dismiss the non-dipole hypothesis on phenomenological

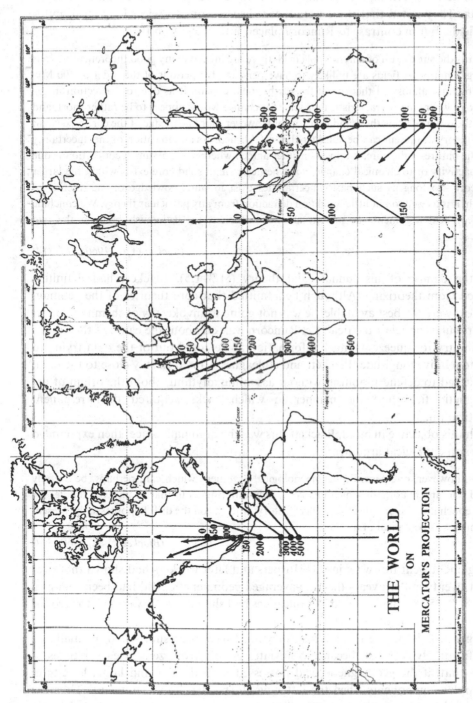

Figure 5.23 Blackett, Clegg, and Stubbs' Figure 8 (1960: 311; diagram on page 312). The points show the paleomagnetically derived latitude and azimuth for Paris, Denver, Alice Springs, and Nagpur, for ages extending back to 500 million years. The number by each point indicates its age in millions of years. Each reference town is placed on the calculated latitude $\lambda_m$ for each epoch at its current meridian. The arrows represent the rotation $\psi$ that brings back the orientation to its ancient value relative to the north geographical pole of the past.

grounds because non-dipolar geomagnetic fields could always be constructed to fit the data. They noted that there were theoretical reasons to reject long-term multipolar fields, but in contrast to Runcorn placed little weight on them.

As regards the suggestion of a non-dipole field, there are indeed many good theoretical reasons why large multipolar fields are unlikely to have persisted for periods of the order of 100 My. This is because almost all theories of the earth's field explain it as due to convection in the liquid core, where the characteristic time scale for change would appear to be rather short, and only of the order of a few thousand years or so. However, our knowledge of the core is so slight that no theory at present can be considered as established. As an indication of the uncertainty about the nature of the core, it appears neither agreed whether the assumed convection is due to radioactivity or to chemical change, nor whether a small solid core exists within the larger liquid core. So long as such basic uncertainties exist in our knowledge of the interior of the earth, it is as well to put little weight on deduction from any particular theory. We conclude that theoretical arguments at present are not sufficient to distinguish between the two hypotheses.

*(Blackett* et al.*, 1960: 313–314)*

Since the demise of his fundamental theory (§1.3–1.5), Blackett had definitely retreated from theorizing. Although he admitted that core theories of the geomagnetic field were the best available, he was not going to invoke any of them to support any particular model of the field. Let Runcorn wax on about the cause of the dipole; let Irving try to squeeze as much information as he could out of the data trying to estimate relative longitude. Blackett and colleagues wanted to stay close to the data; they wanted to defend mobilism without assuming anything about the origin of the geomagnetic field and the manner in which rocks acquired their remanent magnetization.

With this objective in mind, Blackett (his was the dominant voice) then expounded his phenomenological argument.

There is, however, one general phenomenological argument which can, perhaps, be usefully invoked. On any hypothesis which invokes a field differing from that of an axial dipole, the present epoch must be unique in geological history; whereas on the drift hypothesis, the present epoch may well be typical of all time.

*(Blackett* et al.*, 1960: 314)*

His argument went this way: just as Hospers and later workers had shown that over the last twenty million years the time-averaged geomagnetic field had been that of a GAD (§2.9, §5.18), Blackett and his colleagues had shown that $\phi$ was close to zero for each continent during this interval; thus, if there were non-dipolar components in the geomagnetic field during this interval, they must have been short-lived or small and insignificant. By contrast, because $\phi$ deviates greatly from zero in earlier times, it follows that, if the geomagnetic field possessed both dipolar and large, long-term non-dipolar components, then the latter would be the more significant. Thus the geomagnetic field over the past 20 million years must be atypical unless $\phi$ deviates during earlier times because of continental drift, which they argued occurred at a fairly constant rate because $\phi$ increases steadily over time.

Thus if we assume an axial dipole field, and interpret the palaeomagnetic results obtained so far as being due to continental drift, it seems plausible to assume on the basis of the present data that the rate of movement of the main land masses today is very much the same as it has been throughout the geological history of the earth.

*(Blackett et al., 1960: 314)*

This argument was generally but not entirely phenomenological, for Blackett relied on the methodological principle of uniformitarianism. Continuous and steady continental drift is, he argued, methodologically preferable to a geomagnetic field that in earlier times was forever undergoing substantial rapid changes. Thus, even Blackett, the phenomenologist, was in 1960 willing to move slightly beyond the data. Seemingly, appealing to this methodological principle was to him less distasteful than appealing to a substantive theory of the origin of the geomagnetic field.

Stepping back from these arguments that mobilist-minded paleomagnetists were having amongst themselves, Blackett then suggested that paleoclimatology offered the best prospect for independently testing the paleomagnetic results; in his next paper he did just this.

For historically the suggestion of continental drift long preceded rock magnetic studies. It was based entirely on a multitude of terrestrial surface phenomena, geometrical, geological, palaeontological, and palaeo-climatological. Accepting for the present what seems to be the general view that such data alone neither conclusively support nor refute this hypothesis, the obvious procedure is to test numerically the quantitative deductions made from palaeomagnetic measurements ... against the other set of facts. Without doubt, the most suitable of these is the climatological evidence. For the terrestrial climate is now, and to some degree must always have been, markedly zoned according to geographical latitude. Now, it is just the geographical latitude which is the main quantity yielded by the rock magnetic data, on the assumption, of course, of an axial dipole field. Thus it is clearly possible to test in a numerical manner whether the rock magnetic latitudes are or are not consistent with the ancient climates. This comparison will be made in another paper ...[35]

*(Blackett et al., 1960: 315)*

An excellent idea, but hardly new. As Blackett knew, Irving, Creer, Runcorn, Opdyke, and others had been busily working on it and had been for several years (§3.12, §5.10–§5.14).

Although the Imperial College group took a different approach from that of the Newcastle and Canberra groups in marshaling the paleomagnetic data to test mobilism, the two approaches were complementary. As Irving later suggested in commenting about the different approaches:

I always felt that where you had common movement then you should attribute that to polar wandering, and this was the basis of my attempt to produce reconstructions. It is also analytically correct in the sense that one should try and sort out the general effects first, that is polar wandering from the special effects particular to each continent that is continental drift. Now I think that tended to be the Cambridge point of view. The London people of course didn't take that point of view. They thought that one could only think about the latitude changes and changes in the azimuths of continents through time individually. There was certainly a bit of right on both sides. The two points of view are not mutually exclusive.

*(Irving, 1981 interview with author)*

Moreover, all groups vigorously and unequivocally supported mobilism. The Cambridge group, and the Newcastle and ANU groups that grew from it, invoked polar wandering to account for what they perceived as the common movements among continents. But Blackett and company did not reject polar wander outright. Of course, they and Creer, Irving, Runcorn, and others who worked with them all agreed that polar wandering alone would not suffice. Even though Blackett, Clegg, and Stubbs concluded in 1960 that the paleomagnetic results could best be explained by continental drift, they did not exclude the possibility that some polar wandering had occurred as well.

Although there were several disagreements between members of the Imperial College and Newcastle/ANU groups, this one, about the best way to express the paleomagnetic case for mobilism, was the only one that surfaced in the literature. Runcorn (1961) and Irving (1964) briefly responded.

Blackett, Clegg and Stubbs (1960) have argued that, because both these methods assume that the mean geomagnetic field has been dipolar through geological time, the primary observational data is obscured by assumptions about the entirely distinct problem of the theory of the main geomagnetic field. They, therefore, attempt to present the data for one site (or sites close together in one continent) by plotting the angle $\phi$ between the mean palaeomagnetic directions for each geological period and the observed field directions in late Tertiary times (that corresponding to a dipole along the present geographical axis). They show that $\phi$ increases with geological time, measured from the present, for each continent, a result parallel to that of Creer, Irving, and Runcorn ... who show that the pole ... moves gradually along a smooth path over the earth's surface through geological time. Both approaches are therefore equivalent in indicating that the phenomena of rock magnetization are not likely to have been produced by causes of local origin. The quantity $\phi$ however has no simple relation to any parameter of world-wide significance, so that we hold that either one of the above two methods [used at Cambridge, Newcastle, and Canberra] is more useful as a method of examining rock magnetism data [than the one used by Blackett].

*(Runcorn, 1961: 290; my bracketed additions)*

Runcorn's defense of mobilism was not divorced from causal explanations for the origin of the geomagnetic field. Blackett sought to minimize causal explanations; he wanted to minimize the assumptions made when presenting the paleomagnetic case for mobilism even if it meant inventing a quantity that although amenable to statistical analysis had, in Runcorn's words, "no simple relation to any parameter of world-wide significance."

Irving thinks that Blackett's approach did not amount to very much. When asked about its significance and Blackett's attack, Irving responded retrospectively in characteristic forthright manner:

Now the Blackett business. Blackett's attack on the analysis of paleomagnetic data. Frankly, I thought that it was a rather retrograde piece of work. You see what had happened in the Cambridge group. The main steps had been taken. Those steps were to show which rock type was the right one for giving you the signals you wanted, the fine grained red sandstones; Fisher's statistics, you had to package the data and then calculate poles. And, once you had poles you could then compare them from different places. And that was the correct analytic procedure, which has stood the test of time. Now it was the Cambridge group in which this was created. It was not created by Blackett's people, and so Blackett's people, I felt, had to have some sort of comeback to all this, which they did by saying that we won't look at the poles, we'll look at directions. What this did, in Runcorn's words, and I agree with this entirely, was in looking at the separate polar coordinates, the declination and inclination, rather than the total direction, and rather than look at it globally in terms of poles, one divorced the data from its appropriate spherical environment. And, that is really the crux of the matter.

But the ironical point is that when Blackett's group came, or when Blackett himself came in 1961 or 1962 in his paleoclimatic paper, to produce results of interest geophysically they had to calculate latitudes, and they had to calculate latitudes by first calculating the pole. You cannot calculate latitudes, except in a very specific instance, until you have first calculated the pole. And so they were thrown back on the concept of paleomagnetic poles. "Paleopole" was invented in the Cambridge group in the 1953–1954 period as a result of Hospers' work on showing the axial dipole field, and as a result of work by Creer, mainly, and myself on the directions in Britain.

Blackett's attack was an attack against our analytic procedures. In the long run I don't think it has amounted to much.

(Irving, 1981 interview with author, and slightly revised 2002)

Irving is certainly correct about the accomplishments of the original Cambridge team; it made all the key early moves he mentions. Moreover, as I have shown, Blackett and company even used the lack of coincidence among APW paths from different landmasses to argue that polar wandering alone could not account for the paleomagnetic data. Nobody nowadays uses the Imperial College methods. The Cambridge-invented procedures remain the main analytical basis for presenting and analyzing paleomagnetic data. But Irving's assessment is unfair, because, judging by the way it was received, the Imperial College approach, coming as it did from so authoritative a person as Blackett, actually did convince many at the time who were unfamiliar with the details of the subject that paleomagnetism was a promising new way of studying continental drift.

Blackett and colleagues' figure showing latitude changes of major continents was a simple and effective representation, driving them along the meridional highway as it were (Figure 5.23). Theirs was a simple, essentially drift-only picture, although they never denied that polar wander could have occurred; theirs was a picture that accords well with the modern view about the status of strict polar wander in the later Phanerozoic. The APW path representation of Creer, Irving, and Runcorn led naturally to consideration of polar wander as at least a contributing factor; subsequently this possibility was retained by them, perhaps longer than it ought to have been.

## Notes

1 Some paleomagnetic surveying was done in regions where previous work had been done, and comparisons were made with APW paths from different regions. For instance, one of Runcorn's last Cambridge students, Du Bois (1957), analyzed Triassic samples from the Connecticut Valley, combined them and his earlier work from the Keweenawan, and compared them with the APW path from Great Britain. Presenting his results at the November 1956 meeting on rock magnetism held in the Physics Department at Imperial College, he argued in favor of mobilism. Du Bois (1959a) extended global support for the GAD hypothesis by showing that paleopoles from North America coincided with the present geographic pole.

The fundamental assumption made in interpreting palaeomagnetic results is that the average geomagnetic field corresponds to that of a central axial dipole. This communication reports on some palaeomagnetic results which support this assumption and which were made on Late Tertiary volcanic rocks, which were collected during the summer of 1955 from north-western Canada by field geologists of the Geological Survey of Canada from flat-lying, undisturbed, vesicular basalts.

*(Du Bois, 1959a: 1617)*

Du Bois (1959b) also found that Newfoundland had not rotated appreciably relative to Canada's mainland Maritime Provinces since Carboniferous times.
2 At the time, the Rajmahal Traps were considered with some uncertainty to be Jurassic. Later radiometric dating established that they are Cretaceous, 105 Ma (McDougall and McElhinny, 1970). This change, although important in other respects (§5.2), did not impair the general arguments of Clegg *et al.*, 1958.
3 It is now known that the Deccan Traps only span a few million years, and that they contain substantial present Earth's field overprints that can be cleanly demagnetized (Vandamme *et al.*, 1991), so option (b) is correct and there is no significant motion during their rapid deposition. So (c) is incorrect.
4 McDougall and McElhinny (1970) showed that the basalts of the Rajmahal Traps are very stable magnetically and have negligible present field overprints, confirming that (b) as applied to the Rajmahal Traps, is incorrect.
5 This correlation of supposed Late Precambrian high paleolatitudes and the Adelaidean glaciations has now been shown to be erroneous, the effect of late Phanerozoic overprinting (Schmidt *et al.*, 1991).
6 In 2008 on the fiftieth anniversary of the publication of Irving and Green's APW path for Australia, the ANU held a celebration and a plaque was placed on a remnant of the paleomagnetic laboratory that remarkably had survived encroachment of University buildings. The plaque commemorates not just their work but also that of others on the early development of a reversal timescale before it was moved off-campus six years later. It reads:

On this site in 1955 a wooden building was built for the Department of Geophysics as a non-magnetic laboratory for the measurement of the weak remanent magnetization of rocks. At that time there were no other buildings nearby, which was important as the instruments housed were sensitive to magnetic and ground disturbances such as those generated by the passage of cars. An east-west wing was added in 1958, and a concrete-block addition in 1963.

It was here that the first polar wander path for Australia was determined. A great difference was found between this path and those for northern hemisphere continents. The significance of this difference was that it provided a most important demonstration that continental drift occurred. With subsequent developments, this demonstration led to the now accepted theory of plate tectonics. It was here also that measurements were made which related to the history of reversals of the geomagnetic field. Establishing this history was a fundamental discovery, which also played a most important part in the development of plate tectonics.

The building was vacated for geophysical purposes in 1964 when a new laboratory was built in an old quarry on the eastern slopes of Black Mountain. The wooden buildings were removed after 1969 and the concrete-block addition became part of the Research School of Biological Sciences.

This plaque was unveiled on 22 October 2008 by Professor Kurt Lambeck FRS, President of the Australian Academy of Sciences.

7 Gough (1989) has described paleomagnetic work in Africa during the 1950s and early 1960s.

8 At the time of Gough's study the Pilansberg dykes were thought to be Paleozoic. Subsequent radiometric studies showed the dykes to be much older, mid-Proterozoic. Notwithstanding, this general conclusion reached by Gough is valid.

9 Karroo basalts and dolerites are widespread across southern Africa. The southern Karroo had recently yielded ages of 183.7 ± 0.6 Ma, which is not significantly different from ages of Ferrar dolerites of Antarctica (183.6 ± 1.0) (Encarnación *et al.*, 1996).

10 According to Clegg, Nairn also remarked, "The extension of the work to continents other than Europe and North America is generally handicapped by the imperfectly known geology ..." Nairn had du Toit's work as a guide to African geology. But Nairn's remark surely does not apply to Australia, where Irving not only had David and Browne's monumental work, *The Geology of the Commonwealth of Australia*, but even had the very good fortune to have Browne as his personal guide.

11 These procedures for removing unwanted VRMs of recent origin generally proved very satisfactory during the mobilism debate. Later much older, more stable secondary magnetizations were discovered and found to be widespread, especially in orogenic belts and at plate margins (Chamalaun and Creer, 1964; Irving, 1964: 99; Irving and Opdyke, 1965). They are now known to be commonly caused by burial and diagenesis (chemical remanent magnetization). They played little part in the mobilism debate, creating only minor difficulties.

12 I say "almost completely unsurveyed" because Irving (1957b, 1958a) actually got some orientated rock samples from South America and published the results before Creer's first results appeared. Diehard mobilist Maack of the Instituto de Biologia e Pesquisas, Curitiba, Brazil, at Carey's instigation sent Irving (1958a: 53) some oriented samples taken from the Parana dykes. Irving plotted the results on Figure 13 (1957b: 214) and Figure 13 (1958a: 51). Only the second plot had a circle of confidence, and Irving (1958a: 52) noted that the "result from South America is based on a small number of measurements and may be in error." Moreover, he used it only to illustrate a limitation involved with construction of paleogeographic maps based solely on paleomagnetic data. Thus, Creer was the first to obtain and publish fully described paleomagnetic data from South America.

13 S. W. Carey asked Maack to send Irving the samples. Irving was never consulted by Carey, and Irving was not pleased because he knew that Creer was going to South America to collect samples. "Sam Carey arranged this. I was a bit embarrassed by this because I knew Ken was thinking of going there. Carey never consulted me. I got the samples after such great effort on Maack's part, then I could not study them. Also Carey was breathing down my neck" (Irving, March 4, 1999 note to author).

14  The Serra Geral volcanics are part of the vast Paraná Large Igneous Province, which at the time of Creer's work was considered to be Late Triassic or Early Jurassic. Radiometric studies have shown it to be Cretaceous (120 Ma) and associated with the initial opening of the South Atlantic.

15  See I, §6.10 and §7.4–§7.7 for work by King, Long, and Hamilton on Antarctica.

16  There is an excellent photo of Bull's first collecting site on Wikipedia, under "Bull Pass." He collected about a dozen specimens from top to bottom of the sheet as it is exposed in Bull Pass. The pass was named after Bull (Bull, January 10, 2008 email to author).

17  Since 1960 there has been a great deal of radiometric and magnetic work. K and F of Figures 5.9 and 5.10 have been dated as Early Jurassic (Encarnación *et al.*, 1996), R, T and SG as Cretaceous. The slightly more northwesterly position of pole SG likely marks the initial rupture of Gondwana, the rifting of East Gondwana (Antarctica, India, and Australia) from West Gondwana (South America and Africa). What a difference a handful of accurate radiometric ages made!

18  In comparison, Irving (1964: 363–384) cites thirty publications by paleomagnetists in Blackett's group with Blackett, Clegg, Almond, Stubbs, Deutsch, Everitt, Leng, or Fuller as first author, and eighty-four works with Hospers, Creer, Runcorn, Opdyke, Du Bois, Nairn, or himself as first author.

19  See McElhinny (1994), Khramov (1994), and Irving (1960) for background information about Khramov's early interest in paleomagnetism.

20  Khramov was particularly impressed with Hospers' work on reversals. His work showed that sequential (that is stratigraphically one above the other) reversals occurred in sedimentary rocks independently of the process involved, which was quite different from that of igneous rocks, thus satisfying the first and most important requirement of the hypothesis of geomagnetic field reversals. The second requirement is that reversals should be synchronous worldwide, which was satisfied about half-a-dozen years later.

21  I should note that Khramov and others began to obtain divergent earlier paleopoles from the east and west of the Soviet Union. They found that older Paleozoic paleopoles derived from rocks west of the Urals were in keeping with European paleopoles, but those obtained from rocks east of the Urals suggest that current-day Russia had been divided into at least two blocks, with the Urals representing their collision. See McElhinny, 1973: 243–246 for a good summary.

22  Historian Mott Greene told me that the use of fossilized sand dunes to determine paleowind directions had been used in the late nineteenth century by German geologists; the procedure was discussed in a popular high school textbook in geology, but was later forgotten (conversation with Greene, January 1999).

23  Bagnold was elected to the Royal Society in 1944, three years after the publication of his *The Physics of Blown Sand and Desert Dunes*. Bagnold, a professional soldier, cherished his election to the Royal Society above earning the rank of brigadier general in the army. He recalled his surprise upon hearing of his election.

> On my return to England [in 1944] I was astonished to find that I had been elected a Fellow of the Royal Society. I had had no inkling that I had been proposed or even considered. Thus my vague wish as a young officer to become an FRS rather than a general had come true. It was more surprising because I was merely an amateur scientist with no academic standing.
>
> *(Bagnold, 1990: 145; my bracketed addition)*

Bagnold, who took an undergraduate degree in engineering at Gonville and Caius College, Cambridge, while on leave from the British army, worked on the physics of dune formation and the migration of dunes while stationed in Sudan between the first and second world wars. He continued his work in England, built a wind tunnel to obtain needed data, and put forth a quantitative account of the action of winds on sand and the formation and migration of sand dunes. Bagnold later suggested that his lack of geological knowledge may not have been a disadvantage.

Looking back, I am astonished that I managed to do so much in the five years from 1935 to 1939, and in spite of interruptions, including three overseas expeditions that took me away for over a year, and in spite of being a complete novice at any kind of scientific research. I had two advantages. As an amateur with no academic background, my mind was uncluttered by any previous unproved and possibly misleading ideas. So I reasoned simply from the well-established principles of physics. Further, I had an almost unique field experience of unimpeded behavior of wind and sand in the entire absence of interference by rainfall or any living matter. I also had the ability to design and make my own bits of experimental equipment as I needed them, thus saving a lot of time.

*(Bagnold, 1990: 107)*

24 McKee initially argued that there was no dominant direction of movement for the Coconino sandstone, but changed his mind by 1938 as reported by Reiche (1938: 919, Note 23).

25 See Miknulic and Kluessendorf (2001) for a biography of Raasch.

26 Runcorn gave a talk on April 7, 1958 to the Society, and it was published in the proceedings of the symposium (Runcorn, 1958).

27 With the benefit of another decade of data, McElhinny (1973: 245–255) compiled a beautiful example of fixing Europe and North America relative to each other using APW segments without longitudinal ambiguity.

28 A good example of what can happen when hemispheric ambiguity is not recognized is when Graham assumed the Silurian Rose Hill Formation of Maryland was reversely magnetized and had not moved, when in fact it was then in the Southern Hemisphere and normally magnetized (Note 9, §1.8).

29 The remainder of this section was rewritten largely as a result of editing by E. Irving.

30 Upon seeing this passage once again, Irving (June 2000 comment to author) remarked, "I am fairly sure Keith [Runcorn] wrote this. It was pretty silly of us not to follow up this use of Euler's Point Theorem."

31 His paper was submitted about a year after Creer, Irving, Nairn, and Runcorn had submitted their pivot point method. Like them, he made no attempt to carry out analytically or numerically Eulerian rotations. Irving did the rotations graphically.

32 This secondary intersection (5) is very close to that originally obtained from NRM data without demagnetization (Irving, 1958b). Clearly (5) is strongly affected by secondary overprinting. This does not mean that the results of Irving and Green (1958) did not indicate large-scale motion of Australia, only that the newer data do so more convincingly and fix the position of Australia relative to Europe more accurately.

33 Studies of the dispersion of the ancient geomagnetic field and their comparison with models expanded over the following decades (Merrill and McElhinny, 1983) and have since become a major topic of paleomagnetic research. I am concerned here only with their birth and infancy. See also Evans (1976, 2005) and Evans and Hoye (2007).

34 Blackett apparently forgot that paleomagnetism began before Wegener first proposed continental displacement in 1912. See, for example, David (1904) and Brunhes (1906).

35 As described in §5.5 and §5.6 the age of the Serra Geral volcanics and the magnetization of the Tasmanian dolerites are Cretaceous (Note14), thus this graph refers to a wider time interval. However, this does not affect the general relationship claimed by Gough *et al.* (1964) between dispersion and latitude.

# 6

# Earth expansion enters the mobilist controversy

## 6.1 Outline

Before continuing with the developing paleomagnetic case for mobilism, I examine the concomitant revival of Earth expansion. Although now almost forgotten, it was, in its heyday, intimately intertwined with the developing case for global mobilism; some, for example, even thought it provided the mechanism that moved continents and opened oceans. In fact, once fixism was weakened by the maturing paleomagnetic case, and defeated by the confirmation of the Vine–Matthews hypothesis (IV, §5.5) and of Wilson's (and Coode's) hypothesis of transform faults (IV, §5.11), expansion became, for a few years, the only real competition to seafloor spreading, the hypothesis that oceanic crust was created at oceanic ridges and spread sideways from them.

Several workers had suggested Earth expansion to be the mechanism for continental drift during the 1930s, although nobody then seemed to care.[1] However, in the second half of the 1950s the physicist Laszlo Egyed, the geologist Samuel Warren Carey, and the oceanographer Bruce Heezen, independently of one another, proposed expansion afresh, and it began to attract attention. Arthur Holmes, for example, supplemented mantle convection with modest expansion in the second edition of his *Principles of Physical Geology*, and Creer and Wilson flirted with it. Several workers who opposed expansion thought it had attracted enough general interest for them to take the trouble to present their objections to it.

Egyed proposed Earth expansion in a series of papers published in the mid-1950s. Before, he had not supported mobilism, but he now realized that expansion might provide a mechanism for it. His last expansion paper was published in 1969. He died two years later. Carey, who became a mobilist in the 1930s, proposed expansion shortly after Egyed, and first published on it in the 1958 proceedings of the Hobart symposium. The symposium, held two years earlier in 1956, was the first major international meeting on continental drift during the 1950s. Carey continued vehemently to believe in Earth expansion until his death in 2002 at the age of ninety. Heezen, a leading oceanographer at Lamont Geological Observatory (Lamont), first proposed expansion in 1957 (III, §6.9), and continued to support it until the middle

1960s when he abandoned it in favor of seafloor spreading (IV, §6.15). Prior to adopting expansionism, Heezen had not been a supporter of mobilism; like Egyed, he was an expansionist first and a mobilist later.

Egyed advocated expansion at a constant slow rate, Carey and Heezen at faster and accelerating rates. Egyed believed that Earth began expanding as soon as it formed and has continued to do so at about 0.5 mm per year in radius. Carey and Heezen maintained that Earth underwent rapid expansion only after the partition of Pangea and did so at an average rate of radial increase of roughly ten times Egyed's; Earth's surface area had roughly doubled since the Paleozoic.

Egyed, Carey, and Heezen all became interested in paleomagnetism. Heezen began commenting favorably on its support for mobilism when developing his thoughts on Earth expansion. Carey already had become very interested in paleomagnetism as it began to favor mobilism; he had, recall, helped Irving find suitable samples in Tasmania (§3.12, §6.5), invited him to speak at Hobart (§5.16), and had encouraged Blackett to see if there was paleomagnetic support for a counterclockwise rotation of Iberia relative to the rest of Europe, an idea that Argand (1924:145) had proposed (I, §8.7) and which Carey developed. Egyed was impressed with the paleomagnetic case for mobilism and, in 1960, invented a way to use it to test for Earth expansion. That same year Cox and Doell applied Egyed's test (§8.14). Workers continued to use paleomagnetic data to test Earth expansion well into the 1970s. Of these, only Ward and van Hilten fall within the time-frame of this book and I shall deal with them later (IV, §3.10).

Heezen proposed expansion primarily to explain the origin of oceanic ridges. I shall make only passing reference to him until I discuss marine geology in Volume III. I shall use Egyed's correspondence with Arthur Holmes to get an idea of Egyed's views and to learn more about Holmes' thoughts on expansion, and his openness to novel ideas. For Carey I shall pay as much attention to his pre- as to his post-expansionist mobilist views.

## 6.2 Laszlo Egyed and his version of Earth expansion

Laszlo Egyed was born on February 12, 1914, in Fogaras, Hungary. He studied mathematics and physics at Budapest University, obtained his Ph.D. in 1938, and became a lecturer in mathematics. Two years later he took up a position in the Hungarian–American Oil Company, working in exploration geophysics, especially gravity and geomagnetism. He returned to Budapest University in 1947, and was senior lecturer in geophysics until 1951 when the university asked him to reorganize and head the Department of Geophysics. In 1956 he moved to Eötvös Loránd University as professor of geophysics and Director of the Geophysical Institute. He also reorganized the Seismology Observatory at the university. Egyed was awarded the D.Sc. degree in 1953, elected a corresponding member of the Hungarian Academy of Sciences in 1960 and a regular member shortly before his death. In 1957 he was given the highest Hungarian State award, the Kossuth Prize.

Egyed developed slow expansion in a series of papers in the mid-1950s (1956a, b, c, 1957). During the late 1950s and early 1960s (1959, 1960a, b, 1961, and 1963) he modified his ideas, and proposed various methods for estimating the rate of expansion. After the advent of plate tectonics, he still continued to defend slow expansion or dilatation (1969). Egyed began by developing a dynamic model of Earth's interior. His model implied expansion, and he found that it offered solutions to various problems, including a mechanism for continental drift. He also developed several ways to estimate the expansion rate, and found that they agreed, amounting to an increase of Earth's radius of roughly 0.5 mm per year. Taking this consilience as significant, he argued that it improved the overall problem-solving effectiveness of his dilatation theory (RS1) and in 1957 he commented on its development in a letter to Holmes:

Here I cannot but insert a purely subjective remark: the idea of dilatation has arisen quite independently of my own will, as a result of investigations on the internal constitution of the Earth. At first, I myself was in doubt as to take it seriously or not, investigated the consequences of this idea only to detect contradictions in the line of thought that led to it. I had then no knowledge of a number of results which may be derived from dilatation ... I am convinced that my proofs are weak if regarded one by one. I did not publish a number of them. Others are not fully evaluated as yet. On the basis of the collective force of weak proofs as well as by the ability to predict a number of facts of which I had no former knowledge, on the basis of my theory, I dare to make the statement that expansion is the most important factor in Earth development ...

*(Egyed, October 15, 1957 letter to Holmes)*

Originally having severe doubts, he came to think that expansion was fundamental once he realized that it solved problems of which at first he was unaware.

That same year he outlined his global theory in this way:

A new conception of dynamic character is given for the internal constitution of the Earth. The main feature of the conception is that the material composition of the Earth becomes more and more homogeneous within its interior. The Earth's core consists of matter in ultra high-pressure state. This ultraphase state is, however, unstable and, therefore, a steady increase of the Earth's volume is going on. It is shown, furthermore, that the expansion of the Earth is able to account for the formation of the crust and oceanic basins, the energies of tectonic forces and earthquakes, the origin of deep-focus earthquakes, the periodicity of geological phenomena, the continental drift and mountain building, and is supported also by paleogeographical data.

*(Egyed, 1957: 101)*

I now consider his theory in more detail, especially the solutions that he thought it offered.

Egyed's theory of Earth expansion depended on a model of Earth's interior proposed in the late 1940s by the British geophysicist W. H. Ramsey. Ramsey (1949) speculated that the core and mantle are different phases of the same ultramafic material. Phase differences arose because of pressure changes. Super-pressurized material in the inner and outer core remains unchanged so long as pressure remains unchanged, but as pressure is reduced, the solid inner core transforms into liquid outer core, and finally into solid mantle. Consequently he argued that overall density decreases, Earth expands, and the primordial continental crust, which initially continuously covered the Earth, fractured and separated into the present continents, with intervening oceans. Assuming the present area of the sialic continents equaled the surface area of the primordial Earth, he proposed that Earth's radius has increased at a rate of 0.55 mm per year (1956a: 59).

His second estimate of the rate of expansion was based on paleogeographical maps constructed by the Russian N. M. Strahow (1948) and the French team of H. and G. Termier (1952), who had argued that continental areas covered by water had decreased through time. Egyed assumed that the volume of the hydrosphere has not increased by more than 4% since the Early Cambrian, from which he calculated an expansion rate of about 0.5 mm per year since then (500 Ma). Egyed presented this paleogeographical procedure in two short papers (1956a, b).

Egyed's third way of calculating the rate of expansion began by rejecting the widely held belief that the decrease in Earth's rate of rotation is caused by tidal friction.

Observations have shown that the angular velocity of the Earth's rotation has been steadily decreasing. The value of it amounts to approximately $5.2 \times 10^{-6}$ sec/year ... Formerly, the decrease of velocity has been assigned to the effect of tidal phenomena, but the latest investigations have shown that their effect is compensated by the effect of the atmosphere on the whole. However the decrease in the velocity of rotation is a natural consequence of the Earth's expansion.

*(Egyed, 1956a: 61)*

Expansion increased Earth's moment of inertia thereby decreasing its rate of rotation. He calculated that an expansion of the radius of 0.38 mm per year would by itself account for the known decrease in the rate of rotation. Egyed (1956a: 62) concluded that his value, obtained by ignoring tidal friction, was "in very good agreement with the values determined by [his] other methods." He settled on an average increase of 0.5 mm per year.

Although Egyed's letter to Holmes above indicates that he did not come up with Earth expansion because he wanted to find a drift mechanism, but afterwards realized that he had inadvertently found a way to move continents, so he grafted continental drift onto his slow expansion model. He wrote:

The continental drift was the idea by which the Taylor–Wegener theory claimed to explain the distribution of the present continents and oceans. The most serious problem of the Wegener theory was, however, the origin of forces producing the rupture and drifting apart of the continental blocks. In our model these forces are obvious consequences of the expansion of the Earth. On this basis, the explanation for the distribution of continental masses is the following.

After the development of Pangea in consequence of the expansion of the Earth, stresses were produced in the crust, that was also bent owing to the growth of the radius. These stresses surpassed, at a certain limit, the strength of the rocks and the crust was disrupted in the same way as in the case of the formation of the Pacific ocean.

*(Egyed, 1957: 118)*

For Egyed an unexpected bonus of his theory was its ability to offer a solution to the second stage problem, the mechanism for drifting continents.

### 6.3 Holmes assesses Egyed's expansion theory

Egyed sent Holmes several of his expansion papers. Although I don't know when or who began the correspondence, Holmes' letter of April 6, 1957, and Egyed's response of May 6, 1957, reveal that they had begun discussing expansion earlier. Egyed noted differences between his and Ramsey's views, and discussed the transition of matter under very high pressure in the core to lower pressure in the mantle. Egyed closed his letter by emphasizing that his theory was in the developmental stage. He added that he hoped to continue their friendly exchange.

There is no doubt that a great number of details are still to be investigated and cleared in connection with these problems. I must point out, that my paper on the internal structure of the Earth is but a rough sketch of a theory; to develop it in detail will be the program of a long train of future investigations.

*(Egyed, May 6, 1957 letter to Holmes)*

After having "had time to give a little consideration to" Egyed's papers, Holmes replied on August 22, 1957, that he thought Egyed's ideas important.

I have now had time to give a little consideration to your really fascinating papers and to the letter you so kindly sent on May 6th. Your hypothesis is so important that it deserves to be shown as tenable or untenable, so perhaps you will not mind if I tell you of some difficulties I feel personally.

*(Holmes, August 22, 1957 letter to Egyed)*

Holmes raised two theoretical difficulties with Egyed's estimation of radius increase of 0.5 mm per year from the secular decrease of continental areas covered by shallow seas (RS2). He found dubious Egyed's assumption that the average height of continents relative to ocean floors remained unchanged.

On p. 60 [of 1956a] you say "It may be assumed that the level-distribution on the Earth was comparatively the same 500 million years ago as it is today." Your whole estimate of the rate of increase of the Earth's radius (by this method) depends on this assumption. But, I must say as a geologist, that it is an assumption I could not accept, since the present level-distribution is highly exceptional and not to be regarded as average. During the Cambrian I should have thought the lands were very low-lying in most places.

*(Holmes, August 22, 1957 letter to Egyed; my bracketed addition)*

He also found dubious Egyed's assumption that subaerial continental areas had increased over time, and disputed his extrapolation of the rate of the increase beyond the time limits of the published paleogeographical maps.

Returning to p. 60 [of your 1956a]: if the land area [uncovered by water] is now twice what it was 500 m.y. ago, it must have been extremely small (1/16th) 2000 m.y. ago; this expansion and rising of the land areas through geological time is not at all what one should expect from the effects of denudation. Indeed I find [one] of the most fundamental problems to be the reason for the continents keeping their heads above water. What continually raises them, or lowers the sea floor? There are various hypothetical possibilities, but expansion of the Earth as a whole cannot, I think, be more than a minor factor.

*(Holmes, August 22, 1957 letter to Egyed; my bracketed additions)*

Holmes thought that expansion alone was insufficient to keep the continents from drowning; it was likely only a "minor factor."

Holmes raised a further theoretical difficulty, questioning Egyed's proposal that as Earth expands the rotational rate decreases because the moment of inertia increases. He was baffled by Egyed's appeal to the atmosphere to dismiss tidal effects as the cause of the lengthening of the day.

On p. 61 [of your 1956a] you ignore the braking effect of the tides on the ground that "their effect is compensated by the effect of the atmosphere on the whole." I do not know about this, and find it difficult to understand how tidal effects in the atmosphere could speed up the earth's rotation so as to balance the slowing down due to the shallow water tides. Surely any tidal effect must act as a brake, and therefore only a small proportion of the slowing down (if any) could be ascribed to the expansion of the Earth.

*(Holmes, August 22, 1957 letter to Egyed; my bracketed addition)*

Egyed thought that ocean trenches were formed by tension owing to expansion. Holmes thought otherwise, and his response recalls his earlier debates with Jeffreys and shows his preference for mantle convection.

I also feel very doubtful about explaining deep-sea trenches as due to tension. The photograph you sent showing rupture in a foam-rubber strip under tension would demonstrate a perfect explanation if the mantle could be regarded as a rigid solid. But the phenomena of isostasy surely show that it behaves like a fluid for processes of very long duration. So neither the rupture nor the trench at the surface could persist for long in the real earth: indeed the trench would not have time to be formed. I know that Ewing and his colleagues also think that tension is the cause of trenches, but the trenches are so badly out of gravitational equilibrium that some process must be at work holding down their floors – otherwise they would "bob-up" again until isostasy was restored. Tension could not keep them down, but downward turning currents might suck them down and hold them down.

*(Holmes, August 22, 1957 letter to Egyed)*

This "bob-up" tendency was a real difficulty. Although Holmes' comment indicates that he was uncomfortable with Egyed's proposed cause for the decrease of pressure within Earth, he was even more perplexed with Egyed's assumption that proto-Earth was comprised of entirely ultra-high pressure matter.

P. 56 [of your 1956a]. The Earth began with its whole mass in the ultra-high pressure condition. Your letter presents possible modes of transition to the normal state, but my chief difficulty is to understand how the ultra-high pressure matter had not become normal before the Earth was formed, or during the process of the Earth's origin. However, that is a matter we obviously can know nothing about in the present state of knowledge.

*(Holmes, August 22, 1957 letter to Egyed; my bracketed addition)*

Holmes finally raised a related anomaly difficulty. Thinking about ways to test Egyed's idea that the whole Earth had once been denser, he reasoned that if gravity at Earth's surface is now less than in the past, then the angle of cross-bedding (repose) in sands should be shallower.[2] As *G* decreases there would be less vertical component to the movement of sand in a depositing environment. Holmes continued:

What can be tested is the inference from our theory that gravity must have been much higher at the Earth's surface in the Precambrian. The angle of cross-bedding in sandstones etc. seems to have been the same throughout geological history, so far as I can judge from personal experience, photographs and information from other geologists. A good example is to be seen in the Lorraine (Huronian) of Canada, the age of which is well over 1800 million years ... This seems to indicate that gravity has not seriously varied over the past. I understand from friends in Sierra Leone that metamorphic sediments older than 3000 m.y. show cross bedding of the same kind as recent deposits.

*(Holmes, August 22, 1957 letter to Egyed)*

Holmes, veteran of scientific disputes, told Egyed, "Now it is your turn, and I hope you will be able to refute some of the arguments I have used." He added, with evident sincerity, "I very much look forward to your comments."

Egyed attended the September 1957 IUGG meeting in Toronto, and found Holmes' letter waiting for him upon returning to Budapest. He wrote two responses. The first, dated October 15, 1957, responded to the difficulties Holmes had raised against Egyed's explanation for the decrease in Earth's rate of rotation, and his assumption that proto-Earth remained in an ultra-high pressure condition (RS1). He made several suggestions, but they had little effect on the ensuing expansion controversy. Egyed was also intrigued with Holmes' mention of paleogravity, and in his second response (November 14) he attempted to refute Holmes' anomaly difficulty, the unchanging angle of cross-bedding with time (RS1). He was eager to test his theory.

On the other hand, you certainly brought home a point in mentioning the variations of gravity. I have thought of this problem some time ago – in what connection, I will tell below. It is clear that if it should be possible to determine ancient values of the gravity force – analogously to palaeomagnetic investigations – absolute proof for or against my theory could be obtained. Of course the determination of such values would be rather hard, as on the assumption of half a millimetre of radius increases per annum no more than ten per cent of gravity increase would have resulted since the beginning of the Cambrian and no more than 40 to 50 per cent since the early Precambrian (2000 million years ago).

*(Egyed, October 15, 1957 letter to Holmes)*

Egyed continued to think about Holmes' cross-bedding criticism, and met it this way (RS1).

Your remark concerning the connection of cross bedding and gravity has kept on humming in my ear. I have to some extent referred to this point in my last letter, but I myself have found that answer unsatisfactory. Now, I hope I can give a better one.

The angle of cross bedding can be regarded as the angle of internal friction of a grained sediment. Thus, principally, it may be regarded as a slope where the gravitational attraction acting upon any particle is insufficient to overcome friction. However, friction is solely dependent upon perpendicular pressure: $S = uN$, $S$ being the force of friction, $N$ the normal pressure and $u$ a constant depending on the nature of the sediment. Using the notations of the annexed figure [which is lost], $S = mg\sin\alpha$, $N = mg\cos\alpha$, and thus t[an]g[ent]$\alpha = u$ independently of the value of $g$.

In any other medium other than thin air the amount of $m$ will seem smaller because of buoyancy, but as seen from the above equations that does not alter the angle of cross bedding either.

Thus, it seems that the constancy of the angle of cross bedding over the Earth's history may be a correct observation, but it has no meaning as regards the variations of gravity in the past.

*(Egyed, November 14, 1957 letter to Holmes; my bracketed additions)*

Egyed had found an answer to Holmes. Their courteous correspondence gave him confidence and helped him articulate and develop slow expansion.

### 6.4 Egyed develops his expansion theory and proposes a paleomagnetic test

In July 1959, Egyed sent Holmes an optimistic letter explaining that he had invented a paleomagnetic test of Earth expansion, and also had modified his theory by appealing to P. M. Dirac's idea that $G$ increases over time, thus removing the need for an ultra-high pressure primordial Earth (Dirac, 1937, 1938). Egyed was at first doubtful because he thought Dirac's suggestion clashed with the general theory of relativity. Writing to Holmes, Egyed said:

It is known that around 1939 Dirac reached the conclusion that the value of the gravity "constant" varies inversely with the "age of the Universe." I must confess that, at that time, I shared the doubts of most physicists concerning this hypothesis.

*(Egyed, July 31, 1959 letter to Holmes)*

A paper by C. Gilbert (1956) caused Egyed to change his mind. He welcomed Gilbert's results, and was quite emphatic about it, writing Holmes on July 31, 1959, that Gilbert "has succeeded in proving [Dirac's] idea quite irrefutably." In print he was more restrained, remarking (1960b: 186) that Gilbert "showed ... that Dirac's hypothesis is a corollary of the general theory of relativity." With $G$

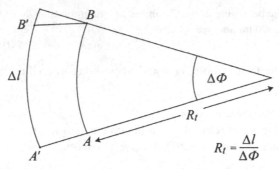

Figure 6.1 Egyed sent Holmes this sketch of his paleomagnetic test of expansion. *A* and *B* are paleomagnetically determined sites on an undeformed landmass along an ancient meridian, $\Delta\Phi$ is the latitude distance between them at $t$, $R_t$ is the ancient radius at $t$, and $A'$ and $B'$ are the current positions of $A$ and $B$, $\Delta l$ is the distance between them.

decreasing over time, he no longer had to postulate a denser primordial Earth. Egyed (1959, 1960b) also grafted an account of the solar system onto his Earth expansion, broadening its application.

Turning to his proposed use of paleomagnetism, he sent Holmes a sketch (Figure 6.1), which he included at the end of his July 31, 1959 letter.

Egyed's idea was straightforward. As will be explained momentarily, he needed determinations of paleolatitude from rocks of similar age located close to the same meridian on the same undeformed landmass.

Pleased with what he had done, Egyed discussed the possibility of putting his views into book form, and writing it in a language other than his native Hungarian for a wider readership.

As for myself, I am very occupied these days. The number of our crew is restricted, so that more work falls to one person. I cuddle the plan of putting my studies concerning the development of the Earth into book form. I should like to give a synthesis of geological and geophysical results on the hand of most up-to-date data, e.g., under the title "Dynamics of the Earth." It is a pity, however, that our mother language, otherwise so sweet to us, presents a sometimes insurmountable barrier to the propagation of our ideas.

*(Egyed, August 30, 1959 letter to Holmes)*

Holmes was taken with Egyed's ingenuity. He started his reply with advice and an offer to help; the book should be in English, and Holmes would offer help in contacting publishers and revising Egyed's English.

It was a great pleasure to receive your splendid letter of 31st July with all its bright ideas. But I am sorry you have increased departmental duties to occupy your time; I know only too well how very frustrating that can be. At the same time, I do hope you will be able to write your proposed book. Geology, as well as geophysics, is crying out for new and fertilizing ideas, and you can provide us with so many. It is essential, however, that it should be written in English and published in Britain or the U.S.A. If I can do anything to help, either by revising your

English (which is already remarkably good) or by finding you a publisher, you have only to ask and it will be a pleasure and an honour to do whatever I can.

*(Holmes, August 30, 1959 letter to Egyed)*

Holmes' remark about the lack of progress in geology and geophysics likely was prompted by reading Jeffreys' new edition of *The Earth* (1959), for he told Egyed:

I have just been going through the new edition of Jeffreys' "Earth" and find it very disappointing. He does not mention either you or Carey. In fact he seems to ignore everything he does not like, throughout the book. So, far from being helpful, it is quite misleading in places. I was hoping to find some information about tidal friction and the opposing effect of the Sun's attraction on the atmospheric tides (to which you drew my attention). He agrees that they are of the same order, but then leaves the whole subject in the air – most unsatisfactory. Most geologists know little or nothing about these topics and if you could present them clearly with a minimum of mathematics in your book it would serve a most useful purpose.

*(Holmes, August 30, 1959 letter to Egyed)*

Holmes was speaking from much experience; from his battles over continental drift and Earth's age he knew about intransigency among Earth scientists. Turning to Egyed's appeal to Dirac and Gilbert, Holmes expressed caution, but was on the whole optimistic. He also let Egyed know his gut preference for a "steady state" view of the universe.

I am grateful to you for drawing my attention to the paper of Gilbert in the MNRAS – a publication I do not see. I hardly think he has proved that G was formerly greater. If the "steady state" hypothesis of the Universe turned out to be correct (which heaven forbid!) surely G would remain constant? However, apart from that very doubtful possibility, it seems reasonably certain that G is decreasing with time. It may even turn out that all the other evidence for an expanding earth is also evidence that G is decreasing and that the "steady state" concept is wrong. But that is probably looking too far ahead. Meanwhile a varying G provides much that we need, though I wonder if it would be enough, by itself. The rate of expansion appears to have accelerated during geological time, and some additional factor may be necessary to account for that.

*(Holmes, August 30, 1959 letter to Egyed)*

Holmes' remarks indicate that he was becoming favorably inclined toward expansion, and he eventually supported Egyed's slow expansion in the second edition of his *Principles* (1965). However, it is unclear whether in the late 1950s he had yet decided between slow or fast expansion because Holmes was by now corresponding with Carey, and well aware of his less moderate views.

Holmes was intrigued with the way Egyed had grafted his ideas about the origin of the solar system onto his expanding Earth theory, and applauded his rejection of Urey's cold-origin Earth:

I am glad that you cannot accept a cold origin for the earth. Perhaps I have been too influenced by Urey's recent papers. Your suggested origin of the Solar System is wonderful and I hope you will soon publish it in English. Would an article in *Science Progress* or in the *Scientific American* not be worthwhile?

*(Holmes, August 30, 1959 letter to Egyed)*

Egyed apparently agreed with Holmes, but not completely, for he published (1960b) a short paper on the origin of the solar system, in *Nature*.

Holmes was impressed with Egyed's paleomagnetic test and urged him to publish a short note so paleomagnetists could get started.

> I must congratulate you most warmly on your fine idea for determining palaeo-radii. This should be sent as a Letter to *Nature* so that everyone interested can get started. There are certainly grave difficulties. Africa for example has a number of N-S trending orogenic belts of Precambrian age and I think the members of the various Geological Surveys could suggest localities and material suitable for testing. But the trouble is, would it be certain that these belts were N-S trending at the time of their formation? With polar wandering, "drift" and possibly tilting of the axis of rotation which would result from expansion not symmetrical about the axis, the whole surface seems to have been in a state of flux. However, while some of the results might be ambiguous as a consequence of these disturbing possibilities, there would still be a few that would prove expansion and provide a limit for its rate.
>
> *(Holmes, August 30, 1959 letter to Egyed)*

Once again, Holmes' fertile mind was at work offering helpful advice. Egyed should quickly publish his idea.

Egyed sent his idea to *Geofisica Pura e Applicata* (note received on April 5, 1960), combining it with a brief discussion of the new unusually high heat-flow values from the East Pacific Rise, of the new paleomagnetic support for drifting continents, and of expansion as a mechanism for continental drift.

> All and sundry of the geophysical arguments against continental drift have made the point that it seems to be impossible for the continents to swim forth on the surface of the mantle ... However, paleomagnetic data suggest that in the geological past the continents were much closer to each other and have even been connected. Now let us seek a way out of this apparent contradiction. I have proposed in a number of papers ... the hypothesis of large-scale Earth expansion, taking place at a radius increase rate of 0.4–0.8 millimeters per year. However, in case the Earth is expanding, continental drift is nothing more than the formation of new ocean basins along the gaping rifts which come to exist between continents. In that case, it was not necessary for the continents to have moved with respect to the mantle, as an expansion of the mantle, as a whole must have taken place, with the substances of the subcrustal earth-shells surging to fill the newly-formed fissures.
>
> *(Egyed, 1960a: 115)*

Egyed noted Heezen and Marie Tharp's discovery of a "central graben feature along" the East Pacific Rise "coinciding with the highest values of heat flow and also with the line of earthquake epicenters." But his idea was that "the formation of new ocean basins along the gaping rifts which come to exist between continents" was not seafloor spreading, it was similar to what Heezen had given during a 1959 symposium at Columbia University (III, §6.11) and Carey (1958) had set out in his contribution to the Hobart symposium proceedings (§6.13). At the time, Egyed was unaware of his fellow expansionists.

Egyed introduced his paleomagnetic test of expansion in his closing paragraph.

Lastly, I wish to make another point concerning palaeomagnetic measurements. The most striking argument in favour of the continental drift, or, stated more correctly, of the relative displacements of the continents, is the set of palaeomagnetic measurement results. Now there is a theoretical possibility, no use of which was unfortunately made as yet, to check by palaeomagnetic measurements the reality of Earth expansion, i.e. the validity of the above-given explanation for continental drift. If we have a contiguous ancient shield, at two points situated on the same ancient meridian and being a distance $\Delta t$ apart, the palaeo-latitudes, $\varphi_1$ and $\varphi_2$, as determined by palaeomagnetic methods on rocks of the identical age $t$, then dividing their distance by the latitude difference we have the palaeoradius corresponding to the time $t$: $R_t = \Delta t / \varphi_2 - \varphi_1$. In case the Earth is expanding, $R_t$ must be smaller than the present radius. To make the insecurities arising in paleomagnetic measurements and in age determinations as small as possible, it would be necessary to estimate beforehand the geological age at which the error would be likely to be a minimum: nevertheless, if the expansion rates obtained by other methods (0.4–0.8 millimeters per year) are somewhere near the truth, it is by no means impossible to obtain results by the outlined method.

*(Egyed, 1960a: 116)*

Egyed believed that the paleomagnetic case provided "the most striking argument" in favor of drifting continents. He was beginning to learn about the classical case, but found the new paleomagnetic support for mobilism more significant.

Holmes encouraged Egyed to try to teach a course in the United States. He thought it would help him put his thoughts into book form. Holmes also mentioned that Carey was going to lecture at Yale during the upcoming year.

I expect you know the work of Carey, who has reached the idea of an expanding earth by way of trying to solve the continental drift problem. He is now going to Yale as an Exchange-Professor for the coming session, and I am sure this (and many lectures he will give in other Universities) will go far towards making the idea familiar and breaking down old prejudices. If you could also go to an American University for a session, or even for a term, it would make an opportunity for giving the book in the form of a course of lectures, from which it would afterwards be easily written. But it may not be easy to find someone to fill your position at the Institute.

*(Holmes, August 30, 1959 letter to Egyed)*

In his reply to Holmes, Egyed thanked him for the offer of help. But, he asked for more.

However, I have the temerity to ask you in excess of the help you offered, to consent to criticizing my thoughts in the course of the development of the book most severely, as it can be expected only in this way to produce something that is useful to the geologist as well as to the geophysicist. Unfortunately, the field I wish to write about is in our country cultivated by some immediate co-workers of mine only.

*(Egyed, October 7, 1959 letter to Holmes)*

Egyed also thanked Holmes for telling him about Carey.

I thank you for having my attention called to Carey. Unfortunately I know nothing about him. I should be very grateful if you would inform me as to where his works were published.

*(Egyed, October 7, 1959 letter to Holmes)*

Holmes answered both points.

As to helping you with criticism I shall be honoured to be able to do whatever I can. Help will be mainly geological, of course, particularly pointing out where more detailed exposition or clarity of expression may be desirable from a geological point of view. For the geophysical side, no doubt you will have your own colleagues to indicate special difficulties that have to be faced and, if possible, overcome. In this connection I should very much like to know whether you have met with any serious opposition to your conclusions about the expanding earth? You have met most of the leading workers and it would be interesting to know if any of them, e.g. Heiskanen, Vening Meinesz, have raised objections. Every possible objection should, of course, be clearly stated and answered in the book. Here in Britain I cannot say the theory has aroused much interest as yet – but it will grow as people begin to talk about it.

*(Holmes, undated letter to Egyed)*

Egyed was fortunate to have Holmes; it is hard to imagine a better sparring partner. Clever fellow that he was, Egyed had come to cherish his criticisms.

Holmes, upon explaining the practical difficulty Egyed would face in trying to obtain Carey's work on Earth expansion, arranged for him to receive a copy.

The only published work of Carey's in which he reaches the conclusion that the earth must have expanded is in a long Symposium on Continental Drift held last year in Tasmania. The papers have been issued only in a duplicated form by the University, and cannot, I think, be purchased in the ordinary way. I have written to the University of Tasmania to ask them to be good enough to send you a copy.

*(Holmes, undated letter to Egyed)*

Holmes also told Egyed about Heezen's work.

You may also have noticed that Dr. B. C. Heezen of the Lamont Geological Observatory (Torrey Cliff, Palisades, New York, USA) has announced his adherence to the expanding earth theory at the recent meetings of the Oceanographic congress. He arrived at it independently from the evidences of tension, especially on the ocean floors. He will probably be sending you his papers, if he has not done so already.

*(Holmes, undated letter to Egyed)*

And now, I leave Egyed for the moment and will return to him later (IV, §3.10). As for Holmes, he was also corresponding with Heezen about his version of Earth expansion (III, §6.10). It is time to introduce Carey, his strong support of continental drift, his initial advocacy of mantle convection as its cause, and his late shift to Earth expansion.

### 6.5 Carey, the man and his views

Samuel Warren Carey (1911–2002) was colorful, appearing larger than life.[3] To get the flavor of the man, I begin with comments by Carey himself and by others who knew him and with whom we are already acquainted.

... We had Carey here for several days. What vitality! He tired me out, but his ideas are very interesting.

*(Arthur Holmes, postcard to a Dr. Suter, written in 1960, after Carey*
*and Holmes finally met after many years of corresponding, and*
*after Carey finally visited Holmes in Edinburgh.)*

I went to make these collections (of Tasmanian rocks) in the southern summer of 1955 – about February 1955 – almost before I had unpacked my bags in Australia after I arrived at the end of 1954. But the first thing I had to do being instructed by Jaeger (he who must be obeyed) was to meet Carey. When I got to the University of Tasmania, of course, Carey wasn't there. He left me a long message about how I had to meet him at the top of the hill at Tungatinah power station that was being built in central Tasmania. Big hydro developments were going on at the time. Carey had been hired by the Tasmanian Hydro to do various geological reconnaissance work, and he was doing it from a helicopter. I had this message from him in Hobart saying that I had to meet him at a certain time – remember that Tungatinah is in central Tasmania at least a hundred miles northwest of Hobart – on top of this hill which was used as a helicopter landing pad. I suppose the altitude was around 3000 feet. But you could drive to the top. So I drove there at the appointed hour – I cannot remember the details precisely. We were in the clouds essentially, but the clouds broke and down came this helicopter. But it didn't even land; you had to come down a ladder. And down came Carey dressed in raingear because he had been out in the Western Tiers where it rains almost incessantly. He said, "Ah, you're Irving. Let's go and have a cup of tea." So we went down to the power station that was being constructed where there was a canteen. We had many cups of tea, and he gave me a beautiful account of the geology of Tasmania, which I wrote down furiously. And then some time later, it was not more than an hour-and-a-half or so, he said, "Well I've got to get back." So we drove up to this pad. There were clouds all over the place. Finally the heavens opened and the helicopter came down and picked up Carey and he disappeared into the blue. He was obviously doing many jobs at high speed the whole time. But he had taken this short time out – probably at considerable expense for the hydro company – to come and give me these instructions about where I could get Tasmanian dolerites, where best to see them, and so on and so forth. And that's how I came to meet him. But, of course, the lecture consisted of a non-stop torrent of Tasmanian geology from the mouth of someone who knew more about it than anyone else. And, that was, of course, very helpful to me.

*(Irving, January 3, 2000 interview with author)*

As you are a philosopher, may I mention another matter Americans do not understand? Whereas I have frequently been invited to America, and to Britain, I certainly would not want to stay there too long. The confining walls of conformist dogma are too dominating. To think originally you must go alone into the wilderness. One of my greatest good fortunes was to spend six years in New Guinea (very primitive then) when I would go many weeks at a time without seeing another white man. (The natives were excellent people, but entirely neolithic.) With no journals or peer pressure, I could think! I welcome visits to the USA and bubble with the contemporary froth, but return to Tasmania to think. Nobody comes to Tasmania to go anywhere else, because Tasmania is not on the way to anywhere else.

*(Carey, November 25, 1981 letter to author; also see Carey, 1988: 120)*

For my part, public acclaim means little, and at 70 I don't suppose many years remain, and that matters little too. Twice I have declined nomination to the Australian Academy of Science, and although I was elected Honorary Life Fellow of the Indian Academy, and of the Geological Society of London, and of the Geological Society of America, in none of these cases was I consulted beforehand, but simply told after the event that I had been so honoured. However one private honour, I did appreciate: Half a century ago I translated Argand's paper at the Brussels International Geological Congress, "La tectonique de l'Asie". Decades later a large package arrived in the mail from Wegmann (Argand's successor in Neuchatel) containing all the original coloured manuscript drawings for Argand's paper. After studying my continental drift symposium he wrote: "Herewith I pass to you the Mantle of Argand."

*(Carey, November 25, 1981 letter to author)*

Carey was very difficult to have an argument with. He stated his point of view, and then that was rather the end of the matter. Your data was made to support it or cast aside. It was always very entertaining, very interesting to talk to Carey, but [he was not] a very fruitful person to have long discussions with.

*(Irving, 1981 interview with author; my bracketed addition)*

But Carey, you know, was always difficult to talk with because he was always very full of geological facts which he fitted into his scheme, and in later years, of course, fit into the idea of an expanding Earth. So it was a bit difficult to have, as it were, a detailed scientific discussion with Carey. But, of course, we knew each other. You know, I don't think he influenced me as it were.

*(Runcorn, 1984 interview with author)*

Carey was a great talker, and while he was at Newcastle, communication was essentially one way, i.e. outward from him. He did not seem to be greatly interested in receiving information about the origins of the palaeomagnetic data, instability thereof, nor methods and procedures nor anything else. He was a great "missionary" for his cause – expansion.

*(Creer, February 15, 2000 email to author)*

But, of course, Carey's expositions were very difficult to come to grips with because at any rate for a person like me who is a physicist, if you will, likes simple arguments. But from Carey you got a mass of geological data, and he used to go on regularly for two hours in giving a lecture. When he came to Newcastle he gave two two-hour lectures. Then in the discussion period it was very difficult to have a discussion. You just triggered off an evangelical outpour from him. So, I think, in a way Carey hasn't been given the recognition that he deserves. Although I think people increasingly do look back at his work with approval and interest. But, of course, you see, through being such a strong advocate of expansion his reputation was damaged because people do think that the kind of expansion he talks about is completely inadmissible on a number of grounds.

*(Runcorn, 1984 interview with author)*

Carey was a man of unbounded energy and drive, and a creative thinker. He viewed himself as a lifelong iconoclast, fighting the tyranny of orthodoxy. Although he generally shunned public acclaim for his professional accomplishments, he welcomed the praise of those he respected and considered fellow travelers. He was an excellent teacher, and interesting to talk with. He could be very convincing, and enthusiastically helped those who might supply him with data supportive of his views. But, at the same time, he could be very dominating, impatient of those with whom he disagreed. He seemed almost incapable of imaging that he could ever be wrong, unless, he himself raised the criticism.

Against the current of the times, he accepted mobilism during the 1930s. After du Toit's death in 1949, Carey became the noisiest, most insistent defender of continental drift. Aided by his incredible energy and enthusiasm, his thorough knowledge of field geology, and an overwhelming confidence in his point of view, he tried to convince anyone who would listen that continents had drifted. Although Carey was a supporter of mobilism throughout his career, his allegiance to it splits into two, the periods before and after he accepted expansion. He originally favored continental drift without expansion, first agreeing with Wegener's reconstruction of a single supercontinent, Pangea, and then adopting du Toit's twin supercontinents Laurasia and Gondwana. Finally, in mid-life he first returned to Wegener's single Pangea and then embraced Earth expansion, very rapid expansion. As he recalled:

From 1930 until 1937 I had followed the Wegener model (underpinned by mantle convection), thence until 1956 the du Toit model of Laurasia and Gondwanaland, twin *Urkotinente*, and thereafter the expansion model wherein continents dispersed from a pan-global continental crust while each continent remained attached to its own subjacent mantle.

*(Carey, 1988: 105)*

Upon switching to Earth expansion he instantly became its staunchest and most vehement proponent, arguing in its favor in three books and articles (1976, 1983, and 1988).

Although almost all Earth scientists held Carey's fast expansion as highly improbable, he did attract a small following. When plate tectonics became orthodoxy, he, like Jeffreys, his longtime fixist foe, rejected it, and continued to champion Earth expansion. When he died, in his nineties, he was still being ridiculed by some for his unyielding support of expansion, but derision never weakened his commitment.

Carey was born on November 1, 1911, on a small farm near Campbelltown, New South Wales, Australia. Campbelltown is approximately twenty-five miles southwest of Sydney. Carey, the fifth of nine children, had a dramatic entrance into the world.

I was born in Australia, on a tiny farm a few miles east of Campbelltown N.S.W., not much of a farm, only a few sheep and cattle, and my mother grew household crops, but my father was a printer and our limited income came mostly from that. My birth was violent. My mother was driving a sulky on a brush track when the wind blew a dead branch from high up, struck the horse on the snout so it bolted, with my mother desperately hanging on to the reins, until the axle caught on a stump and was stopped dead. My mother was pitched over a fence into an orchard – and I was born.

*(Carey, January 25, 2000 letter to author)*

He began school at five. His mother wanted her children educated.

I started school as soon as I turned five, and had to walk nearly four miles back and forth each day. My mother believed deeply in education, which was the most fortunate thing of my life.

*(Carey, January 25, 2000 letter to author)*

Carey's mother was a strong individual. She had to be, for her husband deserted her, leaving her to raise the children. She opened "a mixed shop – a formidable task for an inexperienced country woman," and Carey recalled, "As a teenager I regularly delivered customer orders in my billycart." Carey and his siblings "all got after-school jobs and for the vacations" (January 25, 2000 letter to author).

Carey attended the University of Sydney. In 1932 he earned the Honors B.Sc., which required an additional year's work after getting the B.Sc. He then took his M.Sc. in 1933 in one year. Carey described how he found a way to save money while traveling to and from his assigned area of fieldwork.

For my honors years, I was assigned a field area more than 300 miles away, but was always able to travel free on the railway because when a grazier is sending three or more railway truckloads of sheep or cattle to Sydney market, a drover is entitled to travel with them to check them on long stops lest some have got down and would be dead on arrival, then unload and water them. He is entitled to return free within three months on a normal passenger train. So for all my field years I was officially a drover and interleaved three months at the university and three months in the field.

*(Carey, January 25, 2000 letter to author)*

Once Carey obtained his M.Sc., he planned to go to the University of Cambridge for his Ph.D. He changed his mind when an opportunity arose in 1934 to go to New Guinea.

I had intended to go on to Cambridge and was assured of the necessary scholarship. But I was essentially a field man so when the opportunity was sprung on me, I went to New Guinea for Oil Search Ltd., where I would be months without seeing another white man, and when I did, it would be a surveyor, Jack Fryer, of our own company whose job was to tie together our stream traverses, the only outcrops were in the streams, air photos were still unthought of, and the river pattern wholly unknown as was the stratigraphy. I had no library access whatever, and could not transport more than one textbook as I shifted camp almost daily.

*(Carey, January 25, 2000 letter to author)*

He then decided to work for a D.Sc. degree, and in preparation he needed to take a year off from his employment. Finances were no problem.

My first two years were in far northwest New Guinea near the then Dutch border. At the end of 1937 after a year or more in the jungles of western Papua, I decided to take a year off at my own expense to prepare a doctoral thesis. I chartered aircraft to take me to islands I did not know enough about. In New Guinea I had been completely maintained by my company, so my salary went untouched into the bank, except for a regular allotment paid to my mother.

*(Carey, January 25, 2000 letter to author)*

Carey received his D.Sc. from the University of Sydney in 1939. His thesis was on the tectonic evolution of New Guinea and Melanesia. Holmes was one of the outside examiners. Leo A. Cotton, who was then Chair of Geology at the University of Sydney had met Holmes in Honolulu at the 1920 Pacific Science Congress, and had asked Holmes.[4] Holmes liked Carey's thesis, and his report was extremely complementary. He began by making several general points.

23rd January, 1939

REPORT ON THE WORK OF Mr. S. W. CAREY SUBMITTED IN CANDIDATURE FOR THE DEGREE OF DOCTOR OF SCIENCE

The record of original research work submitted by Mr. S. W. Carey, M. Sc., candidate for the University of Sydney Degree of Doctor of Science, consists of (a) an unpublished Thesis entitled "The Tectonic Evolution of New Guinea and Melanesia", supported by (b) five published papers, four of which make original contributions to the Geology of the Werrie Basin, while the fifth, written in collaboration with Dr. W. R. Browne, is a valuable review of the Carboniferous stratigraphy of New South Wales and Queensland.

In assessing the value of the Candidate's work I have been especially impressed by (1) the high quality and substantial amount of his fieldwork (stratigraphical and structural) and the energetic and enterprising way in which he has taken full advantage of his opportunities, despite the very real dangers and difficulties of the country in which most of his work has been carried out. (2) his independent capacity to recognize the far-reaching application of his local observations to regional problems and geophysical hypotheses; to state these problems and hypotheses accurately; and to devise and carry out scientifically valid tests of his own further inferences. (3) his wide knowledge of the relevant literature, which is extensive and scattered and published in many languages. (4) the good English and excellent style in which the record of his work and its results has been presented.

Holmes then turned to the application of Carey's work to continental drift. He was pleased with Carey's findings and interpretation.

It has been a personal pleasure to read the Thesis and to learn so much that is new and fruitful about an area in which I have long been interested. Ever since Wegener introduced his ideas of continental drift, New Guinea has been recognised as a critical area in tectonic geology. Now, for the first time, a wealth of new knowledge has been gathered which is sufficient in its scope to justify a synthesis of the structure of the island as a whole, and an objective discussion of the relationships of this structure to that of the adjoining areas. For the first time, too, it has become possible to realise the true bearing of the evidence provided by New Guinea on the many problems raised by the conception of continental drift. The Candidate has achieved these aims with conspicuous success. In particular I would draw attention to the important fact that his own contributions consist not only of observations in hitherto unexplored territory (geologically, if not topographically), but also of observations, confirmed again and again, which have disclosed and corrected certain errors in the geological work of earlier pioneers. As these errors are at present a definite stumbling block in current presentations of the evidence for and against continental drift, it is much to be desired that the Candidate should be allowed to publish his thesis as soon as possible.

I have no hesitation in stating that the Thesis makes an original contribution to geology of substantial value and distinguished merit and that it is, in my opinion, sufficiently outstanding in these respects to warrant the recommendation that the Degree of Doctor of Science should be granted to its author, Mr. S. W. Carey, without further examination.

Carey could not have had a more knowledgeable and appreciative examiner, and Holmes remained a valuable ally for Carey.

It took much longer to hear from the second reviewer, and Carey assumed that he had not passed.

Because my thesis was entitled "Tectonic Evolution of New Guinea and Melanesia" a Dutchman was appointed examiner along with Arthur Holmes because my thesis raised fundamental tectonic problems, including the evolution of the East Indies, but when my thesis traveled by sea to him in Holland, he was in Java. So my thesis went back by sea to him there, but by the time it got to Java, he was in Timor so it was sent there by sea, only to reach Timor after he had returned to Java, so back to Batavia [now Jakarta] it went there after he had returned to Amsterdam. Meanwhile deep in the New Guinea jungle, I assumed I had been failed. I had no way of knowing that Holmes had reported months before in glowing terms, but it could not be submitted to the Faculty until all examiners had reported. The third examiner was Dr. Woolnough, who was then Chief Commonwealth Geologist.

*(Carey, January 25, 2000 letter to author; my bracketed addition)*

Carey passed, but did not find out until he had returned to New Guinea, where he worked as a senior geologist for the Australasian Petroleum Company Papua until the outbreak of World War II. Carey then served as a special unit paratroop commando with the Australian Imperial Forces.

I was in "Z Special Unit" whose function was commando attacks deep beyond the front lines. My younger brother and one of my best Z officers, who had been training to be a doctor, were ceremoniously beheaded months after their capture near Singapore. But don't make a hero of me, because the major operation I was scheduled to command to blow up Rabaul [on New Britain] never got away, because the Americans refused to allow a submarine to drop us in enemy waters hundreds of miles away, whence we could travel by foldboats to our intended target. But when the Americans started island-hopping northward in the west Pacific, the role of Z became redundant, so I left the army to become Chief Geologist of Tasmania.

*(Carey, January 25, 2000 letter to author; my bracketed addition)*

In 1942 he became chief government geologist in Tasmania, holding the position until 1948 when he was invited to start up a geology department at the University of Tasmania. He remained there as professor of geology until retirement in 1979. He served as President, Geological Society of Australia for 1977–8. He was named Officer of the Order of Australia in 1977, received the Gondwanaland Medal in 1963 from the Geological Society of South Africa, the Clarke Medal from the Royal Society of New South Wales in 1969, and its Johnston Medal in 1976. For many years he would not allow his nomination to go forward, but finally he did and was elected to the Australian Academy of Sciences in 1989. In 1998 he received the Science Medal of the Australian and New Zealand Association for the Advancement of Science.

I shall begin by examining the development of Carey's ideas about mobilism before he adopted expansion.[5] The harm that Carey's reputation suffered during the many years after he adopted and continued to support expansion has, I believe, obscured the originality and great value of his earlier ideas – the quality of his early pro-mobilism arguments, and their closeness to seafloor spreading. He also made several major contributions to structural geology. I believe that it is time to say that if instead of stubbornly holding fast to expansion he had, in the late sixties, accepted plate tectonics, he would now be viewed as one of the most important geologists of the twentieth century.

## 6.6 Carey's defense of mobilism

Carey defended mobilism in four ways. First, he devised a new way to put the continents back together. Working in the tradition of Suess and Argand, he (1955a) used trend-line analysis, basing his method on the idea that many orogenic regions had been bent in plan section as a result of continental drift. He unbent them, moving continents accordingly, and so developed a reconstruction similar to Wegener's Pangea.

Second, Carey (1954) sought to free geotectonic theorizing from restrictive models of Earth's interior that did not allow for crystalline flow (he called it rheidity) and mantle convection. He emphasized the need for structural geologists to imagine flow of crystalline rock, and argued against Jeffrey's rejection of large-scale mantle convection.

Third, he made a major contribution to the question of the congruency of facing continental edges. Jeffreys claimed that the purported fit between South America and Africa was in error by 15° (RS2). Carey (1955b: 196) showed that the fit was very good; Jeffreys' difficulty had no merit, being based on no method (RS1).

Finally, Carey submitted a paper in 1953 to *JGR*, in which he developed a mechanism for continental drift that, like Holmes' mechanism, invoked convection in the mantle (§6.12). It was rejected. Only a single figure illustrating his use of convection has survived. He later claimed that he had anticipated much of what came to be known as the new global tectonics. Without the original paper, it is difficult to evaluate Carey's claim, but I shall try to do so.

It is difficult to determine which of these four steps Carey took first, because some, as he recalled, were taken many years before being published. I shall deal with them in the order above.

## 6.7 Carey's oroclines

As a student in the early thirties it seemed to me that if continental shields had moved the long distances visualised by Wegener or the convectionists, then the orogenic belts between them, which were mobile and deforming at the relevant times, must record indelibly the relative motions ... And so it was. Even the most primitive physical maps of the earth's surface confirmed these inductions. These disturbed belts certainly looked as though they had been bent and stretched and disrupted. I knew by 1938 that if you straightened the obvious bends, restored the visible stretches and reunited the dislocations, these processes alone reproduced a Pangea essentially the same as Wegener had deduced from wholly different grounds. Most of what I published in the 1954 orocline paper was in the 1937 draft of my doctoral thesis but were omitted at the eleventh hour because I realised that these concepts were too radical for acceptance then and would have cost me my degree. Even in 1954, publication was refused by referees of the Geological Society of Australia (a work which was later to receive the award of the Gondwana Medal and the Clark Medal).

> *(Carey, 1970 Presidential Address to the Australian and New Zealand Association*
> *for the Advancement of Science. Carey, 1970: 178–189)*

In 1938 I wrote: "New Guinea has been sheared westward under a colossal shear system on a scale grander than has been demonstrated anywhere else on the globe" ... "The stresses which are responsible for this great westerly displacement are of continental dimension. They are probably related to the main architectural pattern of the globe."

> *(Carey, 1963: 371)*

Carey contended retrospectively that he conceived his orocline method of reconstructing Pangea at least fifteen years before he published it. His orocline paper (1955a) is now regarded as a *tour de force*. It is ironic that he should have excised this novel idea from his dissertation, even though the open-minded Arthur Holmes served as an outside examiner. But Holmes was only one member of the committee.[6] Perhaps Carey was right, for he still had trouble getting his paper accepted when he submitted it for publication in 1953. His ideas were far reaching and controversial.

Carey's orocline concept owed much to Argand. He read *La Tectonique de l'Asie* (Argand, 1924) in the early 1930s, which soon led him to find a new way of reassembling continents. Argand was a master at trend-line analysis, used so successfully by Suess. Carey became the new master, and at the outset he acknowledged his debt to Suess and Argand.

Sir James Hall first made the induction that observed contortions in strata were actual deformations of beds originally flat, and therewith structural geology began. A century and a half has seen much progress, but most of it has been toward increasing our understanding of the internal anatomy of orogens. Little has been done about the deformation of the orogenic belt as a whole. Again, with the outstanding exception of Suess and his disciple Argand, the bias of structural nomenclature is towards the deformation of the profile cross-section of the orogen, not its shape in plan. There are adequate terms for the deformation and dislocation of strata in the vertical plane, and even for the further deformation in the vertical plane of such folded structures but there is no term for deformation of structures in plan, and still less for the impressed bending of the orogenic belt itself, or for its longitudinal stretching. Yet in view of the horizontal compression involved in orogenesis, and its longitudinal variation, there must be substantial deformation in plan.

> *(Carey, 1955a: 255)*

Evidently Carey was impressed with mobilism as far back as 1930, when he began thinking that if continental drift had occurred, it should have left indelible tracks in orogenic belts. Primed to see tracks, he saw them while doing fieldwork in

New Guinea, where he found "a colossal shear system on a scale grander than" had "been demonstrated anywhere else on the globe." Carey had gone to the right place to see what he thought he might see.

In his paper, twenty-five years later, on the basis of "indelible tracks" left by continental drift, Carey introduced the *orocline* concept:

The face of the globe shows many areas where orogenic belts wheel in trend through large angles, sometimes as much as 180°. Such a form could have one of two origins. Either the orogenic zone had that shape from the beginning, or the bend represents an impressed strain (here defined as an orocline). Classical geology has always made the first assumption, explicitly or implicitly, using the concept of cratons around which the orogens were moulded.

*(Carey, 1955a: 255)*

Carey needed some way to decide whether curvature was original or impressed. He straightened the bends, determining how major structures would have been affected. If there were (were not) structural matches, he accepted (rejected) them as oroclines. He identified twenty-five oroclines.

In every case the major structures of the region not only agree with assumption, but lesser compressional and tensional structures fall where they would be predicted by the impressed strain theory, and quite unexpected solutions of other major tectonic problems result.

*(Carey, 1955a: 255)*

Carey argued that his "orocline" solutions gained further support because they were consistent with or solved additional problems about the structure of the region in question (RS1). He noted the similarity between his and du Toit's reconstructions of Laurasia, arguing that support for both reconstructions increased because they agreed even though developed through different methods (RS3).

Oroclines are best understood by considering examples, and I begin with the rotation of Iberia. Carey reviewed an idea of Argand's (1924), which was seconded by du Toit (1937), regarding the clockwise rotation of the Iberian peninsula relative to the rest of Europe (I, §8.7); Clegg and company were to begin testing this paleomagnetically in 1957 (§3.13). Carey first identified the differential compression of the Pyrenees, and argued that they were caused by this rotation. He then recognized the Bay of Biscay as the complementary tensional wedge (Figure 6.2). In Carey's words:

If the Pyrenean compression is smoothed out by rotation of OB about O away from OA the movement would (qualitatively) close the Bay of Biscay by bringing the Spanish continental shelf into contact with the shelf off Brittany. The simplest interpretation of the morphology of this region is therefore to regard the Bay of Biscay as the tensional rift demanded by the geometry of the Pyrenean compression, since there is no other feature in the region which has the required characteristics, and the magnitude for the tension called for is certainly great enough to have permanent physiographic expression. It seems therefore reasonable to conclude that the Spanish Peninsula was rotated through some 35° during the Pyrenean folding.

*(Carey, 1955a: 261; references are to his Figure 2)*

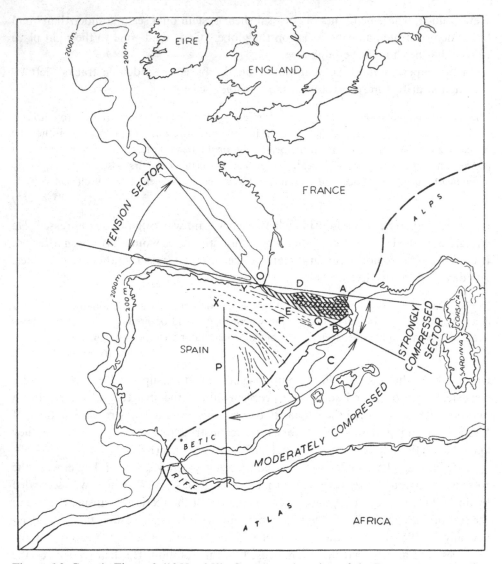

Figure 6.2 Carey's Figure 2 (1955a: 258). Carey's explanation of the Pyrenean compression sector was the complementary tension sector, the Bay of Biscay. O is the rotational center for both movements. The moderately compressing wedge, comprising the Ebro bundle of folds, rotates XP to YB. Carey maintained that its center of rotation was offset somewhat to the west of the apex of the Biscay and Pyrenean sector. He hypothesized an easterly translation of Iberia accompanying its rotation.

Carey referred to the Bay of Biscay as a "tensional rift" in-filled by newly formed ocean floor. He singled out the 200- and 2000-meter isobaths as the opposing divergent sides; they define the steep slope of continental shelves of Spain to the south, and of France and southwestern Ireland and Britain to the north. He was

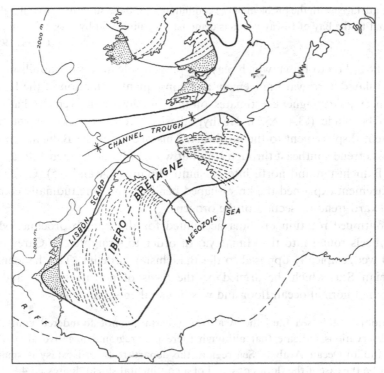

Figure 6.3 Carey's Figure 4 (1955a: 260) : matching of structural trend links across the closed Biscay rift.

familiar with the new seismic profiling (Bullard and Gaskell, 1941; Hill and King, 1942; and Hill and Laughton, 1954) which indicated:

> The steep continental slope running from south-west of Ireland to the head of the Bay of Biscay marks the boundary between typical continental and oceanic profiles with the Mohorovicic discontinuity rising from depths of 30 to 40 km. to very shallow depths.
>
> *(Carey, 1955a: 260)*

Carey identified complementary secondary structures. He recognized the moderate compression of the Ebro bundle of folds, moving line XP through arc C to YB (Figure 6.2). He noted that the center of rotation about which their compression occurred was somewhat displaced from that of Pyrenean compression as a whole and suggested that slight transcurrent motion accompanied the rotations.

Carey then sought older (pre-Pyrenean) structures that matched across the Bay of Biscay, which he argued were a classic structural disjunct (Figure 6.3).

> ... the following structural units all match up in order across the restored Biscay rift: the Liassic sea of southern France and eastern Spain, the Ibero-Bretagne fold bundles of Palaeozoic and crystalline rocks, the Lisbon Scarp and the submarine flexure off Brittany, and the Mesozoic sediments of the Channel trough and of the coast plain of Portugal. Unless we close the Bay of

Biscay, we are faced with the problem of the continuation of these structures, for all eight broken ends run out to the Bay of Biscay which is demonstrably underlain by very different rock.

*(Carey, 1955a: 265)*

He then moved on to India, which played a pivotal role in testing mobilism, because of its postulated northward movement and consequent formation of the Himalayas, and because paleomagnetic estimates of this northward movement had recently begun to be made (§3.4, §5.2). He hypothesized counterclockwise rotation and transcurrent displacement to the northeast. This produced the Baluchistan orocline whose folds trend southeast through Iran but which wheels through 120° as it passes through Baluchistan and north into Kashmir (Figures 6.4 and 6.5). Carey thought the displacement explained the knee-shaped bends in the longitudinally compressed northeastward trending section of the orogen.

The continued rotation of India accounted for the Punjab orocline, where the orogen wheels round into the Himalaya. In order to straighten it, Carey needed a tensional wedge apically opposed to the Baluchistan orocline. This he identified as the Arabian Sea, which, he argued on the basis of marine gravity and seismic profiling, had normal ocean floor and was not sunken sialic crust.

Yet the floor of this sea [i.e., the Arabian] is certainly not foundered continent. The gravity observations indicate that although there is a regional negative anomaly in the northern Indian Ocean–Arabian Sea region, the departure from isostasy is small, which would not be the case if the floor consisted of a continental shield depressed 4 km. For the foundering of a block similar to the Deccan Peninsula through 4 km. of material from beneath, which would result in a mass deficiency of three thousand million tons per square km. (Hess and Maxwell, 1953, for the Caribbean, and Officer, 1954, for the south-west Pacific).

*(Carey, 1955a; my bracketed addition)*

Carey viewed the Arabian Sea as a triangular rift whose formation was a necessary accompaniment to the rotation of peninsular India and the Baluchistan orocline. By closing the Arabian Sea, Carey (1955a: 267) referred to Figure 6.6, and claimed, "the postulated Arabian Sea rift becomes an integral part of the greatest rift system of the globe."

Carey next examined a second rotation at Y in Figure 6.5, the locus of the Punjab orocline, and the westward tapering of the Tibetan plateau. The plateau stands approximately 4 km above the level of the Indian peninsula and other continents generally, which, in turn, stand approximately 4 km above ocean floors. Carey attributed the doubly high plateau to either rotational thrusting of peninsular India under the Asian block or to face-to-face crumbling of the two blocks until the thickness doubled. Transcurrent movement accounted for the tensional opening of the Red Sea and Persian Gulf, the shift in rotational centers from X to Y, the movement of the Indian block into or under the Asian block, and the stress-induced double-knee bending of the orogen between X and Y. As for consilience with

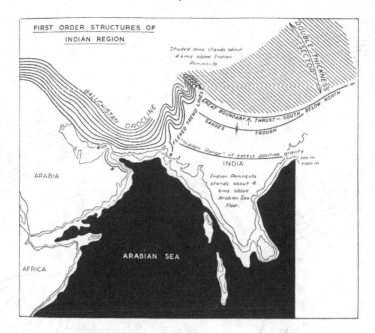

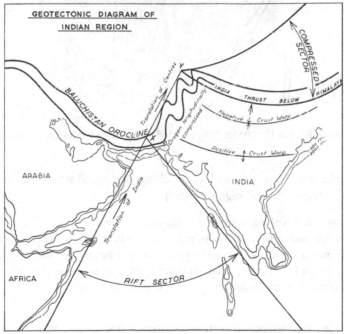

Figures 6.4 and 6.5 Carey's Figures 9 and 10 (1955a: 265 and 268). Figure 6.4 shows the Punjab and Baluchistan oroclines. The Punjab orocline begins where the Tibetan plateau, the double thickness sector, meets the flexed trend lines of the northward trending section of the Baluchistan orocline. X in Figure 6.5 is the center of rotation for the Baluchistan orocline and the pivoting point for the rotation of India, which formed the Arabian Sea. Y is the pivot point of the Punjab orocline. Translation occurs along line XY and its southward extension.

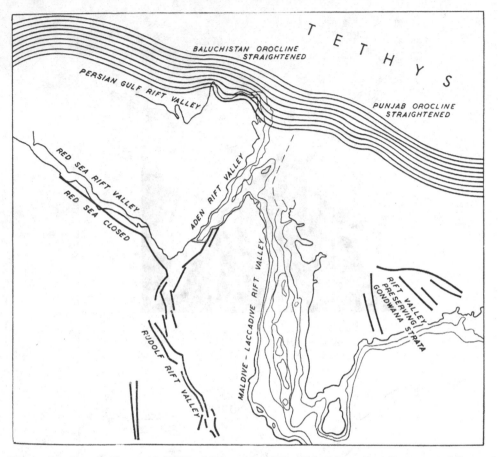

Figure 6.6 Carey's Figure 12 (1955a: 269) entitled "The Baluchistan Orocline restored."

secondary structures, Carey pointed to matching geological structures on the Indian, Arabian, and African coastal regions (RS1).

There are no structures striking transverse to the coasts which are not reasonably matched on the coast set in apposition to it. It is reasonable to conclude that whereas the geology of the opposed coasts may not, on available evidence, prove their former juxtaposition in the manner proposed from the study of the first order strains, the geology as known is quite consistent with this interpretation, and forms a coherent palaeogeographical picture.

*(Carey, 1955a: 268–269)*

I now turn to his reconstruction of Laurasia based primarily on his unbending of the Alaskan orocline and the concomitant opening of the Arctic and North Atlantic Basins. As Carey's method will now be clear, I shall restrict discussion to aspects of Figures 6.7 and 6.8, in which the Alaskan orocline is key: the bend of the Western Cordillera on North America as it extends northwards through Canada into Alaska. Carey (1955a:

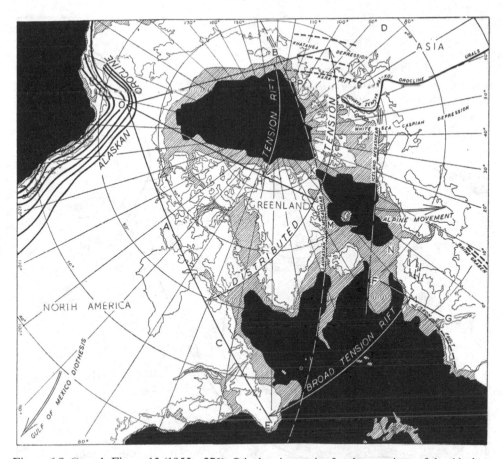

Figure 6.7 Carey's Figure 13 (1955a: 270). O is the pivot point for the rotations of the Alaskan orocline and three tensional sections. The simple tension rift, the Arctic Basin, is OHB. The broad tension rift, the North Atlantic Basin, is ECO, GFO. The wider distributed zone of tension is ABCD. The black areas are true ocean basins. The basin of the East Greenland Sea, surrounded by points K, M, N, and L, is between the Arctic and Atlantic Basins. The Laurasian and Iceland Megashears are shown as broken lines with their direction of motion indicated by arrows. Minor rift areas are shown with parallel broken lines. They include the Khatanga Depression, Kara Rift, White Sea – Caspian Depression, and Rhine Graben. The White Sea – Caspian Depression cuts across the Barents Sea (unlabeled), the Baffin Bay trough (unlabeled) extends southward into the Davis Strait (unlabeled), and both separate the western coast of Greenland from the Baffin Islands. The Taimyr Peninsula (unlabeled) is crossed over by the Kara Rift near P.

271–272) identified a smaller tension rift, the Arctic Basin; a much larger tension rift, the North Atlantic Basin; and a zone of distributed tension, "the Greenland, Barents and Kara Seas, and the Khatanga, Davis Strait and Baffin Bay troughs." With the restoration of the orocline, the two ocean basins, three seas, and three smaller troughs all close.

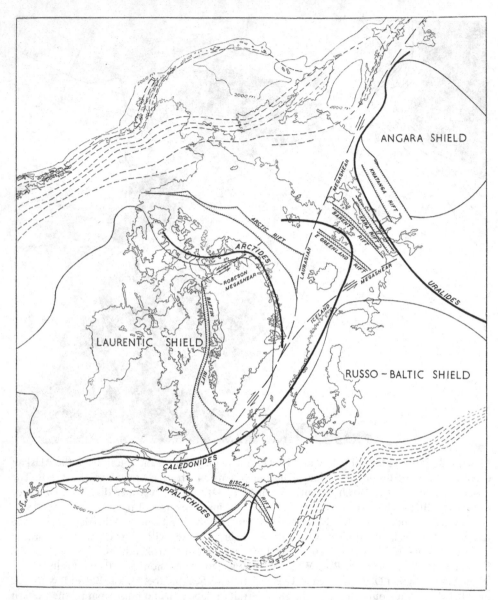

Figure 6.8 Carey's Figure 19 (1955a: 278). The Alaskan orocline is restored and Laurasia is revealed. Carey matched up four mountain systems that had become disjoined with the breakup of Laurasia. Also shown are the unbending six oroclines he identified in the Alpine system. Compare with Figure 6.9.

Carey identified accompanying displacements along huge transcurrent faults, which he designated *megashears* if they exceeded 100 km in length. The largest of these was the 1000 km, right lateral Laurasian Megashear, and he hinted at things to come.

The Pyrenean and Baluchistan oroclines were found to be associated with transcurrent movement in the same sense as the rotation. Similar evidence is present in the case of the Alaskan orocline. I shall discuss elsewhere the whole question of this and other megashears. It will then be shown that this megashear follows the south-eastern side of the Greenland block, and the northern edge of the Barents-Kara shelf from Spitsbergen to the Taimyr Peninsula, then continues across east Siberia to the *Fossa magna* of Honshu then onto the Mariannas. This will be found to be the principal line of slippage between the Asian and North American masses.

*(Carey, 1955a: 276; the Taimyr Peninsula is at 76° N, 103° E*
*in Russia just north of Khatanga)*

Unlike most contemporary structural geologists, Carey thought in global as well as regional terms. He also identified what he called the Iceland megashear, which converged with the Laurasian megashear and along which displacement was left lateral. He thought it significant that the Iceland megashear and the Great Glen fault in Scotland, which runs nearby and is parallel, had the same sense of movement. Another proposed left lateral megashear, already known in the literature as the Robeson megashear, extended across the northern edge of Greenland.

Carey was able to reassemble Laurasia by straightening the Alaskan orocline, closing the Atlantic and Arctic Oceans, and collapsing many small rifts (Figure 6.8). He noted that his Laurasia was constructed on grounds that were entirely independent of those du Toit had used; their consilience provided mutual support (RS1).

[My analysis] reproduces Du Toit's Laurasia without Schuchert's North-Atlantic misfit and without Wegener's dilemma of widening the Bering Strait as he moved America towards Europe. The palaeogeographic arguments used by Du Toit to build his Laurasia have not been used in the present synthesis. The palaeogeographic and tectonic analyses stand therefore as independent corroborative testimonies.

*(Carey, 1955a: 277; my bracketed addition)*

His orocline unwinding solution kept the relative positions of Siberia and Alaska essentially unchanged. With no widening of the Bering Strait, Carey had an answer to one of the difficulties that Schuchert had raised almost thirty years before at the AAPG symposium on continental drift (I, §3.4). Carey viewed his restoration of the Alaskan orocline as providing a unified explanation of the initiation and subsequent evolution of the many geological structures. Moreover, he claimed that his synthesis explained much more than the alternative did, which assumed that curvature in plan was original (RS3).

The adoption of the orocline alternative and the restoration which is implied, ipso facto results in comprehensive and unexpected integration of diverse facts. Structures such as the Rhine

Graben, Lisbon Fault, White Sea-Caspian depression, the inflections of Novaya Zemlya, Khatanga rift valley, the Arctic and North Atlantic basins, and Robeson Channel Megashear, all become integrated as parts of a single movement. The Urals become part of a structure twice as large. Schuchert's two classic criticisms of this part of the Wegener hypothesis, the "Atlantic misfit," and the objection that if America is moved closer to Europe the Bering Strait must be widened equally, both vanish.

*(Carey, 1955a: 278–279)*

In his grandiloquent style Carey argued that the features common to his analyses of the Pyrenean, Baluchistan, and Alaskan oroclines increased the support for each (RS1).

The Pyrenean, Baluchistan and Alaskan oroclines each independently reproduces parallel phenomena – an inflection in the orogens, apical rift sector, consistent major and minor tension phenomena, transcurrent displacement in the same sense, coherent palaeogeography, integration of all regional structures large and small, and automatic unexpected solution of diverse anomalies. It is surely in the highest degree improbable that any one such concurrence should be fortuitous. The triple repetition of such concurrences is surely strong ground for accepting the induction that all three bends are in fact oroclines.

*(Carey, 1955a: 279)*

Carey then went on to identify six more oroclines, all in the European Alpine system (Figure 6.9). He admitted that his analysis raised difficulties (RS2), but he was particularly pleased that one of its consequences was that it revealed the ancient Tethys Sea.

The single induction that the six great wheels in the trend lines of Europe are impressed strains produces tectonic coherence from tectonic chaos ... The great Tethys emerges from the straightened Alpine folds ...

*(Carey, 1955a: 286)*

Carey's discussion of the Alpine orogeny shows his debt to Suess and particularly Argand. Carey may have pickaxed samples from the "Alps" of New Guinea, but although not one of them, he, like any good continental European structural geologist, had to prove his mettle by trying to make sense of the Alpine system. But Carey did more – he used his analysis of the European and Baluchistan oroclines to position Africa, Arabia, and India together to form the shore of Tethys (Figure 6.10).

   In all these examples, Carey identified and reversed the implied strains in plan, looked for and found complementary tensional structures, and then sought corroboration from regional geology. Using spherical transparent overlays (§6.11) he conducted his structural analyses on a large globe, and accurately took account of movement over the sphere. Carey's tectonic analysis, as I have said, was a *tour de force*. The Clarke and Gondwana medals he received for it were surely well deserved. His development and application of the idea of oroclines was original and fruitful; it is now generally accepted. Following Argand, Carey returned geotectonics to the study of first-order structures. Straightening oroclines allowed him to reconstruct

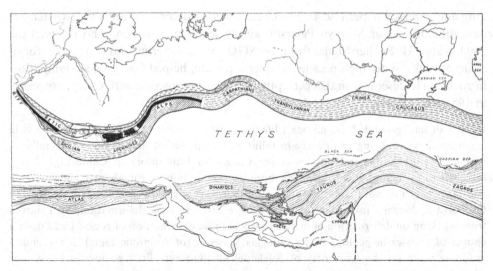

Figure 6.9 Carey's Figure 21 (1955a: 281) showing the result of restoring the six oroclines of Europe with the Tethys Sea exposed.

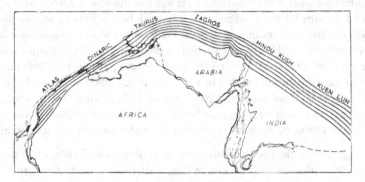

Figure 6.10 Carey's Figure 22 (1955a: 285) showing the Tethyan portion of Gondwana resulting from reversing European and Baluchistan oroclines.

continents. He postulated specific rotations that could be tested. One does not have to stretch one's imagination to see similarities between Carey's ideas about the rotation of crustal blocks and plate tectonics.

## 6.8 Carey's solid but flowing mantle

Carey recognized that his geotectonic reconstructions, notably his oroclines, required a mantle that on long timescales behaves as a fluid. To envisage this he developed what he called his rheid concept. Carey defined a "rheid" as a substance that under long-term stress deforms as a fluid yet remains solid in the short term. Rheidity is a measure of the tendency to flow under given conditions. He described his ideas in a long and complex paper (1954). The circumstances under which this was written are described in a letter Mervyn Paterson kindly sent me. Paterson, who received the 2004 Walter H. Bucher Medal from the AGU for his experimental work on deformation of rocks under high pressure and temperature, helped Carey understand "flow in rocks." He described in this perceptive letter his interaction with Carey, especially in 1953 when Carey was at ANU and had this to say:

By way of background, I did not start life as an earth scientist but my first degree was in engineering, specializing in primary metallurgy. I then moved into secondary metallurgy and took a PhD in metal physics at the Cavendish Laboratory in Cambridge. I was attached to an aeronautical research laboratory in Melbourne for about eight years, on and off between doctoral and post-doctoral sojourns overseas. In 1953 I moved to ANU in Canberra at Jaeger's invitation, to undertake research in rock deformation. So I moved from working on deformation of metals to working on deformation of rocks. I had done a couple of courses in geology under Douglas Mawson (of Antarctic fame) in my under-graduate years [at the University of Adelaide] but had very little geological knowledge when I moved to Canberra.

When I joined Jaeger's department of geophysics in Canberra in June 1953, Sam Carey was spending a sabbatical year in the department. He had had made in the ANU workshops a large sphere, around perhaps 1.5 metres in diameter, on which he was moving around his continents etc. If I recall correctly, his orocline concept was taking shape at this time. Almost immediately I became involved in many discussions with him on rheology of rocks. My impression is that by this time he was well aware that moving the land masses in continental drift implied plastic deformation in the sub-stratum but that he was still a little baffled about mechanism and how rocks could undergo such deformation, having in mind Jeffreys' strictures. Nevertheless I think he already had a concept that the rocks must be flowing. I don't remember when he started using the term rheid but I think he probably invented it during his sojourn in Canberra at that time.

With my background in the plastic deformation of crystalline materials, I had no problem in conceiving of plastic flow in rocks (it was, in fact, what I had come to Canberra to work on). So Sam was very keen to discuss ideas about flow in rocks with me. It was clear that what one was concerned with was *creep* in rock. I clearly remember introducing him to the Andrade creep equation … which I think was quite novel to him at the time. It helped him to put a concept of time dependence into deformation and, I believe, played quite a role in his rheid concept. I always felt that the rheid concept was a bit of overkill and an unnecessary term in dealing with what, to my mind, was a natural matter of creep in rock. But it seems to have helped him overcome a conceptual difficulty involving time scales in the plastic deformation in the Earth. So I feel that my discussions with him in 1953 were of considerable significance for his rheid paper. As a classical geologist Carey evidently did not have much of a grounding in mechanics and so in writing his rheid material he was effectively self-teaching himself things that might have been more obvious to someone with a classical mechanics background. His relative naivety in this area is illustrated by your quotation in §6.9 from his rheid paper: "The only strain during flow is the viscous shear strain between successive laminae. There is no flow across the flow laminae." This seems to imply an almost geological-stratification concept of "laminar" rather than the mathematical concept of laminar flow in a continuum.

I had no preconceptions about continental drift when I came to Canberra and so I was immediately strongly influenced by Carey. I therefore from that time had no problem with the idea of continental drift, perhaps too uncritically so. I remember seeing Jaeger's appointment of Irving as a step towards testing the idea of relative movement of the continents but almost with it being a foregone conclusion (of course, memory from this distance can be very unreliable). By the time that I went on a three-month leave in 1957 to Griggs' lab in UCLA, already aware of Irving's early results, I had no doubt in my mind that continental drift was an established concept. So I was mightily surprised to find that this was utter heresy in the US (this may not have applied to Griggs, who I think was probably a fence-sitter at that time, although I don't recall any discussions with him on the subject).

I didn't have a lot of interaction with Carey after the time he was in residence in Canberra but I met him casually on many occasions and felt that I knew him well. He was the only Australian geology professor who attended the meeting on structural geology and rheology that I organized in Canberra in 1962 (a series that is still going); E. S. Hills did not attend although invited. So he took a keen interest in structural geology in the Australian scene. I always got along affably with him on our casual encounters but certainly he had an ego of mountainous proportions and a tendency to play the "grand old man."

*(Mervyn Paterson, April 12, 2007 email to author; my bracketed addition)*

Carey himself recalled:

Two decades earlier, the American Association of Petroleum Geologists had convened a continental drift symposium in 1926 in New York in which the consensus had discredited continental drift as physically impossible. So during my early years as a professor I pondered deeply the physics of the crystalline flow of glaciers, of rock salt, and of the rocks of the earth's crust, the forces available in the earth to make them flow, and the effect of the scale of size and time on these phenomena. From these meditations came the rheid concept, the geotectonic scale concept, the mechanics of glacier flow, and the basic principles of folding. Gradually these insights became the foundations of my developing philosophy of global tectonics and led me to abandon the generally accepted canons of compressional tectonics.

*(Carey, 1988, viii)*

E. N. da C. Andrade (1910, 1914), a pioneering rheologist, identified the following four sorts of deformation occurring in solids that are under stress long enough to deform by viscous flow: elastic deformation whereby they return to their original shape when stress is released; time-independent non-elastic deformation or plasticity whereby they do not return to their original shape; transient creep; and viscous flow. Only the last two increase over time; viscous flow increases directly, transient creep as the cube root of time. If stress is applied over a very long duration, viscous flow or creep, as Paterson called it, becomes by far the largest component of deformation.

Given the preeminent effect of time in large-scale tectonic processes, Carey argued that the appropriate Earth model is not derived from seismology. The seismologist's Earth has a "solid inner core, a fluid outer core, and a solid mantle and crust, because the rheidity for all but the outer core is much greater than the period of seismic waves," whereas "the geologist's earth for [long-term] tectonic phenomena should be a thin crust and a completely fluid interior" (Carey, 1954: 73; my bracketed addition). Although Jeffreys, whom Carey surely had in mind, acknowledged the different effects of short- and long-term loads, he claimed:

A preoccupation with elastic phenomena in geological processes is still apparent in Jeffreys' latest book (1952) which opens with the statement: "In studying the constitution and history of the Earth, we are largely concerned with the phenomena of change of dimensions, especially of shape. Hence the physical properties that we most need to know, are the elastic properties."

*(Carey, 1954: 75; the Jefferys quotation is from the third edition of* The Earth*)*

Carey disagreed, believing that such attitudes had hampered progress.

I suggest that, for loads of longer duration than ten thousand years, the elastic properties are of little relevance in changes of shape, and that a good deal of wrong thinking in tectonics by geologists during the last generation might have been avoided had they adhered to the earth model appropriate to the time scale of the processes they contemplated.

*(Carey, 1954: 75)*

### 6.9 Examples of rheid flow

Carey went on to describe rheid flow first of homogeneous materials such as ice and salt and then more typical heterogeneous rock formations that involved flow of several different materials each with different rheidities. He described crystalline schists from Broken Hill in Australia, which, because of its economic importance as one of the most productive silver–lead–zinc bodies in the world, had been well studied, leading

to the identification of the original strata and their systematic tracing in great detail, it is found that the beds have been folded and extruded into flow patterns resembling in every way those of a viscous fluid. Individual beds thicken to many times their original thickness and elsewhere are attenuated to mere films of thousands of feet ... It is commonly agreed by geologists who have worked on this field and on crystalline schists generally that such deformations are flow phenomena.

*(Carey, 1954: 85)*

He reproduced a diagram (Figure 6.11) showing a vertical cross section of crystalline schists at Broken Hill.

Carey emphasized the similarity in the rheid flow (creep) between the heterogeneous Broken Hill rocks and homogeneous ice and salt.

Observe, however, that here [at Broken Hill] we are dealing with a heterogeneous aggregate of different minerals, each with its own rheidity, perhaps differing widely from each other, whereas in the previous examples [glacier, salt dome, extruded thrust-sheets of gypsum, and re-intrusion of serpentine], the flowing material has been largely mono-mineralic. However, in such cases the aggregate as a whole has a viscosity, rigidity and rheidity, and may be treated as homogeneous when bulk behavior is being considered. The flow of such material is not

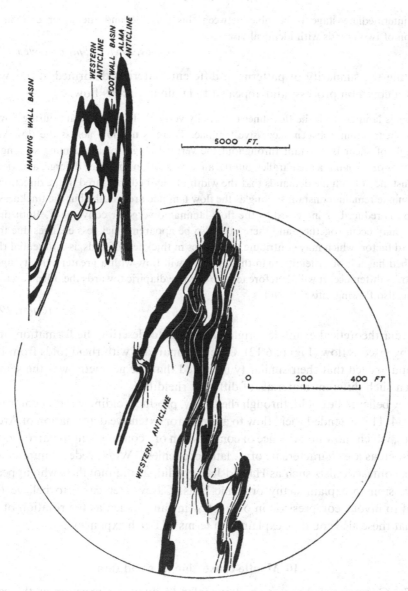

Figure 6.11 Carey's Figure 9 (1954: 86).Vertical cross section of crystalline schists in South Mine, Broken Hill (after Gustafson *et al.*, 1950). The larger circle is an eight times enlargement of the smaller and shows detail of second-order folds. There is a strong similarity in patterns regardless of scale. Minor anticlinal axes are shown by continuous lines and minor synclines by broken lines.

different mechanically from the flow of milk, which is a disperse of one fluid in another with a viscosity different by many orders of magnitude, or the flow of toothpaste or concrete or other suspensions where the viscosity of some of the suspended particles is so much greater than that of the suspending medium that the whole of the flow takes place in the medium.

Every intermediate stage is possible between this extreme and the other extreme of an emulsion of two liquids with identical viscosities.

*(Carey, 1954: 85; my bracketed additions)*

Accenting the similarity of patterns in different materials deformed in this way, he sought a common process, and appealed to laminar viscous flow.

The flow is laminar with the flow lines essentially vertical. The only strain during flow is the viscous shear strain between successive laminae. There is no flow across the flow laminae. The angle of shear is constant throughout the vertical length of any lamina so long as the boundaries of the lamina are parallel, but the angle will diminish if they diverge, and vice versa. The constancy of volume demands that the width of any bed measured in the direction of the flow laminae remains constant so long as the flow laminae are parallel but the thickness in this direction is reduced or increased as the flow laminae diverge or converge ... Both thickness changes may occur together, and there may also be apparent thickness changes due to pitch. The third factor, which may contribute to changes in thickness of beds, is differential rheidity. If one bed has a lower rheidity than the others, it will flow with a greater velocity under the same stress difference. It will therefore tend to become diapiric towards the others even though they are also flowing into rheid folds.

*(Carey, 1954: 92)*

He gave a theoretical example, originally meant to describe the formation of a salt dome by viscous flow (Figure 6.12). Carey compared it with rheid folds from Broken Hill, and argued that their similarity indicated that the geometry was the same even between such different materials of different rheidity.

Carey believed he could, through rheid flow, produce folding without compression. He (1954: 115) extended rheid flow to account for extensive deformation of Archaean gneisses, which show no evidence of compression or "confinement to narrow orogenic zones such as are characteristic of ... later orogenies." Who needed compression? Of course, contractionists such as Harold Jeffreys did, as did mobilists who appealed to compression to explain many orogenic belts. Carey later came to believe that he needed to invoke compression in only a few instances such as the rotation of Spain, and that these also could be explained in terms of Earth expansion.

## 6.10 Mantle convection as rheid flow

In 1976, Carey recalled that he had appealed to mantle convection as the cause of continental drift in his 1953 paper that had been rejected by *JGR* (§6.12) (Carey, 1976: 9). However, in his later 1954 "rheidity" paper (Carey, 1954), although he argued for large-scale mantle convection, he did not mention it as a mechanism for continental drift. He did not deny it, he just did not mention continental drift in connection with his rheid model. Perhaps he wanted workers to appreciate and use his rheid model regardless of whether they were fixists or mobilists. Perhaps if he linked his mantle convection and continental drift he feared it might prejudice acceptance of his orocline concept, such were the toxic connotations of drift.

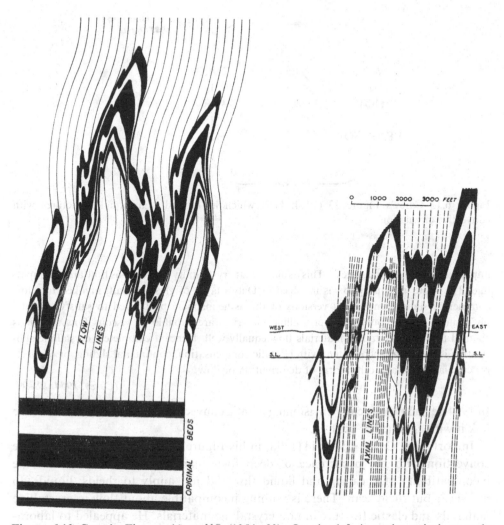

Figure 6.12 Carey's Figures 16 and17 (1954: 93). On the left is a theoretical example of rheid folds; on the right, a field example showing a cross section of rheid folds from Broken Hill.

The rheidity of the mantle allowed for convection and, unlike Holmes, Carey did not have to worry about the need for a partially liquid mantle; rheidity, especially that of gneisses and schists, paved the way conceptually for flow in a crystalline mantle. Nor did Carey have to follow Daly, who argued for a layer of glassy basalt to allow continental sliding. Mentioning Daly and continental sliding was the closest Carey got to continental drift in his "rheidity" paper.

The flow of glaciers, salt, gypsum, and gneisses, described above, all occur in holocrystalline materials. Yet there is still a widespread belief that evidence of fluid behaviour within the earth

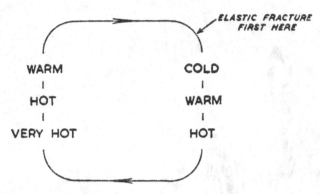

Figure 6.13 Carey's Figure 23 (1954: 111), which he labeled "Earthquake fractures with convective circulation."

implies a non-crystalline state ... This assumes that crystallinity and viscous flow are incompatible. The same misconception is involved in Daly's layer of glassy basalt, invoked to allow continental sliding, and many versions of the asthenosphere postulated to satisfy isostatic phenomena ... Of course, non-crystalline materials are capable of viscous flow, but it is claimed that holocrystalline materials flow equally well. Hence a completely crystalline substratum is in no way inconsistent with isostatic compensation, continental sliding, or convective circulation, or any other form of deformation or flow.

*(Carey, 1954: 98–99)*

In 1954 he had no difficulty envisioning mantle convection, but he did not at the time link it with continental drift.

Importantly, Carey (1954: 111) did, in his Figure 23 (Figure 6.13), link mantle convection with the occurrence of deep focus earthquakes, claiming that the incompatibility of shearing and liquid flow did not apply to rheids; liquids do not shear but rheids can. There is nothing incompatible about flow in crystalline materials and elastic fracture in non-crystalline materials. He appealed to laboratory and field evidence for fracture in cases where flow predominates; structures at Broken Hill and salt domes provided instances where fracture occurred but flow dominated. Bridgman and Griggs had performed the relevant laboratory studies. With these theoretical difficulties removed, Carey (1954: 100–101) linked together deep focus earthquakes and large-scale convection in the mantle, maintaining that "deep focus earthquakes are not inconsistent with convective circulation in the same material."

In homily fashion Carey suggested convection as an explanation of the origin of Vening Meinesz's negative gravity anomalies.

Dynamic movements within the earth could themselves cause large anomaly. Where convection currents turn downwards the vertical component of the viscous shear may support very substantial loads which would be expressed as negative anomalies with no support whatever from strength. This is the cause of the depressed water surface above the drain hole of a bathtub. The "Meinesz" negative anomaly strips are, it is suggested, of such an origin.

*(Carey, 1954: 103–104)*

With his recognition of the importance of creep (as Paterson preferred to called it) in geotectonics, he showed, with help from Paterson and, I suspect, also from Jaeger whom he thanked for his guidance as well as encouragement (1954: 116), that he could delve into rheology and provide a model of Earth that provided workers in geotectonics with enough flow to make sense out of the deformation they observed in outcrops. He appreciated developments in the growing field of rheology, knew about laboratory work in solid state physics, and understood the need to proceed cautiously when extrapolating short-term laboratory to geological phenomena, and he challenged Jeffreys. I move now to the other occasion when he challenged the old fixist, this time with complete success.

## 6.11 Carey's fit of Africa and South America

Carey attributed his initial interest in continental drift and coastline congruencies to Sir Edgeworth David, who Carey described in a 1984 letter to me as his "old beloved master." He subsequently remarked that it

was not until the third edition (1922) of Wegener's book was translated into English [in 1924], and burst like a bombshell on incredulous Americans and Britishers, that the status of the match across the South Atlantic was critically examined. My own personal inspiration, Sir Edgeworth David, who became deeply but cautiously interested, cut out the continents from a globe and pieced them together: "and would you believe it," he said, "the astonishing thing is that they do fit. David delivered a public address on June 12, 1928, in which he said that the whole theory looks fantastic at first sight, but a contemplation of the facts includes the hypothesis within the realm of probability.

*(Carey, 1988: 95; my bracketed addition)*

Carey recalled that it was a suggestion of Cotton (I, §9.2) that prompted him to begin reassembling continents, and it was he who, with a background in physics, apparently suggested that Carey use a process akin to Euler's point theorem; Carey chose another method.

As a student, on the suggestion of Professor Leo A. Cotton, I made cartographically valid stereographic projections, using both the 100- and 1000-fathom isobaths, not only of the South Atlantic, and Australia-India with the Bay of Bengal against Australia's northwest shelf. Cotton had suggested plotting on a normal equatorial stereographic net, and rotating the outline about an appropriate Euler pole, which is relatively easy on a stereographic projection, but I preferred to take the more arduous route of calculating a series of oblique stereographic projects, because I could work on a much larger scale, and Cotton's suggestions, although theoretically valid, accumulated progressive plotting errors.

*(Carey, 1988: 95)*

Jeffreys (1929: 322) had argued in the second edition of *The Earth* that Wegener's fit of the continents was mistaken – "the alleged fit of South America into the angle of Africa is seen on a moment's examination of a globe to be really a misfit of about 15°." Carey took Jeffreys at his word, that he had made only a cursory examination, and thought others would quickly realize Jeffreys' error. But Jeffreys' criticism, which he (1952: 349) repeated in the third edition of *The Earth*, remained unanswered; and, what is even more, Carey's former boss at the Anglo-Persian Oil Company, G. M. Lees (1953), cited Jeffreys' difficulty as a reason for rejecting continental drift in his Presidential Address to the Geological Society of London in which he defended contractionism. Carey (1988) later recalled these events and his reaction to them, adding that it was Lees who arranged for publication of the improved fit.

In 1929 appeared Sir Harold Jeffreys' prestigious book, *The Earth* – quite the most authorita-tive treatise ever on the physics of the earth … However, Jeffreys was completely opposed to Wegener's hypothesis, and in regard to the alleged fit of South America into the angle of Africa, he wrote: "On a moment's examination of the globe, this is seen to be really a misfit by almost 15°. The coasts along the arms could not be brought within hundreds of kilometers of each other without distortion. The widths of the shallow margins of the oceans lend no support to the idea that the forms have been greatly altered by denudation and deposition …"

From my many "moments" of accurate examination of this question that I had done, I knew this statement to be incorrect. I considered that the matter was rather trivial, that the true position would be generally realized, and that this criticism would fade away. But Jeffreys' prestige was so great that most workers accepted his pronouncement as final. Jeffreys repeated the statement in the second edition of his book, in 1952, and to rub salt on the wound, Dr. George Martin Lees (my former chief in the Anglo-Persian Oil Company), in his 1953 presidential address to the Geological Society of London, listed this as one of his three crucial reasons for rejecting the Wegener hypothesis. So I sent Lees my stereographic projections of two decades earlier … I added that "whether the continental drift hypothesis be true or false, this argument should never be used against it again." I asked Lees to arrange publication of this rebuttal, which he did.

*(Carey, 1988: 103–104)*

Carey's paper, which appeared in *Geological Magazine*, provided a detailed account of his technique of making reconstructions using his thirty-inch globe and continents molded out of sheets of polyvinyl chloride with a curvature matching its surface. In this way he hoped to allay doubts about his method.

Lest there should be any doubt about the validity of the methods, and also because it should make it easier for others who would make such comparisons, it is necessary to describe in some detail the methods which I have developed for such global morphological studies.

*(Carey, 1955a: 197)*

If readers doubted him, a relative unknown outside Australia, and assumed that Jeffreys could never have made such a careless blunder, they could try it for themselves. His map (Figure 6.14) showed that the fit was very good, the most significant overlap

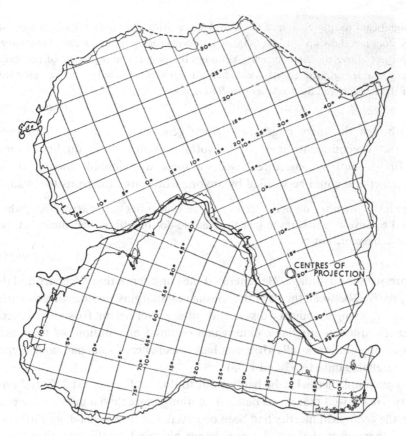

Figure 6.14 Carey's Text-Fig 1 (1955b: 198) The continental edges of Africa and South America fitted together at the 2000-meter isobath halfway down the continental slope. The largest overlap is at the Niger delta (approximately 4° N, 6° E on the African grid).

being the Niger delta. Turning this apparent anomaly difficulty on its head, he argued that it should be expected because the Niger delta had formed after partition. There also were tiny overlaps where the Walvis and Martin Vaz ridges leave both coasts.

Carey compared the facing edges of the two continents at the 2000-meter isobath.

Which isobath should be used in comparing such coasts? All will agree without argument that the present coast should be rejected and that some level on the slope from the continental shelf would give a more significant comparison. Examination of the problem shows that the 2000 m. isobath is the logical boundary. The mean level of the ocean floors is a little more than 4000 m. below sea-level and the mean level of the continental plateaus is a little above sea-level. The 2000 m. isobath is therefore a close approximation to half the "freeboard" of the continental rafts above the sea floor.

*(Carey, 1955b: 199)*

His choice of this isobath had an important consequence: it made the present Mid-Atlantic Ridge irrelevant to determining the goodness of fit.

This is significant to the relation of the Mid-Atlantic Ridge to the fitting of these coasts. For this ridge does not show up until the 3000 m. and deeper isobaths are drawn. Only a few small islands project above the 2000 m. line. Although the position and pattern of this ridge may have considerable significance, its width has no significance whatsoever in comparing the forms of the African and South American continents.

*(Carey, 1955b: 199)*

Thus although the ridge might have some genetic relation to current processes moving the two continents apart, it had nothing to do with the quality of the fit.

His fit was not a conjecture based on some questionable theory. It was an objective fact and could be verified by anyone who took "the pains to do so."

The diagrams constructed do not express opinions, or theories but objective facts, which may be verified by anyone who cares to take the pains to do so. Jeffreys' statement that there is a 15° misfit is therefore untrue.

*(Carey, 1955b: 199)*

Carey pressed his advantage. He reiterated the consilience that Wegener and du Toit had emphasized between the fit of the continental margins, geological structures, and stratigraphy, giving mobilism a distinct advantage over fixism; fixists attributed these consiliences either to unreliable data which required no solution, or to coincidence which was no solution at all. Mobilists had consilience; fixists had coincidence and plea of unreliable data (RS1 and RS3).

Carey accomplished what he had set out to do. He laid to rest Jeffreys' anomaly difficulty. He showed that it was based on nothing more than a momentary glance at a globe; the geological niceties had been observed, the fit was good, and others could check for themselves. Jeffreys did not change his mind on this question, remaining stubbornly wrong.

### 6.12  Carey's views in the 1950s prior to embracing expansionism: his appeal to mantle convection

*12$^{th}$ October, 1971*

The Secretary,
American Geophysical Union,
2100 Pennsylvania Avenue,
N. W. Washington, D. C. 20037 U.S.A.

*Dear Sir,*

*Twenty years ago at the time when the gross separation of the continents was a heresy to be ridiculed, I prepared a note to answer one of Jeffreys' criticisms of continental drift and submitted it to your predecessor for publication. It was rejected with a note from a referee stating that it was naïve and unsuitable for publication ...*

*This note contained perhaps the earliest viable exposition of subduction, which is now the accepted dogma, particularly in America. You will also find in my continental drift symposium (published in 1957, although taught by me for many years previously), the first viable exposition of ocean floor growth at the mid-ocean ridges. This is reproduced on page 179 and figure 6 of Search, vol. 1, enclosed.*

*A visiting American geologist recently suggested that I should now send back to you the 1953 note with the suggestion that you might choose to publish it now as an historical document. I therefore enclose a copy of the original manuscript.*

*Although I worked with subduction models for more years than any of the new generation of subducers has yet done, I have since moved on to what I think are more probable models. American thinking has now arrived pretty much at where I was twenty years ago.*

*Yours sincerely,*
*S. Warren Carey Professor of Geology*
*(Carey, 1976: 10. Carey's italics. The reference to* Search *is to Carey (1970: 181). The Proceedings of Carey's continental drift symposium were actually published in 1958.)*

Carey wrote this letter in 1971 to *JGR*, suggesting that the editors might now want to publish his paper (originally submitted in 1953) and make good its former error. *JGR* declined and Carey's 18-year-old manuscript remained unpublished.[7] The text of his submission is lost. Apparently in it, he, like Holmes over two decades earlier, proposed mantle convection as the cause of continental drift. Carey's letter is interesting because he retrospectively tells, at least in part, why he wrote the paper, something about his pre-expansionist ideas on the kinematics of mobilism and how they compare with modern views about seafloor generation and subduction.

Using Carey's letter, other retrospective comments by him, a diagram from his original 1953 paper he later found and published (1976: 11), and his account of seafloor generation at oceanic ridges in the 1958 proceedings of the 1956 Hobart symposium, I shall attempt to reconstruct his convection solution to the mechanism problem, and place it within the mobilist view he held in the mid-1950s (1954, 1955a, 1955b) prior to his conversion to expansion. However, because his manuscript is lost, it is particularly difficult to separate his early ideas about the generation of seafloor from his later adoption of Earth expansion.

Carey had twice challenged Jeffreys; the fit across the South Atlantic was not poor as Jeffreys contended but was almost perfect. Mindful of developments in rheology, Carey had, evidently with help from Paterson, proposed a notion of mantle creep that might give mobilists the sort of Earth they needed to explain flow of long duration; flow that he contended could be observed under the microscope, and at local, regional, and global scales. This was his answer to Jeffreys' unyielding Earth. Moreover, Carey's mantle in his 1954 rheidity paper allowed for large-scale convection, which could even extend through the asthenosphere into the lithosphere. It seems that in 1953 he felt he had laid the groundwork for enabling convection to become the long sought after mechanism of continental drift.

Carey accompanied his invocation of convection with an account of the destruction of oceanic crust. In addition to his 1971 letter to *JGR*, there is the figure from the lost 1953 manuscript that he later found and reproduced in Carey (1976). There also are retrospective comments, including the one below, written twenty-five years later.

The tectonic confirmation of Wegener's continental dispersion acerbated its complementary difficulty: – If vast new oceans developed by widening rifts between continents, equal area had to vanish elsewhere. By the early fifties I prepared a paper which developed the transport of continental blocks on the back of convection cells, which advanced against the ocean, became the subcrust over the downgoing limb of the convection cell tended to be drawn into the downflow, as a highly viscous fluid must. This continually sapped the foundations of the more brittle crust above, which therefore progressively subsided into the sink. In this way oceanic crust was steadily excised and swallowed, so that the sialic crustal block advanced against the ocean floor, notwithstanding equal or perhaps greater strength in the oceanic crust.

*(Carey, 1976: 9)*

His resurrected figure (Figure 6.15) shows the formation of a rift basin off the eastern coast of Asia. I assume, in keeping with Carey's remarks, that his explanation could be applied with appropriate changes to the formation of the Red Sea, Atlantic, or Indian basins.

He believed that in his rheid paper (1954) he had made mantle convection plausible. Because viscosity and subsequent lowering of rheidity decreases exponentially with temperature, mantle convection currents can, he argued, rise upward to just below continents, where they increase the temperature of the upper mantle, asthenosphere, and subcrust. As the rising currents split, the crust is subject to tension, fractures, and eventually splits apart. New oceans form in the wake of drifting continents. Once currents move horizontally just below continents, there is sufficient viscous drag to propel them too. When material cools, it falls back into the mantle. Carey (1954) accounted for pronounced negative gravity anomalies over ocean trenches by appealing to descending convection currents. Figure 6.15 indicates that he thought the same in 1953. He had to find a way to get rid of oceanic crust in front of drifting continents. Like Holmes, Carey claimed that oceanic crust descended into the mantle at ocean trenches, but the likeness stops there. Holmes postulated intense *compression* at trenches, which transforms basaltic oceanic crust into denser eclogite, which sinks, merging into the descending currents. But Carey rejected compression. He pictured ocean trenches as regions of *tension*, where oceanic crust sinks not because of increased density but by a removal of its subcrustal foundation. This "sapping" of ocean crust occurs because the rheid subcrust gets caught up in the descending currents, putting the oceanic crust under great tension; the crust, being brittle, collapses into the mantle below. The gap created provides room for advancing continents. Although Carey owed a large debt to Holmes, their accounts differed in these key ways.

I have said enough about Carey's novel geotectonic ideas prior to his sudden conversion to Earth expansion, about his attempts to liberate tectonics from the straightjacket of an unyielding Earth, about his fit between South America and Africa, and about his appeal to convection and to the destruction and creation of oceanic crust.

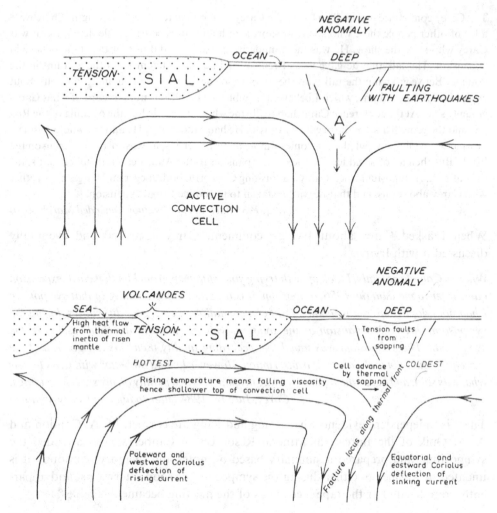

Figure 6.15 Figure 2 from Carey's now lost 1953 manuscript, reproduced with the following caption as Figure 5 in Carey (1976: 11). Migration of continents through sapping and consumption of oceanic crust.

## 6.13 Carey switches to expansionism

Carey didn't, in my recollection, present his expanding earth theory at the meeting ...

*(Lester King, December 29, 1981 letter to author)*

There is no question that Carey was an expansionist when he *wrote* his contribution to the proceedings of the March 1956 Hobart symposium. The proceedings were not published until 1958. When did he adopt expansion? Carey claimed it was during the symposium. King, Irving, and others, who were also there, believe that Carey did not discuss expansion at the meeting.

The Carey conference was really a forum for Carey to annunciate his ideas at length. There were a lot of other people there who spoke at short length and some considerable detail, but it was Carey who stole the show. He was the dominating person. He did not, at that time believe in expansion. The talk that he gave is, as I remember it, quite different from the account in the volume. Between giving the talk and the account in the volume, he convinced himself about expansion. So the volume, which I believe was published in 1958, is not a record of what Carey actually said. At the conference Carey had really the whole thing. He had the opening of the Red Sea and the growth lines in the ocean, and so on. He had subduction. He had the whole business of what we would now call plate tectonics – more at his fingertips, he was sort of teetering on the brink. But then all of a sudden, he took the expansion route. And, I can remember that some time after the symposium when Carey was visiting Canberra, both Jaeger and I arguing together with Carey about this, and that he was mistaken to introduce global expansion.

*(Irving, 1981 interview with author; amended March 2000)*

When I asked Carey about Irving's comment, Carey disagreed and eventually discussed it with Irving.

*When in Canada recently, I took up with Irving your statement from him that earth expansion came later to me than the 1956 symposium. It is true that at the opening of that symposium I had not adopted the radius change. But contemplation of emergent difficulties during the symposium threw up expansion as the solution, which I stated in the final session of the symposium. The symposium was taped, but recording quality then was not good, and the secretary could not make a useful transcription. However I could do better with it, as I knew what was said, and I used the tapes in writing my paper, based mainly on my speech headings.*

*(Carey, July 18, 1984 letter to author; Carey's italics)*

There is independent evidence that Irving and King are correct. M. A. Condon and A. A. Öpik of the Bureau of Mineral Resources in Canberra both attended the symposium, and prepared a summary based on notes and memory. Although it is undated, it was begun either during the symposium or soon afterwards, and apparently completed after the tape recordings of the meeting became available.[8]

A brief summary of the papers and discussion is presented. Discussions during the meeting were tape-recorded; the records will be distributed and contributors will be able to produce a concise summary. The following notes are preliminary, composed from brief notes and memory, and will be completed later when the tape-recorded material is available.

*(Condon and Öpik, 1956, "Summary," p. 1)*

Their summary of Carey's talk made no mention of Earth expansion. Nor did they mention Earth expansion in their summary of the discussions. To the contrary, they stated that he appealed to convection as the mechanism for continental drift.

Professor Carey replied to criticisms already published and presented at the symposium ... It has been objected that the sima is too strong relative to the sial to allow passage of continental matter: under similar stress conditions the sial would be expected to fail before the sima. Carey stated that the likely mechanism – convection in the crust and mantle – is most likely to develop at the continental margin where rates of heat loss are most different. The surface layer of the sima and the whole of the sial would together tend to move and as it moved the convection current would move with it to continue the movement with it.

*(Condon and Öpik, 1956, "Summary," p. 10)*

Finally, I shall show (§6.16) that Longwell was the only contributor to the published proceedings of the Hobart symposium other than Carey himself who mentioned expansion, but he did so *only* in his final summary, *after* reading the papers as prepared for publication. In addition, Condon and Öpik made no mention of expansion in their account of Longwell's summary *as he presented it at the symposium.* I find it hard to believe that King, Irving, Condon, and Öpik would all have missed such a startling idea as expansion as presented by Carey; it was not in his character to mention the idea softly in passing.[9]

Regardless of when he switched to expansion, why did he so suddenly do so? He explained that in his original reconstruction of Pangea there were some troublesome residual misfits or gaps, but these disappeared when he reassembled continents on a smaller globe; the fit of his reconstructions was improved. And so he began taking expansion seriously.[10]

Carey (1958: 346) postulated an accelerating rate of expansion; since the Paleozoic he proposed the rate of increase in the Earth's radius had averaged 0.5 cm per year, which was ten times faster than Egyed's non-accelerating rate (§6.2–§6.4).

A sialic crust covered the young Earth whose diameter was less than half, its surface area less than a quarter, and its density eight times greater than the present values. He proposed an increasing rate of expansion fairly late in the Pre-Cambrian, which continued through the Paleozoic, but it was the later breakup of Pangea whose expansion became very rapid that received most of his attention because it allowed him to bring into play his orocline concept. It was also Pangea's fragmentation and his proposed rapid expansion that later caught the attention of paleomagnetists (§8.14).

Convection no longer played the central role it had in Carey's pre-expansionist account in which he needed mantle convection to get rid of oceanic crust in front of advancing continents. With a growing Earth, Carey had all he wanted, an increase in Earth's surface area to explain the increasing separation of continents.

Unlike convection, the generation of seafloor by dilatation played a central role in Carey's expansion account. Expansion caused tensional fractures into which material from Earth's mantle and sediments flowed. Rift valleys widened to become rift seas and eventually rift oceans. Expansion, dilatation, and variable inward flow became Carey's ways of creating not only oceanic crust, rift valleys, seas, and oceans, but also orogenic belts and mountains. He argued that the final product depended on the amount and rate of inward sedimentation versus the amount and rate of dilatation. If sedimentation was meager, an oceanic rift developed. If sedimentation was abundant, an orogenic belt developed; thus he proposed a theory of mountain building without compression; rheid folding in orogens, and upflow of the less dense and hotter mixture of sediment and rheid material from the mantle caused the uplift of mountain belts. Carey (1958: 331) closed his account of dilatational orogens with evident glee; he did not need compression to form mountains, even for the Alps, the hallowed ground of compressionism and contractionism.

I have now reached the very citadel of the compressional dogma – hundreds of kilometers of crustal shortening demanded in the Alps. My claim is that careful argument is necessary for the granting of even one metre of rapprochement of the jaws of the Alpine Geosyncline. Even the Jura with its decollement need be no more than the forward push of the central zone of the regurgitated Alps! ... not a march of Africa of 500 or 1000 km on to Europe during the Alpine revolution, but a dilatation of 700 km!

On his expansion theory, he did not have to appeal to contractionism as he had in 1955 in order to explain the Pyrenees and Himalayas as resulting from the rotation of Iberia and India and the concomitant openings of the Bay of Biscay and Arabian Sea; growing rift oceans could rotate continental blocks toward or away from each other.

Carey recognized that finding the cause of Earth expansion, especially at an accelerating rate, was a serious external difficulty, because any suggestion was almost bound to contradict the laws of physics and theories then current about the origin of Earth and the solar system. Carey thought that only a phase change deep in Earth's interior could provide the forty-fold increase in volume he required. He speculated that common elements might form heavy isotopes under great pressure and very low temperatures, much as Egyed had proposed before he began to favor Dirac's decrease in $G$ over time (§6.4). Instead of Egyed's hot Earth under great pressure, Carey appealed to a dense, cold Earth.

Carey did not care if his expansion created a severe external difficulty. He favored rapid expansion, and would do so unless faced with an ultimate absurdity.

As a geologist my duty is therefore to follow fearlessly where the orocline interpretation leads me. It might lead ultimately to an absurdity whereupon I would abandon it, but it has not done so yet. On the contrary it promotes me to ask the physicist to seek an earth model, which will expand at an increasing rate with time.

*(Carey, 1958: 349)*

Was Jeffreys listening? If so, as Holmes noted to Egyed (§6.4), he never bothered to respond; in fact by his own admission he was not even reading Carey (III, §1.18). Carey never encountered an absurdity that carried sufficient weight with him to abandon expansion. Nowadays, many accept oroclines but few, if any, believe that unwinding oroclines requires expansion. His orocline concept has had a lasting influence on tectonics.

## 6.14 Carey's account of seafloor generation after he embraced Earth expansion

... it is essential to my thesis to postulate that dilatational rifts may develop in either continental or oceanic crust, and that dilatational tectonic ascent of crystalline sima may occur along such rifts and result in new simatic crust separating the former rift margins. According to this view, the Great Rift Valleys of Africa, the Red Sea, and the Atlantic Ocean are similar phenomena, but differ in degree of dilatation.

*(Carey, 1958: 179)*

In the publication of the 1956 Hobart symposium, Carey proposed that the creation of new oceanic crust begins with the formation of a dilatation crack in continental or oceanic crust (Figure 6.16a). It is vertical and normal to the tension. He did not identify the cause of the tension. Shear failure occurs at the depth where overburden pressure exceeds shear strength. One or two shear fractures develop at an angle of 45° plus half the angle of internal friction (Figure 6.16b). Creep relaxes stresses and does so increasingly with temperature, so flow increases with depth. As depth increases, brittle shears become ductile shears, and ductile shears are transformed into laminar flow without fracture (Figures 6.16b, c). The depth where this transformation occurs is the lower limit of shallow focus earthquakes. As widening continues, secondary fracturing of crust occurs to prevent the formation of voids along fracture surfaces; the situation of Figure 6.16d is isostatically unstable because of the resulting mass deficiency, and the area YY rises (Figure 6.16e).

This produces a typical rift valley with raised but actively rising rim, and a negative gravity anomaly due to still incomplete adjustment (Fig. 3e [my 6.18e]). The floor may be sinking or rising according as the subsidence caused by continuing dilatation is currently more or less rapid than the isostatic rise.

*(Carey, 1958: 185; my bracketed addition)*

So far Carey has taken us from a shallow tensional fracture to the formation of a rift valley. To this point the only newly created oceanic crust is at Y between the secondary fractures in the sialic crust (Figure 6.16e). He explained how the imaginary rift valley evolves into a rift sea, and a rift ocean.

It has already been pointed out that isostasy causes the line XX' of Fig. 3d [my 6.16d] to bulge upwards to XX' of Fig. 3e [my 6.16e]. This means that level for level the material at YY between and below the faults is warmer than the otherwise similar material outside and above the faults. Hence if stretching continues, the relief by flow extends inwards *below* the original faults and faults such as PP and P'P' extend upwards until they meet, while the material above them is drawn away, allowing their junction to rise isostatically to reach the surface. Thus the rift widens by flow in depth with repeated slices developing below the older faults. This process can go on indefinitely to produce an ever widening ocean (Fig. 3f [my 6.16f]). At all stages while this process continues, the marginal coasts may long since have ceased activity, but there will be an active median line of faulting and seismicity following a ridge composed of two tilted rims with an intervening narrow trough on the site of the latest extension. This trough has a

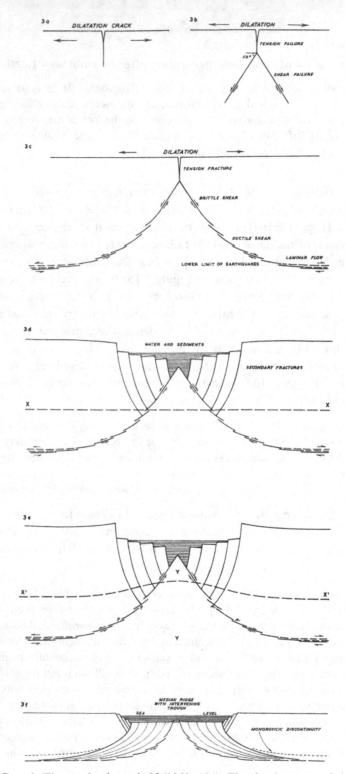

Figure 6.16 Carey's Figures 3a through 3f (1958: 184). The development of rift valleys and rift oceans.

mass deficiency and will be in process of rising isostatically. The raised rims will be due to the fact that regional isostatic adjustment is reached quite rapidly, but more local isostatic adjustment follows much more slowly and may never be fully attained. As local adjustment becomes more complete, the raised rim subsides. In any case it would be eliminated whenever progressive dilatation caused a further fracture to appear (such as PP in Fig. 3e [my 6.16e]) since the former raised surface would then be in the area of slump settling (as in Fig. 3c [my 6.16c]). Hence there will at any time be only one pair of raised rims even in a wide rift.

*(Carey, 1958: 186–187; my bracketed additions)*

In this way Carey explained not only rifts, rift seas, and rift oceans, but also the origin of ocean ridges and creation of new seafloor within the narrow rift valley. Raised rims border the trough. The rims continue to rise isostatically and subsequently subside. So there is a broad ridge with a central rift valley. As long as dilatation continues, new seafloor flows into the rift valley. The ocean basin continues to widen, but the ridge itself does not widen. New seafloor flows into the central valley between slices of older seafloor, which migrates away from the rift valley, making way for new seafloor. Along the upper flanks of the ridge, slices of seafloor migrate to the lower flanks and away from the ridge axis. As I understand him when he wrote his 1958 paper, Carey believed that seafloor is composed of successive narrow crustal strips arranged parallel to the ridge axis, which get older with distance from the axis.

For anyone familiar with the Vine–Matthews hypothesis (IV, §2.13), it is hard to encounter Carey's account of seafloor generation without exclaiming, "Vine and Matthews could have just as easily begun with Carey or Holmes instead of Hess." In Holmes' account, as I understand it, there are no clearly defined slices of oceanic crust running parallel to ridge axes of ever-increasing age outwards, although they can be read into his account; there is room for them, but Holmes does not draw them out. For Carey, at this stage in the development of his ideas when writing his 1958 paper, the formation of growth strips of seafloor running parallel to ridge axes and increasing in age with distance from the axis is central. I have already compared Holmes and Hess (I, §5.6). The similarity between Carey's and Hess's accounts, I believe, is greater with regard to formation of new blocks of seafloor, even though Carey's account as it appeared in the 1958 Hobart volume, unlike that of Holmes and Hess, was embedded with Earth expansion instead of mantle convection.

Carey did not at the time stress the age–distance relation and striped nature of the seafloor for he had no pressing reason to do so. However, he did argue that his hypothesis not only explained the seismicity associated with oceanic ridges but also their newly discovered central rift valleys and the mid-oceanic position of ridges that had formed as a consequence of the splitting apart of continental blocks (Carey, 1958: 187).

Carey believed that he had greatly influenced Hess. When I sent him my paper (Frankel, 1981) in which I argued, as did Hess, that his own hypothesis of seafloor spreading offered mobilists a solution to the mechanism problem, Carey vociferously claimed that I had exaggerated the importance of Hess's contribution, and informed me of his own role both in solving the mechanism problem and in converting Hess to continental drift and seafloor spreading.

*Harry Hess certainly warrants an eulogy. He was one of the "greats" of our generation, exceeded perhaps by R. A. Daly and Arthur Holmes. Arthur and Harry were two of my closest personal friends. But your account is extra-ordinarily biased and inaccurate.*

*On the first page you write of "the most serious problem faced by drifters, namely how the continents could plow their way through the sea-floor and remain intact ... [Hess] offered a solution to the engineering problem by providing an adequate mechanism for the drift of continents." RUBBISH! Read again p. 179 of my continental drift symposium (copy attached) and Figure 3, which gives the mechanism of ocean floor spreading by continued insertion of successive strips of new sea floor. I explained this in lectures in Princeton in 1959 and 1960, and had long discussions with Harry while a guest at his home on several occasions, and during these converted Harry to continental drift and ocean-floor spreading ... Incidentally, I had already dealt with this specific question in my paper submitted to the American Geophysical Union (see pages 9–11 of my book [Carey 1976]).*
    *(Carey, 1981 letter to author; my bracketed addition; Carey's italics. Figure 3 is my*
    *Figure 6.16, and the appropriate passage from Carey (1958: 179) has been discussed above.)*

Carey also publicly noted this similarity, and the influence he may have had on Hess.

Meanwhile in 1959 I presented these concepts and illustrations [that had been in my rejected 1953 manuscript I sent off to the American Geophysical Union] in lectures in Yale, Columbia, Princeton, and a score of universities through North America – so perhaps some seeds fall on fertile ground. My 1953 naiveté became the 1962 approbation of my friend Harry Hess as "a more acceptable mechanism whereby continents ride passively on convecting mantle instead of having to plow through ocean crust."
    *(Carey, 1976: 9; the quoted passage is from Hess, 1962: 599)*

Although Hess was a mobilist before Carey visited Princeton in 1959 (III, §3.12) and had rejected Earth expansion (III, §3.15), Carey's hypothesis of seafloor generation may have served as a catalyst in getting Hess to seafloor spreading, something that he could disentangle from Earth expansion. Perhaps Carey simply served as a catalyst for Hess's rethinking about the origin of ridges (III, §3.12).

    Carey also provided a detailed account of the formation of the Red Sea. Heezen shortly thereafter did the same (III, §6.9), and their accounts have many similarities. Moreover, Egyed already had independently suggested that new ocean basins form from rifts between continents, and had noted Heezen and Tharp's recognition of a central rift valley along the axis of ocean ridges with their associated high heat flow and seismicity (III, §6.7).

### 6.15  Carey's appeal to paleomagnetism

Carey may have found out about paleomagnetism and its possible use in testing mobilism from Blackett, who visited Australia in 1953. Ronald Green recalled:

Blackett visited Australia in 1953 and lectured on the construction of an astatic magnetometer and how it could be used for measuring the magnetization of rocks. Carey persuaded Blackett to work on a bore core that passed through the Jurassic dolerites from Tasmania. The bore core showed nearly vertical magnetization and there was one reversed specimen. At the time there was some doubt whether this represented a true reversal or merely a mistake in indicating the upward direction of the bore core. Jaeger subsequently did more work and showed that it was a true reversal. Irving thoroughly sampled the Tasmanian dolerites and showed that they were magnetized almost vertically (85°). Carey, of course, regarded this result that during the Jurassic Tasmania was near the pole.

*(Green, February 5, 1979 letter to author)*

Green, who did not go to the University of Tasmania (on staff in the Geology Department) until at least five years after these events, probably heard this story retrospectively from Carey. Carey also talked retrospectively about the bore cores, and he even thought that work on them led to Jaeger's inviting Irving to ANU.

At this time [i.e., the early 1950s], a vertical, 300m bore into the Jurassic dolerite on the eastern shore of Great Lake had just been completed by the Tasmanian Hydro-Electric Commission, so I air-freighted to Blackett cores 30m apart, and predicted that they should prove to have been relatively near the pole of the time. In due course Blackett cabled that except for the surface sample, the magnetisation of each core was close to its axis, confirming the tectonic prediction. This led, through the co-operation of Professors Jaeger and Runcorn, to the invitation to E. Irving, then a graduate student at Cambridge, to take up a research fellowship at the Australian National University, where he commenced a systematic study of the remanent magnetisation of the Tasmanian Jurassic dolerite. This resulted in Irving's paper (1956b) which established a Tasmanian Jurassic palaeolatitude of 80° S.

*(Carey, 1976: 181; my bracketed addition)*

Although Carey may have helped arrange sending the cores to Blackett's laboratory, Carey's and Green's versions of what happened are misleading. Jaeger was the key person, not Carey. There are no firmly established reversals in the dolerites themselves. There is a reversal, in a Triassic aged volcanic tuff adjacent to a sill. Although reversed magnetization was found in a section of one of the cores, the possibility of it being caused by incorrect labeling was raised from the beginning. At the time, Irving wrote of the dolerites as Mesozoic because they were then only known to be Jurassic or Cretaceous (Irving, 1956b).

Blackett actually saw the cores while he was in Australia in 1953. Jaeger and Germaine Joplin, a petrologist at ANU, were the first to work on them magnetically and found near-vertical magnetizations. Because the cores had not been oriented azimuthally, no information could be obtained about the ancient declination of the field. Jaeger and Joplin also identified a possible reversed section.

Blackett and Clegg found a number of cases of reversed magnetization in samples selected from core No. 5001. As remarked earlier, it has been shown that these reversals were not caused by stray magnetic fields either in the laboratory or in transit to England. There remains the possibility of mismarking, either when the core was drawn or at some subsequent time. The practice in Tasmania is that all cores are put in boxes immediately on drawing and that red arrows pointing downwards are then marked on the core. Only samples which have an actual arrowhead on them have been used. Even with the greatest care, there is the possibility that one or two pieces may be dropped in transferring from the core barrel to the box; in this case these pieces might be inserted in the wrong place in the box and mismarked in direction.

*(Jaeger and Joplin, 1955: 18)*

Carey is not mentioned. The Hydroelectric Commission of Tasmania supplied the core, Carey and Jaeger were consultants.

Almond, Clegg, and Jaeger stated in the next paper on the cores that the Hydroelectric Commission supplied Jaeger with them, adding (1956: 771–772), "A number of samples from different sill complexes were taken from bore holes and were shipped to London for measurement at various times." Again, there is no mention of Carey.[11] Almond and her co-authors concluded that the very high inclinations found in the Tasmanian samples support apparent polar wandering relative to Tasmania since the Mesozoic period. They also discussed the apparent reversal in the sills, but remained uncommitted as to whether it was real or due to mislabeling. They also found another reversed magnetization. One core did pass through a dolerite sill and penetrated a Triassic volcanic tuff, which was reversely magnetized, and which the authors (1956: 776) argued was a legitimate reversal. Likely it is this Triassic reversal that Green had in mind.

Carey is wrong about Irving. There were no cooperative efforts between Jaeger and Runcorn to hire or arrange for Irving to apply for the position. As explained already, Irving, after seeing an advertisement in the *Cambridge Reporter* (§3.4), applied for the position for which Runcorn did not give him a recommendation; his recommendations were from Blackett and Fisher. The advertisement was for research fellowships at ANU generally and did not specify paleomagnetism. Moreover, Carey's comment intimates that he was the one who was behind ANU's beginning paleomagnetism. This is to Jaeger's credit. Once again, Carey overstated his influence.

However, Carey certainly was eager for paleomagnetic data. He helped Irving locate Tasmanian dolerites (§6.5), and encouraged Blackett to test the rotation of Iberia. Carey, recognizing the global improvement of getting paleomagnetic results from the vast Mesozoic mafic complexes (now called Large Igneous Provinces) associated with the partition of Pangea, and without mentioning it to Irving, arranged for samples of the Serra Geral Volcanics from southern Brazil to be sent to Irving at ANU. Maack, who later helped Creer collect samples there (§5.6), sent them. Carey also got King and a mining company to send samples of Karroo dolerite from Natal from mines in Transvaal, South Africa; M. R. Banks, a member of Carey's department on sabbatical in the United States, sent samples from the Upper Triassic lavas of Massachusetts (Du Bois *et al.*, 1957; Irving and Banks, 1961). However, as far as Irving was concerned, Carey was a little too eager for paleomagnetic data.

Now when I arrived in Australia, Carey was enormously helpful to me. He spent a lot of time explaining to me the geology of Tasmania, and he helped me a great deal. But, he was a very domineering person, and he rather expected that I would sort of provide the data and he would then sort of use it to prove his ideas. And, we did get into a little fracas about this because there was really no way in which I was going to let this sort of thing happen. I was certainly going to get my own information and have a crack at it, and after it was described and published, well then that was the time for other people to look at it. But, if I was going to spend a lot of time and effort getting data, then I had the right of having the first go at interpreting it. And that led to some sort of separation between Carey and me, but it was one that was readily healed, we kept discussing things over the years and never let this matter sour our relationship.

*(Irving, 1981 interview with author)*

One way to get to see Irving's data quickly was to invite him to give a paper at the symposium on continental drift (see following section). Irving agreed, used the meeting to compare various mobilist reconstructions to see which one best fit the available paleomagnetic data, and cautiously concluded that Carey's reconstruction was the best fit (§5.16). Carey, however, did not just welcome the paleomagnetic data; he embraced it, arguing that it removed any doubts about his method. He used the data to defend his reconstructions and the validity of his orocline concept (RS1), for example, for the Alaskan orocline.

The locus of palaeomagnetic poles of Britain and North America which have been measured wholly independently of this hypothesis, and have no theoretical connections with it, shows an angular rotation and offset which agrees in magnitude and sense with the rotation demanded by the Alaskan Orocline. No other hypothesis yet submitted reduces this anomaly to zero within the statistical limits.

*(Carey, 1958: 215)*

Carey did more. When he came to write his paper for the Hobart proceedings, he had already developed a reconstruction somewhat different from that presented at the meeting; recall that the latter (the one Irving tested (§6.16)) was based on an Earth of present size. Carey tested his new Pangea and had more data; he likely had Irving (1957a), Irving and Green (1957a, b), and possibly Green and Irving (1958).[12] Carey (1958: 281) compared the range of poles from four periods with continents fixed and reconstructed. His range was the diameter of a small circle containing all poles for each period. For the Carboniferous this reduced from 69.5° to 35°, for the Permian from 86° to 45°, for the Triassic from 61° to 21°, and for the Jurassic, from 60° to 13.5°. The fit for the Carboniferous and Permian certainly was not tighter. Carey accented the fit for the Jurassic, arguing that it was the best one could expect.

This [scatter of 13.5°] is the same order of magnitude as the scatter of the poles found for any one period from rocks from a single continent. We should not expect a smaller scatter for Pangea under the most favourable circumstances.

*(Carey, 1958: 281; my bracketed addition)*

He claimed that the paleomagnetic results vindicated his orocline-based reconstruction of Pangea.

For the Jurassic, which represents approximately the date for which the assembly submitted might be expected to be valid, the 75% reduction of scatter is very strong support for the general validity of the procedure. The residual must include the inherent scatter due to any short-term oscillation of the magnetic pole not neutralised by the sampling procedure (the Quaternary remanent magnetism leaves no doubt that such fluctuations are real). It also includes any experimental errors (the measurements are inherently subject to errors of a few degrees), and any departure from theory, either in relation to the geocentric dipole and its perturbations, or in its geometrical relations to the earth as we know it at present (this includes any changes in the earth's diameter). In addition to all of these, the error includes any misinterpretations of the deformations which have been used in the assembly, either qualitative or quantitative. I do not pretend that the assembly I have given is final. Even fairly substantial misinterpretations might be obscured by the bluntness of the palaeomagnetic check. But that the overall guiding principles are valid I no longer have doubt.

*(Carey, 1958: 281–282)*

Carey was much more certain than Irving. Carey (1958: 282) added, "The orocline technique is, as Irving has shown, the only hypothesis which so far has achieved zero error in this test." But Irving had also pointed out outstanding misfits with Carey's hypothesis.

In figure 13 [which showed the reconstruction Carey presented at the meeting from the Northern Hemisphere] these poles are compared on Carey's reconstruction. Again there is a discrepancy between the Tasmanian dolerites and the Drakensberg Basalts suggesting that Africa and Australia are wrongly placed relative to each other [see §5.7 and note where it is shown to be an age problem]. The Deccan Trap pole is now well to the westward. Figure 13 also includes the poles given by the European and North American data as in figure 10b [which showed the reconstruction Carey presented at the meeting from the Southern Hemisphere]. The Tasmanian dolerite pole is now only 5° from the Triassic poles of the North Hemisphere; the separation on the present continental distribution ... is 50° separation. This gives support to the relative positions of Australia, North America and Europe postulated by Carey. However, there is 50° separation between the Cretaceous poles of North America and the approximately contemporaneous Deccan Traps, and the palaeomagnetic evidence is not consistent with the relative positions, assigned by Carey to North America and India.

*(Irving, 1958a: 50; my bracketed additions)*

Carey addressed none of these concerns; he was working with his revised reconstruction and had more data. But even with the new reconstruction there was no substantial difference in the placement of India relative to North America and no new relevant data from India. The new data were from Australia, and Carey changed the position of Australia relative to North America. But Irving did not raise an anomaly difficulty with the previous fit between North America and Australia; he, as shown (§5.16), expressed the need for caution. I doubt that Irving would have changed his mind if he had tested Carey's revised reconstruction. Irving, for example, was bothered by the poorness of fit for the Permian and Carboniferous, which was much less good than that for the Jurassic, as Carey had shown.[13] Yet Carey believed that paleomagnetism had removed any doubts about his method; in view of his expression of caution, Irving surely would have disagreed. Moreover, there is the additional important but obvious point. Even if Carey's reconstruction of Pangea fit very well with the paleomagnetic data from the Jurassic, it did not mean that Earth expansion gained support because he could not, as he maintained, achieve a satisfying reconstruction when he circumnavigated his globe, unbending his oroclines as he went. Those who refused to follow him to his expanding Earth could agree with him about the good fit between Africa and South America, and the rotation of Iberia and India. They could reject some of his oroclines, and still remain fixed-radius mobilists. In fact, in years to come this is precisely what happened. Many of his proposed oroclinal rotations were corroborated by paleomagnetic tests. His fit between South America and Africa, especially in its later manifestation by Bullard, Everitt, and Smith (IV, §3.4), was adopted as one of the cornerstones of the new global tectonics by old and newly declared mobilists who, however, rejected Earth expansion.

Carey was ingenious and enthusiastic. He developed a highly original view, integrating results from tectonics, structural geology, rheology, marine geology, and paleomagnetism. He gave tectonicists the prospect of an Earth that would flow if subjected to forces of the duration needed to move continents, make ocean basins, and build mountains. He offered ideas that pushed to the limits what was known in the 1950s. A decade prior to the development of plate tectonics, he provided a mobilistic synthesis with premonitions of some aspects of plate tectonics. But, on the negative side, he seemed incapable of self-doubt, and was unable to change his mind when difficulties accumulated against expansion, difficulties that left plate tectonics unscathed.

## 6.16  Other contributions to the 1956 Hobart symposium

When organizing that [1956] symposium, I decided to go right to the heart of the opposition to drift, which was Yale, led by Schuchert. Schuchert was dead, so I invited the Yale Chairman, Longwell, to come as principal guest.

*(Carey, July 18, 1984 letter to author; my bracketed addition)*

Among the symposium participants were Professor Lester King, of Durban, who had assumed the mantle of du Toit, Dr. J. W. Evans, who had long been interested in continental drift from the view point of the distribution of insects, Dr. Reinhard Maack, of Curitiba, the leading authority on the regional geology of Brazil, Professor Kenneth E. Castor, of Cincinnati, who had spent many years studying the Paleozoic stratigraphy of South America and Africa, Edward Irving, who was already establishing leadership in paleomagnetism, Professor J. C. Jaeger, who had done so much to stimulate investigation of new fields in geophysics, Dr. Rudolf O. Brunnschweiler, who had become a mobilist under Staub of Zurich, Dr. James

M. Dickins and Dr. G. A. Thomas, who had compared the Permian faunas of India and northwest Australia, Professor Alan H. Voisey, originally sympathetic to continental drift, but who had absorbed the American skepticism during his teaching assignments there, and Dr. Armin A. Opik, the Estonian world authority on early Paleozoic trilobites, who was opposed to continental drift.

*(Carey, 1988: 117)*

Of participants mentioned by Carey, all but Castor, Jaeger, and Öpik had either their paper or an abstract of their talk included in the proceedings. Of the symposiasts who published their contributions, four (Brunnschweiler, Carey, Irving, and King) favored mobilism, two (Voisey and Stirton) supported fixism, one (Evans) preferred former land connections between Australia and Antarctica but left unanswered whether or not mobilism supplied the connections, and four (Longwell, Gill, Dickens, and Thomas) tended to be neutral. Condon and Öpik's (1956) summary included remarks by Castor, which indicated that he favored mobilism (I, §6.2), by Jaeger who was sympathetic to paleomagnetism's support of mobilism (see next section), and by Öpik (I, §9.3) who firmly rejected mobilism.

Longwell gave the opening address and also a final summary. In his opening address, he had described his former views about the mobilism controversy, and examined the recent paleomagnetic evidence and Carey's fit of South America and Africa. In his final summary written after reading the other written contributions, he restricted his attention to those of Irving and Carey, the two he thought particularly important. Because Longwell was a longtime and relatively restrained player in the controversy and had something to say about both paleomagnetism and Carey, his contributions describe well the attitude of an interested worker who, in the mid-1950s, had no strong commitment to either fixism or mobilism (I, §3.7; I, §6.8). Before discussing Longwell's comments, I shall remark first on other published contributions, summarizing their attitude toward mobilism, and what, if anything, they said about Carey's views and about paleomagnetism. Besides Irving only Longwell, Carey, and Voisey discussed the paleomagnetic support for mobilism.

Although Brunnschweiler, King, and Irving supported mobilism, nobody said anything about Carey's expansion, as would be expected, had he not discussed it at the meeting. However, Brunnschweiler, King, and Irving referred to Carey's reconstruction or offered alternative reconstructions. Irving, as discussed above, argued that Carey's Laurasia was in broad agreement with the paleomagnetic data, because it fitted them better than King's or du Toit's. King, who probably stood below only Carey in the intensity of his support for mobilism during the 1950s (I, §6.10), gave two presentations. He updated du Toit's reconstruction of the continents, included a slightly different arrangement of Gondwana, and disagreed with Carey over the positioning of continents before their dispersal. In his second paper he argued that the distribution of oceanic ridges supported his reconstruction of Gondwana.

Brunnschweiler, who wrote about Indo-Pacific faunal relations during the Meso-zoic, favored mobilism. But he objected to Carey's version because he believed (as is now known to be so) that continental drift had happened more than once.

Before viewing S. W. Carey's new re-assembly of Pangaea in the light of Mesozoic geographies it should perhaps be stated where and why the writer's opinions on the drift problem differ from those held by Carey. Professor Carey's Pangaea has, I think, little to do with the primordial Pangaea and it resembles only in parts such subsequent Pangaeas as were formed before the worldwide Late Palaeozoic orogeny. In other words, his re-assembly can only be taken as giving the position of the continents and related fragments as they stood when the Late Palaeozoic orogenic cycle had ended, i.e. before the Alpine orogeny had entered its first major phases. It seems to me *a priori* incompatible with the implications of the drift hypothesis that the break-up of a Pangaea should have been a unique event of Mesozoic times ... I cannot comprehend how any of the great orogenies in the history of the Earth's surface can be satisfactorily explained without assuming a major or minor component of lateral movement of crustal fragments, i.e., without continental drift.

*(Brunnschweiler, 1958: 129)*

Brunnschweiler made no reference to Carey's expansion in his attack, which, if he had known of it, he might have recast. Regardless, I shall recast it anyway and in doing so raise a general difficulty with Carey's idea of rapid expansion as the cause of drift. Carey invoked expansion to account for the dispersion of Pangea and bending of certain mountain belts. If continental drift was not a single happening (there were earlier episodes during which other older mountain belts were bent), then Carey would have had to argue for earlier phases of large-scale expansion. But the present area of continental crust fixed the minimum size that the Earth could have been, and its current size set its greatest size. Carey already needed most of the available expansion to account for the dispersion of the continents in his onetime breakup of Pangea and formation of oroclines. Thus, if continental drift was not a unique happening, Carey either would have to offer a different explanation for earlier episodes of continental drift and for earlier oroclines in order to lower his estimate of the amount of expansion needed to account for the rupture and dispersal of Pangea, or he would have to find some way to shrink Earth in earlier times.

The entomologist J. W. Evans, Director of the Australian Museum, Sydney, one of the few Australian scientists sympathetic toward mobilism, argued for land connections among Gondwana fragments entirely on paleobiogeographical grounds. He analyzed the distribution of various modern terrestrial biota in addition to insects, and argued that it requires that Australia, New Zealand, South Africa, Antarctica, South Africa, and Madagascar formerly had land connections. Evans left to geologists the question of what such land connections were and how they had disappeared. He offered no overall reconstruction of his own (see I, §9.4 for further discussion of Evans' view).

Others, like Longwell, tended to be neutral. Edmund D. Gill, a paleontologist from the National Museum of Victoria, Australia, wrote about Australian Early Devonian biota and ended:

It is concluded that it is not possible on present palaeobiological evidence to state whether in Lower Devonian time there was migration of such degree as to explain the faunas in the geographical positions they now occupy, or whether continental drift has taken place, or both.

(Gill, 1958: 115)

He said nothing about Carey's expansion.

Dickens and Thomas, Australian paleontologists, examined Permian fauna associated with glacial deposits, which have a close affinity with faunas in India, South Africa, and South America. They remained neutral.

The faunal relationships are consistent with reconstructions of Gondwana made in accordance with the theory of Continental Drift, especially that of Carey (this volume). But, it is not necessary to postulate drift to explain the relationships, although communication by shallow sea-way would appear to be necessary.

(Dickens and Thomas, 1958: 126–127)

They apparently thought Carey believed his reconstruction was compatible with these faunal distributions.

R. A. Stirton and Alan H. Voisey supported fixism. Stirton had obtained his Ph.D. at the University of California, Berkeley, having studied under W. D. Matthew, the staunch fixist and mentor to G. G. Simpson (I, §3.6). He contributed a two and a half page abstract on Australian monotremes (the duck-billed platypus and echidna) and marsupials, which reads as if G. G. Simpson himself had dictated it. He believed that the ancestors of monotremes came from Southeast Asia, and that they had arrived entangled in the roots of mangroves, drifting to Australia on ocean currents. He rejected landbridges because they would have brought placental mammals too. These opinions, he wrote, have "long been held by many but not all American vertebrate paleontologists." Despite the purpose of the symposium, he said nothing about continental drift, perhaps he thought it unworthy of mention.

Voisey, from the University of New England, Armidale, New South Wales, took on both mobilism and Carey. Voisey was up front about his recent switch to fixism. Sitting on the fence was not an option for him.

Every student of tectonics has to make up his mind whether he is prepared to accept such enormous movements or whether he will regard the continents and ocean basins as being more or less fixed in position. Either he is a "drifter" or he is not. Having changed sides in the last five years the writer will now endeavour to explain why he is no longer content to drift and why he has now decided to take a firm stand with the continental blocks.[14]

(Voisey, 1958: 162)

As perhaps expected from Carey's comment about him losing his way in America, Voisey had come to adopt the idea that continents had grown from original nuclei through accretion of mountains that evolved from marginal geosynclines and island arcs as championed by Marshall Kay (I, §7.3). He also emphasized the importance of compressional forces in the formation of mountains, and noted the tiny role that Carey accorded them. Voisey repeated many of the standard difficulties raised against classical mobilistic arguments, but not for the absence of a satisfactory mechanism, which seemed not to bother him.

Voisey dismissed the continental fits of Wegener, King, and du Toit because they invoked too much distortion. He acknowledged the greater rigor of Carey's fit of Africa and South America but argued that it was less good than he maintained (RS2).

The most impressive argument favouring continental drift, the fit between Brazil and Africa, has recently been emphasised by Carey (1955a). He showed that "the fit" taking the 2000 m. isobath is remarkably good and very much closer than that accepted by many critics of the drift hypothesis. In spite of the case made for the fit one cannot help wondering whether too much is not being made of what is after all only a coincidence of the angles made by the meeting of the coasts in each case. The lengths of the coasts do not coincide. While one feels that the suggestion that this is merely an accidental circumstance is not a fair and full explanation it is questionable whether the fact of "the fit" should of itself force acceptance of all the other principles involved in the hypothesis.

*(Voisey, 1958: 164; Carey (1955a) is my Carey (1955b))*

Voisey claimed that mobilists had overestimated structural similarities between continents and appealed to similarities of process.

The writer just cannot regard as significant the sorts of comparison made by Wegener ... du Toit ... and King between Africa and South America. It is being more and more recognized that geosynclinal belts the world over have analogous histories. Under these circumstances claims of mathematical probability for the connections break down. Facts too are so varied and personal selection from them so differently made.

*(Voisey, 1958: 166)*

Lumping together the paleontological and paleoclimatological arguments of mobilists, he dismissed the former by appeal to isthmian links, and cited trans-Pacific instances of disjunctive biota that mobilism, he believed, left unexplained. However, he said only this about mobilism's solution to the origin of Permo-Carboniferous glaciation.

There seems to be some case for polar movement relative to the continents when the distribution of glacial deposits, red beds and corals are considered but it is questionable whether relative movements between the continents would do much to remove these anomalies.

*(Voisey, 1958: 168)*

Given the strength of mobilism's solution to the glaciation problem in Voisey's home base in northern New South Wales amidst well-exposed Carboniferous glacial phenomena, his argument is puzzling: it was his weakest argument against the classical case for mobilism.

Unlike most others, Voisey did discuss recent developments in paleomagnetism; his cautionary remark, that we should wait for more results before evaluating its support for mobilism, will, as this account proceeds, become a familiar fixist refrain.

It would appear to be too soon to evaluate the contribution made by palaeomagnetic investigations up to the present and we must wait for further results.

*(Voisey, 1958: 168)*

Voisey considered recent work at sea, showing the effect of his American sabbatical. He noted that sediment cores from near Bermuda indicated that the Atlantic Ocean there is at least as old as Early Cretaceous, and Maurice Ewing and others thought that there were still older sediments beneath, which meant that the Atlantic Basin was older than mobilists maintained.

Longwell began his opening address reminiscing about the classical period of the mobilism controversy and his role as a fairly neutral participant in it. He next assessed the current status of the classical evidence for mobilism. His comments reflected the views he had expressed almost thirty years earlier at the AAPG symposium, his involvement in the debate between du Toit and Simpson during the 1940s, and his later trenchant criticism of the geodetic support for mobilism (I, §3.13). He then turned to paleomagnetism and Carey's fit.

His analysis of paleomagnetism reads much like Runcorn's before he switched from polar wandering only to continental drift. Longwell cited three papers, all from the same volume of *Nature*. In the first, Day and Runcorn (1955) summarized the 1955 Cambridge meeting (§3.10) at which Runcorn and Gold argued for polar wandering without drift. In the second, Runcorn (1955b) presented more of his results and concluded:

... it seems possible that the motion of the pole was not smooth: in fact, it may have been rather like a random walk. It is thus impossible to tell at this stage whether the discrepancies between the pole positions inferred from "contemporaneous" strata in Great Britain and Arizona are due to systematic errors or to the strata not being truly contemporaneous.

*(Runcorn, 1955b: 505)*

The third paper, Du Bois (1955), reported results from the Precambrian Keweenawan rocks of Michigan, and also argued in favor of polar wandering alone as an explanation for the paleomagnetic results.

Longwell had read these papers with great care. He summarized Runcorn's and Du Bois' results, noted that they agreed as to interpretation, and asked:

What can this mean? To me the most plausible suggestion is that the dynamic axis of rotation has stayed essentially fixed, the body of the Earth has slowly turned through the dynamic axis by as much as 90° since late Precambrian time, and 50° or more since the Carboniferous period. T. Gold (1955), a British scientist, reasons that a plastic sphere in rapid rotation will make this kind of adjustment normally and rapidly. The Earth's bulge is a stabilizing element, but even so there seems to have been slow turning through the axis, perhaps because of some asymmetry in distribution of mass caused by orogeny. If this is true, a slow shift has occurred in location of the bulge, with a corresponding shift in climatic zones.

*(Longwell, 1958a: 6)*

He was impressed with Runcorn's, Du Bois' and Gold's case for polar wandering alone.

Longwell then considered whether polar wandering without drift explains Permo-Carboniferous glaciation.

Assume that at a certain late Paleozoic date the position of the north magnetic pole was near the island of Formosa. This corresponds to the average value found by Runcorn for Permian rocks in the Colorado Plateau. If the magnetic pole at that time had its present latitude, about 73°, the geographic pole may have had its location a little north of New Guinea. A large part of Australia then would have been within the Arctic Circle, the rest of it high in the temperate zone. The corresponding position of the south geographic pole, in northeastern Brazil, would have given a frigid climate to much of South America and the South Atlantic.

*(Longwell, 1958a: 7)*

So far so good, but there was a difficulty (RS2).

But, someone may object, those polar locations would give India and South Africa rather low latitudes; we must have a credible explanation for late Palaeozoic glaciation in the Indian Peninsula and the three southern lands.

*(Longwell, 1958a: 7)*

Consequently, he appealed to rapid polar wandering without drift (RS1).

A logical answer to this point, at least in general terms, is not difficult. The assumed positions of the poles, not far from the present equator at 140° E and 40° W longitude, would associate Australia and India in the northern hemisphere, South America and Africa in the southern. A moderate polar wandering with a short geologic interval of time would then provide widespread glaciation in all four lands, especially if one or more occurrences of abnormal climate like that of the Pleistocene coincided with critical polar positions.

*(Longwell, 1958a: 7)*

The longtime skeptic realized that this solution might stretch things a bit too far for the less credulous, but he found rapid polar wander easier to swallow than changing India's position.

Such a scheme to explain the glaciation in India seems more reasonable than the grotesque distortions to which Wegener and others have subjected the Indian Peninsula in moving it from a supposed earlier berth alongside South Africa.

*(Longwell, 1958a: 7)*

Longwell then raised the possibility that glacial and paleomagnetic results might not be dated accurately enough to warrant claims of synchronicity.

Exactly how close is the correlation among Indian, African, and South American glacial beds? Even the pronouncement Late Carboniferous sounds deceptively simple. And this matter of correlation looms large in our comparisons of values in paleomagnetism. Already, with research in that field just well started, apparent discrepancies are being pointed out and variously interpreted.

*(Longwell, 1958a: 7–8)*

Considering that APW paths for Europe and North America showed that it was quite possible that the discrepancies in them could have arisen from a combination of rapid polar wandering only and non-synchronicity of the corresponding poles, he deemed drift unnecessary.

Longwell admitted that the fixist still had to explain the glacial flow directions onto landmasses from present day neighboring oceans. Where did the ice come from? Longwell resorted to sunken sialic landmasses.

One potent argument for drift remains in the distribution of glacial features in the southern continents. Australian geologists cite evidence for movement of the late Palaeozoic ice northward, across the present southern coast. Where is the land that nourished and sent out these ice sheets? Likewise South Africa and South America had ice centers off their present east coasts. These facts present grave problems ... This problem of late Palaeozoic ice centres may now be the strongest single argument for continental drift. There is strong evidence that a land mass with sialic composition lay west of the California coast in early Mesozoic time ... and the surface of an old land that supplied sediments to the Appalachian geosyncline is now thousands of feet beneath the continental shelf.

*(Longwell, 1958a: 7–8)*

Sunken borderlands had a long history, and were incompatible with isostasy, but he did not insist that this explanation was difficulty-free. His assessment of the developments in paleomagnetism through 1955 to the beginning of 1956 was generally constructive, he had read the papers carefully.

Longwell acknowledged the closeness of Carey's fit of South America and Africa, but did not directly come out and say so, diverting the argument by referring to the origin of the Moon.

The most elementary example of physical evidence is of course the "fit" between eastern South America and western Africa, presumably the very germ of the drift concept. Carey recently (1955) emphasized the close articulation of these opposing edges, and one of my countrymen, H. B. Baker (1914), judging the fit too accurate to have persisted since the Mesozoic era, suggests birth of the Moon from the Pacific basin and consequent sliding of the continents as late as Miocene time. Aside from the physical improbability of this concept, catastrophism of such major proportions at that date surely would have left a record in interrupted development of living things.[15]

*(Longwell, 1958a: 5; Carey (1955) is my Carey (1955b))*

The closeness of Carey's fit did not make him less skeptical of mobilism, which makes at least some sense in light of his preference for interpreting the paleomagnetic results in terms of polar wandering alone.

These critical comments may suggest that I am an incurable sceptic toward continental drift. I admit to scepticism which, it seems to me, is the soul of scientific inquiry; but I have no unfriendly feeling for the concept of drifting continents – on the contrary I find it attractive and in several respects credible. The Atlantic basin looks like a gigantic rift; if the fit between South America and Africa is not genetic, surely it is a device of Satan for our frustration; the east and west coasts of Africa are strangely bare – we may say raw – in comparison with continental margins generally; pairing of some structural trends on opposite sides of the ocean is at least suggestive of former continuity. In spite of this appeal, I remain incredulous on two main grounds. (1) Evidence known to me is suggestive and circumstantial only – no clinching argument has appeared. (2) Quite aside from the problem of a propelling force, the supposed horizontal creep of sialic blocks through sima seems to be highly improbable. Rigidity in the crust presumably decays downward and at some depth disappears. In that deep zone the principles of rheology must apply, as indeed they must in more limited degree at higher levels. But modes of tectonic failure in the outer, visible zone suggest strongly that some long-term rigidity is there a reality. No basic principle known to me seems compatible with the assumed horizontal movement of a sialic mass, as a unit, through sima. But developments may change this point of view. Not long ago I could not see a logical mechanism for polar wandering, yet this concept is now given a respectful hearing.

*(Longwell, 1958a: 10)*

He remained unsatisfied with mobilist solutions; they were not without their difficulties. Earlier, Holmes had not convinced him to change his mind. Although he acknowledged that the mobilist explanation of Permo-Carboniferous glaciation was persuasive, he thought polar wandering without drift was sufficient. The need to posit continental drift depended on establishing the synchroneity of the relevant poles and glaciations, which was not always evident. He wrote off the close fit between South America and Africa as the work of the devil. Nonetheless, by-and-large Longwell's response was reasonable, being made in March 1956 before Runcorn's conversion, and before he had learned much about paleomagnetic results from other landmasses and before he had heard Irving's reinterpretation of the APW paths in terms of continental drift. Finally, he may have had Graham in mind when noting that serious difficulties were beginning to arise in paleomagnetism.

The explorations in palaeomagnetism, now only well started, may open new vistas in Earth history. Workers in this new field are finding serious difficulties, and some students express doubts that results can be dependable. Others hail the magnetic records as the long-sought Rosetta stone, a key to geological puzzles. Gold (1955) is confident that eventually the magnetic data will tell us whether polar wandering, or continental drift, or both together made the tangled skein of climatic records that seem insoluble on the basis of present world geography. If the sum of evidence finally convinces us that the continents have drifted, no doubt we can then come to agreement of geophysical principles that make this possible.

*(Longwell, 1958a: 10–11)*

It was simply too early to tell. He would wait and see. He did not have long to wait, for Irving gave his talk a few hours later. (I have already reviewed Irving's contribution to the Hobart symposium elsewhere (§5.16) along with the mid-1950s contributions of other paleomagnetists to paleogeography.)

Longwell's short epilogue, devoted almost entirely to Carey and Irving, differed considerably from his opening comments. It was written after reading preprints of the papers that were to be included in the proceedings. He was impressed with Carey's careful empirical approach, and his use of a globe to work out his reconstructions in their correct spherical environment, but he disagreed with the interpretation of some of his structures as oroclines, although he thought the concept fruitful.

Continental drift must involve major crustal deformation, and Carey's "tectonic approach" properly forms a major unit in this symposium. Readers of his text and visitors to his workshop will be impressed with the careful empirical studies on which his concepts are based. He works out all geographic relationships on a large globe, and thus avoids distortion of form and scale that are inherent in flat maps. His idea of the orocline as a major tectonic feature introduces a new element that surely will have a stimulating effect in future global tectonics. No doubt there will be critical disagreement with Carey's proposed explanations of specific orogenic features; I am not in full sympathy with some of his suggestions on North American structure. But his is a new concept that has global applications, and no one geologist can start with full knowledge of all mountain zones, many of which are puzzling to the best informed workers in a given area. Scientific research must proceed by trial and error, and the orocline offers a new approach that merits thorough testing.

*(Longwell, 1958b: 356)*

Longwell appreciated Carey's global approach, and recommended that structural geologists test his orocline-based reconstructions.

Not surprisingly, Carey's fast expansion appeared absurd to most Earth scientists. Longwell raised obvious difficulties, but welcomed Carey's guess because, as he emphasized, the whole issue of crustal deformation was wide open.

Many who find the idea of oroclines merely *novel* will no doubt pronounce Carey's suggestion of an expanding Earth *radical*, even shocking. A number of major questions are posed. How did the complete sialic shell postulated for the small primitive Earth originate? As only a fraction of the present water must have made a universal ocean during a considerable part of the assumed history, has a major part of the water been added since late Precambrian time? What is the exact mechanism by which sialic plates have receded great distances across a basaltic floor from a mid-Atlantic fracture zone that is pronounced still active? The truth is, of course, we are completely ignorant about the Earth's early history and the fundamental cause of orogeny. For a long time geologists were content with the concept of a cooling, shrinking Earth. After this was questioned seriously, many students seized on the suggestion of convection currents as the motive force for crustal deformation. Still the subject is wide open, and another major guess will do no harm – in fact it certainly will stimulate useful thinking. We should welcome any key that may help unlock our greatest puzzle.[16]

*(Longwell, 1958b: 357–358)*

Longwell's open-minded but skeptical assessment of expansion matched his temperament.

In his epilogue, he separated Carey's expansion from his excellent fit of the continents, a practice that others would follow. He characterized the fit in glowing terms but it alone was insufficient to require mobilism; matching of trans-oceanic structures was not unambiguous, and he used as an example Carey's new and unorthodox linking of the Appalachians not with the Hercynides of Europe but with the Urals.

Many physical features of the Earth are cited in support of the concept that some landmasses now far apart were once united. A remarkable example is the close correspondence in outline, at the 2000-meter isobase, between the facing margins of South America and Africa ... The Mid-Atlantic Ridge is indeed a strange accident if it has no genetic relation to the margins of the Atlantic basin. Arguments based solely on such features can be only speculative, as emphasized by a suggestion in this symposium that the Appalachian belt was once continuous with the Urals, not with the Hercynian orogen of western Europe.

*(Longwell, 1958b: 356)*

Longwell was not objecting to Carey's attempt to find consilience, but thought the quantitative checks offered by paleomagnetism more worthwhile. Longwell was impressed with Irving's presentation, but unfortunately did not go into much detail. He did not, for example, discuss Irving's Indian and Tasmanian results nor his linking of paleoclimatological and paleomagnetic findings that he presented at the meeting. He did mention Irving's support in favor of Carey and Wegener's Pangea, but, perhaps more interestingly, he praised Irving for his cautious attitude, in tune with his own skeptical approach. Longwell's comments even suggest that he was seriously willing to entertain Irving's (and Creer's, and by the time he was writing his epilogue, Runcorn's) view that the separate APW paths for Great Britain and North America support continental drift and not polar wandering without drift.

We seek quantitative checks on any proposed re-assembly of lands, and such a check is attempted by Irving through the data from studies of rock magnetism. His results favor the concept of wide continental drift since Palaeozoic time, and suggest an earlier integration of lands similar to the map of Pangea outlined by Wegener and indorsed in its main features by Carey. Some readers may accept these results as firm proof that the continents have drifted. Irving points out that quantities of diverse data still require study, and that the present relation of the geomagnetic field to the axis of rotation is not assuredly genetic and permanent. This cautious attitude is properly scientific; but the divergence between ancient polar locations as indicated at stations on opposite sides of the Atlantic seems too consistent to be accidental. Readers will see much merit in Irving's analysis and will regard rock magnetism as a highly promising source of information on ancient geographies. Further developments in this field of study will command wide interest.

*(Longwell, 1958b: 357)*

Longwell's high praise for the paleomagnetic case for mobilism was not to be repeated for some time. Paleomagnetism had not knocked Longwell clean off his fence-sitting position but it had certainly administered a jolt; he understood and appreciated the care taken by the original developers of mobilism's paleomagnetic test.

### 6.17  Jaeger favors mobilism because of its paleomagnetic support

Jaeger saw drift as a big unsolved problem but something he could not do anything about himself in his own specialty. In the 50s and early 60s I cannot recall him making any committed statement either way and he certainly never sought to influence anyone at ANU. As you know we wrote a mobilist paper together that was published in Madagascar [Jaeger and Irving, 1957]. He had sat through geology courses of Carey's in Hobart before he went to ANU, so he was well aware of what it was all about. His attitude as departmental head was exactly right for the times because his department was split on the matter. Ringwood, who was a powerful member of his ANU staff, was a fixist till quite late and Jaeger was tactful – he never for instance mentioned our differences in public or anywhere so far as I remember. Staff peace was his objective, and that everyone should get on with it. He did give a talk to the Australian Academy when it was quite new (probably about 1959) in which he gave a general review of work going on at ANU in geophysics. As I recall, it was a public evening meeting. I remember the following clearly. He gave a very brief review of paleomagnetic work and showed some diagrams of mine in which I had reconstructed movements of Australia – Australia moving all over the place. I sat with Mrs. Jaeger who whispered to me when she saw it – "Ah, Waltzing Matilda." Mrs. Jaeger's whispers were really stage whispers so many people heard it, and there were ripples of laughter. Clearly he was making an effort to publicise work on drift in his department at a high level in Australia at a time when few bothered with it. He supported me all the way. He saw the paleomagnetic data as something that had to be explained and it was his job to see that more data were obtained.

*(Irving, October 20, 2002 email to author; my bracketed addition)*

As I shall explain, most of Irving's reminiscence is correct, some is not: the Academy meeting was in 1956, his portrayal of Jaeger as sitting on the fence is wrong, and when he spoke of Australia "moving all over the place" he was likely thinking of his own reconstruction efforts two years later (§5.16).

Jaeger was well informed on continental drift. Before going to the ANU he was Head of the Department of Mathematics at the University of Tasmania, Hobart, and he had audited courses given by Carey in the Department of Geology. In his early years at ANU, he hired Irving to develop paleomagnetics (§3.11), helped him publish his early work (§3.12), co-authored papers on the magnetism of the Tasmanian dolerite (§3.12), worked with Blackett's group arguing for movement of Tasmania relative to the geographical pole (§3.13), along with Paterson helped Carey discuss mantle convection (§6.10), and soon would argue with Carey about Earth expansion (§6.13).

In April 1956, one month after the Hobart symposium (§6.16), he gave a lecture at the Australian Academy of Sciences describing achievements of the recently formed

Department of Geophysics. Included in the published report was a concise, two and a half page summary of the paleomagnetic case for mobilism (Jaeger, 1956). After briefly reviewing methods, field tests, and reversals, in much the same way as Irving had done at Hobart (§6.16), he went straight to the heart of the matter, acknowledging that the GAD hypothesis was both needed and reasonable.

Assuming, now, that a direction of magnetization can be assigned to a rock formation, it is necessary to consider what information can be deduced from this about the Earth's magnetic field at the time when the formation was laid down. It is known that the Earth's magnetic field during the past few hundred years has been that of a dipole at its centre with its axis inclined at the small angle to the axis of rotation, together with, in addition, a smaller non-dipole field. Both the non-dipole field, and the direction of the dipole vary rapidly. Measurements of the magnetization of rocks from historic lava flows and from earlier flows as far back as the Miocene (Hospers, 1955) have shown that the Earth's magnetic field, when averaged over a period of a few thousand years, was, during the past 20 000 000 years, that of a dipole with its axis along the axis of rotation of the Earth. The fundamental assumption is now made ... that this has been the case throughout geological time. Clearly it is a reasonable assumption on any dynamo theory of the Earth's magnetic field in which the magnetic and rotational axes may be assumed to be closely coupled, but it has to be remembered that it is an assumption.

*(Jaeger, 1956; the Hospers reference is the same as mine)*

Referring to Figure 6.17, he argued for continental drift. He described the APW path for Britain, and then noted that all six poles from North America were situated to the west of it; but if the North Atlantic is "closed" in the manner required by Carey, the results agreed very well. Poles from Gondwana were much more discordant, with British poles and with one another. Those from India (Deccan Traps, §3.12) and Australia (Tasmanian dolerites, §3.12) being more than 70°, and that from South America being 25° distant from the positions expected at corresponding times on the British path.[17] Thus he argued that the present magnetic results cannot be explained by a simple wandering of the pole relative to the existing distribution of continents, but that substantial relative movement between continents is also necessary. This relative movement is, "in the direction postulated by the exponents of continental drift ..." (Jaeger, 1956: 102).

In this way, in the published account he endorsed mobilism, and seems to have done so largely because of its paleomagnetic support. His was a straightforward reading of the empirical and theoretical evidence. Fully introduced to mobilism, familiar with drift reconstructions, and having up-to-the-moment paleomagnetic data at his disposal, he was in a unique position to appreciate its relevance and strength. But his strong support of mobilism is inconsistent with Irving's retrospective comment, "In the 50s and early 60s I cannot recall him making any committed statement either way." Surely Jaeger's paper is a firm endorsement of mobilism, there are no conditions – no mechanism difficulties, no irregular geomagnetic fields. These are possible contributory reasons for the conflict. Irving later wrote:

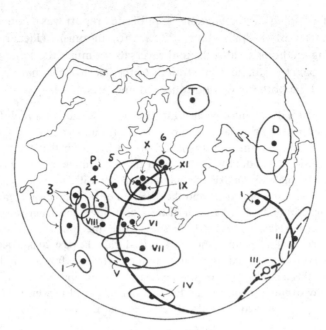

Figure 6.17 Jaeger's Figure 1 (1956: 102). Poles as known in 1956 taken from Irving, 1956 (§3.12). Poles in the Northern Hemisphere are indicated by dots and open circles when on the lower hemisphere. The APW path for Britain is drawn. Results from Europe are denoted by Roman numerals and results from North American rocks by Arabic: I, Lower Torridonian; II, Longmyndian; III, Upper Torridonian; IV, Cambrian; 1, Silurian; V, Devonian; VI and 2, Carboniferous; VII and 3, Permian; VIII and 4, Triassic; 5, Cretaceous; IX, Eocene; X, Oligocene; 6, Miocene; XI, Miocene to present; T, Tasmanian silts (Jurassic or Cretaceous); D, Deccan Traps (Cretaceous to Eocene); P, Parana dykes (same age as Serra Geral Volcanics then considered Jurassic §5.6).

The tone of Jaeger's lecture was likely less firmly driftist than his written account. This meeting of the newly formed Academy was attended, as I remember, by the founding fellows including the geologists E. S. Hills and W. R. Brown, influential representatives of predominantly, fixist, Australian opinion (I, §9.6). Jaeger as head of a new "upstart" institution and not a bone fide earth scientist in the eyes of many could not afford to appear to give a biased account, he had to add all the caveats.

*(Irving, April 2011 email to author)*

A half century later, Irving was very surprised when I told him that Jaeger's lecture had been published. He does not remember Jaeger telling him about it, it is not in the extensive bibliography of his book (1964), and it is absent from his reprints collection: he never saw it; he had only the forty-five-year-old memory of a lecture.

 In addition to matters of tone, Jaeger's lecture did differ in one substantial way from the published version – there is little or nothing in it that is likely to have caused Mrs. Jaeger to exclaim, "Ah, Waltzing Matilda." What Jaeger might have shown

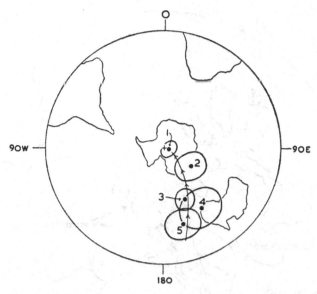

Figure 6.18 First attempt, 1956, at a polar wander path for Australia by Green and Irving (Figure 3 from Irving, 1957b: 198). Pole 1 (Recent and Pleistocene) is from the Newer Volcanics of Victoria; 2 (Lower Tertiary) is from the Older Volcanics of Victoria; 3 (Jurassic or Cretaceous) is from the Tasmanian dolerite sills; 4 (Upper Carboniferous) is from the Kuttung Volcanics; 5 (Upper Carboniferous) is from Upper Kuttung sediments. Poles 4 and 5 appeared in Irving (1957b), 3 in Irving (1956b), and 2 and 1 in Green and Irving (1958). "Records show that samples of the above had been measured in Canberra by early 1956" (Irving, May 16, 2003 email to author).

during his talk was the unpublished early version (Figure 6.18) of what later became Irving and Green's APW path for Australia (§5.3). In fact, Irving believes,

... it is entirely possible that I made available to him for his lecture something like Figure 6.18 and it was this that prompted Mrs. Jaeger to exclaim "Waltzing Matilda." It is likely he asked me for this because he wanted something really new and fully Australian for this meeting of the fledgling Academy.

*(Irving, May 16, 2003 email to author)*

A little later, Jaeger was invited to speak at the third Pacific and Indian Ocean Congress at Antananarivo, Madagascar. In his lecture (Jaeger and Irving, 1957), he reviewed methods and the paleomagnetic case for continental drift, updating his lecture to the Australian Academy of Sciences. In it, however, there was something new – the first paleomagnetic test of a well-constructed map of all the reassembled continents, Pangea. This is worth a comment because it was the only instance in which paleomagnetic results were shown to be systematically not conformable with Carey's reconstruction. (Du Toit and King presented Laurasia and Gondwana separately and Wegener's maps were not very detailed.) Carey's maps were

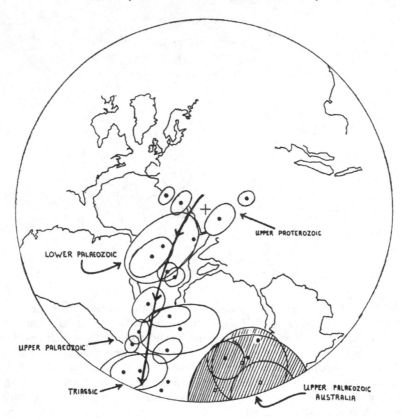

Figure 6.19 Pole positions recalculated on Carey's reconstruction (Jaeger and Irving, 1957). The reconstruction is from Carey (1958, Figure 39d) as presented at the Hobart symposium. Note the wrong positioning of Madagascar which Carey quickly corrected. Stereographic projection, the center of projection is indicated by a cross; knowing of Carey's new reconstruction, Irving was able to recalculate poles and errors relative to Pangea; except for two or three from the Triassic all are Paleozoic or Precambrian.

constructed accurately on a sphere. At about the time of the Hobart symposium, Carey sent Jaeger a large blue-print copy of the reconstruction he presented there (Carey, 1958, Figure 39d). Jaeger passed it to Irving who updated his global compilation of poles and recalculated them relative to it. He did not do so in time for the symposium, otherwise he would surely have presented it. There was a spectacular reduction in dispersion of poles, which now fitted "very well into a single curve, except for the Upper Paleozoic of Australia" (Figure 6.19). Jaeger and Irving (1957: 242) remarked that this discrepancy could be made to disappear "by moving Australia [clockwise] around Antarctica into a position nearer to South America"; they implied that Carey had placed Australia incorrectly within Gondwana. As Irving (2004: 16) later remarked, "We were wrong and it looks as if Carey was correct ..."; the reason being that the discordance was later (beyond the scope of this book) found to be not just an Australian peculiarity, but Gondwana-wide: it was a general paleomagnetic discordance between Laurasia and Gondwana that was related to pre-Mesozoic events and hence not applicable to the paleomagnetic test of Wegenerian continental drift.[18] It is noteworthy that the discovery of this discordance was made from measurements of natural remanent magnetization before the development of magnetic cleaning methods.

Jaeger never became a vocal mobilist. There is no evidence that he attempted to influence anyone at ANU to adopt mobilism. Ringwood, who joined Jaeger's department in 1958 after these events, was strongly opposed to mobilism. A major researcher in geochemistry of the mantle, he earned his Ph.D. in 1953 at the University of Melbourne, supervised by the staunch anti-drifter E. S. Hills (I, §9.3). He went on to study with Francis Birch at Harvard University (1957–8), another anti-drifter, and was himself, as D. H. Green (1998: 352) notes, "an unforgiving opponent of … continental drift." Irving believes that Jaeger, as Head of the Department of Geophysics at ANU, wisely said little about mobilism around the department because he wanted to maintain peace and it was perhaps because of this that Irving felt that Jaeger was luke-warm on drift. At this time, nothing that he could have said to Ringwood would have changed his mind.

## Notes

1 Early works on expansion are by O. C. Hilgenberg (1933), a German physicist, and J. K. E. Halm (1935), a South African astronomer; both used it to explain continental drift. Carey rediscovered them after he came up with his own version. Hilgenberg's version, or at least his rate of expansion, was sometimes tested by paleomagnetists along with the rates proposed by Carey, Egyed, and Heezen. Runcorn invited Hilgenberg to present a paper at a NATO sponsored conference he held at Newcastle in March 1967. Hilgenberg's paper was not included in the proceedings, but the author privately arranged for its publication. It was entitled "The red shift of the Sun's light in relation to the swallowing constant of the ether stream hypothesis" (See Carey (1976: 23–26), Holmes (1965: 965–966) and Scalera and Braun (2003: 25–41) for comments about these and other early expansionists.

2 Experiments during parabolic flight have shown that the effect of varying gravity on the angle of repose is negligible, far less than other factors (Nakashima *et al.*, 2011).

3 See Elliston (2002) for more on Carey's life.

4 I want to thank Cherry Lewis for sending me Holmes' report of Carey's work and for telling me that Cotton nominated Holmes. Her finding Holmes' report was wonderful.

5 Much of my discussion here is based on Carey's autobiographical comments about his early work, and its rejection by various editors. I hope that some of his rejected manuscripts are included among his papers.

6 However, Carey might not have known that Holmes was going to be his outside examiner; Holmes might not have been asked until Carey had finished his dissertation.

7 Carey's remark that his 1953 paper is of historical interest is definitely correct. I asked him for a copy; he told me that he did not have one. I hope one is found in his papers.

8 Homer Le Grand (1988: 168) cited Condon and Öpik's summary, and I learned of it from him. Le Grand, knowing that there was some disagreement about whether Carey presented his version of Earth expansion at the meeting, concluded, based on the summary of Carey's talk presented by Condon and Öpik, that he did not.

9 Carey (1958: 316) has discussed his struggles in attempting to reconstruct Pangea on a globe before embracing expansion. He has also said that Figure 39d, which appears on p. 280 of the symposium volume (1958), was the one he submitted at the Hobart meeting. Indeed, it was; it is pre-expansionist, and it was the one Irving and others used when discussing and testing Carey's reconstruction of Pangea at the symposium. Irving (March 2000 note to author) added that Carey "sent it to Jaeger before the meeting, and I used it to do the fits that I presented at the meeting." Carey also provided an interesting reason for omitting from his 1938 dissertation much of the material he included in his 1955 paper on oroclines. It was, he said (1958: 316), because "I was painfully aware that there was a crucial link missing from the global synthesis." But, elsewhere, as already noted, he gave another reason: it was "omitted at the eleventh hour because I realized that these concepts were too radical for acceptance then and would have cost me my degree" (Carey, 1970: 179; quoted in its entirety in §6.7).

10 Irving (September 2000 note to author) does not see these "misfits" as being real misfits. With Carey's re-assembly, Australia and Indonesia were supposed to be adjacent to each other, yet he could not get rid of a gap between them. But, Australia and Indonesia were not anywhere near each other in Pangean times. There were no unambiguous connecting links between Australia and most of Indonesia and Southeast Asia before the Tertiary, and Carey never gave any evidence that there were. It appears that Carey simply said that they were connected, and that was the end of the matter.

11 Although Carey played no significant role in getting Blackett, Almond, and Clegg to undertake a paleomagnetic investigation of the cores, he did send some core samples to Fredrik Walker, a petrologist at the University of Cape Town. Walker, thoroughly acquainted with the very similar dolerites of the Karroo in southern Africa and the Palisades in northeastern United States, wrote a critical review of Jaeger and Joplin's hypothesis about the formation of the Tasmanian dolerite sills. He thanked Carey for sending him samples.

The author's qualifications for embarking on such a discussion are an acquaintance with the contemporaneous and very similar intrusives of the Karroo and Palisadan dolerite provinces; the kind gift by Professor W. W. Carey of a series of buttons from the 1050-foot core, 180 of which have been sliced and petrographically examined; and lastly the privilege of friendly and stimulating discussion with Professor Jaeger.

*(Walker, 1956: 436)*

12 Carey certainly could have seen Irving and Green (1957a) in which they gave Lower Tertiary and Pleistocene poles for Australia. Their paper appeared in *Nature* on May 25, 1957. In addition, their paper, Green and Irving (1958) where they also presented an APW path for Australia was published in the *Proceedings of the Royal Society of Victoria* early in 1958. However as Irving (March 2000 note to author) noted, "Of course I likely sent him a preprint of Green & Irving (1958) which took some time to come out; it contained the earliest version of our APW path."

13 The poorness of fit of the Carboniferous and Permian paleomagnetic results to standard reconstructions of Pangea compared to that of Triassic and Jurassic results has now been repeatedly confirmed, and has been interpreted as indication that Late Paleozoic and Triassic Pangea were substantially different (Irving, 1977; Muttoni *et al.*, 2003).

14 Carey tells us in the lengthy passage quoted at the beginning of this section that Voisey changed his mind and supported fixism because "he had absorbed the American skepticism during his teaching assignments there." However, Carey gave me a different explanation. After sending him excerpts from a paper in which I wrote that Voisey "had the audacity to argue against continental drift at Carey's symposium", Carey said that I was mistaken, and responded with the following:

You were wider of the mark with Voisey, who "had the audacity to argue against continental drift at Carey's continental drift symposium." Utter crap! Alan Voisey and I were inseparable mates throughout our university years and throughout our lives. He was

best man at my wedding, and I would have been at his, but I was then in the interior of
New Guinea (long before the first airlines!). Alan quite deliberately took an opposing
stance, mainly at my suggestion, because I was afraid the symposium would be too
one-sided, in spite of my efforts to balance it.

*(Carey, July 18, 1984 letter to author)*

However, Carey's characterization differs from what Voisey (1991) said about himself in
his *Memoirs*. Although he did not even discuss Carey's conference – but included a
photograph of the participants – it is clear that he did not believe in continental drift during
the 1950s. According to Voisey, J. Tuzo Wilson got him to favor geosynclinal theories in
1950, and he soon became a proponent of Marshall Kay's geosynclinal views. Kay (I, §7.3)
strongly favored fixism.

[In 1950] Tuzo Wilson visited each Australian University, gave lectures and discussed
his views with staff and students ... He left me hooked on "geosynclines" with a
commission to write up the Australian scene which he said was quite unclear to overseas
geologists ... I spent the next few years working on Australian geosynclines and
became enamoured with the ideas of Marshall Kay.

*(Voisey, 1991: 45)*

Voisey also stated that he did not become a proponent of continental drift until 1970, when
he first saw the Dwyka Tillite in South Africa during a Gondwana symposium.

The arrival of the participants in the Gondwana symposium heralded a marked change
in tempo ... The Pakhuis Pass excursion ... gave me my first view of the Dwyka Tillite,
which resembled the Kullatine tillites of the Macleay and Manning areas very closely. Thus
I began to believe in continental drift.

*(Voisey, 1991: 77)*

It seems to me that Carey is wrong about what Voisey believed in the 1950s. However,
Voisey speaks very fondly of Carey; it is evident that they were great friends, first meeting
as students at the University of Sydney in 1929. Voisey, in fact, tells some moving and
entertaining stories about Carey. Carey criticised another of my assertions, this time
correctly. I had claimed that T. W. E. David was anti-mobilist, basing my assessment
on the generally anti-drift opinions expressed in David's *The Geology of the Commonwealth
of Australia*. However, David had been sympathetic to mobilism, and the anti-mobilist
views were actually those of W. R. Browne, who completed the manuscript after David's
death. Once David realized he would be unable to complete the manuscript, he asked
Browne to do so. Browne agreed to the monumental task, and it took him fifteen years.
Carey recalled an amusing anecdote about Browne and himself: "Some forty years ago
I criticised Browne as 'a rooted continentalist' but he flashed back that I was 'an
uprooted continentalist!'" Carey could have asked Browne to speak at his symposium.
Actually, there were many Australian Earth scientists he could have asked. Perhaps he
did ask some of them. I do not know.

15 Longwell's reference to H. W. Baker (1912–14) is curious. Baker's first note, "Origin of the
Moon," was published in the *Detroit Free Press* on April 23, 1911; his "The origin of
continental forms" appeared in the *Annual Reports of the Michigan Academy of Science* from
1912 through 1914, and he privately published his *The Atlantic Rift and its Meaning* in 1932.

16 Note Longwell's use of *plates*.

17 Irving obtained this South American pole from rocks that Maack sent him from the Parana
dykes (1957a).

18 By the mid-1970s, the discordance was known to be Gondwana-wide and had been
confirmed repeatedly (McElhinny and McFadden, 2000: 299). An explanation was given
by Irving (1977), Kanasewich *et al.* (1978), and Morel and Irving (1981), which, 20 years
later with benefit of much more data, was modified and greatly improved by Muttoni *et al.*
(2003). The solution can be envisaged in this general way. Imagine, as we now know,

that the Australian Late Paleozoic poles of Figure 6.19 referred to *all* of Gondwana. In order to unify them with the Late Paleozoic poles of Laurasia, *all* of Gondwana (not just Australia as Jaeger and Irving had proposed) must be rotated clockwise relative to Laurasia about a pivot in Antarctica. Minimization of the rotation required to satisfy the data places northwestern South America (Columbia) alongside eastern North America (New York) and northeastern Africa (Morocco) immediately beneath central Europe. According to Muttoni *et al.* (2003), this rotation occurred toward the end of the Paleozoic along a ~3500 km shear zone in the Appalachian-Hercynian fold belt. The new assembly is called Pangea B. On this view Wegener's Pangea (Pangea A) lasted intact during the Triassic for only about 50 million years before fragmentation began in the Early Jurassic. In contrast, Köppen and Wegener (1924) wanted it formed by the Devonian and little changed until fragmentation in the later Mesozoic and Cenozoic. An alternative, reflecting earlier doubts, was to assume the geomagnetic field had long-term non-dipole components (Briden *et al.*, 1971; Van der Voo and Torsvik, 2001). There is a review by Irving (2004). I thank Irving for this flash-forward note. He also noted that most Earth scientists continue to reject Pangea B.

# 7

# Development and criticism of the paleomagnetic case for mobilism: late 1950s and early 1960s

## 7.1 Removing difficulties during the development and enlargement of the paleomagnetic case for mobilism

I want to avoid being deceived by what I'm trying to find out in the experiments I'm doing. Everybody criticizes everybody for everything. It's part of being a scientist who is doing things that are novel.

*(Janis Giorgi, immunologist at UCLA, working on a novel approach for an AIDS vaccine. See Cohen, 1998: 35, 38.)*

I shall now trace the continuing development of the paleomagnetic case for mobilism, and examine the more important attacks against it and paleomagnetists' responses to them during the late 1950s and early 1960s. I shall argue that participants acted in accordance with the three standard research strategies, RS1, RS2, and RS3, by criticizing, defending, and enlarging paleomagnetic support for mobilism; the critics raised difficulties, and paleomagnetists attempted to remove them, thereby improving their case for mobilism. Retrospectively, I think that paleomagnetists' behavior may be profitably described in terms of these research strategies, even though they did not at the time think about their work in these terms; I am not claiming that they developed and deliberately applied them.

Research Strategy 1 (RS1) was used by paleomagnetists to expand the problem-solving effectiveness of their solutions and theories (I, §1.13). They improved their solutions by removing difficulties, and sometimes they developed new solutions to replace earlier ones that were plagued with what they saw as insurmountable difficulties. They improved their solutions by showing their consilience with others. They also used RS1 to solve problems that they had not yet addressed. RS2 was an attack strategy; paleomagnetists used it to raise difficulties against their opponents' solutions. They raised empirical difficulties by showing that an opponent's solution faced anomalies, lacked significant supporting data, or depended on unreliable data; they raised theoretical difficulties against a solution by arguing that it depended on unfounded assumptions, clashed with established theories, or contained inconsistencies. Typically they used RS2 to raise difficulties against their opponents' solutions, but sometimes tellingly used it preemptively, identifying difficulties possessed by their own solutions before their opponents did so. RS3 was used by paleomagnetists who supported drift to identify and emphasize solutions that had a decided advantage over their competitors. When a solution explained something that was anomalous or previously unexplained by competing solutions, they emphasized their solution's advantage. If a fixist solution was plagued by difficulties, they emphasized these difficulties and the superiority of their own solution.

This and the following chapter are primarily concerned with the difficulties that, in the late 1950s and early 1960s, were raised against the paleomagnetic case for continental drift, and the measures paleomagnetists took to dispose of them.

Good researchers, especially when working in a new area or on a controversial topic, raise potential difficulties with their own research plans and results. They critically assess their own work by raising difficulties that it might later encounter. They may introduce new solutions to compete with their own. Moreover, if their new work shows promise and supports old controversial ideas such as continental drift, they may immediately begin addressing long-standing seemingly invulnerable difficulties that were raised long ago against these old ideas. Good researchers criticize their own plans and early results, attempting thereby to deprive others of the opportunity do so. They cannot anticipate every difficulty or prepare for every attack, especially if their work is novel. Having recognized difficulties as best they can, they may attempt to remove them, or propose a means of removing them. Good researchers do not present solutions themselves that have obvious, severe difficulties, because they do not want to be thought incompetent, nor do they want, as a result of carelessness, to overlook significant difficulties and be fooled into thinking they have the correct answer. Good researchers not only critically examine the work of those with whom they disagree, they critically examine their own.

A matter of special concern to researchers in a newly developing field is the need for reliable data without which any solution would be dismissed, and without which there would be no sound basis to continue. It is no surprise therefore that very early in their work, paleomagnetists became aware of and addressed unreliability difficulties. With paleomagnetism in its infancy, there were at first no established procedures and little agreement on criteria for judging results. Blackett and those at the Carnegie Institution had sufficiently sensitive magnetometers, a necessary starting point. Irving was concerned about getting reliable data from rock specimens whose magnetic constituents were not necessarily uniformly distributed. Ways had to be found to identify stable remanent magnetizations; as early as the first decade of the twentieth century, P. David (1904) and B. Brunhes (1906) established the igneous contact test (§1.11); Graham in 1949 developed the conglomerate and fold or tilt tests (§1.8), which conceptually owed much to the earlier work of G. Folgerhaiter (1899) and Brunhes and David. Irving made studies of soft sediment slump beds (§2.5) which, like the igneous contact test, offered the prospect of tracing magnetization right back to the time of their deposition. Hospers realized in 1950 that his data from his first Icelandic collection needed statistical treatment, and Fisher provided the method that allowed data to be conveniently packaged (§1.13) and enabled powerful consistency tests to be made (§1.14). In mid-1951, Irving and Runcorn, searching for the lithologies that provided a good record of the geomagnetic field, discovered that fine-grained red sandstones and siltstones did so, and searched them out in later work (§1.17). All these tasks were completed before paleomagnetic testing of mobilism could begin, and were made by researchers to protect their work from unreliability difficulties. Although not consciously acting in accordance with RS2 and RS1, raising and then removing difficulties, they attempted, preemptively, to prevent unreliability difficulties from plaguing their solutions, trying to make their solutions less vulnerable, not giving potential criticism a chance to become a dangerous threat.

Paleomagnetists worried endlessly about the reliability of their data. Creer and Clegg confirmed and extended Irving and Runcorn's discovery that fine-grained red beds generally provide reliable results. Irving checked the reliability of Creer's new magnetometer by re-measuring samples he already had measured on Blackett's magnetometer and later did the same at ANU, comparing standard specimens that he had earlier measured on both Blackett's and Creer's magnetometers. At the January 1954 meeting on paleomagnetism in Birmingham, R. F. King discussed his experiments on the settling of sediments to determine how well they recorded the geomagnetic field, as did Clegg. Using ideas from his 1954 Ph.D. thesis, Irving began, in 1956, using paleoclimatology to check the reliability of paleomagnetic data by seeing how well they agreed; Creer did the same for South America, and Nairn for China. Opdyke and Runcorn came up with the idea of checking paleomagnetic results against evidence of planetary air circulation, comparing them with paleowind directions determined from aeolian sandstones. By 1958, Creer, and As and Zijderveld developed magnetic cleaning methods as a response to concerns about the reliability of data. Well-dated rocks were preferred to blunt the charge that the age basis of results was unreliable. Paleomagnetists also realized that they needed more than just a few samples; for example, Clegg and colleagues boosted Irving's early Indian work on the Deccan Traps, noting that their results were very similar. To average out secular variation and demonstrate repeatability, a sufficient number of samples from a rock formation spaced through a substantial thickness and spread over a large area were required. Contemporaneous sedimentary and igneous rocks magnetized in different ways were sought because agreement between them would enhance reliability. These examples, and there are many more, show that paleomagnetists acted in a manner consistent with RS1 and RS2, planning their work so as to minimize or answer in advance potential unreliability difficulties.

Once paleomagnetists had embarked on their mobilism test, they soon realized it was vulnerable to a potentially fatal theoretical difficulty. They needed to find support for the GAD hypothesis, which was the basis of their mobilism test. Had they not done so, their test would have failed. In late 1952 Hospers marshaled his own data from Icelandic lavas, and that of others mainly from Europe provided evidence in support of the GAD hypothesis for the Neogene and Quaternary.

He (1953a) found that the mean directions of magnetization he calculated agreed within error with Earth's GAD. Later Hospers (1955) expressed this result in terms of poles of this age and they clustered about the present geographic pole. Campbell and Runcorn (1956) offered the same defense, providing new data from Late Tertiary lavas from Washington and Oregon, which they collected during summer 1953, and Opdyke and Runcorn (1956) did the same based on Pliocene and Pleistocene lavas from northern Arizona, which they collected during summer 1956. Irving and Green (1957a, b) presented the same defense, providing data from Australian rocks ranging in age from Pliocene to Recent, as did Creer (1958b) from Early Quaternary basalt flows in Argentina. Du Bois (1959a) provided a similar defense from Late Cenozoic volcanic rocks in northwestern Canada. Turnbull (1959) did so too, based on Cenozoic aged volcanics from Antarctica.[1] Runcorn (1954) provided theoretical support for the GAD hypothesis by incorporating, as Elsasser had done, the Coriolis force into his solution to the origin of the geomagnetic field. Irving (1956a) argued that if the GAD hypothesis were correct then paleolatitudes determined from paleo-magnetic and paleoclimatological data for the same region should agree, and, finding considerable agreement, claimed that he had found new independent support for GAD. Creer, Irving, Nairn, and Runcorn (Creer *et al.*, 1958) pointed out that the interpretation of the paleomagnetic data in terms of continental drift did not necessarily require that the axes of Earth's rotation and the geomagnetic field coincide; the less restrictive hypothesis that the mean geomagnetic field has been that of a geocentric dipole was sufficient. Blackett, Clegg, and Stubbs (1960) appealed to the methodological principle of uniformitarianism to defend the GAD hypothesis.

British paleomagnetists, having at first only Creer's APW path for Britain, were faced with a vast self-evident missing-data difficulty: they needed paths for other landmasses. Thus, Runcorn's APW path for North America; Clegg and his co-workers' findings in India and France; Irving's initial Tasmanian work and then his and Green's APW path for Australia; Gough's, Nairn's, Graham and Hales' work in Africa; Creer's work in South America; and the investigations by Turnbull, Blundell, and Bull in Antarctica, may retrospectively be viewed as attempts to remove or address this obvious requirement. Although some paleomagnetists noted the possibility of drift as an explanation, none argued strongly for it until poles from other landmasses had been obtained; if they had, they would have been vulnerable to this missing-data difficulty.

Once data from other continents had been obtained, and paleomagnetists in Britain began to favor mobilism, the missing-data difficulty was much reduced but certain observational requirements remained. Green (1958) and Runcorn (1957) noted that rapid polar wandering without continental drift remained a possibility because poles compared from different landmasses may not be of comparable age. Without such well-dated poles, the paleomagnetic case for continental drift still faced a missing-data difficulty; paleomagnetists had to find samples that could be dated sufficiently well to establish intercontinental time-equivalence and diminish the possibility of very rapid polar wander. The cleaning techniques of Creer (1958a, b), and As and Zijderveld (1958) allowed paleomagnetists to use a wider range of lithologies without sacrificing reliability, assisting this search for time-equivalent poles.

These preemptive actions of paleomagnetists during the design and development of their case for mobilism, born as they were of protective self-criticism (their implicit use of RS1 and RS2), certainly reduced the number and effectiveness of potential difficulties. But there were no "silver bullets." Many results from many different geological settings in places far removed from Britain were required. In practice, the global test turned out to be a messy but efficient process, moving from an inchoate idea, through laboratory building and field collection campaigns, to the global implementation – a search for better more widely based science. There was no international project or plan. It worked through the initiative of individuals and groups. In a very, very controversial debate, their close attention to potential difficulties helped paleomagnetists maintain their credibility and, against the predictions of their critics, they, for the most part, avoided being fooled by their own astounding results.

### 7.2 Maintaining standards: quarrels among paleomagnetists supportive of mobilism

Before describing fixists' attacks on the paleomagnetic case for mobilism, I want to examine a disagreement, a disagreement within the paleomagnetic community between Irving and Nairn. It illustrates Irving's nervous acute concern that paleomagnetic data should be reliable. Nairn (1960) had reported results from a variety of rocks collected between 1955 and 1957 from Britain, France, and Germany. He submitted his paper at the end of 1958, but it was not published until the first half of 1960. There was nothing surprising about his results. They were in general agreement with previous paleomagnetic findings and with known paleoclimatic results from Western Europe, including contemporaneous paleowind directions. Irving (1961) wrote a critique of Nairn's work, and Nairn (1961) responded. Irving's discomfort was not with Nairn's results but with his methods.

Irving, as the following letter he sent to Cox and Doell reveals, thought that Nairn's work was sloppy. Somebody had to begin policing paleomagnetic work. Irving had somewhat mixed feelings about acting, but decided it had to be done. Irving wanted paleomagnetic data to grow, but not by adding unreliable results, results that could not stand alone, like work that he and Nairn and Creer had just completed on the Great Whin Sill (Creer, Irving, and Nairn, 1959).

25th July 1960

I have been very worried about the poor quality of a lot of palaeomagnetic work and have attacked Nairn. The discussion, sent to the *Journal of Geology*, is enclosed. In the early days this sort of reconnaissance work was acceptable – but not now – at least to me. A pity, because I thought that we had agreed what constituted a reasonable job of work when we did the Whin Sill, but it appears not. I have discussed these results of Nairn's with him over the 2 years both before and since publication but get nowhere – at least he says yes to all my criticism and then publishes this frightful stuff.

I have concentrated on the broad issues, but God knows, I could write a book on all the mistakes in detail you probably have noticed. Please tell me if I am being stupid and irascible. Will be very grateful for your comments. I have discussed it in great detail with Jaeger and geologists here and have their support. I have only just sent it to the Journal so there is no question of it definitely being published yet, but you may if you wish discuss it with colleagues such as Verhoogen & Longwell [who had retired from Yale and was at Menlo Park] etc. using discretion, of course …

> Very best wishes,
> Ted
> *(my bracketed addition)*

The *Journal of Geology* received Irving's critique on July 29, just four days after he sent the above letter. The editor swiftly accepted (August 5). Irving received a letter from Cox on August 9, which unfortunately he did not keep, and there is no copy in Cox's papers. He replied to both the *Journal* and Cox on the 9th. He thanked Cox for his comments, told them what the *Journal* had suggested he might want to change in his paper, and explained why he was not going to make them, reflecting that he enjoyed arguing. For him, enjoying an argument was just enjoying science.

My grateful thanks for the remarks on the Nairn letter. The *Journal of Geology* has agreed to publish it as it stands but suggest that I should reference your review [i.e., Cox and Doell's 1960 review of Paleomagnetism] and perhaps shorten my letter as a result. I shall not do this as I explained in my letter to them (a copy is enclosed), written before I got your letter this morning, for these reasons: 1) I have only today read your review. So our remarks which agree in general have been prepared separately. This is important. 2) I have no intention of referencing a review when there are perfectly good originals which reveal the points years earlier. 3) You do not discuss Nairn's results … You say the broadsword is unnecessary and then say I have been over generous regarding statistics! As, a matter of fact, it may sound a bit pompous and self-righteous and I have struggled hard with this, but for years now these points have been made and yet this bloody awful stuff still gets published. I'm sorry it's the broadsword for me. As a matter of fact I rather enjoy arguments.

> *(Irving, August 9, 1960 letter to Cox; my bracketed addition)*

I now consider Irving's critique. I shall return (§8.12) to his correspondence with Cox when examining Irving's response to Cox and Doell's review, and see just how well Irving described himself in his final remark.[2]

Irving believed that much recent work fell below standard. Nairn was not the only one. If paleomagnetism was to earn and keep the respect of Earth scientists, standards had to improve, not fall; practitioners had to obtain reliable results.

More recently, however, despite many technical advances, there has been a tendency to publish paleomagnetic pole positions accompanied by little or no evidence of the underlying assumptions and without a realistic assessment of the errors. For instance, since the beginning of 1959 about 17 papers have appeared containing 54 pole determinations, and in only 16 of these determinations is any direct evidence of stability (that is, fold tests, conglomerate tests, igneous contact studies, a.c. and thermal demagnetization, etc.) presented. In the remaining 38 results no evidence is given other than the fact that the directions of natural remanent magnetization ... differ significantly from the present field direction, and the results are repeatable after storage of specimens in the earth's field for periods of a year or so. But such evidence provides no information about the stability in the geological time scale. In view of the developing interests of geologists, many of whom may not be in a position to assess the reliability of the methods, a discussion of the basic requirements of this work is needed.

*(Irving, 1961: 226–227)*

As shown from his July 25 letter to Cox and Doell, he was particularly upset with Nairn whom he thought understood what was at stake. Irving attributed twenty-two of the thirty-eight (58%) unreliable paleomagnetic poles to Nairn. Irving also chose to make an example out of Nairn because his results made up about 20% of the pole determinations for Europe, and Irving did not want geologists to be misled into thinking that Nairn's results were unquestionably reliable.

The results chosen are the 22 new pole positions derived by Nairn (1960) from the magnetic results from rocks of Devonian to Tertiary age in western Europe. Nairn has measured the direction of natural remanent magnetization ... in specimens from certain rock formations, and he interprets these directly in terms of paleomagnetic pole positions with little regard to these requirements. Nairn's results have been chosen because they represent a substantial proportion (about 20 per cent numerically) of the paleomagnetic pole determinations from Europe, and it is therefore possible that geologists will look to these determinations for support for their own theories without questioning the basis on which they were derived.

*(Irving, 1961: 227)*

Irving began by summarizing the method for determining poles and highlighted three conditions that must be met.

The basic argument in this work is that some rocks acquire a component of magnetization (usually referred to as the primary magnetization, $m_p$) parallel to the earth's magnetic field at the time they were formed, so that the variations in directions of the field throughout geological history may, in principle, be obtained from a laboratory study of orientated rock samples; and, if these components, $m_p$, in a set of samples representative of a particular rock unit are acquired over a period of time longer than $10^4$ yet not greatly exceeding $10^7$ years, then some average of these will be an estimate of the geocentric dipole field freed of the secular variation yet not substantially affected by movements of the dipole axis (polar wandering). In each determination it is necessary (a) to provide evidence for believing the directions obtained in the laboratory are those of the geomagnetic field at the time the rocks were laid down; (b) to insure that there is an adequate time coverage; and (c) to see that the results are obtained and analyzed in a consistent fashion appropriate to the assumptions made.

*(Irving, 1961: 226)*

He then proceeded to show how Nairn failed to satisfy these conditions. As for (a) providing evidence for believing the remanence directions are those of the geomagnetic field when the rocks were laid down, Irving argued that although some of Nairn's samples had been remagnetized, he had not bothered to rid them of their secondary magnetization by magnetic cleaning as spelled out by Creer (1958a, b), and As and Zijderveld (1958). Regarding (b) time coverage, Nairn failed to obtain samples with sufficient vertical separation to average out secular variation, or sufficient lateral spread to reduce errors caused by local disturbances of the Earth's magnetic field or inaccuracies in determining structural tilts.

Nairn also had incorrectly applied Fisher's statistics. Irving argued that he had inflated the number of supposed independent data-points, artificially reducing estimates of dispersion and errors. He described the hierarchical nature of paleomagnetic surveys; several samples ($s$) are taken from each of several sites, and several specimens ($n$), the unit actually measured in the magnetometer, are cut from each sample. At each site the time taken for samples or specimens to be deposited or cooled through their Curie point should be negligible compared to the time elapsed between sites. At each site, the $sn$ observations are combined to obtain a site direction ($M_s$). The mean formation direction $M_f$ is an estimate of the dipole field direction, and is obtained by summing the $M_s$ directions giving each unit weight. But what Nairn actually did in some cases was to sum values for each specimen instead of each site. This greatly increased the number of data points, because there were more specimens than samples, and samples than sites. As a result, according to Irving, Nairn's data points were not independent. His results appeared more accurate than they really were.

The editor of the *Journal of Geology* asked Irving if he had any objection to Nairn's response appearing in the same issue as Irving's attack. Irving said he had "no objection at all ... providing of course, publication is not held up unduly." It was not; Nairn completed his reply within two months. He stood by his results, did some of the things that Irving wanted, and explained his selection procedures. First he responded to Irving's criticisms about his failure to clean his samples. He appealed to authority. He cleverly cited Irving, repeating Irving's (1959b) response to As and Zijderveld's charge about the unreliability of paleomagnetic results based on uncleaned rocks (§5.5). Similarly, Nairn argued that his results were not suspect because they agreed with earlier results. He even quoted Irving's.

In fact this agreement between many determinations from Permian rocks in Europe would seem to be a most convincing confirmation of procedures in palaeomagnetism
> *(Nairn, 1961: 233; the original passage is from Irving, 1959b: 140)*

Nairn (1961: 233) then treated some of his samples in alternating magnetic fields, and reported no significant changes. Turning to his alleged misuse of Fisher's statistics, he redid some of the calculations in the manner Irving recommended and found no substantial difference. His authority in this case was Runcorn.

In this connection Runcorn (1957) described a more realistic scheme which takes into account small sample sizes; this was the method adopted in collections from Europe. There is no agreed method for the statistical treatment of results. Irving describes one method; another, that of Runcorn (1957), is the one I adopted (see, also Runcorn, 1957, and Collinson and Runcorn,

1960), and it is similar to the way in which some of the American results [of Runcorn's] in Creer *et al.* (1957) were calculated.

*(Nairn, 1961: 235; the other references correspond to mine; my bracketed addition)*

Nairn also explained that he was careful to use results only from sites where dispersion was not great; of the 277 sites he sampled he used results from 81 (29%), rejecting the widely scattered results from the rest (71%).

The following year Irving asked Cox what he thought of Nairn's response.

What do you think about the Nairn correspondence? I would very much like to know whether you think it has done any good. My own feeling is that it hasn't, since geologists can't see through his reply. One thing appalls me – we are now told that 2/3 of his results were not used on the grounds that they are no good, that is, highly scattered. It is therefore not surprising that the remainder of his results agree. Perhaps one is allowed to neglect 10% of one's results on intuitive grounds but not 60% particularly for reasons that would automatically almost give you the result desired. I am contemplating a 2 sentence reply to this effect. Do you consider it worthwhile? I have discussed all these points with Alan Nairn and he says yes to everything I say (as he has done for the past 5 years) but then when I see his current work he is doing exactly the same again. A lot of all this derives from the Runcorn methodology.

*(Irving, July 13, 1961 letter to Cox)*

Cox was of two minds, but had advice for Irving.

Concerning your new letter about Nairn's article, my own reactions are mixed and I'm not sure they'll be of much value to you. Certainly Nairn's work could have been more careful and the reporting of it more complete; but then a lot of other careless work has been [done in] our field and reported selectively, and doubtless this state of affairs will continue. It would be a Herculean task for you to try to take each one of these to task as they come up. A quite analogous situation exists in the field of radiometric dating, and the unreliable data seems to fall by the way-side soon enough. I suspect the only effective way to raise and maintain standards is to do careful work and report it completely. The only other approach I can think of that's apt to have any substantial effect would be for you to write a rigorous text-book on the subject. The time is ripe for a good, complete one and if rigorous standards were set out with great clarity, as there's plenty of space for in a book, it would make the judging of subsequent paleomagnetic research much easier for the informed non-specialist.

*(Cox, letter to Irving, September 1, 1961; my bracketed insertion because the typed material has been rendered illegible)*

Irving (1964) wrote the textbook.

Wiley invited me to write the book. As I understand, it was at Bill Bonini's (Princeton) suggestion. Whether Cox was consulted I do not know. The book was not a best seller, but it became a citation classic.

*(Irving, October 2000 comment to author)*

In fact, he already had begun writing about a month before he received Cox's letter (Irving, September 2000 letter to author).

### 7.3 Preview of the attacks against paleomagnetic support for mobilism and their defeat

Paleomagnetists, geophysicists, and geologists alike raised difficulties. Graham was the most vociferous paleomagnetist to do so. He suggested in 1956 that even stable remanent magnetism might not record the ancient geomagnetic field, a prospect with the potential to undermine seriously the paleomagnetic case for mobilism and reversals. Cox and Doell also attacked the paleomagnetic case for mobilism. As early as 1957, Cox raised an anomaly difficulty, and he and Doell raised others in their extensive 1960 "Review of Paleomagnetism."[3] Jeffreys, Munk, and MacDonald, prominent geophysicists, joined in: Jeffreys in the fourth (1959) edition of *The Earth*, and Munk and MacDonald in their 1960 *The Rotation of the Earth*, arguing that the paleomagnetic case for continental drift should be shunned. Several geologists argued that paleomagnetic results were inconsistent with paleontological findings, and suggested that they should not be taken seriously partly because of these difficulties and partly because paleomagnetism was a juvenile study. Irving, Creer, and Runcorn were especially diligent in responding to criticisms raised before 1960; although in the 1960s they did continue to deal with difficulties raised during that decade, and Irving would discuss some of them in his 1964 textbook, they devoted less time to doing so than to expanding paleomagnetic support. Blackett and Clegg also addressed some of the difficulties.

Viewed from a general perspective, paleomagnetists who advocated mobilism were a beleaguered group. Although they were not to know it at the time, their ordeal was to be short-lived. By the mid-1960s, the focus of the mobilism controversy shifted to the oceans, and developments there eventually led to mobilism's worldwide acceptance. Continental paleomagnetists supportive of drift had less need to defend their work; criticisms fell away or lost their bite as their spectacular results of the 1950s were followed by the even more spectacular results from oceans. The success of the hypotheses of seafloor spreading and its two major corollaries, the Vine–Matthews hypothesis and Wilson's idea of transform faults, showed how justified paleomagnetic support for mobilism was. Hence the early 1960s marked the conclusion of the continent-based paleomagnetic test of Wegenerian continental drift. It also marked a change in the main objectives of continental paleomagnetism from tracking the paleogeography (especially the role of latitude) of the Mesozoic and Cenozoic to tracking the paleogeography of the Paleozoic and Precambrian, and to the study of the tectonic motions in orogenic belts which continues today.

### 7.4 Graham's magnetostriction difficulty[4]

In June 1955, John Graham submitted a paper describing results obtained during the prior three months from over 300 samples of Permian and Triassic sedimentary rocks from the southwestern United States (Graham, 1955: 330). Comparing them with results from rocks of similar age from Britain, he argued that their general agreement supported polar wandering alone, favorably citing Gold and Vening Meinesz as offering explanations of its cause. He found no support for continental drift, essentially offering the same interpretation as Runcorn (1955b) at first did (§3.10).

Thus, it follows that the Wegener hypothesis of continental drift finds no support in these observations. If it should turn out that measurements from other continents confirm the pole location determined from these measurements, then the Wegener hypothesis will have ceased to be a concept worth serious consideration.

(Graham, 1955: 345)

Unlike Runcorn, however, Graham did not soon change his mind and become a mobilist, but became ever more the skeptic, to the point of even doubting his earlier support of polar wandering.

In three papers, Graham questioned the very basis of paleomagnetism, that magnetically stable samples could provide an accurate record of the direction of the ancient geomagnetic field (paleofield) at the time they acquired their magnetization (RS2). The first, submitted in November 1956 (Graham, 1956), was immediately followed by his presentation at the November 1956 meeting on rock magnetism held at Imperial College, London (Graham, 1957). The third was submitted eight months after that meeting (Graham et al., 1957). He appealed to magnetostriction, the effect of stress on magnetization, which he thought had been ignored by mobilist paleomagnetists. Many rocks acquire their primary remanent magnetization while under stress, or suffer stress later; igneous rocks undergo changes in stress as they cool; many sedimentary rocks had been deeply buried before uplift and erosion exposed them. The removal of such stress might change the directions of the magnetization, which would not then reliably record the paleofield.

Graham began thinking about the role of magnetostriction when visiting the Bernard Price Institute in South Africa in 1955, while puzzling over the widely scattered magnetization directions (later shown to be caused by lightning) often observed by Anton Hales, Ian Gough, K. W. T. Graham, and Nairn in their work on surface outcrops of diabase intrusions in southern Africa. Graham reasoned that the intrusions had been cooled and magnetized under vertical compression at depth, and suggested they acquired a new magnetization once this stress was released during uplift and erosional unloading. In support, he noted that in laboratory experiments various igneous rocks had behaved in this way when subjected to compression, that the strength of their newly acquired magnetization was in proportion to the initial applied stress, and that it disappeared once pressure was released. The impermanence of the stress-induced magnetization did not bother him, and he concluded:

The writer's present feeling is that, although many of the available rock magnetism data are consistent with various (unrelated) versions of the hypotheses of continental drift and polar wandering, they by no means prove that such important processes have taken place. The data are tantalizing, but they must be greatly amplified and strengthened. We must look for wide regional consistencies in the magnetization patterns of rocks of varied types in varied settings, and we must obtain accurate and ample knowledge of the processes by which rocks are magnetized.

*(Graham, 1956: 739)*

At the Imperial College conference, Graham extended his criticism to include red beds, the source of many pro-drift results, proposing that their remanent magnetization may be altered as burial stress was released upon uplifting and exposure by erosion. He invoked a two-step process, with the second occurring as hematite grains change shape during uplift; their magnetization could be systematically changed, yet they would retain their coherence.

It is generally assumed (and the mechanism can be demonstrated in the laboratory) that the initial magnetization direction is parallel to the magnetic field applied during the deposition of the hematite, but it is not certain that this is the only step in the magnetization process. It is possible that the following may take place: the original sediments are deposited and subsequently buried by many thousands of feet of other sediments. Hematite is then deposited in the pore spaces and becomes "permanently" magnetized in the direction of the prevailing earth's magnetic field. While at depth, the bulk sample is in static equilibrium, but owing to the strength of the sediments, it is not in hydrostatic equilibrium. Thus, on becoming exposed at the surface by erosion, the sample experiences an unloading and upward elongation. To the hematite which was deposited under hydrostatic equilibrium in the pore spaces at depth, this appears like a tension, and its magnetic moment is subject to change by magnetostriction at the time of unloading. Hence the final measured direction of magnetization may have no simple relationship to the magnetic field applied during the deposition of the hematite. A regionally consistent bias of the magnetization directions could result from the fact that the unloading everywhere is primarily upwards. The magnitude of the effect should be subject to the mineralogic constitution of the sediments because elastic constants depend on composition.

*(Graham, 1957: 362–363)*

His attitude was deeply pessimistic. He seemed to be saying that unless we knew everything we could understand nothing. In contrast, the British paleomagnetists to whom he was at the time speaking, while admitting that there were complications, adopted a more optimistic view; they recognized that they had some new quite remarkable results which they believed had global significance, and they simply forged ahead and tried to make some general tectonic sense of them.

As he developed his magnetostriction difficulty, Graham became more convinced of its seriousness, especially as a result of work he carried out with A. F. Buddington, from Princeton University, and J. R. Balsley, from the USGS. They showed in the laboratory that some rocks magnetized under stress did not, after stress was removed, return to their original magnetic state, either in direction or intensity. Their data were obtained from notably anisotropic very high grade metamorphic Precambrian rocks of the Adirondack Mountains from upper New York State. They subjected them to changing stress ranging from 350 to 2650 psi, and they claimed:

Inasmuch as many rocks acquire magnetization while under stress and are relatively stress-free when their magnetizations are measured, these results are offered as support of the opinion that many conclusions that have been offered on the basis of rock-magnetism data, relating to polar wandering, continental drift, secular variation and reversal of the earth's magnetic field, are subject to serious doubt.

*(Graham et al., 1957: 465)*

They thought this difficulty would be very hard to overcome. Their experiments indicated that small changes in stress brought about large changes in magnetization that were not systematic. Moreover, even if the changes were shown to vary with lithology and the stress history inferred from that lithology, they argued that it still would be very hard to infer anything about the orientation of the paleofield when the rocks were formed because it would be next to impossible to determine the individual stress history of each sample.

This is so for two simple reasons: that we can give neither an accurate account of the stress history of the rock nor of the chemical and physical evolution of the magnetic species. It is, of course, clear that the older the rock, the greater will be the uncertainty. The only obvious way out of these difficulties is to have so many field observations from so many rocks of so many types in so many settings that the insidious influence of magnetostriction, either taken alone or in conjunction with (other?) time-dependent parameters, is eliminated. This is hardly an encouraging prospect.

*(Graham et al., 1957: 474)*

Although they fell short of saying that the effects of magnetostriction entirely undermined the paleomagnetic case for mobilism, they certainly believed that the latter's chance of success was slight.

The ultimate geophysical implications of these observations are not yet known, but they certainly do not further the hope that has prevailed for decades that by way of the techniques of rock magnetism it will be possible to deal effectively with such major geophysical questions as continental drift and polar wandering. The present observations do not assure that such hopes are beyond reach; however, they call for great caution in accepting rock-magnetism data, of the sort usually presented, as evidence bearing satisfactorily on major geophysical phenomena.

*(Graham et al., 1957: 474)*

## 7.5 Removing the magnetostriction difficulty

This attack by Graham and colleagues generated responses from seasoned paleo-magnetists. Du Bois *et al.* (1957) argued that effects of magnetostriction in rocks used for the drift test were insignificant. They proposed the following:

A test relevant to this discussion is to determine whether rocks of the same age within one continent give pole positions in agreement, although it must be remembered that the geological age of those rocks which are most strongly magnetized, lavas and red sandstones, cannot usually be established with precision because diagnostic fossils are often scarce or lacking.

*(Du Bois et al., 1957: 1186)*

They described the general consistency of poles across the United States, from Triassic red sandstones from Utah, and Triassic mafic lava intrusions and red sandstones from the Connecticut Valley and New Jersey.

Results from both igneous and sedimentary rocks laid down at approximately the same geological time but at points thousands of miles apart within the same continent are in agreement, and indicate the high degree of certainty with which these "fossil" directions may be identified with the Earth's magnetic field during the Upper Triassic so that in these red sandstones and lavas it seems that the special effects [i.e., magnetostriction] mentioned above are of negligible importance.

*(Du Bois et al., 1957: 1186; my bracketed addition)*

Their results were consistent with these rocks being good recorders of the paleofield and were inconsistent with Graham's magnetostriction hypothesis (RS2). In principle, some rocks may behave as Graham proposed, but not those used by paleomagnetists for their drift test. Some years later Blackett, Clegg, and Stubbs (1960: 305–306) extended this general argument by appealing to data worldwide, and arguing that the agreement between results from similar-age igneous and sedimentary rocks with their differing stress histories from the same region was inconsistent with Graham's proposal (RS2).

Clegg and two Indian colleagues, Radakrishnamurty and Sahasrabudhe, in their analysis of the Deccan Traps, rather than singling out similar-age rocks with different stress histories but similar paleomagnetic characteristics, drew attention to two very similar sequences of plateau basalts with the same stress history but different paleomagnetic characteristics (Clegg et al., 1958: 831). They rejected magnetostriction as a cause for this difference (RS2), attributing it instead to the northward drift of India between the times the two sequences were deposited (RS1).

Telling responses came from P. M. Stott, an Irving student, and F. D. Stacey, a rock magnetist in the Geophysics Department at ANU, and from J. W. Kern, a student of Verhoogen's at the University of California, Berkeley. Stacey (1958) first offered a theoretical analysis of the effect of isothermal stress on the direction of thermo-remanent magnetization, arguing that although theoretically possible, it is not necessarily irreversible. Stott and Stacey (1959) then performed a clever experiment. Taking a varied selection of igneous rocks, mainly basalts and dolerites ranging in age from Silurian to Quaternary that had been used in the Australia drift test, they cut two specimens from each. All samples were magnetically isotropic. They thermally magnetized one specimen under stress and the other without stress. Upon removing the stress and cooling, they found no significant difference in the directions of magnetization of stressed and unstressed specimens from the same sample. Emphasizing again that Graham and company had observed irreversible effects only in magnetically anisotropic metamorphic rocks that were not relevant to the paleomagnetic drift test, and citing Du Bois *et al.* (1957) on the absence of field evidence, they (1959: 385) concluded, "large systematic errors due to magneto-striction are most improbable in igneous rocks of types normally used for palaeo-magnetic work" (RS2).

Somewhat repetitively, Irving (1959a) offered the next rebuttal. Summarizing the field evidence of Du Bois et al. (1957) and Stott and Stacey's first experiments, he agreed with both, and labeled Graham's difficulty presumptive and irrelevant.

Evidence ... has been put forward by Graham & others (1957) who have shown in the laboratory that linear compressions of 250kg/cm$^2$ can change the directions of magnetiza-tion in certain metamorphic rocks by as much as 30°. This is presumptive evidence only since the deflections are not permanent; with few exceptions the magnetizations resume their original orientations after the stress is removed. In any case the material used is irrelevant.

*(Irving, 1959a: 73)*

Because Graham and company had found stress-induced deflections only in magnet-ite or ilmenomagnetite-bearing gneisses, and few in gneisses containing hematite as their chief magnetic component, Irving raised an anomaly difficulty by singling out several instances where good agreement was found between lavas containing mag-netite and contemporaneous red hematitic sediments from the same region (RS2). He dismissed Graham's difficulty.

It is clear that there is no substantial basis in the observations available at present from either the field or laboratory for supposing that magnetostrictive effects are responsible for the contrast in the observed variation [in pole positions] within and between continents.

*(Irving, 1959a: 74; my bracketed addition)*

Graham and company, however, were not finished. They raised a theoretical diffi-culty (RS2), claiming that it was the laboratory experiments of Scott and Stacey that were irrelevant; they had assumed wrongly that the conditions under which their samples had been magnetized and cooled in the laboratory accurately modeled the complicated stress histories of rocks.

Stott and Stacey ignore these facts [about the relevant complicated stress and temperature history of magnetized igneous rocks], assuming that their experiments on natural samples which have been heat-treated 15 min. at 650°C. and then cooled over a period of several hours while under axial stress give realistic replicas of typical conditions in Nature. Clearly, this assumption is not safe, except possibly for the case of some young rocks which were cooled quickly; it does not permit generalizations, for example, about the possible role of magnetostriction in influencing the directions of residual magnetization of samples of a dolerite sill which cooled over a period of hundreds of years while under load, remained under a load that decreased over millions of years while magnetic phases were segregating from original simpler phases, and finally were brought to the laboratory for measurement.

*(Graham et al., 1959: 1318; my bracketed addition)*

Unable to resist this opportunity to lampoon their opponents, they concluded:

We feel that this oversight by Stott and Stacey should be pointed out for, in our opinion, there are already too many instances where conclusions founded improperly have given the subject of palaeomagnetism a questionable status.

*(Graham et al., 1959: 1318)*

Stott and Stacey responded with experiments and Stacey with a further theoretical paper. First, and again using only isotropic rocks that had been used in the drift test, Stott and Stacey (1960) found no significant differences between stressed and unstressed igneous rocks. Then, suspecting that Graham and company had found irreversible results because they had worked with magnetically anisotropic metamorphic rocks, Stott and Stacy now carried out their own test on anisotropic rocks and found, as expected, that they acquired a magnetization that diverged from the field's direction. They concluded that only highly anisotropic rocks are magnetically affected by stress, and added yet again that magnetostriction was irrelevant to the drift test since such rocks were not being used in the test.

Nevertheless, they were unwilling to dismiss entirely Graham and company's concern. They acknowledged that some ferromagnetic materials could acquire an irreversible stress-induced viscous magnetization when under considerable stress applied over long intervals. This, however, they argued would be removed in paleomagnetic work by routine magnetic cleaning.

We have subjected a variety of rocks to stresses up to 1000 kg/cm$^2$ for short periods (minutes to hours) in the presence of the earth's field without measurably changing their natural moments, but when rocks are stressed for thousands or millions of years the stress-aided viscous magnetizations could well become important. Any viscous magnetization, whether acquired with or without stress, is less stable than thermo-remanent magnetization and its relative instability becomes apparent when rocks are subjected to alternating-field demagnetization. Stress-aided viscous magnetization may well be one cause of "unstable" magnetic moments which have been found to be superimposed upon the stable primary moments of some rocks.

*(Stott and Stacey, 1960: 2424)*

Graham offered no reply.

John W. Kern (1961a, b, c) took the next step while working on his Ph.D. Obtaining samples of the Adirondack metamorphic rocks that Graham and company used in their original experiment, Kern divided them into three groups, cooling them in an applied field with no stress, with an applied stress of 175 bars, and with an applied stress of 350 bars. Divergences in direction of magnetization from that of the ambient field occurred with specimens from each group. Then, he subjected his specimens to magnetic cleaning, and found that the scatter was much reduced for all specimens regardless of whether or not they were cooled under stress. Even anisotropic rocks may accurately reflect the paleofield, if they are cleaned. He showed that Graham and company could have got good data if only they had cleaned their specimens. By way of an explanation, Kern also suggested that stress effects are greatest in large magnetic grains, which are also less magnetically stable than fine grains. Thus stress effects can be preferentially removed by cleaning in alternating fields.

By the early 1960s paleomagnetists no longer took Graham's difficulty seriously. The accumulating field evidence, the laboratory work of Stott, Stacey, and Kern, the theoretical work of Kern and Stacey, and the now general use of magnetic cleaning were sufficient counters. For example, Cox and Doell, who were critical of the paleomagnetic support for mobilism, did not find Graham's difficulty bothersome in their reviews of 1960 or 1961 (Cox and Doell, 1960: 655; Doell and Cox, 1961a: 235–236).

In closing this discussion of Graham's appeal to magnetostriction, it is worth recalling his classic fold test.[5] As Irving put it:

The irony of all this is that Graham's own fold test (if the test is positive) shows that stress is negligible. If it were not, then the test wouldn't work. This is because angular relation of stress to bedding differs along a fold. He never did tell us how a fold test worked if stress strongly affected remanence directions. He seemed to want to escape from the consequences of what he had done.

*(Irving, November 1999 note to author)*

Graham had already given up his fold test in 1951 when he realized there was no way to keep it and avoid field reversals, continental drift or polar wandering alone. He then readopted his test once he thought that the idea of self-reversing rocks would save him from mobilism (§1.10). Perhaps this time he thought magnetostriction would save him. It was not to be.

This ended Graham's involvement in the mobilism debate. He had imagined magnetostriction as a widespread, possibly fatal, flaw in the paleomagnetic case for continental drift. That he was wrong, and could be shown to be wrong on field and laboratory evidence, and on grounds of the physical theory of rock magnetism, signified the depth and maturity of that case. By now an acknowledged expert in the physics of the magnetization of rocks, Stacey is reported to have said to Irving at the time "that paleomagnetism had fulfilled all the requirements of physics and that we could proceed without fear of further attacks from that direction" (Irving, 1988: 1013). And so it was to prove. Stacey documented his case in his comprehensive 1963 review.

## 7.6 Cox's troublesome Siletz River Volcanics

I turn now to something very different, the story of how Allan Cox came to study the Siletz Volcanics and what happened when he did; it is an interesting and telling story. It likely began with John Verhoogen, Cox's supervisor at Berkeley, and Cox learning through him of work in Britain by Irving on the Deccan Traps (§1.20). Verhoogen, a highly respected Earth scientist, was one of very few senior workers in the United

States who, while not a supporter, was not strongly antagonistic to mobilism. Upon returning to Berkeley from the September 1954 IUGG meeting in Rome, where Runcorn had described Creer and Irving's, and Clegg's results from Britain, Verhoogen discussed the new findings in a talk that Cox later characterized as "one of the most lucid talks I've ever heard on the larger issues in geophysics" (Glen, 1982: 166).

How could Verhoogen have learned of Irving's Deccan Trap results? Prior to the Rome meeting, Verhoogen visited Cambridge. Irving remembers talking with him there and answering his insistent questions about the Torridonian, on its magnetic stability and the processes by which it acquired its magnetization, but has no recollection of talking to him about his Deccan Trap results. Twenty-five years later Irving asked Cox how he came to know of his early Deccan results. Cox replied:

I've checked with John Verhoogen and we're certain that we had not seen your thesis at the time I started working on the Siletz River Volcanics. John doesn't remember suggesting that I work on them and thinks I got the idea on my own. I seem to remember that John had received a letter from someone at Cambridge, possibly P. M. Du Bois, mentioning your result and that this had stimulated me to work on the Eocene from North America. My reason for working on marine rocks is that these could be dated paleontologically with some precision. As you will remember, in 1955 there were as yet no radiometric dates from early Tertiary volcanic rocks. It was Parke Snavely at the USGS who told me about the Siletz.

I don't know how you feel about it, but to me it is amazing to find that work done as a somewhat confused graduate student should end up being of even the slightest historical interest. Yet the historians of science sure seem to be sniffing around our early traces!

*(Cox, August 12, 1980 letter to Irving)*

Whatever the means by which he knew of Irving's result, Cox needed results from Early Tertiary rocks of North America for comparison, and Snavely, who had mapped and named the Siletz River Volcanics of the Oregon Coast Range in 1948, told him about them (Snavely and Baldwin, 1948; see also Glen, 1982: 167). Du Bois and Verhoogen had met at the August 1954 meeting organized by UCLA's Institute of Geophysics – the meeting at which Runcorn and Graham argued about reversals and polar wandering (§3.8).[6]

Cox began collecting samples in 1955, and submitted his results to *Nature* in January 1957 (Cox, 1957). It was a very convincing study. He had fifty-seven oriented samples from eight lava flows distributed over a distance of 38 miles across the Oregon Coast Range. The declination and inclination of the basalt flows were respectively 70° east and 55° down with an error ($P = 0.05$) of 7°. From this he calculated a pole in the North Atlantic southeast of Newfoundland. He greatly reduced scatter of magnetization directions between flows by heating samples to 225 °C, followed by cooling in the absence of a magnetic field, and by correcting for tilt using interbedded sedimentary strata. The tilt test was strongly positive. Their early to middle Eocene age was also securely fixed by over thirty species of fossils in interbedded sedimentary strata. He found both normally and reversely magnetized

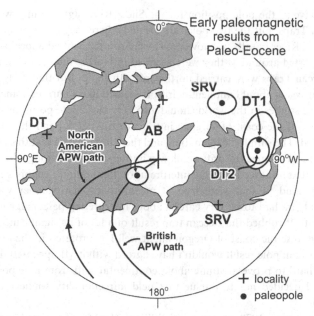

Figure 7.1 Poles from Paleo-Eocene igneous rocks known to Cox in 1957. AB, Antrim Basalts, Northern Ireland (Hospers and Charlesworth, 1954); DT, Deccan Traps; DT1 (Irving, 1954), DT2 (Clegg *et al.*, 1956); SRV = Siletz River Volcanics (Cox, 1957). APW paths for Britain and North America are from Figure 3.6 in §3.7.

flows, which were interbedded in hundreds of feet of marine sediments. So secular variation was likely fully averaged out. Cox's Eocene aged paleopole from Oregon was certainly reliable.

Cox noted that his Eocene aged Oregon pole (SRV, Figure 7.1) "lies about 30° north-east of the average pole found by Clegg *et al.* from the Deccan Traps [DT2, Figure 7.1], which are probably slightly older. It differs by a greater distance from the Eocene poles computed from European data [AB, Figure 7.1]" and from the APW paths of Britain and North America. He went on to remark:

These measurements suggest the desirability of additional measurements in the early Tertiary in order to establish the rate of variation of the average field. From this we may be able to determine the accuracy in dating which is necessary for intercontinental comparisons of directions of remanent magnetization.

*(Cox, 1957: 686; my bracketed additions)*

There was no mention of any tectonic explanation such as that already offered by Clegg *et al.* (1956) for their Deccan results.

Why did the Siletz pole become so troublesome to the paleomagnetic case for drift, and what was it that really links it with the Deccan Traps? Consider these retrospective comments by Cox and Irving.

As luck would have it, the pole position for the Siletz River agrees fairly well with the pole from the Deccan Traps, but the two poles are located where they are for completely different reasons. The Siletz River basalts, we now know, were rotated around a local vertical axis; they were somehow twisted around as they were accreted in North America or shortly afterwards, whereas the Deccan Traps were carried northward with India. But that early result led me to become less enthusiastic about continental drift. It seemed like a curious coincidence, that two of the first Eocene studies in the world should give very close pole positions out in the South Atlantic [actually, off the coast of Virginia in the North Atlantic] and that neither India nor Oregon had drifted. The simplest explanation, and the wrong one, is that this represented polar wander and not any evidence for drift at all. My impression of Verhoogen's reaction to all of this, I think, was that he agreed with my interpretation that it didn't supply strong support for continental drift. And, if that were true, it meant that polar wander was a very rapid phenomenon, and you had to be very careful about interpreting paleomagnetic data in terms of continental drift. It robbed the Deccan trap result of a lot of its significance. We now know it was a fluke, because the coast of Oregon might well have rotated in the other direction, in which case the Deccan pole result wouldn't have agreed with it. [It appeared that] it was going to be very, very hard to prove anything about continental drift from pole positions, and that was one reason I decided not to pursue that field [directionality studies], and to work on reversals instead.

*(Allan Cox, from a 1978 interview with William Glen; first bracketed addition is mine, the second is Glen's. See Glen, 1982: 167–168.)*

Then there comes along Allan Cox, a most remarkable and very influential character. Allan had heard this result from the Deccan Traps. He was the second paleomagnetism student of Verhoogen's. Doell was the first student – how a man can be so lucky as to get two bright students like that consecutively is beyond mind boggling. What Allan did was to sample from the early Tertiary Siletz Volcanics, which are very well dated. It was a beautifully observed piece of work, which is still absolutely sound. It agreed approximately with the Deccan pole. Now this has got to be the biggest fluke in history. Cox, of course, was not a drifter, and he attributed his results to the fact that the field was varying very, very rapidly. Hence this was absolutely vital. If it was correct, we had no hope of comparing results from continent to continent because the geological correlation simply was not that accurate.[7]

*(Irving, taped address at 1981 AAAS symposium on "What happened after Wegener")*

But when the odd weird result did turn up, and that was, of course, the Siletz volcanics ... What were we going to do about that? How could we accommodate it without having very complex and very rapid polar wander paths – the rates of motion would have been ten times faster than would have been conceivable? How were we going to get around that? At this stage I simply threw up my hands and essentially said, "Well there is likely a rotation." And it was from Cox's original beautifully observed result that we first thought magnetically about the enormous mobility that there was in orogenic belts. At least *I* thought of it for the first time; somebody else might have seen it. I became conscious of it in 1959 when I wrote that review of the enormous mobility that is to be found in orogenic zones. Argand had foreshadowed the rotations. Of course, as I've said before, I'd been conditioned by reading Argand as a student. It was obviously one's conditioning.[8]

*(Irving, 1987 interview with author)*

Paleomagnetists attempted to distinguish continental drift from polar wandering by comparing equivalent paleopoles from different continents – poles that they hoped were sufficiently close in age to discount effects of rapid polar wandering. Lack of coincidence could be attributed to drift, but also, if intercontinental correlations were sufficiently inaccurate, to very rapid polar wandering; the more rapid it was, the higher the dating accuracy would need to be. It was not easy to find formations suitable for paleomagnetic investigation that could also be correlated from continent to continent with such high accuracy.

Cox (1957) seemed to be calling into question the general comparisons that mobilist British paleomagnetists had very recently been making between what appeared to be the rather smooth APW paths from Britain and North America. Nothing like the huge 60° departure of the Siletz pole had ever been found within a single continent. Moreover, the relative proximity of the Siletz and somewhat older Deccan Trap poles indicated that in the offing there may be confirmation of this aberration from far-away India.

## 7.7 Irving explains the aberrant Siletz paleopole

Cox's paper, published on March 30, 1957, immediately caught Irving's attention, and on April 17 he wrote Runcorn, sending him a preprint of his and Green's paper on the Newer Volcanics of Victoria (Irving and Green, 1957b). In it, as noted already (§5.3), they argued that their results increased support for GAD, for the pole of the Newer Volcanics coincided with the present geographic pole. They also compared roughly equivalent pole positions in Lower Tertiary and Cretaceous times obtained from North America, Europe, Australia, and India, which they illustrated in their Figure 12 (Figure 7.2), arguing that Australia and India had undergone changes in latitude relative to one another and to Europe and North America since Cretaceous and Lower Tertiary times, and that Europe and North America had not undergone such latitude changes (continental drift) relative to each other during that time. They rejected wild swings of rapid polar wandering without continental drift as an explanation because such wandering would likely have caused differences in pole positions from North America and Europe, which had not been observed. But Irving quickly realized that Cox had observed exactly that; he had created a difficulty, which he explained to Runcorn:

Cox's letter in Nature raises a bit of a problem ... but this paper [with Green] was written before I knew of his results. In any case, results are coming out so quickly that interpretations are bound to change and one has to draw the line somewhere. I'm prepared to leave this section as it stands, but if you think some reference to his work is needed, I suggest the following –

"The effect of such polar wandering, if it occurred, should be present in the European and North American results; from the earlier data (figure 12) this does not appear to be so, but some recent work by Cox (26) suggests that it might have occurred."

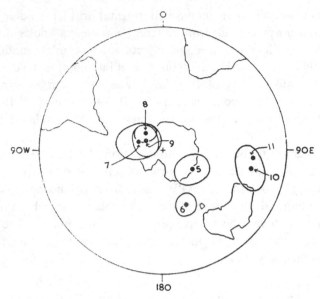

Figure 7.2   Irving and Green's Figure 12 (1958b: 356). The numbered poles are as follows: (5) Irving and Green's Older Volcanics of Victoria; (6) Irving's Tasmanian dolerites (Jurassic or Cretaceous); (7) Roche's Oligocene paleopole of France; (8) Hospers and Charlesworth's Eocene paleopole of Northern Ireland; (9) Runcorn's Cretaceous paleopole of the United States; (10) Irving's Cretaceous to Eocene paleopole of India's Deccan Traps; (11) Clegg, Deutsch, and Griffiths' Deccan Trap paleopole.

There is no need to alter figure 12. On the whole I prefer to omit any reference as I think this rather curious result merits a more careful discussion; conversely, it may be a red-herring and best left stranded.

*(Irving, April 17, 1957 letter to Runcorn;*
*Irving's December 2010 bracketed addition; Cox (26) is my Cox, 1957)*

Irving and Green (1957b: 358) did add a reference to Cox's paper, writing, "The effect of such polar wandering, if it occurred, should be present in the European and North American results; in fact, it is not (Fig. 12) (see note added in proof)," where they added,

Recent work by Cox ... on lavas of Eocene age in the Western United States suggest that comparatively rapid polar wandering of the type mentioned ... may have occurred in lower Tertiary times.

*(Irving and Green, 1957b: 359)*

This shows that Irving initially found no way to avoid Cox's difficulty; rapid polar wandering was a real possibility, which prejudiced the case he and Green had made.

Irving (1959a) returned to Cox's result two years later, after giving it the "careful discussion" it "merit[ed]" in his major review of 1959 in which he developed his argument for mobilism by showing that the results from the same continent are in broad agreement with one another but not with those from other continents (§5.17). Cox's Siletz pole from Oregon remained the one glaring exception. Irving accepted Cox's results as sound, and began as follows:

Apart from the work of Cox (1957) on the Siletz volcanics of Oregon few results are available from the Palaeogene of North America. Early work by Torreson & others (1949) on Tertiary sediments in the Western United States suffered from lack of adequate stability evidence (Hospers 1955). However, for comparison with Cox's work a result from the Laney shale is given. This pole coincides with the present geographic pole whereas Cox's pole is in mid-Atlantic. Cox has studied 8 flows spread through about 4000 feet, and reversals occur. It is somewhat unlikely therefore that its divergence from the present pole is due to a failure to average out the secular variation.

*(Irving, 1959a: 65)*

Cox's result required an explanation, and Irving proposed a tectonic clockwise rotation of the Siletz region.

It is noteworthy that the distance of [Cox's] pole from Oregon, 55° + or − 7°, is not greatly different from the present co-latitude of 45° and the large observed divergence could, in part at least, have resulted from a clockwise rotation of this particular part of Oregon since the middle Eocene. The region has undergone orogenic movement since that time so this possibility has to be borne in mind. Measurements on undisturbed Eocene rocks from other regions of North America should settle the matter.

*(Irving, 1959a: 65; my bracketed addition)*

Although the Laney shale pole had not been shown to be reliable, it had not been shown to be unreliable either, and it was all there was at the time. Irving was willing to use it as a basis of comparison. Moreover, the Laney shale pole was more or less where expected by interpolation in the North American APW path. If correct, it placed Oregon on about the same co-latitude (45°) as at present. Cox's samples had a mean inclination of 55° ± 7°, which yields a co-latitude, as Irving put it, "not greatly different from the present co-latitude of 45°." (The co-latitude of Cox's samples differs from Oregon's present co-latitude within a range of probability of 2% to 18%. The co-latitude for inclination 55° is 54°, for inclination 55° + 7° it is 47°, and for inclination 55° − 7° it is 61°.) By this argument Irving reduced the anomaly to one in declination only, requiring a 70° clockwise rotation of the Oregon Coast Range relative to the rest of North America since the Eocene.

Irving agreed that Cox's data recorded the direction of the geomagnetic field in the Oregon Coast Range during Eocene times, but, he argued, did not do so for the rest of North America. As Irving later recalled, it seemed unlikely that the discrepancy could be reasonably attributed to rapid polar wandering because the effect was apparent in only one coordinate.

There was only one way out of this. And, that was to assume that this was due to a 70° rotation of the Coast Mountain Range. That in fact is the true interpretation. It comes out of the fact that the result is a change in declination but does not affect the inclination. Therefore the chances that this was in fact due to an aberration of the field – it would affect one of the polar coordinates but not the other – seemed a rather extreme thing. And therefore in 1959 I suggested that this in fact was due to a tectonic rotation. I think that what we as paleomagnetists saw for the first time was the kind of mobilism Argand talked about in 1923. We saw for the first time these extreme rotations that can occur in orogenic belts.

*(Irving, taped address at a 1981 AAAS symposium on "What happened after Wegener")*

Irving's explanation, which he elaborated in his book (1964: 249) was not generally accepted for over a dozen years, indeed initially it was ignored as too extreme, but for him it had become less extreme than supposing very rapid polar wandering. Why on Earth would rapid polar wandering cause a change in declination only and not also in inclination? For him that would have been too great a fluke.[9] He understood that his solution, if it were to remain credible, could not be invoked too often, otherwise it would destroy his general finding that there was better overall agreement of equivalent poles within than between continents. There must not be too many Siletz-type poles.

Relative rotations of regions within a continent would result in differing pole results, and the position of the Eocene pole from Oregon would be explained in this way. The better agreement of poles within a continent compared with that between continents is evidence against the occurrence of such rotations, unless these have contrived to bring previously dispersed poles into agreement. It seems much more likely that tectonic rotations would increase dispersion.

*(Irving, 1959a: 75)*

Cox and Doell (1960) later responded to Irving's proposed 70° clockwise rotation of the Oregon Coast Range; they recognized that rotation removed the difficulty, but they did not agree that it had happened, placing their rebuttal within the context of their generally negative assessment of the paleomagnetic case for mobilism. I shall consider their response in its broader context later (§8.3).

Irving's proposal is now generally accepted as correct and is one of the cornerstones of present understanding of the Tertiary tectonics of the Western Cordillera, illustrating, as it does so graphically, the general clockwise shear between the Pacific and North American plates (Beck, 1976).

### 7.8 Other rotations

Following the early study (§3.13) of Clegg and company there was a renewal of work on the rotation of the Iberian Peninsula undertaken by the Dutch paleomagnetic group led by Rutten and Veldkamp during the second half of the 1950s. Veldkamp (1984) gave a short overview of paleomagnetic research in the Netherlands and his and Rutten's involvement. In 1955, Rutten, inspired by Hospers' work, wanted to form a group at the State University at Utrecht, with the hope of using paleomagnetism as a stratigraphic tool, although tectonic rotations became its principal contribution. Veldkamp was then professor of geophysics at Utrecht and Director of the Geophysical Department at the Royal Netherlands Magnetic and Meteorological Institute. He oversaw the construction of the paleomagnetic laboratory. It remained in the institute for five years before moving to the State University at Utrecht. As built an astatic magnetometer and demagnetizer, and he and Zijderveld pioneered work on magnetic stability (§5.5). Van Bemmelen also became involved, and he and Rutten, and later Zijderveld, working in cooperation with Veldkamp, had their geology Ph.D. students undertake paleomagnetic investigations as part of their research projects in many places in Western Europe (Rutten and Veldkamp, 1958; Veldkamp, 1984). G. J. Van der Lingen (1960) and E. J. Schwarz (1963) sampled Permian rocks of Huesca Province, northeastern Spain, and argued that the findings supported the counterclockwise rotation of the Iberian Peninsula; both favored mobilism. Dutch workers also sampled Permian rocks from l'Estérel of southern France (Rutten, Van Everdingen, and Zijderveld, 1957; Rutten and Veldkamp, 1958), from the Italian Alps (Dietzel, 1960), and from the Oslo Graben, Norway (Rutten et al., 1957). Van Hilten (1962) interpreted the results as indicating substantial counterclockwise rotation and northward displacement of the Italian Alps and counterclockwise rotation of the Iberian Peninsula relative to the northern European stable block.

Irving and Tarling (1961) argued for a 7° counterclockwise rotation of the Arabian Peninsula relative to Africa closing the Red Sea and placing opposing shore lines together. Irving wanted to study statistically the secular variation as a function of latitude (§5.18), and variations in low latitudes were few. The Pleistocene to Recent Aden Volcanics, which he had seen when in the army and then again en route to Australia, were a possibility. He asked Tarling (who planned to pursue his Ph.D. with Irving) to sample them on his way from Britain to ANU. They did not intend to engage in a tectonic project, but they quickly realized that their results implied rotation of the Arabian Peninsula and remarked that "the paleomagnetic results are consistent with this movement which is ... similar to that required by Carey [1958], but they do not, of course, prove that it did occur" (Irving and Tarling, 1961: 555–556).

Kawai, Ito, and Kume (1961) showed that the mean declinations of Late Paleozoic and Mesozoic rocks from the southwest and northwest trending arms of Japan differed by 58°. The declinations of Cenozoic rocks are in agreement. They attributed this to bending in plan of the Japanese arc of almost 60°. Although the authors said nothing about Carey, Irving (1964: 250) noted in his discussion of their results, "Carey had previously postulated bending of Japan on tectonic evidence, and his reconstruction gives good agreement for the pre-Tertiary declinations."

Aberrant directions in Phanerozoic rocks like that of the Siletz Volcanics did not occur on cratonic shields, they always occurred in orogenic zones, and by the early

1960s had provided evidence of the high degree of mobility within these zones. Some of the above results were later criticized, but always on technical grounds; unlike the Siletz, they were not frequently invoked as evidence for very rapid polar wander – a means of avoiding continental drift.

## 7.9 Hibberd's rapidly spiraling polar wander paths

Frank. H. Hibberd of the physics department at the University of New England, Armidale, New South Wales, Australia, offered an ingenious interpretation that did not require continental drift. His paper (1962: 221), first submitted in November 1960, required revision, and was resubmitted the following April. Creer in the United Kingdom (Creer, 1962a) and Irving in Australia (Irving, 1962) wrote independent rebuttals, which illustrates the propensity for researchers to raise difficulties, and their implicit use of RS2. Hibberd did not respond.

Hibberd directed his attack against Creer, Irving, and Runcorn, especially at Irving's claim that the greater dispersion of poles between than within continents required continental drift. He (1962: 230) wrote that it "is difficult to see in terms of continental drift and more or less steady polar wandering how apparently contemporaneous poles from the same continent can be widely separated." Sifting out a half-dozen examples, he argued that they could not all be dismissed as unreliable without undermining the general reliability of paleomagnetic data (RS2).

The general difficulty is that in any period the separation between the poles deduced from a single continent is often greater than the separation between poles from different continents. Any one of the particular difficulties when considered alone might perhaps not be decisive, but when taken together they constitute a serious objection to the continental drift hypothesis. It might be possible to argue each of the difficulties as special cases by suggesting that some of the data are inaccurate or unreliable because of magnetic instability, or inadequate sampling, or inadequate correction for geological dip, or incorrect age, and so on. If such procedures are resorted to for all of the "difficult" pole positions one begins to doubt the reliability of all the data and of any conclusions drawn from them.

(Hibberd, 1962: 231)

Hibberd argued that the paleomagnetic directional data indicated that the north geomagnetic pole slowly spiraled northward and westward (Figure 7.3).

For each geological period he identified two groups of poles. The major group fit his path; the minor group did not, which he attributed to secondary overprinting. He claimed moderate success for his solution.

Creer and Irving counterattacked. Creer began with a summary of Hibberd's scheme and criticized his overprinting hypothesis.

The essence of Hibberd's argument is that the palaeomagnetic data can be explained without resort to the theory of continental drift if it is assumed that certain of the data include the effects of a secondary magnetization in the direction of an axial dipolar geomagnetic field in

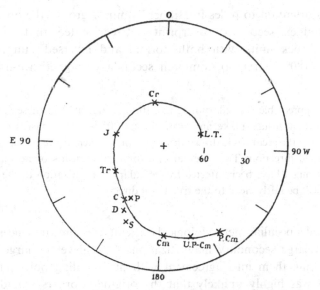

Figure 7.3  Hibberd's Figure 5 (1962: 230). Hibberd's spiral polar wander path. L.T., Late Tertiary; Cr, Cretaceous; J, Jurassic; Tr, Triassic; P, Permian; C, Carboniferous; D, Devonian; S, Silurian; Cm, Cambrian; U.P-Cm, Upper Pre-Cambrian; P-Cm, Pre-Cambrian.

addition to the fossil component. As Hibberd points out, the effect of this is to move the calculated pole position along a meridian away from the true pole position appropriate to the fossil component of magnetization alone (provided the bedding is flat). This is a reasonable argument to test but I disagree with Hibberd's conclusion that it provides an adequate explanation of the data.

*(Creer, 1962a: 275)*

Appealing to the very different grouping of poles from Laurasia and Gondwana, he showed that mobilism offered a superior solution (RS3).

When polar wandering curves ... are drawn for the different continents for which adequate data are at present available it is immediately obvious that those for North America and Europe approach the present north pole from the Pacific while those from Australia and India do so from the Atlantic. In fact the longitudes of the pole positions from the two Laurasian continents are seen to differ from those from the Gondwanic continents by about 180° ... More recent data from S. America and Africa indicate that the polar wandering curves for these two continents also converge along paths from the Atlantic rather than the Pacific. Thus the major and minor groups formed by Hibberd will be explained in terms of the theory of continental drift rather than the hypothesis of secondary magnetization if a correlation can be shown between each of his groups and the places from which the palaeomagnetic data were obtained.

*(Creer, 1962a: 275)*

Creer found a strong correlation between grouping of poles and their place of derivation, and concluded that it "appears that the theory of continental drift fits the data very well, anomalous results ... forming only 7.5 per cent of the total."

Turning his attention to poles in Hibberd's minor group (the ones purportedly affected by hidden secondary overprints), Creer noted that over half were obtained from rock units with both normal and reversed samples (directions approximately 180° apart), so dominant secondary magnetization could not be present (RS2).

> The absence of appreciable secondary magnetization is indicated in these cases because the means of the normal groups make angles of about 180° with the means of corresponding reversed groups and the great circle drawn through pairs of such means would pass close to the present dipole field direction. The external bisector of the means of normal and reversed directions which might have been affected by secondary magnetization to the extent likely in the above data will be fairly close to the true fossil directions.
>
> *(Creer, 1962a: 276)*

Creer also raised a peculiar methodological difficulty; in some instances Hibberd had to invoke very large secondary magnetizations to achieve the large pole changes necessary to bring them into agreement with his spiraling pole path, and Creer argued that it was highly unlikely that the original workers would have missed overprints of such magnitude (RS2).

> If secondary remanent magnetization had been as strong as this I believe the original workers would have noticed it, first because the angle between pairs of normal and reversed directions of magnetization of a particular age and from a particular place would differ greatly from 180°, and second because streaking of directions would have been noticed (Creer, 1957).
>
> *(Creer, 1962a: 277)*

Paleomagnetists could not have been as incompetent as Hibberd implied, Creer being one of them! Finally Creer (1962a: 277–278) separated the results into geological periods, and showed that the percentage of anomalous results for each period was less for the hypothesis of continental drift than for Hibberd's spiral (RS2 and RS3).

Irving began by showing that the Australian results diverged significantly from Hibberd's spiral (RS2).

> It is claimed that this [Hibberd's] path is based on palaeomagnetic pole determinations from all continents ... [But] the Hibberd hypothesis bears no relation to the Australian data. The Australian Cretaceous determination is 20° from the appropriate point on Hibberd's curve, the Jurassic determinations 60° and the Permian and Upper Carboniferous determinations 90°.
>
> *(Irving, 1962: 279; my bracketed additions)*

He then proceeded to dispose of several arguments that could be used against what he had just said. His criticism of Hibberd's spiral could not be dismissed on the grounds that it was based on unreliable data. Fold, conglomerate, and other stability tests had been made. Samples had been demagnetized for removal of secondary magnetizations, many had been radiometrically dated, and others stratigraphically. All magnetizations systematically diverged from the present field.

Hibberd, in his spiraling path, had rejected some Australian poles either because of their "inaccuracy," or because of regional rotations, and Irving raised difficulties against both (RS2).

Hibberd does not consider poles "accurate" if a given component of the polar error exceeds 10° ($P = 0.05$). To reject a result on these grounds when divergences from Hibberd's hypothetical curve often reach 90° and commonly exceed 50° is a quibble. In any case the criterion is incorrect in principle, since it depends on the theory assumed for the calculation of polar errors ... not on the reliability of observations. What really matters is the physical evidence for magnetic stability and the precision and number of the observations. Judged by any contemporary standards these from Australian results stand quite firm.

*(Irving, 1962: 281)*

It was futile to appeal to rotations because the magnetization directions were near vertical. Just as the vertical axis of a spinning top is unaffected by its spinning, so near vertical magnetizations would not be much affected by tectonic rotations.

A further reason suggested by Hibberd (p. 240) for the departure of the Australian results from his curve is that "this region has suffered considerable tectonic movements, perhaps including local rotations in Tertiary times." This is contrary to the geological evidence – Australia has been one of the quietest tectonically of all continents in Tertiary times, major orogenies being completed by the end of the Palaeozoic. In fact, one of the most striking features of the topography of the sampling region is the sequence of Tertiary peneplain surfaces which can be traced unbroken for hundreds of miles. In any case tectonic rotations would not alter the case since the directions of magnetization are approximately vertical!

*(Irving, 1962: 281)*

Irving then identified certain African and North American results that were inconsistent with Hibberd's spiral (RS2). Finally, he raised a methodological difficulty – Hibberd had made too hasty a generalization (RS2).

Finally, some comments ... about Hibberd's statement (p. 231) "that in any period the separation between the poles deduced from a single continent is often greater than the separation between poles from different continents." This statement is supported by a few particular instances. But if all the pole results available to Hibberd at the time of writing are taken and their separations actually measured the following result is found ... Taking the results from the periods Permian, Triassic, Jurassic, Cretaceous and Palaeogene and measuring the separation of poles obtained from the same continent taken two at a time, there are 121 values (88 per cent) which are less than 30° and only 17 greater than 30°. For the values for poles from different continents only 40 (or 16 per cent) are less than 30° whereas 215 exceed 30°. For every instance in which the differences within continents exceed those between continents, there are at least 6 cases for which the reverse is correct ... They show how misleading Hibberd's statement really is.

*(Irving, 1962: 282)*

No more was heard of Hibberd's spiraling polar wander path.

### 7.10 Stehli raises a Permian paleobiogeographic difficulty and Runcorn and Irving counter

In 1957 Francis Stehli, a paleontologist then with the Pan American Petroleum Corporation, argued that neither continents nor poles had moved. Stehli took the Permian pole for Europe/North America that Runcorn had given in 1955 (before his conversion to mobilism; see Figure 3.5) and plotted it on a Mercator projection with continents in their present position. He added occurrences of warm climate brachiopods and fusulinids, and noted that there was a sharp northern limit to their distribution at 55° N present latitude. Because these Permian marine invertebrates likely would not have survived in a temperate climate, he argued that the present 55° latitude roughly corresponded to the subtropical-temperate zone boundary during the Permian; hence globally the Permian climate was considerably milder than that today, and the northern continents had not changed their position relative either to each other or to the pole. Runcorn's Permian pole was wrong (RS2); the geographic pole had not wandered, the distribution of Permian marine invertebrates showed that.

If the interpretation that the faunal discontinuity is the result of temperature is correct, the position of the break [at 55° N present latitude] may be equated with the position of a Permian isotherm whether it be the 15°C isotherm or some other, and must therefore approximately parallel the Permian equator ... this boundary is parallel to the present equator as well. This being so, it follows if the argument is accepted, that the positions of the earth's poles cannot have shifted with respect to northern hemisphere land masses since Permian time. The inference from these data that the poles occupied approximately their present positions in the Permian is in complete disagreement with the interpretation of remanent magnetism data by Runcorn (1956) who suggests that the Permian north pole was in a position within what is here believed to have been the tropical and subtropical temperature belt of Permian time.

(Stehli, 1957: 617; my bracketed addition. Runcorn (1956) is my Runcorn, 1956a)

Runcorn, who by now had not only accepted continental drift but had become a strong and vocal advocate of it, responded in measured tones. Broadening the discussion, he argued that mobilism offered the better solution (RS3).

In view of Dr. Stehli's statements it is surprising that his paper does not include a comparison of the paleontological data with the paleomagnetic data. It is the purpose of this note to show that the distribution of the fusulinid and brachiopod groups fit the predictions of palaeomagnetic data slightly better than they fit the present distribution of the continents; the paleontological data is not decisive on the point but provides no justification for Dr. Stehli's conclusion.

(Runcorn, 1959c: 235–236)

He (1959c: 238) repositioned Stehli's Permian marine fossils relative to the paleomagnetically determined poles and equators for the Carboniferous, Permian, and Triassic, and showed that they were distributed "with roughly equal frequency on either side of the equator in Europe and North America." However, Runcorn admitted that his solution faced an anomaly difficulty – many brachiopod and fusulinids in eastern Asia were situated near the paleomagnetically determined Permian North Pole.

An apparent discrepancy is the relative abundance of these fossils in Asia near the position of the Permian pole plotted on the present map of the world on the basis of the North American and European paleomagnetic results.

(Runcorn, 1959c: 237)

He suggested that eastern Asia might have drifted relative to Europe (RS1). After all, India had moved relative to Europe since the Mesozoic, why not China too? There was no continuous Precambrian or Early Paleozoic basement connecting the three places. He also noted that other paleoclimatic findings suggest that Asia had not been near the North Pole during the late Paleozoic and early Mesozoic.

But it was pointed out by writers on paleomagnetism (Runcorn, 1956c, and Irving, 1956) that, as the paleomagnetic evidence suggested that continental drift as well as polar wandering had occurred, the position of the continent relative to the pole could not be determined from paleomagnetic observation made in other continents. Thus the present distribution of the continents is not the suitable basis on which to consider paleoclimatic data: instead the paleomagnetic and paleoclimatic data for each landmass must be separately compared. In fact, of course, there is evidence to suggest that India has moved relative to Europe since the Mesozoic (Irving, 1956; Clegg, Deutsch, and Griffiths, 1956). We have as yet no paleomagnetic observations from the rest of Asia and hence no evidence concerning its possible motion in respect to Europe. But the abundance of evaporites and dune sandstones of the early Mesozoic and late Paleozoic in Northern China hardly suggest that this part of Asia had a position close to the North Pole at that time. The abundance of warm water invertebrates is in agreement with this surmise. That Asia could have moved relative to Europe appears not to have been discussed previously but its possibility must be entertained in continental drift theories.[10]

(Runcorn, 1959c: 237. Irving, 1956 is my Irving, 1956a. Other references match my bibliography.)

Runcorn, only a few months old convert to drift but speaking as if he had never thought otherwise, ended his paper by referring to the Permo-Carboniferous glaciation, and paleomagnetic support for placing the southern continents close to the South Pole. But he had more in mind. First, he noted that the consilience between the positioning of the southern continents close to the Permian pole and the lack of Permian-aged cold-sensitive marine invertebrate fossils in South America, South Africa, India, and Australia provided continental drift with an advantage over fixism in explaining the distribution of Stehli's marine invertebrates. And then, by reminding the reader of the existence of extensive Permo-Carboniferous glaciation, he called into question Stehli's view that the climate during the Permian was milder than at present. Of course, Stehli knew about the evidence of glaciation, but in his original paper he had attempted to diminish its relevance by claiming that little was known about it.

The climate of the Permian period has long aroused interest and speculation not only for itself but for its bearing on continental relationships. Attention has focused primarily on the climatic conditions responsible for "Permo-Carboniferous" glaciation in the southern hemisphere. The causes of glaciation are imperfectly understood, however, and the climatic implications of an ancient glacial interval cannot yet be adequately interpreted ... It is thus necessary to turn to some other means of evaluating Permian climate. Of those available, zoogeographic study appears most promising.

*(Stehli, 1957: 607)*

His world view at stake, Stehli (1959) responded quickly. First, he claimed that paleomagnetists simply avoided facing up to what he regarded as anomalous results by introducing each time a new interpretation; they presented him with a moving target. Although not putting it in Popperian terms, he implied that, in the hands of paleomagnetists, mobilism was unfalsifiable (RS2). Then he proposed an alternative viewpoint: either the paleomagnetic data were unreliable or they were being misinterpreted. I shall examine what Stehli said in detail because it had a great deal in common with the lines of attack employed against mobilism, with varying degrees of sophistication, by other fixists.

Runcorn points out in the opening sentence of his discussion of Permian climatic zonation and paleomagnetism, that one can interpret paleomagnetic measurements "as demonstrating that the positions of the continents have moved relative to one another and to the axis of rotation through geological time." If it is possible not only to drift the continents but also to shift the axis of rotation, the resulting system is one of great latitude, hence almost any set of data can be satisfied by one continental and/or axial configuration or another. For instance, on the evidence of paleomagnetic measurements from the northern hemisphere, Runcorn places the Permian north pole in Japan. On the basis of similar measurements made in Australia, Irving and Green (1958) place the Permian pole in northwestern Africa or in Spain. This divergence in results is cited as evidence that Australia has undergone relative continental drift at least sufficient to explain the variation in results. An alternative explanation may be presented, however, which should be considered: this is the possibility that paleomagnetic data are either unreliable or not being properly interpreted.

*(Stehli, 1959: 239; the cited paper is the same as my Irving and Green, 1958)*

Stehli was partially right. Some paleomagnetists, notably Runcorn, had initially rejected continental drift, arguing for polar wandering only. Others suggested one or the other, or both in differing amounts. However, unless paleomagnetic data were rejected altogether, continental drift or rapid polar wandering were needed, and rapid polar wander such as Cox (§7.6) and Hibbert (§7.9) had invoked was not very plausible. Stehli ignored, did not know, or simply did not understand the painstaking work paleomagnetists already had done to identify reliable data and distinguish them from the unreliable, and to show that continental drift was needed to explain the reliable data without rapid polar wandering. Notwithstanding, he confidently urged that no firm conclusions be drawn from paleomagnetism.

Because paleomagnetic measurements can be used to support almost any continental configuration it is difficult to see how they can be of more than limited value until consistent results can be obtained from different areas or until a sufficient body of results is available to show that all of the data for a given geological time support a single model. For the present, light can be shed on the validity of the method only by comparing its results with those of independent lines of evidence indicating continental configurations.

*(Stehli, 1959: 239–240)*

Like so many scientists, Stehli believed that his own data and his interpretation of them were superior to those of others. His confidence in the correctness of his own position went beyond the evidence and hinted at excessive pride. Evidence of widespread glaciation in the Southern Hemisphere and peninsular India during the Permian was abundant on five continents, and was considered very reliable by competent observers *who had seen the evidence*, even fixists such as Coleman (I, §3.11). The evidence was at least as compelling as that provided by Stehli; it was more abundant than and at least as well based as Stehli's own view of the paleogeographic implications of his Permian marine invertebrates. Moreover, the observational requirements for sea-level glaciation surely were as well known as those for the intolerance of Stehli's marine invertebrates of cold climates.

Stehli's second response concerned Runcorn's cavalier treatment of the tectonics of eastern Asia. Runcorn suggested that the anomaly could be removed if eastern Asia had moved relative to Europe (RS2).

Accordingly, I presented some paleogeographic data which appeared to me to indicate that no major change in continental configurations were required since Permian time. Runcorn in his discussion has stated that these data better fit an earth model based on paleomagnetic measurements than the present-day earth model, which I employed ... It is difficult for me to see the basis for Runcorn's conclusions that the data fit his paleomagnetic earth model. What Runcorn has termed "an apparent discrepancy" in Asia appears to me, if one were to use northern hemisphere data alone, to suggest a Permian equator at a high angle to that proposed by Runcorn as best fitting the data.

*(Stehli, 1959: 240)*

Stehli did not relish the prospect that future paleomagnetic results from eastern Asia might indicate a lower latitude, as indeed they eventually did (§5.8).

Predictably, Irving did not stand on the sidelines. Although he did not refer to Runcorn's rejoinder (Irving, 1964), he used the same approach as Runcorn, albeit with more care. He argued that the distribution of Permian fusulinids not only did not create an anomaly difficulty for the mobilist interpretation of the paleomagnetic data, but actually solved the problem (RS1). Irving thought the fusulinids of special interest because their distribution, as shown by Stehli, appeared to depend on temperature, and importantly their remains were absent from many Permian marine beds.

This group is chosen because its distribution is thought to be temperature controlled (Stehli, 1957); they are regarded as typically warm water forms. Stehli (1957) has given compilations of the distribution of fusulinid groups which provide an excellent basis for this study, since they give not only the locations which have yielded fossils, but also the location of Permian marine beds in which they are not known to occur; this latter information is important for testing whether or not observed effects arise from inadequacies in the geological or paleomagnetic record.

*(Irving, 1964: 217)*

Irving compiled the paleolatitudes of the Permian marine localities irrespective of whether or not they contained fusulinids using paleomagnetic data obtained from Permian rocks from North America, Australia, and Europe. He found the "estimated paleolatitudes for *all* formations ranged from 20°S to 25°N for North America, 0° to 50°N for Europe and northern Asia, and from 60° to 80°S for Australia." However, the latitude range of formations that contained fusulinids was much more limited; "the cumulative picture shows a close grouping of paleolatitudes around the paleoequator with only one occurrence more than 30° away" (Irving, 1964: 215). He concluded:

The paleolatitudes of the barren marine localities have a much wider range, suggesting that the restriction of the paleolatitudes to low values is a real effect not due to inadequacies of the record. These results are consistent with the view that the fusulinids were tropical forms. There is no inconsistency between the available paleomagnetic results and the fusulinid distribution when observations from the *same* region are considered.

*(Irving, 1964: 219)*

Irving, however, was not finished. He had to consider fusulinids from China and Japan.

As already noted, there is an apparent inconsistency between the abundant occurrences in China and Japan and the paleomagnetic poles from North America, but this arises between observations from *different* regions. Paleomagnetic data, and pole positions and paleolatitudes derived from them, refer only to the regions from which the rocks were collected for magnetic study. The paleolatitudes of the fusulinids of east Asia will remain unknown until reliable paleomagnetic results from the Permian rocks of this region are described. It might be noted that recent results from China indicate a low paleolatitude for the Chinese fusulinid occurrences which is entirely consistent with [the results showing that paleolatitudes for fusulinids during the Permian deviate at most 35° from the paleoequator]; however the data do not fulfill the minimum reliability criteria and require confirmation by more detailed studies.

*(Irving, 1964: 219–220; my bracketed addition. Irving referred to a diagram showing the results. The paleomagnetic work cited by Irving is Wang* et al., *1960.)*

There was no anomaly. Stehli had blundered in his analysis by insisting that contemporaneous paleomagnetic and paleoclimatic data from *different* regions should coincide; that is, he insisted that continents were fixed. There was no reason to expect consistency between the two sorts of data within a composite landmass, unless one can rule out any relative displacement between its component parts, which is not the case in Asia to which India had recently been added. Thus, Irving admitted that his mobilist solution to the distribution of Permian marine invertebrates was not without difficulty. It had its own missing-data difficulty, namely reliable Permian paleomagnetic poles were needed for Japan and China; there were preliminary data, but they did not satisfy Irving's minimum criteria.

## 7.11 Munk and MacDonald attack paleomagnetism

I should pause again to mention style ... Style leads to stories. Who, among those present at the time, could ever forget when Walter Munk was talking to a small seminar at Scripps about a century of ideas on the rotation of the earth? He wrote a long equation on the blackboard, described its ancient and honorable antecedents and the history of its use by eminent scientists, and began "And for over a century ..." when Carl Eckart spoke from the floor. "Walter, the sign in the third term is wrong." Walter continued smoothly, "... none but Carl Eckart ever realized that the sign in the third term is wrong." That was style twice over.

*(Menard, 1986: 14)*

Gordon James Fraser MacDonald (1929–[2002]) was born in Mexico City and moved to the United States when he was 12 years old ... He initially majored in mining engineering at Harvard but dropped "that nonsense" and graduated summa cum laude in geophysics. He was a Junior Fellow and received his doctorate in 1954. He advanced from assistant to Associate Professor at MIT from 1954 to 1958. He then transferred at the urging of most of the other former Junior Fellows in geophysics, including [D. T.] Griggs and [G. C.] Kennedy, to their department at UCLA. He was a full Professor and elected to the American Academy of Arts and Sciences at age 29. Soon he was scheduled to speak at Scripps. Francis Shepard was an old friend of Gordon MacDonald, the premier vulcanologist of Hawaii; so when he heard MacDonald was coming, Shepard invited his many friends at Scripps to a cocktail party. The host was surprised when Dave Griggs walked in accompanied by an unknown youth. As the confusion was resolved, Dave spoke of the scientific abilities of "this" Gordon MacDonald in terms that I heard him use only one other time, in referring to the youthful Dan McKenzie.

Gordon and Walter Munk were among the leaders in a new field, geophysical fluid dynamics, that emphasized the unity of the basic equations for fluid flow whether air, ocean or mantle, and utilized the new computers to obtain previously unobtainable numerical solutions for the equations.

*(Menard, 1986: 226–227; my bracketed additions)*

Munk and MacDonald were a brilliant duo, a combination of experience and youth. Together they wrote a monograph entitled *The Rotation of Earth*, published in 1960, dedicating it to Harold Jeffreys. They devoted their last chapter to polar wandering, against which they raised geodynamical and rheological difficulties. They were not persuaded by the paleomagnetic case for either polar wandering or continental drift.

Munk was a leading figure at Scripps Institution of Oceanography (Scripps), and MacDonald a rising star at the UCLA. Their assessment of paleomagnetism carried great weight with many Earth scientists, often, with the latter, I suspect, regardless of whether they had taken the trouble to understand the paleomagnetic arguments. Menard (1986: 227) likened MacDonald's role in the mobilist debate of the early 1960s to that of Jeffreys during the classical stage of the debate.

At first, the evolution of Munk and MacDonald's thinking followed developments in paleomagnetics with a time lag of about two years. Munk began looking at paleomagnetism before MacDonald, and it was reasonable of him to attribute the only APW path then available (from Britain) to polar wandering, the contribution of continental drift to this path then being unknown. By the end of the decade, after paleomagnetists had fanned out across the world, and found very different APW paths from different continents, continental drift had to be reconsidered. But, as seen below, continental drift was anathema to Munk and MacDonald, and as the relentless build-up of paleomagnetic results favorable to drift continued, rather than treating them seriously, they abandoned them altogether. What Munk had thought was manageable polar wander had morphed into impossible continental drift. It was all too much for the theoreticians, who envisaged a simpler world, and who, like Jeffreys, came to believe that paleomagnetism had nothing at all to say in the matter.

Before detailing the difficulties that Munk and MacDonald raised against the paleomagnetic case for mobilism, we need to examine Munk's earlier views on polar wandering. Munk began by writing an entertaining paper entitled "Polar wandering: a marathon of errors" while spending part of the 1955–6 academic year at the Department of Geodesy and Geophysics, University of Cambridge, in which he surveyed various possibilities and plausibilities. Here is its introduction.

"The subject of the fixity or mobility of the earth's axis of rotation in that body ... [has] from time to time attracted the notice of mathematicians and geologists. The latter look anxiously for some grand cause capable of producing such an enormous effect as the glacial period. Impressed by the magnitude of the phenomenon, several geologists have postulated ... a wide variability in the position of the poles on the earth; and this, again, they have sought to refer back to the upheaval and subsidence of continents." With these words George Darwin introduced his paper, "On the Influence of Geological Changes on the Earth's Axis of Rotation"; it occupies forty-two pages in the *Philosophical Transactions*, and its mathematical complexity is in proportion to the length.

(Munk, 1956: 551)

Munk reviewed Darwin's work, reproduced his equations, and showed that polar wandering could occur only if Earth is sufficiently plastic. Darwin believed that Kelvin had shown that Earth is "sensibly rigid," and therefore had rejected polar wandering. However, regarding Darwin's equations, Munk remarked:

It may have been noticed that the first terms in x and y have the dimensions length/(time)$^2$ instead of length. This is the result of a simple algebraic error made earlier in the derivation. This mistake was discovered by Lambert; the corrected equations are given by Jeffreys.

(Munk, 1956: 552)

Applying this, he showed that polar wandering is impossible no matter how fluid the Earth. Munk then went on to modify the equations in light of new developments in the theory of elasticity, showing that polar wandering now became possible. Comparing his new solution with Gold's hypothesis, and noting, "My solution agrees on all essential points with Gold's conclusions arrived at by non-analytical reasoning," Munk, the geophysicist, ended this way:

In this controversy between physicists and geologists, the physicists, it would seem, have come out second best. They give decisive reasons why polar wandering could not be true when it was weakly supported by palaeoclimatic evidence; and now that rather convincing paleomagnetic evidence has been discovered, they find equally decisive reasons why it could not be otherwise.

*(Munk, 1956: 553–554)*

He referred to Irving's 1954 doctoral thesis, Clegg, Almond, and Stubbs' 1954 paper on Triassic Keuper marls, Creer's 1955 doctoral thesis, and Runcorn's 1955 review. Knowing of the new findings, Munk found the paleomagnetic evidence for polar wandering "rather convincing."

It should come as no surprise that paleomagnetists, who thought that the effects of polar wandering were present in their APW paths, welcomed Munk's theoretical support just as they had Gold's. Within the next two years, Creer, Irving and Runcorn (1957: 148), Green (1958: 382), and Irving (1957b: 205, and 1958b: 224) would reference Munk's paper. Irving even went so far as to return Munk's compliment of providing "convincing" support:

Recently Gold (1955) and Munk (1956) have shown rather convincingly that geological causes could distort the figure of the Earth sufficiently to cause large-scale polar wandering.

*(Irving, 1957: 205)*

But by 1958, however, Munk had changed his mind. He raised theoretical difficulties; the paleomagnetic support was no longer convincing because there were inconsistencies with the paleoclimatic evidence.

Anxious to entertain, he described what he considered to be the usual starting point.

The usual starting point of any discussion on polar wandering is to presume the Earth to be in equilibrium until suddenly disturbed by some implausible rearrangement of matter. The ensuing motion of the pole is then computed for an Earth made of material that can be modeled by appropriate combinations of springs and dashpots. Finally, the computed polar path is found to be in agreement with a bewildering array of paleoclimatic and more recently paleomagnetic evidence.

*(Munk, 1958: 336)*

Rejecting this, Munk (1958: 336) instead thought it "pertinent to inquire whether for any prevailing Earth model the present distribution of mass is consistent with the *present* position of the pole." Using his starting point (which he owed to Birch) he found Gold's theory, and that of others, wanting.[11] He showed that Gold's theory predicted that the *present* pole ought now be south of Hawaii, which it clearly is not. In contrast, the polar wander paths of Milankovitch in 1934 and Jardetsky fifteen years later, based on paleoclimatic evidence, had the pole move *from* south of Hawaii to its present position. Moreover, the paleomagnetically determined APW paths for North America and Europe differed from those of Milankovitch and Jardetsky.

Pinpointing these inconsistencies, Munk (1958: 345) expressed his new dismal assessment of the paleomagnetic (and paleoclimatic) support for polar wandering.

The computed direction of polar wandering is toward the principal pole near Hawaii. On the basis of paleoclimatic evidence which he considered as "unzweideutig," Milankovitch drew the arrow along his orthogonal ... pointing away from the Pacific. Jardetsky seems to have gone along with this interpretation. Recent paleomagnetic evidence also indicates wandering away from the Pacific, but the evidence is far from *unzweideutig*, I think, and the time scales of the climatic and magnetic evidence do not agree at all well.

*(Munk, 1958: 345)*

Within two years Munk, formerly supportive, had come to see the paleomagnetic evidence for polar wandering as "far from unzweideutig." It was no longer convincing; it was disputable and did not agree "at all well" with paleoclimatic evidence. Perhaps, Munk's return to California from England had helped him reevaluate the paleomagnetic support for polar wandering.

I now turn to Munk and MacDonald's attack on paleomagnetism in their co-authored *The Rotation of the Earth*:

It is difficult to write accurately about happenings so long ago, but it is my recollection that those of us who valued precision and good sense, viewed with astonishment Munk and MacDonald's lampooning of the paleomagnetic case for mobilism. These authors admired Jeffreys and Jeffreys had written absurd things about paleomagnetism. As disciples, perhaps they felt obliged to continue his teaching. It is remarkable that people so intelligent did not appreciate that a magnetization observed at a given place maps uniquely into a paleopole relative to that place; all their talk about degrees of freedom and so on obscures this central fact. Provided one has done one's job properly, there is, because of reversals of the field, only one uncertainty – is the paleopole north or south? For the majority of data available at the time when Munk and MacDonald were writing (late 1950s), there were few hemispheric ambiguities because paleopoles could be related to one or other of the continentally based apparent polar wander paths each of which tracked to either the present north or south geographic poles. This unique relationship, between a magnetization direction and its corresponding paleopole, enabled us to make quantitative tests of the correctness or otherwise of given configurations of the continents, to test oroclinal bending of mountain belts, enabled us to begin making paleolatitude analyses of paleoclimatic indicators in order to map ancient climatic zones, and eventually to do many other things we, at the time, only dimly perceived. In the late 1950s, we had this work in hand, and their critique, far from deterring us, urged us forward to develop further our paleomagnetic case for mobilism. But it did make it clear that we had to make a better job of explaining ourselves.

*(Irving, September 2000 note to author)*

Walter Munk and Gordon J. F. MacDonald summarized what I believe was the general reaction to these developments [in paleomagnetism] ... Their book, *The Rotation of the Earth*, received the Monograph Prize of the American Academy of Arts and Sciences for 1959 [for unpublished monographs]. It was dedicated to Sir Harold Jeffreys.

*(Menard, 1986: 89; my bracketed additions)*

The story of polar wandering is varied and complex. Our principal conclusion is that the problem is unsolved. From the point of view of dynamic considerations and rheology the easiest way out is to assign sufficient strength to the Earth to prevent polar wandering, and the empirical evidence, in our view, does not yet compel us to think otherwise. If we must have polar wandering, a number of possibilities present themselves, all somewhat labored ...

*(Munk and MacDonald, 1960: 284)*

Because Munk and MacDonald's theoretical work on the dynamics of polar wandering and their analysis of rheological data led them to believe that it had not occurred, they had to find a way to discount its paleomagnetic support. Borrowing from other recent attacks, they raised difficulties about some of the key assumptions made by the paleomagnetists. They also offered difficulties of their own (RS2).

They began by questioning the two key assumptions, namely, that the natural remanent magnetization records the geomagnetic field, and the GAD hypothesis. They cited (1960: 253) Graham's difficulty about magnetostriction, but without mentioning the work of Stott and Stacey that had shown it irrelevant to the drift test (§7.5).

Magnetostriction may be a factor; laboratory experiments ... have shown that the direction of magnetization can be changed by directed stress. The effect of burial and later distortion of lavas and sediments has not been critically evaluated. However, Graham *et al.* are led by their studies to seriously question the validity of interpreting NRM [natural remanent magnetization] in terms of past positions of the magnetic pole.

*(Munk and MacDonald, 1960: 253; my bracketed addition)*

Questioning (1960: 253–254) the second assumption, they reached far back into the literature to cite Cowling's theorem that a purely axisymmetric convection current could not maintain a geomagnetic field. Elsasser had removed this difficulty in the late 1940s in his self-exciting geomagnetic dynamo (§1.4). They also noted the theoretical argument linking the axes of rotation and the geomagnetic field (§2.13), and the empirical support for the GAD hypothesis obtained from paleomagnetic studies of Miocene and younger rocks (§2.9).

Preparing the ground for their next attack, Munk and MacDonald summarized paleomagnetic results from Europe, North America, Australia, and India, stressing that results from rocks older than 50 million years were not consistent. This lack of agreement is the central argument in the paleomagnetic case for drift, but Munk and MacDonald saw it as evidence against the paleomagnetic method. Turning the paleomagnetic support for continental drift on its head, they remarked:

The consistency of results from a given locality for any geologic time and from different localities for the past 50 million years strongly argues for the method of paleomagnetism. The continuing increase in the degrees of freedom required to account for the paleomagnetic results argue against it. The early measurements of Runcorn in England and America could be explained by a northward motion of the pole through the Pacific. Further data from Great Britain and America now require, in addition to polar wandering, a relative movement of the two continents. The Australian and Indian results are inconsistent with each other and with American and European measurements. They can be brought into line only by further relative motions. Recent European results that Spain be rotated relative to France, Scotland relative to England, and perhaps England with respect to itself. It is usually a bad omen for any method if the degrees of freedom required to interpret measurements grow at the same rate as the number of independent determinations.

(Munk and MacDonald, 1960: 259)

Munk and MacDonald overestimated the rate of increase in the degrees of freedom relative to the number of independent determinations because they did not appreciate the force of the strong concordance among equivalent paleopoles from the same tectonically stable block. The paleomagnetic results were just as expected if intercontinental drift had occurred side-by-side with more local deformation within orogenic zones. Although mobilists would have agreed with Munk and MacDonald's claim that "it is *usually* a bad omen for any method" if the degrees of freedom and independent determinations increase at the same rate, they would have replied that what is usually so is not always so. Tectonics is a complicated matter. Of course, Munk and MacDonald were adamantly against continental drift, and they never hid it from their readers. Stehli made a similar argument, and, as I shall show, Billings (§7.12) would repeat the same argument in his 1959 Presidential Address before the Geological Society of America.

Munk and MacDonald also quarreled with the methods employed by paleomagnetists with their case for polar wandering *and* continental drift while still retaining the possibility of polar wander alone (RS2).

We wish to discuss the rules to be followed if there is continental drift as well as polar wandering. The question concerning the reality of such a drift we can, fortunately, avoid.*

In the analysis of polar wandering, any movement of continents relative to one another enters in two ways: (1) the latitude and longitude of the observatories change over and above the changes resulting from the variable rotation, and (2) the variable distribution of the ocean-continent system leads to a variable excitation function. Under these conditions it is impossible to solve uniquely for the position of the pole without some hypothesis concerning the behavior of the continents relative to the rest of the Earth.[12]

(Munk and MacDonald, 1960: 282)

(1) says that drift introduces another degree of freedom to the movement of land-masses relative to the pole of rotation. (2) implies that drift could bring about polar wandering. Moreover, the introduction of continental drift greatly complicates determination of paleopoles with respect to a given landmass. But why should it be

impossible to obtain a unique position of a paleopole relative to the continent from which it is observed unless assumptions are made about the behavior of the continents relative to the rest of Earth? Munk and MacDonald began their answer by introducing the equations for determining paleopole positions with respect to a given continental block.

A dated rock sample with remanent magnetization serves as our observatory. From this we obtain a record of the inclination, $I$, and declination, $D$, at the time $t$. ... We consider $n$ distinct "continental" blocks. Assume there is no continental rotation and distortion. The hypothesis of no distortion can be checked by comparing the results from many rock samples on one continent. We assume this experiment has been performed and that the results were reasonably consistent.

Let $\theta'(t)$, $\lambda'(t)$ designate the position of the pole of rotation, $\theta_n(t)$, $\lambda_n(t)$ the coordinates of a representative "observatory" on the $n$th continent, and $I_n(t)$, $D_n(t)$ the magnetic readings, all referring to the time, $t$, the sample was magnetized. Let $\theta_n(0)$, $\lambda_n(0)$ designate the position where the sample was collected. Then for any one continent and any one time, the readings $I$, $D$, yield a relation between the positions of pole and continent (Creer, Irving and Runcorn, 1957):

$$\sin \theta' = \sin \theta \cos \phi + \cos \theta \sin \phi \cos D$$
$$\sin(\lambda' - \lambda') = \sec \theta \sin \phi \sin D$$
$$\text{where } \cot \phi = 1/2 \tan I.$$

*(Munk and MacDonald, 1960: 282–283; $\phi$ is the co-latitude of the "observatory" or collecting locality, the angular distance along the great circle from the collecting locality to the north magnetic pole)*

None of this was news to paleomagnetists. Munk and MacDonald were disturbed that the equations had no unique solution, there being more unknowns than equations. But the number of unknowns and equations could be made equal, if either continental drift or polar wandering were assumed absent. If both were present, other assumptions had to be made. Thus these assumptions or rules become endless.

These are $2n$ equations in $2(n + 1)$ unknowns, $\theta_n$, $\lambda_n$, $\theta'$, $\lambda'$. Without further assumptions there can be no solution.

We consider two possible assumptions:

(1)  no continental drift, $\theta_n(t) = \theta_n(0)$, $\lambda_n(t) = \lambda_n(0)$;
(2)  no polar wandering, $\theta'(t) = \theta'(0) = 0$

In the first case we end up with $2n$ equations in 2 unknowns, $\theta'$, $\lambda'$, and can apply the usual statistical techniques to inquire whether $n$ values for the pole position are consistent in light of the (known) observational error. Under the second hypothesis we have $2n$ equations in $2n$ unknowns, and the only test is the one of reasonableness; for example, the overlap of two continents could be considered as unfavorable evidence. A further test for either hypothesis is afforded by a reasonable continuity in time of the motion of the pole or of the continents, respectively.

Suppose we do not wish to exclude either continental drift nor polar wandering. We then need some additional assumptions ... These are some possible rules. There is no end of variation by which this game can be played.

                                                                *(Munk and MacDonald, 1960: 283)*

Again this was familiar ground to paleomagnetists. In fact, Runcorn, notably, had first rejected continental drift, assuming that the sampling localities in Britain and North America were in the same relative position as now when becoming magnetized. At first he explained the lack of close coincidence between equivalent paleopoles in terms of unreliabilities. But he later rejected these unreliabilities, and became favorably inclined toward continental drift. Irving and Creer opted for both polar wandering and continental drift as possible explanations of the poles. All three realized that very rapid and irregular polar wandering was an option consistent with the data, but rejected it. But they also assumed, quite reasonably, that the common changes in latitude observed for Britain and North America arose from polar wandering. Blackett and Clegg also tended to support both. As I shall later show, Deutsch, a Blackett protégé, started thinking in terms of continental drift only (III, §1.8). These paleomagnetists knew they were making some of the very assumptions identified by Munk and MacDonald. Irving rejected reconstructions that involved overlapping continents. Moreover, he made other assumptions when attempting to make paleogeographical reconstructions to deal with the limitations arising from certain longitudinal ambiguities endemic to the paleomagnetic method (§5.16). Munk and MacDonald knew all of this too. They were not saying that paleomagnetists were unaware of what they were doing. Perhaps, they simply wanted those unversed in the methods of paleomagnetism to realize that paleomagnetists could propose many versions of polar wandering, continental drift, or their combination, because all they had to do was make the needed assumptions and their game could be won. What Munk and MacDonald never showed, however, was that continental drift, with or without polar wandering, was an unreasonable interpretation of the paleomagnetic data. They merely showed that no unique solution could be derived without making assumptions. That, of course, is how scientists work, particularly when dealing with historical events hundreds of millions of years ago. Paleomagnetists never claimed they could develop a unique solution. They already had uncovered, and attempted to deal with, the limitations of their method. They had acknowledged the limitation in their various discussions of the ambiguity of longitude, and they had attempted to deal with it as best they could.

   Munk and MacDonald also addressed some of the paleoclimatological and paleontological support, and found it weak. They harked back several years to the 1955 Cambridge symposium (§3.10). Runcorn and Gold spoke on behalf of polar wandering at a time when there was only Creer's APW path for Britain and some as yet inadequately assessed new data from the United States. The paleomagnetic test for drift, although underway, had not yet fully emerged, and it did not do so until the hugely different Southern Hemisphere poles became known, first precociously in

Irving's (1954) thesis then from 1956 through 1960. Munk and MacDonald noted approvingly Durham's fixist stance at this meeting and also Arkell, who had been sympathetic toward mobilism, but had come to reject it during the early 1950s (I, §8.13). Arkell remained mildly sympathetic toward polar wander, and Munk and Macdonald speculated quite erroneously, I am sure, that he might have been pressured by paleomagnetists.[13]

Arkell states that in the Jurassic, the Arctic Ocean was a breeding ground of a rich marine molluscan fauna and could not have been covered by ice to the extent to which it is now. Arkell selects, possibly under pressure from the paleomagnetists, the only possible position of the Jurassic poles if they are assumed as cold as the present poles: the North Pole in the North Pacific and the South Pole in the South Atlantic. He goes on to say that this selection is based more on lack of evidence in the proposed polar regions than anything else.

*(Munk and MacDonald, 1960: 261)*

They also approved of Stehli's arguments but made no reference to Runcorn's reply or the difficulty that he had removed (§7.10).

Stehli (1957) in a detailed study of Permian brachiopods and fusulinids finds a latitudinal distribution. The distributions are obviously inconsistent with a pole in southeast Asia. The advantage of Stehli's treatment is that the temperature dependence of the fauna need not be made explicit. Moreover, since the paleomagnetic rock samples were taken from the same formations, absolute time does not enter into the comparison of the two methods. The scarcity of Southern Hemisphere Permian outcrops does not allow a statement either for or against a northward drift of the southern continents.

*(Munk and MacDonald, 1960: 260)*

Moreover, Munk and MacDonald, clever physicists that they were, had committed a major geological blunder, exposing their ignorance of salient geological facts. The Southern Hemisphere is littered with Permian strata that do not contain Stehli's tropical fauna, but instead, widespread evidence of glaciation.

   They considered polar wandering as an explanation of this glaciation, and found it wanting, concluding:

It appears that there is little positive evidence in the paleoclimatic and paleontological data for polar wandering of the kind suggested by paleomagnetic observations. There is a suggestion of continental drift in the paleoclimatic evidence, but this is not definitive.

*(Munk and MacDonald, 1960: 260)*

They were even less impressed with the support for mobilism offered by paleowinds. Drawing on the Royal Astronomical Society's November 1957 discussion on sand dunes and wind directions (§5.13), they doubted that fossilized dunes could record paleowind directions.

Sand dunes and ash falls may under certain conditions give the prevailing wind direction at the time of deposition. In regions of steady easterlies or westerlies, the pole lies in a direction perpendicular to the prevailing winds. It appears from a recent discussion [reported in

*The Observatory*, 78 (1958): 65–68] that this method for estimating past position of the poles is even more tenuous than the paleomagnetic and paleontologic methods.

*(Munk and MacDonald, 1960: 262; my bracketed addition)*

Munk and MacDonald, presumably drawing on Bagnold's skeptical remarks at the meeting (§5.13), questioned the reliability of the method, and the possibility of identifying ancient barchan sand dunes correctly (RS2). They saw no completely reliable way of distinguishing fluvial from aeolian deposits, and of the slip faces of ancient barchans from those of longitudinal dunes, leading to huge errors in determining ancient wind directions.

A major difficulty is the identification of wind-blown sand and the recognition of the type of dune. Many water-laid deposits closely resemble aeolian sands. The criteria for identifying aeolian material include frosting, excellent sorting, mineral purity and the absence of pebbles, but these features are not uniquely associated with wind-blown deposits. The barchans (crescent-shaped dunes) have a slip face on the lee side; longitudinal dunes develop their slip face at 90° to the average wind direction. Geologists generally interpret ancient dunes as barchans, though the reason for this identification is not always obvious.

*(Munk and MacDonald, 1960: 262)*

Except for their last sentence and the strong implication that these difficulties were insurmountable, I doubt if Bagnold, Shotton, Opdyke, and Runcorn would have disagreed. Munk and MacDonald, who were without personal experience in these historical questions, were not telling those whom they criticized anything they did not already know, had not abundantly acknowledged, and had sought to take account of. Shotton, Opdyke, and geologists who had worked on the aeolian sandstone deposits in the western United States since the 1930s had developed various techniques, which, if used carefully, did allow distinctions to be made with confidence between fluvial and aeolian deposits and between ancient barchan slip faces and those from ancient longitudinal dunes.

Munk and MacDonald ended their monograph by concluding that the problem of polar wandering remained unsolved; Earth appeared to have insufficient strength to prevent it, and the empirical paleomagnetic, paleoclimatic, and paleontologic evidence that had been cited in its favor was not compelling. They did not rule it out, but were strongly inclined against it.

The story of polar wandering is varied and complex. Our principal conclusion is that the problem is unsolved. From the point of view of dynamic considerations and rheology the easiest way out is to assign sufficient strength to the Earth to prevent polar wandering, and the empirical evidence, in our view, does not yet compel us to think otherwise.

*(Munk and MacDonald, 1960: 284)*

As for continental drift, they (1960: 282) brushed it aside, "The question concerning the reality of such a drift we can, fortunately, avoid." It is amazing that two such clever geophysicists, albeit with very little understanding or respect for the geological aspects, imagined that in 1960 they could avoid one of the central questions facing mid-century Earth scientists. They could avoid it only by sticking their heads in the sand.

Although Runcorn, Creer, and Deutsch referred to Munk and MacDonald, no paleomagnetists bothered to rebut them. They really had no reason to do so. It would have been as fruitless as arguing with Jeffreys. Perhaps there are some things to which silence is the only response, and Munk and MacDonald's writings and those of Jeffreys on paleomagnetism were examples. Paleomagnetists inclined toward mobilism already were responding to the difficulties raised by Graham and Stehli that Munk and MacDonald cited. Presumably, they believed that their time was better spent expanding positive support for mobilism. This is not to say that Munk and MacDonald's attack was not influential. If Menard is right, it was.

I do not think, however, that a point-by-point rebuttal would have changed the minds of Earth scientists whose own work gave them no reason to look favorably on mobilism or even to think about it very much. And besides, although some may have thought their attack was reasonable and entertaining, it was already out of date before it was published. But, once again, if Menard is right, and Munk and MacDonald's attack reflected the general reaction to developments in paleomagnetism, then most Earth scientists were not keeping up with these developments and with the rebuttals by paleomagnetists of attacks on their case for mobilism. It seems clear they were not. I also very much suspect that most who failed to keep up with advances in paleomagnetism also knew little about world geology, especially that of the Southern Hemisphere. Overly enamored with theoretical arguments about the alleged unlikelihood of polar wandering and impossibility of continental drift, and supremely confident in their own abilities, Munk and MacDonald, like Jeffreys, could not appreciate the increasing success of the pro-mobilist explanation of the very large intercontinental differences among APW paths. Results, as they came in from different continents, repeatedly confirmed these differences, making fixism less likely and mobilism more and more likely. Munk and MacDonald were too demanding of paleomagnetism, too sure that the empirical arguments of the paleomagnetists would yield to their arguments, too accepting of theoretical arguments based on questionable assumptions about the rheology of Earth's interior, and too confident that their own work was more profound than anything paleomagnetists had to offer. Moreover, they wrote confidently from a milieu in which the great majority of Earth scientists were on their side.

In their non-expert assessment of the paleomagnetism, M&M argued that because it could not itself provide a complete description of continental motions (something no paleomagnetist ever claimed it could) paleomagnetism had not played and could never play a useful part in the mobilism debate. In doing so they held paleomagnetism to a higher standard than other disciplines, which certainly could not do so either. Of course it was not until plate tectonics came along in the later 60s that it became possible to make a satisfactory description by combining data from several disciplines. What M&M did not say was that (1) we had carried out the first rigorous global physical test that could have laid out continental drift flat if all

APW paths had agreed (which they did not), (2) that we had for individual continents confirmed the zonal nature of global climate since the Carboniferous, and (3) that we had shown that latitudes and climate globally were hopelessly conflicted if the continents were kept fixed, but made good sense if they were reconstructed in the manner advocated by Wegener, du Toit etc. I suspect they did not face these arguments because between them they, like Jeffreys, had negligible knowledge of world geology and in their heart of hearts knew they had zero authority to speak about it. Instead they chose to lampoon our arguments, clothing them in fancy language designed, as I am sure they were well aware, to enhance their image of cleverness and to appeal to the fixist sentiments of their day and place. Hubris is the word that best describes their behaviour.

*(Irving, October 2002 note to author)*

## 7.12  Billings attacks paleomagnetic support for mobilism

In his 1959 Presidential Address to the Geological Society of America, Marland P. Billings (1960) reviewed global theories of Earth's major surface features including continental drift and the paleomagnetic evidence for it. A prominent structural geologist and Sturgis Professor at Harvard University, he favored no particular theory, but was definitely hostile to polar wandering and continental drift.

He reviewed the classical arguments for mobilism, and found them wanting. He summarized the paleomagnetic method, relying primarily on Runcorn's five-year-old 1956 pre-conversion paper in the *Bulletin of the Geological Society of America* (Runcorn, 1956a). He noted that at any instant in time the geomagnetic field fails to coincide with the geomagnetic pole and argued therefore that the GAD hypothesis was false (RS2).

Before considering the significance of fossil magnetization, it should be pointed out that within historic times the configuration of the earth's magnetic field has varied greatly. For example at London ... the magnetic declination changed from 11° E. to 24° W. between 1600 and 1800, a change of 35° in 200 years ... Similarly, the magnetic inclination changed from 74° in 1700 to 66° in 1935, a change of 8° in 235 years. At Boston, the inclination changed from 68½° to 74° between 1723 and 1850. If we accept the assumptions generally made in paleomagnetic studies, we would conclude that Boston drifted 300 miles northward in this period.

*(Billings, 1960: 388)*

Billings, rather than raising as he thought a theoretical difficulty against the GAD hypothesis, was making a fool of himself; he mistakenly claimed that the paleomagnetic method requires coincidence of the geomagnetic and rotational axes at all instants in time, whereas it calls for the former to be averaged out over thousands of years. Harboring this profound misconception, it is not surprising that he thought little of the paleomagnetic support for mobilism.

But Billings was not done, adding a litany of, what were becoming, familiar criticisms.

Geologists cannot help viewing with extreme skepticism most of the conclusions reached so far from these paleomagnetic studies. For one thing, the various workers within the field do not agree on their interpretations. Moreover, if slipping of the entire crust is not sufficient to explain their observations, some investigators also invoke drifting and rotating continents. (Creer, 1958: 100, 101). This, of course is an unbeatable combination; anything can be explained. Stehli (1957), from a study of Permian climatic zonation, based on the distribution of certain brachiopods and fusulinids, considered that the Permian climatic zones were similar to those existing at present; in other words, his study suggests that little change in the position of the geographic axis of the earth has occurred since then. Doell and Cox (1959) suggest that during Mesozoic and Early Tertiary time the earth's magnetic field was a comparatively rapidly changing dipole, without requiring extensive displacement or rotation of the continents. Munk (1956), on the basis of his mathematical and physical analysis, doubts the probability of polar wandering.

*(Billings, 1960: 389; the only reference that differs from mine is Creer, 1958a)*

Billings' theoretical difficulty, the "unbeatable combination" by which anything can be explained, echoes those of Stehli (§7.10) and Munk and MacDonald (§7.11); he was claiming erroneously that paleogeographical reconstructions based on paleomagnetism were unfalsifiable. He repeated Stehli's anomaly difficulty (§7.10), and also Doell and Cox's 1959 claim (reviewed below, §8.3) that a rapidly changing geocentric dipole field (polar wander) could explain the paleomagnetic data obtained from Mesozoic and Early Tertiary rocks without invoking continental drift. Moreover, his allusion to Munk's 1956 paper, "Polar wandering: a marathon of errors," was wrong because in it Munk had supported polar wandering (§7.11); perhaps Billings failed to understand the paper; perhaps he never read it and was misled by the title; perhaps he already knew that Munk had changed his mind about polar wandering. Nevertheless, he was certainly right about disagreement among paleomagnetists. But, disagreement among specialists is the rule in new rapidly advancing fields; it was true in Billings' own field of structural geology. However, except for the Americans Graham, Cox, and Doell, by 1959 every major paleomagnetist supported mobilism; they had technical disagreements among themselves, but they did not disagree that continents had drifted. Was his claim, "Geologists cannot help viewing with extreme skepticism most of the conclusions reached so far from these paleomagnetic studies," also correct? Bill Menard (1986: 166) thinks so. If geologists understood paleomagnetism as poorly as Billings, then, sadly, he and Menard were probably right about their fellow geologists.

No paleomagnetist took Billings' misconceived attack on the GAD hypothesis seriously, but Runcorn responded in measured terms in *The Bulletin of the Geological Society of America*, where he was assured of wide readership.

Some authors have doubted these conclusions [in favor of polar wandering and continental drift]. Billings (1960) referred to the absurdity which would result from applying these arguments to the field at the present time (or at any other instant). But the deviations of the field from an axial geocentric dipole change very rapidly compared with geological time, and investigation of the cause of the geomagnetic secular variation shows that Billings' criticism is without substance.

*(Runcorn, 1964: 687; my bracketed addition)*

Runcorn also used this opportunity to defend the GAD hypothesis and its role in the paleomagnetic case for mobilism. He first reintroduced the solution he already had proposed to the second-order problem about the origin of secular variation, saying that Billings had confused timescales.

The relative rotation of the core and mantle, which accounts for the westward drift of the geomagnetic field and the irregular changes in the length of the day, causes the mean geomagnetic field to be symmetrical about the earth's axis of rotation (Runcorn, 1959a), when averaged over periods which are long compared with the secular variation time scale. A geological period is $10^3$ longer; it is therefore not difficult to insure that enough of a formation has been sampled to average out the geomagnetic secular variation in the mean paleomagnetic direction.
*(Runcorn, 1964: 687; reference to Runcorn 1959a is the same as mine)*

Runcorn did, however, admit that the superposition of a multipole geomagnetic field could not be entirely ruled out, but noted that Blackett and company had put forth their phenomenological defense of the paleomagnetic case for mobilism in which they did not invoke hypotheses about the long-term behavior of the geomagnetic field. Thus he argued that geologists should not think that the GAD hypothesis, although much better supported than they may imagine, is absolutely necessary for the paleomagnetic case for mobilism. Finally, Runcorn proposed a general test of the dipole hypothesis, but noted that more data would be needed for the test.

However, the possibility that the field has in remote geological time been an axial-multipole field cannot be rejected a priori … Blackett and others (1960) have tried to present the paleomagnetic evidence for continental drift without any assumptions about the long-term behavior of the geomagnetic field. The comparison of paleomagnetic results for one age cannot yet be made with sufficient accuracy to test directly the dipole character of the mean geomagnetic field.
*(Runcorn, 1964: 688; reference to Blackett et al. is the same as mine)*

### 7.13 Jeffreys attacks paleomagnetism and its support for mobilism

Although I was very friendly with Jeffreys and he used to come to our colloquia, he was always a very difficult person to have an ordinary scientific discussion with because he is a man of very few words. He writes very fluently but talks very hesitantly and it was notorious that if you went to see him and asked him something you might be faced with embarrassing silence.
*(Runcorn, 1984 interview with author)*

After election to the Plumian Chair of Astronomy and Experimental Philosophy at Cambridge in 1946, Jeffreys occasionally visited his old department (Department of Geodesy and Geophysics), where he learned intermittently of the new work in paleomagnetism and its support for continental drift to which he was adamantly opposed.[14] In the fourth edition of *The Earth* (1959), he argued that the mechanism difficulties against polar wandering were bad enough, but those against drift were insurmountable. Hence, the paleomagnetic method must be flawed because the continents could not have drifted.

Evidence from rock magnetism interpreted as in favor of polar wandering or continental drift has accumulated in the last few years ... For some time the results for rocks of the same geological age were consistent with a shift of the magnetic axis, and on the assumption that the magnetic axis was always near to the axis of rotation they were held to show that the axis of rotation has shifted in the Earth. Examination of suggested causes of displacement had led to nothing satisfactory, but the mechanical arguments against polar wandering are nothing like so strong as those against continental drift. However, recent work ... has shown rock magnetization in India and Australia in directions far from the direction of the field indicated by measures in the northern continents, and it has been asserted that they prove continental drift. But the evidence for irregularity of the ocean floor is now stronger than ever; and I shall continue to believe that a viscous material would flatten out under gravity however many physicists treat it as of no importance.

*(Jeffreys, 1959: 369–370)*

Although Jeffreys might have found somebody who claimed that the results from India and Australia "prove continental drift," so far as I know no paleomagnetist ever made any such statement in print.

Jeffreys also raised two specific reliability difficulties against paleomagnetism, one based on his own experience; the other was Graham's magnetostriction difficulty (§7.4). Recalling his previous work in handling magnets, he suggested that remanent magnetization of a sample is probably disturbed when it is collected and transported to the laboratory.

When I last did a magnetic experiment (about 1909) we were warned against careless handling of permanent magnets, and the magnetism was liable to change without much carelessness. In studying the magnetism of rocks the specimen has to be broken off with a geological hammer and then carried to the laboratory. It is supposed that in the process its magnetism does not change to any important extent, and though I have often asked how this comes to be the case I have never received any answer. In fact the modern study started with the announcement that many old rocks showed magnetism in opposite directions to the present field, but later work appeared to show that the magnetism at neighbouring geological dates appeared to concentrate about a direction and its opposite; since then reversals have been ignored, it being usually supposed that the Earth's general field is liable to sudden reversal but not to intermediate shifts. The reason for the reversal, it being supposed genuine, remains unknown.

*(Jeffreys, 1959: 371)*

As far as Jeffreys was concerned, even reversals may not be genuine, but may be caused by removal and transportation of samples from their collection point to the laboratory. Regardless, if the mere handling of permanent magnets can alter their magnetization, smashing sampled rocks with geological hammers was even more likely to alter, even reverse, their remanent magnetization. After summarizing Graham's difficulty based on magnetostriction, Jeffreys concluded:

... it would explain why the directions of magnetization of rocks over a wide area are nearly coincident but depart from the general field, without there needing to be either polar wandering or continental drift. He [Graham] gives examples that appear difficult to explain on either of the latter hypotheses.

*(Jeffreys, 1959: 371; my bracketed addition)*

Either he did not know of the responses that had been made to Graham (§7.5) or thought them not worth mentioning. Paleomagnetists did not respond publicly to Jeffreys, believing him unlikely to change his mind, but also because his influence on the mobilism debate was waning. Irving (September 2000 note to author), for example, "cannot recall anyone taking Jeffreys' statements on drift seriously in the late 1950s. His prejudiced statements were treated as comic relief." Irving also noted that one of the first things done at Cambridge when they began working in paleomagnetism was to determine if hammering samples made any difference.

This was one of the first things we tested. I can remember Hospers hammering specimens repeatedly yet their magnetism remained consistent. Hospers did it in the Earth's field and also in degaussing coils.

*(Irving, May 2000 comment to author)*

Runcorn, typically so quick to respond to criticism of paleomagnetism, probably thought a public response would be pointless.

But a public response may have been helpful. I say this because some did take Jeffreys seriously. John Sclater, who did his graduate work in the Department of Geodesy and Geophysics at Cambridge University in marine geophysics during the first half of the 1960s overlapping with Vine and McKenzie, later discovered a relationship between heat flow of oceanic crust, its subsidence, and its age that was explicable in terms of cooling plates (Sclater and Francheteau, 1970). He was definitely influenced by Jeffreys, retrospectively remarking that he took Jeffreys' criticism of paleomagnetism "very seriously, since he was generally credited with having made prewar geophysics into a respectable discipline" (Sclater, 2001: 142). Sclater certainly could have understood paleomagnetics while a graduate student, but he took Jeffreys at his word and, I suspect, saw no reason or was not encouraged to work through the pro-drift paleomagnetic papers himself. I also suspect that Sclater was not alone. Indeed there was a substantial disconnect during the late 1950s and early 1960s between pro-drift paleomagnetists and marine geologists and geophysicists, be they at Lamont, Scripps, or at Cambridge University in the Department of Geodesy and Geophysics. This may seem surprising, especially at Cambridge; however, once Runcorn left Cambridge for Newcastle and took most of the equipment, paleomagnetism in the Department of Geodesy and Geophysics just limped along, without strong leadership.

### 7.14 Concluding remarks: the prevalence of the three research strategies

If I have shown anything in this and earlier chapters, it is the prevalence of RS1, RS2, and RS3 in the development, extension, attack, and defense of controversial hypotheses related to the drift debate. The paleomagnetists who played key roles directed these strategies *in a self-critical way* against their own work, anticipating difficulties that did or could emerge, thereby strengthening their own position. Hospers realized he needed statistics to deal with the dispersion endemic to paleomagnetic data; R. A. Fisher provided them. Bruhnes, Graham, and Irving developed field tests. Irving, Clegg, and Creer discovered that fine-grained red beds give coherent results and focused on them. R. F. King, Runcorn, and Creer considered the question of whether sediments provide a reliable record of the geomagnetic field. Creer, As, and Zijderveld developed ways to "clean" magnetic samples in the laboratory. Efforts to obtain well-dated samples also reflected the desire to avoid reliability difficulties. Irving, Creer, Nairn, Runcorn, and Opdyke used paleoclimatology to check the reliability of paleolatitudes determined from paleomagnetism on the basis of the GAD model of the geomagnetic field. Hospers, Campbell and Runcorn, Creer, Turnbull, Irving, and Green presented field evidence to buttress the GAD hypothesis, and Runcorn provided theoretical support. Blackett, Clegg, and Stubbs appealed to the general principle of uniformitarianism to justify the GAD hypothesis. Once paleomagnetists turned their attention to the question of relative displacement of the continents, they anticipated and began to address what they realized would become missing-data difficulties. Eventually APW paths of varying completeness and temporal range were developed for different landmasses. Clegg and company worked in India, where they confirmed Irving's initial results. After Creer's pioneering APW path for Britain, Runcorn turned his attention to North America, Irving and Green obtained a path for Australia, Creer traveled to South America, Nairn to Africa, Turnbull, Bull and others to Antarctica, Soviet paleomagnetists got poles from the USSR, and the Japanese from their own country.

By directing research strategies of the second kind (RS2) against their own work, by identifying immediate and potential difficulties, and by employing research strategies of the first kind (RS1) to avoid or remove difficulties before they became dangerous, paleomagnetists showed themselves capable of stepping aside from their main path to inspect what they were doing and to make sure they were not fooling themselves about the progress they thought they were making. At the same time, for the most part, they worked like crazy, pushing ahead with their surveys. Mindful that they might be wrong, their confidence helped drive them to develop a strong case for mobilism. Doubtless there were moments of perhaps not fully justified euphoria, but their confidence did not become over-confidence, nor did they allow their own skepticism to deter them from their task.

My accounts of the early attacks against the paleomagnetic case for mobilism and the ensuing rebuttals show the dominant role played by these three research strategies. Critics used them to attack paleomagnetism and mobilist paleomagnetists acted in accordance with them. Graham raised his magnetostriction difficulty (RS2), which was rebutted by Blackett, Clegg, Du Bois, Irving, Opdyke, and

Runcorn using field results (RS1), and by Stott, Stacey, and Kern based on experiment and theory (RS2). Cox suggested that his results from the Siletz River Volcanics were best explained by very rapid polar wander, thereby raising an anomaly and theoretical difficulty with continental drift (RS2); retaining continental drift, Irving disposed of Cox's difficulty by suggesting (now generally accepted) local tectonic clockwise rotation of the Oregon Coast Ranges (RS1). Hibberd raised a comparable anomaly difficulty with the overall mobilist interpretation of the APW paths, and offered instead his single spiral path and a fixist interpretation (RS2 and RS1). Creer and Irving removed Hibbard's anomaly difficulty (RS1), raised difficulties against his spiral path (RS2), and Creer argued that the mobilist solution was much to be preferred (RS3). Stehli raised an anomaly difficulty based on the distribution of certain warm water Permian marine invertebrates (RS2), arguing the superiority of the fixist solution (RS3). Irving and Runcorn both responded by removing his anomaly difficulty (RS1), raising difficulties with Stehli's solution (RS2), and by arguing that mobilism was altogether preferable (RS3). Stehli then countered by raising a theoretical difficulty, arguing that their pro-drift paleomagnetic view was unfalsifiable because they could adjust their paleogeographical reconstructions to fit any conceivable new data (RS2). Munk and MacDonald argued against polar wandering. But paleomagnetic results demonstrated motion of the paleogeographic pole relative to stable continental blocks (APW), some component of which might be attributed to polar wander. So Munk and MacDonald, reiterating various theoretical difficulties, attempted to discredit the whole paleomagnetic enterprise (RS2). Finally, the staunch fixists Billings and Jeffreys raised their own theoretical difficulties (RS2). Most paleomagnetists left Munk and MacDonald, and Jeffreys in peace, their arguments having been overtaken by events even before they were made, but Runcorn took Billings to task by showing that his doubts about the GAD hypothesis were based on his misunderstanding of timescales (RS1).

In a few short years, paleomagnetists in Britain, Australia, and South Africa had taken their subject from the study of what many considered a miscellany of obscure curiosities to the forefront of Earth science by reopening, globally and regionally, the moribund mobilism debate. They quickly recognized major difficulties and their critics had added others. They quickly answered these difficulties in a manner that the unbiased and knowledgeable reader would find satisfactory. But were paleomagnetists making inroads into the hearts and minds of the majority of Earth scientists who were not to be easily diverted from their own pursuits by the temptations offered by these new and remarkable paleomagnetic data? With a few notable exceptions, the answer is no. I shall later pay special attention to these exceptions, which include Hess (III, §3.12), Dietz (III, §4.6), and Bullard (III, §2.14), but before doing so I need to consider a major critique of paleomagnetism's support for continental drift, which became a stumbling block to its acceptance especially in North America.

# Notes

1 Although at this time it was generally taken that the empirical support for the GAD hypothesis extended globally back to the "mid-Tertiary" this is not, strictly speaking, correct. Going back in time, the question to ask is, "When did continental drift kick in?" That depends on where you are. In Iceland, Europe, and North America, which moved longitudinally mainly, when the tectonic effect kicked in is going to be least easily determined. It is in Iceland (Hospers) and next in Europe (Hospers used mainly Roché's data) and then in North America (Campbell and Runcorn) that empirical support for the GAD hypothesis was first found. Hence there was a tendency to believe that the GAD hypothesis went back globally to "mid-Tertiary" because the oldest rocks were (roughly for the most part) assigned to the Miocene. The Australia data were Plio-Pleistocene, younger than 4 Ma, and there was not enough time coverage to "see" tectonic effects. India, the other continent besides Australia with swift latitudinal changes, had no data from the late Cenozoic. The Antarctic data were Late Cenozoic but were not then accurately dated – also Antarctica does not move much latitudinally. Hence the hard evidence regarding the question of how far back did GAD extend is from Australia: the GAD hypothesis applies for the last 4 Ma. Evidence from elsewhere does not provide an older limit because longitudinal movements predominate. Hence saying globally that the GAD hypothesis is empirically supported back to the mid-Tertiary is not strictly correct, back through the Pliocene would be correct.

2 Irving's remark about preferring the broadsword brings to mind Antonio Pollaiuolo's engraving, *Battle of Ten Naked Men*, with combatants wielding their broadswords. Irving (and Runcorn and Creer) would have fit right in. The cover of David Hull's *Science as Progress* has a cropped section of Pollaiuolo's *Battle* on the cover (Hull, 1988).

3 I shall consider Cox and Doell's major 1960 critique of the drift interpretation of paleomagnetic data in the next chapter. Theirs was the most influential attack, and merits special treatment.

4 Stott and Stacey (1960: 2419) remark, "magnetostriction is the process of change in dimensions which a body undergoes during magnetization" but the term is now generally used in rock magnetism to denote "the converse effect of change in magnetization resulting from mechanically applied elastic deformation." They go on to say, "stress alone cannot induce magnetic moments, it can only change the directions and magnitudes of existing moments."

5 Actually, other paleomagnetists had not ignored magnetostriction. Irving, for example, discussed the idea in his Ph.D. thesis, although he called it the effect of "tectonic stress." He argued that the "effects of tectonic stress" on magnetization of the Torridonian were negligible because of a positive fold test on Ben Liath Mhor (Irving, 1954: 82). In one of his contributions to the suite of papers drawn mostly from Irving's and Creer's Ph.D. dissertations, Irving noted:

Mechanical strains during tectonic activity might have been responsible for the permanent magnetization. Yet the folded beds of Ben Liath Mhor, discussed in the preceding paper, have retained their directions of magnetization in spite of compressive forces sufficient to produce vertical tilt. The beds in the main sequence are horizontal or have gentle tilts. It is, therefore, most unlikely that tectonic movements have had any effect.

*(Irving, 1957c: 103)*

6 Cox told Glen a slightly different story. Cox suggested that Du Bois told Verhoogen about Irving and Deutsch's results from India in the same letter – there is no known surviving copy of this letter – and Glen goes on to suggest that Verhoogen subsequently encouraged Cox to get a North American Eocene paleopole. Here is what Cox said:

Verhoogen got an exciting letter from Phil Du Bois in which he reported that Ernie Deutsch, and I think also Ted Irving, had picked up paleomagnetic evidence [from the trap (basalt) rocks of the Deccan] for the northward drift of India. But at that time, there weren't good

Eocene [paleomagnetic] pole positions from elsewhere in the world. Their pole position for India fell in the South, somewhere off Florida. The simplest interpretation was northward drift of India.

*(Glen, 1982: 166; my bracketed additions)*

Although Verhoogen may have received a letter about Deutsch's Indian results, I do not think it could have been sent until after Cox had decided to get a North American Eocene paleopole. As a student in paleomagnetism in the Department of Geodesy and Geophysics, Cambridge, Du Bois could have been told directly by Irving. There was therefore ample time for Verhoogen to have received a letter from him. They had met in August 1954 at the Edelweiss Conference long before Cox decided on his sampling program. Deutsch's results, referred to in Cox's comment (by which Cox presumably meant the results of the Imperial College group of Clegg, Deutsch, and Griffiths) were not published until 1956; therefore it would seem that Cox could not have heard of them (in those days all journals except for *Nature* came by sea-mail) before embarking on field sampling in 1955. It is, of course, quite possible that Verhoogen received two letters from Du Bois. Du Bois could have told him about Irving's work in the first and about Deutsch's in the second. I think that Cox can only have learned of the results obtained by the Imperial College group after he had begun sampling the Siletz Volcanics. He eventually did know about them, for they were published and he referred to them in his 1957 paper.

7  Irving wonders how Verhoogen could have been so lucky to get Cox and Doell. Runcorn did not do too badly either with Hide, Irving, Creer, and later Opdyke. Hospers was a lucky find for Runcorn, and the same may be said for Irving and Opdyke. However, Runcorn certainly recognized quality when he saw it, getting both Irving and Opdyke as students. He actively pursued Creer to work in paleomagnetics. While Hide was an undergraduate at the University of Manchester, he helped Runcorn with the mine experiments undertaken to test Blackett's theory, and Runcorn became Hide's Ph.D. supervisor at Cambridge (Runcorn, 1984 interview with author; see also Lowes, 1998: 705).

8  A local rotation already had been used by Clegg and company to describe the angular movement of Spain relative to the rest of Europe (§3.13). As explained in Chapter 8, Carey and Argand had made such a prediction, and Clegg *et al.* confirmed it in Clegg *et al.* (1957: 227). Thus, unlike Irving's case, the rotation had already been suggested and was confirmed by paleomagnetic study. Irving invoked rotation to remove difficulties; on any theory, Cox's results were completely unexpected at the time. Irving probably was also influenced by Carey, and both Carey and Argand might have conditioned Irving to think in terms of rotations about local vertical axes. Irving certainly knew of Carey's penchant to postulate rotations. In fact, Carey (1958: 336) even postulated a rotation of Oregon with his Mendocino Orocline, and Irving definitely knew of his proposal while writing his textbook (see, Irving, 1964: 249–250). In a more recent discussion of the issue Irving remarked, "I don't think I [knew of Carey's proposal] when writing my 1959 review. But Carey may well have talked to me about it" (Irving, May 2000 comment to author; my bracketed addition).

9  A much fuller account of the Siletz rotation is given by Cox (1980) based on much subsequent work by him and his students.

10  Runcorn's paper was published in March 1959. Chang and Nairn (1959), which appeared at the end of January, had postulated relative movement between China and Europe with a lower latitude for China. Thus, some work already had been done. Runcorn probably submitted his paper before he knew of the results from China, especially since his paper was not published until Stehli had a chance to write a reply. As I previously explained, Nairn already had suggested, on the basis of paleoclimatological considerations, that Eastern Asia had moved relative to Europe.

11  Birch is credited with the suggestion on p. 275 in Munk and MacDonald (1960).

12  The asterisk referred to the following note about the failure of more recent geodetic measurements from Greenland to confirm the earlier measurements that appeared to

confirm westward movement of Greenland relative to England. Munk and MacDonald found the wonderfully entertaining quotation from Markowitz (1945).

There have been some reports of measured drift. For example, Jelstrup in 1932 found the longitude of Greenland by radio-link to exceed by 5" the longitude determined by Borgen and Copeland in 1870 using lunar culminations and occultations. Markowitz (1945) has examined these and other reported drifts and ascribes them to experimental error. If these older determinations were to be accepted they would mean that "... large shifts occurred until the invention of the telegraph, moderate shifts until the invention of radio, and practically none since then."

*(Munk and MacDonald, 1960: 282)*

I guess the social constructivists were right. Worlds are what we say they are.

13 Irving also considers Munk and MacDonald's suggestion that Arkell was under pressure from the paleomagnetists or anybody to be absurd. "This is a gratuitous statement. Anyone knowing Arkell, even slightly, would recognize that pressure would be the last thing he would have yielded to" (Irving, March 2001 comment to author).

14 Raymond Hide told an interesting anecdote during his acceptance speech upon receiving the 1997 William Bowie Medal from the AGU that reflects great credit on Jeffreys.

At the time of these events, Jeffreys, the first British medalist (of the Bowie Medal), was working in seismology. His brilliant reputation as a writer was never quite matched by his lecturing and conversational skills, but as students we were advised that "his grunts and murmurs should always be taken seriously, because they were likely to contain pearls of wisdom." It was a remark consisting of no more than four words uttered by Jeffreys in 1951 that prompted my own interest and subsequent work in dynamical meteorology, a subject in which Jeffreys had made seminal contributions a quarter of a century earlier. The remark was made as he passed through the large hut where several graduate students in the Department of Geodesy and Geophysics at Cambridge University were engaged in various unrelated laboratory studies. When I showed him some of the flow patterns produced in an apparatus I had designed for investigating thermal convection in a rotating fluid, he muttered "looks like the atmosphere" and wandered off into the field outside the building, leaving me to ponder the implications of what he had said.

*(Hide, 1997)*

# 8

# Major reaction against the paleomagnetic case for mobilism and early work on the radiometric reversal timescale: 1958–1962

Cox and Doell ... compiled all that beautiful data, recognized that it defined polar wander curves, and tiptoed away saying that it could not really mean anything.
(*Warren Hamilton, July 8, 2002 email to Irving, author copied*)

## 8.1 Introduction

Although most early criticisms of the paleomagnetic case for mobilism have been covered in the previous chapter, the most influential, that written by Cox and Doell, has not. Their (1960) "Review of paleomagnetism" (hereafter, *GSA* review) was published in the *Bulletin of the Geological Society of America*, whose prestige and wide circulation meant it was seen by most North American geologists and read by many there and elsewhere, at least the abstract if not the entirety. The *GSA* review provided an unsurpassed summary of paleomagnetic results up to the end of the 1950s, and here I ask the question, did it also provide an adequate review of the application of them to the mobilism debate? It is not unreasonable to expect that it ought because mobilism had, after all, come under active discussion largely because of the flurry of new research in paleomagnetism in the later 1950s.

Doell and Cox (1961a) also wrote a second major review (hereafter, *AG* review) for *Advances in Geophysics*, which most geologists did not read, and which had little or no effect on the evaluation of the paleomagnetic case for continental drift. Many asked for reprints of the *GSA* review, few requested the *AG* review.

All I recall is that we (and GSA) soon ran out of a large number of reprints, but hardly anybody asked for one of the Ad. in Geophys. separates. I guess a lot more read the Bull of the GSA than the other publication.

(*Doell, January 26, 2001 email to author*)

The *GSA* review had a lasting effect; four years later in 1964, because demand was so high, the GSA decided to reprint it as a separate monograph, thus further extending the review's circulation. Although the *GSA* review remains a very good description of methods, its list of results and discussion of continental drift was by 1964 out-of-date in this fast moving field.[1] I shall deal mainly with the *GSA* review because of its greater influence.

Cox and Doell, along with Graham, were, I believe, the only major paleomagnetists who had actually observed oblique magnetization directions (the trademark of latitude change and continental drift) and who had not, by 1960, accepted continental drift or at least become inclined toward it. The *GSA* review did not reject continental drift outright, but inclined toward fixism; generally it argued in favor of polar wandering without continental drift. Given the circumstances of the time, the *GSA* review likely did more than any other publication to discourage Earth scientists from seriously considering the developing paleomagnetic support for mobilism. Because it was well written, and sanctioned by the world's largest geological society, and because the authors were paleomagnetists, the *GSA* review carried real weight with those who were disposed against mobilism, as most were, giving them little reason to reconsider.

Specialists may have their own good reasons for dissenting, and their arguments merit special scrutiny, and I am devoting much of this chapter to those of Doell and Cox. I shall be critical of the *GSA* review, as were several at the time. I shall discuss Runcorn's and Irving's replies to the *GSA* review. Runcorn responded publicly. Irving's responses were both public and private, the latter the more interesting. Irving and Cox corresponded with each other throughout 1960 and 1961. Irving's letters are forthright, Cox's measured. I also shall examine Cox and Doell's application of Egyed's proposed paleomagnetic test of Earth expansion (§6.4); they were the first to take up his suggestion. Then I shall return to the Cox–Irving letters because they reveal much about the initiation, early planning, and development of the radiometrically dated reversal timescale, which, in the mid-1960s, became important to the confirmation of the Vine–Matthews hypothesis and seafloor spreading.

Before I begin, I want to make it quite clear that my critique of the *GSA* review should in no way be viewed as detracting from the later wonderful work by Doell, Dalrymple, and Cox in developing the radiometric reversal timescale. I also want to recognize that Doell kindly answered a number of my questions about their reviews, and disagreed with part of my interpretation. Doell was a very capable scientist; he knew how to get things done, and how to keep a research project going. Cox had the ability to focus on what needed to be done. Both combined excellent field and laboratory work with critical analysis and clear writing.

Some may consider my discussion of the reviews inappropriate, that I pry too deeply, or that my judgment is faulty. Be that as it may. The reviews were important. Here were two highly intelligent, young, newly graduated and, at the time, little known scientists, striving for a place in the world of research, who were, as will shortly be explained, educated and worked deep inside the fixist tradition, and who had assumed the task of reviewing a set of new results that had all the earmarks of being flatly at odds with that tradition and potentially capable of overturning it. The new paleomagnetic results were, after all, the first geophysical evidence of large-scale, long-term motions of continents relative to the geographic pole and to each other. Were they to write favorably of the results and support their implications and so to challenge the prevailing fixist milieu in which they had been educated, worked, and lived, were they to reject them outright, or were they to search for some middle ground? But experience over earlier decades had shown there was no middle ground. One cannot have a little bit of mobilism at some time and not at others, in one place and not everywhere; mobilism did not come in bits. Once large-scale continental drift is accepted in one place and its consequences worked out, little is left of fixism. Perhaps, when they wrote their reviews, Doell and Cox were not fully cognizant of this dilemma, a dilemma that everyone who had extensively examined deeply the ever-expanding global record of biotic, paleogeographic, and tectonic intercontinental disjuncts had had to face (see Volume I). Now there was an entirely new geophysically based set of disjuncts, the intercontinental differences among paleomagnetic polar paths, which had been determined entirely independently of the above, and which, nevertheless, showed close and, by the end of the 1950s, unavoidable parallels with them. Over the previous five years, paleomagnetists had demonstrated the remarkable intracontinental agreement between paleomagnetic results (latitudes) and the evidences of paleoclimate, and with many aspects, general and particular, of the reconstructions of Wegener (and Köppen), du Toit and Carey (§3.12, §5.12–§5.14, §5.16). As I shall explain, all this, the vital underpinnings of the paleomagnetic case for continental drift, went essentially unremarked in their reviews.

## 8.2 Doell and Cox and their milieu and their attitude toward mobilism before 1958

I want now to try to place Cox and Doell's *GSA* review in the context of their education and their earlier assessment of the paleomagnetic evidence for continental drift. First I consider their schooling and early work (details in Glen (1982)). Doell and Cox overlapped at the University of California, Berkeley, from fall 1954 through June 1955. Both were late-bloomers. Each had spent time in the Army, had two false starts in college, and gained field experience by working in exploration geophysics for United Geophysical Co. (Doell) or the USGS (Cox). Doell became a settled student at the age of twenty-seven upon his return to Berkeley in September 1950. He majored in geophysics, and obtained his A.B. degree in June 1952. Remaining at Berkeley, he obtained his Ph.D. in 1955. Cox, after two unsuccessful periods at Berkeley, returned in fall 1954, at the age of twenty-seven. He also majored in geophysics, got his A.B. degree in June 1955 and Ph.D. in June 1959.

Doell summarized what happened to him once he had returned to Berkeley and decided to major in geophysics.

As an undergraduate I intended to return to United Geophysical [the exploratory geophysics company Doell had earlier worked for] after my A.B. degree, so I naturally concentrated on

Seismology (Byerly) and Physics of the Earth – or should I say "all other Geophysics" (Verhoogen). Further, Byerly was my studies advisor and I did a sort of undergraduate research paper under Verhoogen. There were others from whom I learned much and who also no doubt "influenced" me. Having had a somewhat "undistinguished" previous academic experience with Physics and Mathematics majors – fields that really didn't interest me as long time career choices (I later realized), the entire department and atmosphere at Berkeley was a very exciting and revelational experience for me. I sort of look at it as the whole being influential, in that I gave up Petroleum work as a career choice and chose academia.

*(Doell, January 26, 2001 email to author; my bracketed addition)*

In some of his classes, Doell was taught a little about geology outside North America, but most instruction concerned local and regional geology.

In [Francis J.] Turner's Historical Geology course, our examples from the founders gave us a look at Europe – or the British Isles at least. In an Economic Course of [Garniss] Curtis there was stuff on South Africa (gold and diamonds). But by and large I recall most topics being illustrated with more local geology (California and/or North America). But I also recall learning about more general stuff; e.g., the distribution of the great pre-Cambrian shields, the Circum Pacific "Ring of Fire," etc. There is one exception I recall, though, and that was about the recumbent folds of the Alps. I dare say there was some other stuff, but this is what comes to mind.

*(Doell, January 26, 2001 email to author; my bracketed addition)*

Neither continental drift nor polar wandering received much attention. Doell does not think that his teachers had any strong opinions about either of them. His view, as he notes, differs from Glen's (1982), which was that, with the exception of Verhoogen, the Berkeley faculty in geology and paleontology were against mobilism.

Contrary to what Glen reports as coming from Cox, I think that by and large the staff at Berkeley was without strong or fixed opinions about drift and pole wandering. I would describe the general state as one of openness or indecision – at least among those who thought about it much. I sure don't recall any dogmatizing on the subject. It is only in Verhoogen's course [Geophysics 122] that I recall any significant discussion, though I also think it must have at least come up in several others. Geophysics 122 covered a fair amount about gravity, earth tides, rotation, and thermal history (as well as many other subjects). I recall J. V. [John Verhoogen] applying the current thoughts on these subjects to the ideas of drift and wandering, especially his thoughts on the possibility of convection to help with the thermal problems. In general, I think it's fair enough to say that – outside of Geophysics 122 – continental drift may well have been mentioned in some or several other courses but not to any great length.

*(Doell, January 26, 2001 email to author; my bracketed additions)*

In attempting to remember his own attitude toward continental drift and polar wandering, Doell recalled that he studied them in a "Special Studies" course, but cannot remember what he thought of them.

Regarding my own opinion, it also may bear some on the above as well. We had a series of "Special Studies" courses ... These were reading courses wherein the students were given a bunch of stuff to read on a limited – usually controversial subject, as I recall – and were expected to write a report and/or lead a discussion with the course supervisor(s) at the end of the term. Some of us found these sort of fun (and an easy way to get a good grade) and I recall taking several to fill out credit requirements. But of particular interest here is the one on Continental Drift, which I think I shared with one or two others. I recall Verhoogen, Turner and Curtis being the supervisors. Having said this, I can't now remember what I might have concluded, or what side I might have argued, only that I would have been assigned all the pertinent literature on both sides and have studied it in detail. I would guess that I wouldn't have been strongly one way or the other, but whatever the case my then ideas will have been thoroughly obliterated by all that I've thought and studied about the theory since. About what I may then have thought about Polar Wandering, I haven't a clue.

*(Doell, January 26, 2001 email to author)*

Pulling Doell's recollections together, it appears that there was at Berkeley little detailed instruction or discussion about the geology of regions other than North America, that continental drift and polar wandering received little attention in his classes, that nobody warmly endorsed drift, that it was recognized as a long-standing controversial issue, and that Doell familiarized himself with it but had no strong opinion for or against. The classical mobilist texts by Wegener, Argand, and du Toit do not appear to have been recommended reading.

A better fix on Doell's attitudes is obtained from his Ph.D., completed in September 1955. My reading leads me to believe that he did not favor continental drift and was cautiously supportive of polar wandering. After briefly describing some of the paleomagnetic work that had been done on rocks no older than Tertiary, he turned to studies of pre-Tertiary rocks. Doell (1955a: 44) singled out the work by Graham, and Clegg and company as "possibly the most important studies that have appeared recently." Although he was aware of Irving's and Creer's work from reading Runcorn's (1955a) review, he had not seen their dissertations, and apparently had not seen their paper describing Creer's APW path for Britain (Creer, Irving, and Runcorn, 1954). He seems to have known little in detail of their work, likely reflecting the unfortunate tardiness of the Cambridge group to publish (III, §2.21).

Other important studies on Pre-Cambrian and Paleozoic rocks of Great Britain by Irving and Creer have been reported (Runcorn, 1955). However, the details of these studies have not yet appeared.

*(Doell, 1955a: 44; Runcorn, 1955 is my Runcorn, 1955a)*

Doell then described his own study of Paleozoic and Pre-Cambrian rocks from the Grand Canyon, and listed his assumptions.

If conclusions are to be made on the basis of the measurements of this report regarding the earth's magnetic field and direction of the axis of rotation in the past, certain basic assumptions must be made. These are:

1. that the rocks in question received their magnetization at the time they were formed (if this be the period of time for [which] the conclusion is drawn),
2. that this magnetization has been stable since that time, i.e., has not changed its direction with respect to the bedding plane,
3. that there have been no large scale rotations or displacements of the area, from which the rocks were collected with respect to the earth as a whole, and
4. that coincidence of the magnetic dipole field and the axis of rotation was a fact in the geologic past as well as the present.

*(Doell, 1955a: 73 and 75; my bracketed addition)*

The first, second, and fourth assumptions dealt with the paleomagnetic method; taken at its face value, the third seems tantamount to rejecting continental drift. Doell said that each assumption was needed to infer polar wandering. He cited Clegg *et al.* (1954a) in support of the first and second assumptions. He also appealed to Graham (1949).

Graham's studies (1949) on folded sediments and conglomerates have established the validity of the second assumption for some rocks. It is also evident that many of the rocks reported on in this study have not had their directions completely altered to that of the present earth's field, nor have the Triassic rocks studied by Clegg *et al.* (1954).

*(Doell, 1955a: 75; Clegg* et al., *1954 is my Clegg* et al., *1954a)*

Aware of Graham's magnetostriction difficulty (§7.4), he argued that it was not relevant because his own rocks had not been deeply buried. He then turned to (3), noting:

If the third assumption holds, the earth's magnetic field must have had different directions in the Pre-Cambrian and Permian from that which it has at present. If the assumption is not valid, large scale movements of the Grand Canyon area are indicated.

*(Doell, 1955a: 75)*

He rejected continental drift as an explanation of the "different directions in the Pre-Cambrian and Permian," believing it reasonable to do so.

When more data of this nature from various parts of the world have been obtained, the validity of the third assumption will no doubt be determined. At present, it seems reasonable to assume a permanency of the continents and oceans in order to make tentative postulates concerning the position of the poles in the past.

*(Doell, 1955a: 75–76)*

He did not say why it was reasonable to accept fixism.

Turning to (4), essentially the GAD hypothesis, he appealed to Hospers' empirical and the usual theoretical support.

The fourth assumption, made by Hospers (1954) in his review of the measurements on Tertiary rocks, must be generally based on the principle of Uniformitarianism. There is, of course, some theoretical basis for such an assumption in the modern dynamo theories (Runcorn, 1954; Bullard, 1949; Elsasser, 1950).

*(Doell, 1955a: 76; Hospers, 1954 is my Hospers, 1954a)*

While working on his dissertation, Doell seems to have had little sympathy with continental drift, which left him with polar wandering. Assembling poles from both Europe and North America, he constructed a single polar wander curve.

In his thesis, he ended his discussion cautiously, almost retracting his earlier mild support of polar wandering.

These conclusions on polar wandering are not to be taken too seriously, but are presented mainly as an indication of what it will be possible to do with this type of information when a great many more data from all parts of the world have been measured.

*(Doell, 1955a: 78)*

Not only was he unwilling to concede that continental drift was a reasonable possibility, he was unwilling to take polar wandering "too seriously" either, even though he thought the paleomagnetic method was correct, and that his, and the other paleomagnetic results that he cited, were reliable.

In his thesis, Doell gave little consideration to continental drift. For example, he never explicitly distinguished the European from the North American poles, surely a necessary step to take for someone seriously wondering about continental drift. In fact, from North America only his and Graham's Silurian poles were given. In addition, although he (Doell, 1955a: 156) referenced as "in press" Runcorn's (1956a) paper which contains Runcorn's poles from North America, Doell did not include them in his figure. Nor did he reference Creer, Irving, and Runcorn (1954), in which the concept of APW paths was introduced; he seemed to have been unaware of it, yet he does (1955a: 155) refer to a paper by H. Mastuzaki, K. Kobayashi and K. Momose that appeared in the same 1954 volume, albeit an earlier issue, of the *JGG* as the Cambridge paper. Doell referred to none of the papers presented at the Rome meeting, including Clegg *et al.* (1954b).

I therefore think it is fair to say that, at the time of writing his thesis, Doell was not considering continental drift seriously and thought fixed continents "more reasonable"; he thought polar wandering more reasonable than continental drift as an explanation of the paleomagnetic data. His stance was neutral; his statements, carefully qualified. He was content to set aside serious consideration of continental drift until there were more data. His arguments were based solely on paleomagnetic data. He did not discuss the global, tectonic, paleoclimatic, and paleobiogeographic evidence that bore on mobilism (Volume I).

According to Doell, his views about continental drift and polar wandering underwent no substantial change while he was at the University of Toronto (1955–6), where he helped set up a paleomagnetic laboratory, or MIT (1956–8), where he began his compilation of paleomagnetic results. At Toronto, Doell remembered talking about "global ideas" with J. Tuzo Wilson and Jack Jacobs, but does not recollect what was said about continental drift. At that time, Wilson was a vocal fixist.

I do seem to recall that Tuzo, Jack Jacobs and I did a lot of talking about all sorts of "global" kinds of ideas. I certainly found Tuzo to be a very imaginative person and always full of all sorts of thoughts and ideas, many of which Jack and I found quite amusing. Regarding an attitude on C.D. [continental drift] or P.W. [polar wandering], I just plain do not remember.

(Doell, January 26, 2001 email to author; my bracketed additions)

Nor was drift a topic of general discussion among Earth scientists at MIT. However, he did talk about continental drift and polar wandering with Gordon MacDonald, who had become and would remain one of mobilism's most vociferous critics (§7.11; III, §2.7, §2.8).

If there was a general "attitude" at MIT when I was there, I certainly don't recall what it might have been. I would guess there might have been varied private opinions among the staff, but I'd also guess that in general it would not have been a very "hot" topic at all. However, Gordon J. F. MacDonald was a new Assistant Professor there, as was I. We became relatively good friends and regularly played squash together. We certainly did, as I recall, frequently discuss C.D. and P.W. In fact, I recall that he and I did a series of presentations in the department wide seminar. I would guess that I would have reviewed the paleomagnetic data supporting the ideas, but I can't remember if I endorsed it or not, and I remember Gordon doing a tour-de-force in Math in which only he and I had an inkling of what he was doing!

(Doell, January 26, 2001 email to author)

MacDonald did not hide his views. Doell remained neutral.

I sure don't think Gordon McD. and I ever came to any conclusions. As you know he was pretty firm in his ideas, and I think I remained relatively open on what you call the mobilist ideas.

(Doell, February 2, 2001 email to author)

From his time at Berkeley until his return to California to work at the USGS (1959), Doell remained neutral toward mobilism and fixism.

Glen (1982) discussed the attitude toward mobilism of the Berkeley faculty in geology and paleontology at the time Cox was there. Glen (1982: 164) claimed that Earth scientists in both departments, like most in the United States, were fixists, that Verhoogen was the notable exception, and that many of the students favored mobilism. Glen (1982: 163–165) characterized Verhoogen as "an early advocate of drift." Verhoogen told Glen, "I became favorably disposed to it, just on the general principle that you don't believe what your professors tell you." Glen also reported that Verhoogen spoke in favor of mobilism in 1948 at a seminar sponsored by the Paleontology Department at Berkeley, and that Verhoogen believed in convection, and saw no geophysical objections to mobilism. He also told Glen, "'he saw no physical objections to it' and thought that 'the lack of a mechanism did not preclude it.'" But Verhoogen's seeming enthusiasm for mobilism did not extend to his publications during this time.[2] Glen's account also differs from Irving's experience with Verhoogen. Irving recalled:

I talked with Verhoogen at length on two occasions in the 1950s – first in Cambridge in the summer of 1953 or 1954, and in Berkeley in July or August 1957, when I was en route from Australia to the Toronto IUGG meeting, and staying for several days with Allan Cox. Verhoogen was not an active supporter of drift. He had me give a talk to the Geology Department – I gave the first APW (Irving and Green, 1958) for Australia, hugely different from Europe and North America. It was the same talk I gave at Toronto a week or two later. He asked many questions, as did other members of the staff, but neither he nor they, as any drifter would have, shouted "Hurrah! Wonderful! Just what Wegener and du Toit said!" He was not a drifter, and I am pretty sure no one in the audience was. He was not aggressively hostile, but it is my recollection that at the time he had a long, long way to go. Anyway, as a theoretician, and given the climate of the times, it was very unlikely that Verhoogen would have been a supporter of drift, because the evidence for it, both old and new, was field-based especially in the southern continents, and that was not his strength.

*(Irving, March 2001 comment to author)*

Even Cox did not single out Verhoogen when giving the citation for Irving's Bucher Medal. Cox (1979: 659) recalled, "It would have been difficult in 1956 to find 20 earth scientists in North America who agreed with Irving [that continental] drift had occurred]." You would think that Cox would have singled out his mentor Verhoogen, if he had believed in drift.

Glen based his claims about the general attitude of the faculty and students toward mobilism, in part, on what Cox told him had happened at a meeting on continental drift, sponsored by the Berkeley Geology Club. Cox, a co-founder of this student club, told Glen:

During the first series of meetings at which student papers were delivered, they decided to take a vote on the question of drift, and "the ayes had it." The advocates of drift were a strong majority. Some of the faculty were "outraged" on hearing this.

*(Glen, 1982: 166)*

Glen also quoted a passage from a 1958 letter that Cox had sent to Doell.

The Geology club has been a great deal of fun. The reaction of FJT [Francis J. Turner] and the faculty was that they didn't care what we did, as long as we didn't *involve* them in the proceedings ... Several of the faculty were P.O'ed because they weren't invited to the Continental Drift Symposium. One of the reasons the Club has caught on, I suspect, is its slightly revolutionary flavor.

*(Glen, 1982: 166; Glen's emphasis and bracketed addition)*

Were those who were upset, upset because they were not invited, or because "the ayes had it," or because of both? And, which members of the faculty were upset? Elsewhere, Glen (1982: 165) correctly, I believe, noted that the paleontological community was particularly anti-mobilism, "especially those with interest in the later Mesozoic and Cenozoic record of North America." Berkeley had two paleontologists who specialized in Tertiary life and were outspoken critics of mobilism: Ralph Chaney (I, §3.8), an expert on North American Tertiary flora, and J. W. Durham, a specialist in Tertiary marine fauna. Durham (1952, 1959) opposed polar wandering and continental drift throughout the 1950s, and argued against them at the 1955 colloquium on continental drift at Cambridge (§3.10).

The Cox–Glen characterization of the attitude of the Berkeley faculty in the geology and paleontology departments toward mobilism differs from Doell's recollections. Both may be correct. Doell's above remarks pertain to 1952 through 1955, while Cox's comments to Glen pertain to the period from 1955 through 1959 approximately. Doell and Cox overlapped at Berkeley only from fall 1954 until June 1955. The difference between Doell's and Cox's (and Glen's) accounts may therefore reflect a change in faculty outlook toward fixism in the mid-1950s. What could have provoked such a change? The vote by the students in favor of mobilism at the Berkeley Geology Club may have been the catalyst, but negative reaction to the upsurge of paleomagnetic evidence for mobilism could have been the underlying cause. Verhoogen knew about the developing paleomagnetic support for mobilism, and encouraged Doell and Cox to obtain paleomagnetic poles for North America. Chaney and Durham, perhaps even Francis Turner, could have become alarmed.

What about Cox's own attitude toward mobilism? Glen did not say how he voted at the Berkeley Geology Club, and Cox may not have told him. But he did tell Glen (1982: 158) that he "regarded *Principles of Geology* (1951) by Gilluly, Waters, and Woodford as an influential and profound introduction to the subject." Cox read it in 1951 while uncertain as to whether to continue with chemistry or switch to geology. Because the first (1951) and second (1959) editions were popular, I shall digress and describe their rejection of continental drift to get a fix on what its readers were learning, and *not learning*, about continental drift.[3] Indeed, if Gilluly and his co-authors' assessment of continental drift is a fair indicator of their readers' familiarity with the literature on continental drift, it was certainly minimal; they raised a difficulty against the fit of continental margins on opposite sides of the Atlantic: "Indeed, this 'fit' at a depth of 10,000 feet is far less impressive, and there were notable differences in the continental outlines in Cretaceous or even in mid-Tertiary time" (1951: 530). They raised precisely the same difficulty in their 1959 edition; they can hardly have read Carey's 1955 paper in which he showed the good fit at the 2000-meter isobath (§6.11). They raised a difficulty with Wegener's explanation of fold mountain belts, noting (1951: 534), "even if we grant it for the Tertiary mountain chains we are left with the necessity of accounting for the formation of the many far older ones." They attempted to account for the tectonic disjuncts on opposite sides of the Atlantic with a particularly bizarre counterexample (1951: 530–531).

The apparent interruption of these fold lines by the ocean suggests, but does not prove, that they were formerly joined and have since drifted apart. The hazard in assigning that this occurred is shown by the fact that Atlas and Pyrenees ranges also end at the shore, and that although some geologists have tried to identify their western continuation in the Greater Antilles, it is clear from the patterns of the gravity anomalies and island arcs that the Antilles structures never did extend eastward to join the European and African ranges, but curve southward through the nearly drowned arc of the Lesser Antilles to join the Venezuelan ranges.

Argand and du Toit proposed that the counterclockwise rotation of Iberia formed the Pyrenees, which were not therefore a tectonic disjunct. Gilluly and company repeated the usual difficulty that tidal and centrifugal forces were too weak to move continents. On the positive side they did describe the theory of continental drift as "a brilliant *tour-de-force*," and noted:

Indeed, should evidence for continental drift ever become compelling, it may be that subcrustal convection currents will prove to be the driving mechanism that brings about the displacements.

*(Gilluly et al., 1951: 540)*

They correctly claimed that drift's solution to the distribution of Permo-Carboniferous glaciation was perhaps "its strongest argument," but found it unsatisfactory. Emphasizing the purportedly anomalous Squantum Tillite, they did not mention Wegener's and van der Gracht's 1928 suggestion that the Squantum might not be glacial (I, §3.12). They claimed instead that Wegener interpreted the Squantum Tillite as "a mountain glacier," which made no sense because it "was within a few miles of the Rhode Island coals." It was a phantom difficulty arising from their unfamiliarity with Wegener's and van der Gracht's work. They raised another phantom difficulty in their further attempt to find fault with drift's explanation of Permo-Carboniferous glaciation.

But even on this assumption [that the Squantum tillite is a mountainous glacier] both the Russian and Siberian *Glossopteris* localities must have been in the tropics, so climatic zoning of the plants is not an adequate explanation of their distribution.

*(Gilluly et al., 1951: 534; my bracketed addition)*

Again, they showed their ignorance of the mobilism literature. During the 1930s and 1940s, du Toit, Hoeg, and Jongmans had noted that the Russian and Siberian fossils had been mistakenly identified (I, §3.5). When Edwards (1955) discussed the mistaken identification of *Glossopteris*, noting that the mistake had found its "way into textbooks with astonishing rapidity," he should have added, "and where it continues to remain even after having been shown mistaken" (I, §3.9). Gilluly and colleagues (1951: 534) also belittled any attempt to explain Late Paleozoic climates by moving continents rather than climatic zones, and had little confidence in inferring reliable information about such climates based on extinct fauna.

Others besides Wegener have tried to explain the climates of the late Paleozoic by different groupings of the continents. All require special explanations for nearly as many phenomena as they explain. All the plants of the late Paleozoic are extinct; few paleobotanists have much faith in climate inferences from them, for there are many examples of later adaptations of plants to environments that differ widely from those in which they arose.

Their discussion in the second edition (1959) remained unchanged or little changed. They noted in their "Preface to Second Edition" that "all chapters have been rewritten and reorganized to take account of new developments." Either they did not care about new developments favoring continental drift or were unaware of them. They said nothing at all about the mid-1950s surge of paleomagnetic results favorable to mobilism, which surely rated as a "new development."

Cox thought Gilluly and company's *Principles of Geology* profound and influential; I suspect he agreed with the authors' assessment (1951: 444) that continental drift's "support does not seem substantial." If Cox did not read works by paleontologists and paleoclimatologists who favored continental drift, he probably was under the erroneous impression that they had little to offer. Given his interests and high opinion of this text, I suspect that he did not do so, and thus lacked the background to appreciate what Irving, Opdyke, and Runcorn, and later Blackett had written about paleoclimatology to buttress the GAD hypothesis and strengthen the case for drift.

As already noted, Doell himself recalled retrospectively that Cox and he had substantially the same opinion about mobilism, and his recollection is that they were neutral, if not somewhat sympathetic toward mobilism; the evidence from Doell's dissertation is that he himself was not sympathetic. What do their publications say? Both reported poles that were relevant to the mobilism question (Doell, 1955b; Cox, 1957). Doell discussed his poles from the Permian and Precambrian of the Grand Canyon in his Ph.D. dissertation (§3.8), and in his paper pointed out that they coincided with Runcorn's, but made no mention of continental drift; he seems not to have felt strongly about mobilism. Explaining why he said nothing about continental drift, he later noted, "I was certainly a 'very new kid on the block' and would have felt it prudent to defer those conclusions to those who had done so much more in this regard" (January 26, 2001 email to author). But Doell had argued against continental drift in his Ph.D. dissertation, which he had finished three months before his paper appeared in *Nature*. Also, he did show much confidence when discussing reversals.

No "reversals" were found by Dr. Runcorn or me in either the Supai rocks or the Hakatai rocks. Thus if the main geomagnetic field reverses itself as postulated by Hospers for the Tertiary, and by Clegg *et al.* for the Triassic, such reversals are not evident in the periods represented by the Supai and Hakatai rocks.

(Doell, 1955b: 1167)

Bearing in mind their backgrounds and experience and distinguishing what they wrote at the time from what they and others wrote subsequently, these are my conclusions about their attitude to global tectonics when they began facing the task of writing the *GSA* review and deciding how they should address the growing paleomagnetic case for mobilism: (1) from their education and experience they did not have a strong background in those elements of global geology needed for an understanding of the classical case for mobilism; (2) they did not favor continental drift; and (3) they preferred polar wandering as a general explanation of the oblique magnetizations that were then beginning to be recognized as a general characteristic of older rocks. Recall that Cox in his 1957 study of the Siletz River Volcanics and for many years after argued that rapid polar wandering, not regional tectonics, was the explanation of their oblique magnetizations (§7.6).

## 8.3 Genesis of the *GSA* and *AG* reviews[4]

Doell became an assistant professor at the Massachusetts Institute of Technology (MIT) in the fall of 1956, after spending a year at the University of Toronto, where he helped assemble a paleomagnetic laboratory. Once he got to MIT, he began compiling and calculating paleomagnetic poles.

The review(s) really got started independently by both Allan and I. When starting the paleomagnetic program at MIT, and trying to think of local problems that students there might undertake, it became apparent to me that a review of all the paleomagnetic data into one place would be helpful – and that it would also be nice if they were all treated with the same statistical methods – i.e. Fisher's. So I started this compilation and recalculation soon after coming to MIT. Because much of the local exposed geology was of Paleozoic and Mesozoic age, I started with these periods. At the 1958 Spring AGU meeting in Washington, I talked about my progress.

*(Doell, January 26, 2001 email to author)*

In the meantime, Cox, who was then a graduate student in geophysics at the University of California, Berkeley, was compiling his own list. Doell recalled:

Meanwhile, Allan, working on Cenozoic rocks, had come to the same conclusion regarding previous work on rocks of these ages. He also had been compiling and recalculating these data just as I had been doing on the older rocks data. Although we corresponded sporadically during these years – mostly about instrument and technique matters, neither of us had been aware of the others' efforts in this regard.

*(Doell, January 26, 2001 email to author)*

When Doell returned to Berkeley to teach a summer course in geology, he and Cox renewed their friendship. Agreeing so well, they decided to write a book.

When I accepted the offer to teach the Summer Session Geology 1 course at Berkeley, Allan graciously offered to let me crash in his modest apartment until I found a place. Needless to say, being the only two working in the field at that time we had lots to talk about. We soon found that we thought very similarly about things geological (not surprising in that we had pretty much the same teachers) and had very similar goals, and most significant to this discussion, we decided that we wanted to collaborate on a book covering all of paleomagnetism – not just the compiling and recalculating that we had individually been doing. We would also include all the paleomagnetic methods, statistics used, and pertinent knowledge of the geomagnetic field.

*(Doell, January 26, 2001 email to author)*

They got a good start on the book before Doell returned to MIT, and continued to make progress once Doell began teaching.

We spent much time working on this while I was at Berkeley. I think that we got fairly far along on a text draft, and the big Table of data (which always seemed to take the nature of a work in progress as new stuff kept coming in!). I'm pretty sure we had also sketched many of the figures we wanted, but let me assure you that I'd be hard pressed to be at all specific on our progress. We also continued work on the book after I returned to MIT, through correspondence, up until the time I entered duty at Menlo Park the 1st of March '59. However, due to my illness, moving, etc., I imagine there was not a large amount of progress during that period.

*(Doell, January 26, 2001 email to author)*

In November 1958, Doell (Glen, 1982: 153) was diagnosed with a malignant melanoma on both legs. The prognosis was not good, but Doell beat the odds. He returned to California after the operation. They continued with the book as Doell began working at Menlo Park for the USGS. They were anxious to finish and start on new research.

After March 1st of '59 I would have spent lots of effort getting the lab set up, equipment ordered, etc. But there would also have been a lot of waiting time and I recall my office being cluttered with stuff on the book much of the time – certainly when going over things with Allan when he came over from Berkeley to visit. He came on duty in July and I sort of remembered that we were working hard on it hoping to have it out of the way so we could get to all the new studies we wanted to do.

*(Doell, January 26, 2001 email to author)*

Then they got an offer they could not refuse. The GSA had asked Jim Balsley, Chief of the Geophysics Branch of the USGS from 1953 to 1960 and Doell and Cox's senior officer, to write a review of paleomagnetism. Balsley had collaborated with Graham and Buddington on stress effects on remanent magnetization, maintaining that they cast "serious doubts" on mobilist interpretations (§7.4). Their strong supporter, Balsley, asked them to write the review. Balsley was then asked to write a second review, this time for *Advances in Geophysics*, and he passed that to Doell and Cox; he seemed determined that both major US reviews should speak with one voice on the mobilism question.

It must have been sometime that spring or summer that Jim Balsley told us of being asked by GSA to do a review of Paleomagnetism for them. Details I don't recall, but the result was that he said he was too busy (certainly true) and would we do it if he suggested that to GSA and they agreed. At first I kind of recall that Allan and I were upset because if we did that we'd certainly have to put off the book – if not give it up entirely. I dare say we were also flattered to be asked. I'm not sure we'd formally made a decision about this, when exactly the same thing happened with *Advances in Geophysics*: a request to Balsley, bucked on to us.

*(Doell, January 26, 2001 email to author)*

They decided to write both reviews, and Doell described what led them to agree to do it.

I know we agonized a great deal over this, but I hardly recall all the details. In the end I think there were two major thoughts that were decisive: one, we certainly didn't want to displease Jim (although I know he would have supported whatever we did); and second we realized more and more that we'd never get a better format than the *GSA* review to get a large audience of geologists. I can still "hear" our geologic colleagues at the Survey telling us, "you guys keep griping about how stupid we geologists are about things magnetic, well you'll never get another chance like this one!"

*(Doell, January 26, 2001 email to author)*

The second reason is particularly interesting because it shows that Doell and Cox agreed with other paleomagnetists about the difficulty most geologists had in understanding geomagnetism and paleomagnetism.

## 8.4 The *GSA* review

The *GSA* review was erudite and long, over 120 pages. (For brevity, references to the review in this section will be "1960.") It provided a solid introduction to paleomagnetism, basic assumptions, statistical analysis, and the various field and laboratory tests that had been developed to determine its reliability. It contained a huge list of results, updated and more detailed than Irving's (1959a) bare-bones compilation.[5] There were remarks on the reliability of each result, many were re-analyzed, and computational errors were corrected. They assessed the paleomagnetic case for continental drift and it is this with which I am here concerned and which is relevant to the mobilism controversy.

My discussion is arranged under eight headings (a)–(h), beginning with **(a) the abstract**, which reads:

Evaluation of the data leads to the following general conclusions:

(1) The earth's average magnetic field, throughout Oligocene to Recent time, has very closely approximated that due to a dipole at the center of the earth oriented parallel to the present axis of rotation.
(2) Paleomagnetic results for the Mesozoic and early Tertiary might be explained more plausibly by a relatively rapidly changing magnetic field, with or without wandering of the rotational pole, than by large-scale continental drift.
(3) The Carboniferous and especially the Permian magnetic fields were relatively very "steady" and were vastly different from the present configuration of the field.
(4) The Precambrian magnetic field was different from the present field configuration and, considering the time spanned, was remarkably consistent for all continents.

*(Cox and Doell, 1960: 645)*

Those, especially non-paleomagnetists, who read it and it only, would have gathered that there was little or no paleomagnetic support for mobilism. There were no surprises in (1) repeating as it did what others had discovered and said many times: they noted that Oligocene to Recent results supported the GAD hypothesis, poles never being far from the present geographic pole (§2.9, §2.13, §3.12, §5.4, §5.6, §5.7, §5.8, §5.16, §5.18, §5.19). Only in (2) is continental drift mentioned, which is deemed less plausible than "a rapidly changing magnetic field" such as would be produced by polar wandering. Although not actually said in (3), they did explicitly acknowledge some support for mobilism because they were impressed with the quality of the Late Paleozoic Australian data, and the fact that it gave poles differing in position by more than 60° from poles of similar age from North America and Europe. As for the Pre-Cambrian data (4), they were insufficient to support mobilism, the implication being that polar wander could explain the wide range of observed oblique directions of magnetization.

The *GSA* review itself began by explaining **(b) the "field configuration,"** hypotheses used to test the theory of continental drift and to derive paleogeographic reconstructions from paleomagnetic data, especially the need for a time-averaged geocentric dipole field; they insisted that the paleomagnetic case for mobilism was not justified unless this was satisfied for every relevant time period for every continental block.

In order to interpret the results from some period in terms of continental drift we must first have enough data to know the configuration of the magnetic field during that period. Was it, for example, essentially a dipole field as at present? It is further necessary to assess the expected variation between sampling areas that have not been displaced with respect to each other. Only then is it possible to infer that a departure in direction at some given sampling area indicates displacement of that area with respect to others. Polar-wander interpretations require further that there be some connection between the magnetic-field configuration and the earth's axis of rotation. Continental-drift interpretations do not require such a connection, but they do require that the field configuration be changing at a rate which is small in comparison with the degree to which we can establish the contemporaneity of the formations compared.

*(Cox and Doell, 1960: 756)*

They did not make clear that paleomagnetists who had developed their case for mobilism had, from the very beginning, laid bare and tested their basic assumptions, and had not found them wanting. Runcorn (1959a) had distinguished between the need for the GAD hypothesis and less restrictive alternatives for polar wander interpretations (§5.18). Creer and others (1958) had noted that strictly the axial element of the GAD hypothesis was not required for drift interpretations; the idea of a time-averaged geocentric dipole field was all that was needed to determine if relative displacement of continents had occurred (§5.18). Irving (1959a) had compared the variation between intracontinental and intercontinental equivalent poles, and found the latter much greater (§5.17). Indeed, he had already shown that within-continental dispersion of poles considered interval by interval for the Late Carboniferous through Paleogene (Early Tertiary) were very similar to that between continents in the Neogene (Late Tertiary and Quaternary); at the time every geological period for every continent was not, then, individually mapped, as Cox and Doell's quest for perfection required, but grouped together their dispersions were comparable. Hidden in Irving's analysis there were aberrant exceptions (the Siletz Volcanics for instance (§7.7)), but they must have been very few, otherwise overall agreement from period to period within continental blocks simply would not have been observed.

Cox and Doell, however, demanded more stringent criteria before claims about mobilism were warranted. They insisted on knowing what they described as "the configuration of the field," essentially that dispersion of the poles from many formations for *each* geological period from *each* continental block be known before drift between continents could be inferred. To frugal experimentalists anxious to make the most of their hard-won results, the redundancy required by these criteria was nothing short of staggering. For example, if it is shown (as Du Bois *et al.* (1957) had shown for the Late Triassic of North America) that over the sideways spread of one continental craton and through a long interval of time prior to the Neogene the paleomagnetic directions correspond to that of a geocentric dipole (that is, poles agreed), it would be not easy although not impossible to envisage how the global field could be otherwise. If this is demonstrated for a second and then a third continent for second and third intervals of time, the chances of the global field not being a geocentric dipole becomes vanishingly small. But Cox and Doell insisted that for each interval for which continental drift was being entertained, "It is further necessary to assess the expected variation between sampling areas that have not been displaced with respect to each other," that is, all continental cratons. They did not tell paleomagnetists when they could cease; they condemned mobilist paleomagnetists to endless tasks, demonstrating consilience to a geocentric dipole field for all continental cratons for all intervals of time – truly the Twelve Labors of Hercules. Theirs was an impractical, utopian research program with an objective not achieved even now fifty years later, but it allowed them to argue that the paleomagnetic case for continental drift was weak.[6] This was their core position, and it was this that Runcorn later stressed (§8.9) when contrasting their approach with that used by paleomagnetists who supported mobilism. It was their utopian vision of what was required that allowed them, as Irving (§8.12) later said, "to avoid" the main issues.

The *GSA* review treated mobilism in a piecemeal way, isolated from happenings before and after. By contrast, paleomagnetists supportive of mobilism treated APW paths as entities extending from the present back to the Pre-Cambrian, stressing the time-ordered sequence of poles and the increasing divergence of paths backwards through time from the current geographic pole and from one another. Mobilist-minded paleomagnetists were willing to give weight to the continuity of historical processes responsible for these paths: if they found strong support for the GAD hypothesis for one time period, they were willing to take a progressive stance and generalize to other periods, and to go ahead and test it using the paleoclimatic evidence; and if, in this one period, they found strong support for drift from intercontinental comparison of poles, they were willing to conclude that it was a general phenomenon, not something that started in one period and stopped in the next.

The *GSA* review considered the youngest results first, **(c) Neogene**. Results for the Quaternary provided strong support for the coincidence of the time-averaged geomagnetic field with Earth's axis of rotation, the GAD hypothesis. Going further back into the Late Tertiary, they rejected appreciable apparent polar wandering during this period, mirroring the interpretation of European paleomagnetists, notably Hospers (§2.9).

To the extent that we are willing to extrapolate the axial dipole model back into the past, the paleomagnetic evidence indicates that no large shift of the axis of rotation has occurred during the late Tertiary – a conclusion previously arrived at by Hospers ... from analysis of fewer data.

*(Cox and Doell, 1960: 756)*

Cox and Doell then asked whether any pre-Oligocene data satisfied their criteria.

With this information we could then discuss, with some confidence, the application of these paleomagnetic data to possible post-Eocene polar wandering and continental drift. In interpreting the pre-Oligocene data, is there any time for which we may also establish a field configuration and an estimate of the scatter?

*(Cox and Doell, 1960: 760)*

For the **(d) Early Paleozoic and Precambrian** results, their answer was "No," paleomagnetists had not established the configuration of the geomagnetic field for these more remote times (RS2).

In general, there are relatively few measurements available from early Paleozoic rocks, and those that are available show a considerable range in pole positions. Therefore without sufficient data to determine a field configuration or to assess the expected scatter in the results, interpretations for these data involving continental drift or polar wandering are hazardous. Many more Precambrian data are available, and, as might be expected from older rocks with large relative age differences, the scatter is large. As mentioned earlier, many of the virtual geomagnetic poles tend to lie in the eastern Pacific near the equator. Even if this group of poles represents an average field configuration, we must expect a rather large variation in any given measurement. Thus, as for the early Paleozoic, a large uncertainty exists in the basic data available for polar-wandering and continental-drift interpretations, and the paths [for Europe, North America, and Australia, put forth for these periods] might best be regarded, at present, as working hypotheses.

*(Cox and Doell, 1960: 762; my bracketed addition)*

What the *GSA* review did not say was that a wide scatter of poles is just what would be expected from Early Paleozoic and Precambrian rocks, if continental drift had occurred, but the scatter could not be understood with much certainty until subsequent (Mesozoic and Cenozoic) drift had been worked out. Wegenerian drift would have to be deciphered first before earlier episodes could be understood. Apart from Argand's brief mention of his "Proto-Atlantic" (I, §8.7), there was at the time no testable theory of older drift. The paleomagnetic mobilism test of the second half of the 1950s was made on the basis of Late Paleozoic, Mesozoic, and Early Cenozoic rocks, not older ones.[7] The review then raised a general unreliability difficulty with mobilist interpretations of **(e) the Mesozoic and Early Tertiary** results (RS2); although there was some evidence for drift, the sometimes high intracratonic dispersion of poles weakened its support.

Paleomagnetic measurements for the Mesozoic and early Tertiary present an interesting problem; although the preceding Permian and following post-Eocene results are each internally consistent, the geomagnetic pole positions from the Mesozoic and early Tertiary are quite scattered. Some impressively consistent results have been obtained for individual Triassic formations, and the distribution of results from North America and Europe to some extent suggests relative drift. However, the scatter between mean formation directions is quite large and weakens the conclusion that a relative displacement has taken place. The paleomagnetic results from the Jurassic are at least as scattered as those for the Triassic and no conclusive statements concerning drift or wandering can be made.

*(Cox and Doell, 1960: 762)*

Cox's data from the Eocene Siletz Volcanics, they argued, posed problems for mobilism. Although they acknowledged Irving's proposed local rotation solution (§7.7), they argued that Eocene reference poles from cratonic North America would be needed before local rotation could be invoked – a missing-data difficulty; it was not possible to tell whether North America's APW path was simple or complexly curved (RS2).[8] Cretaceous and Eocene results from India, Japan, and Australia have been interpreted as indicative of movement relative to North America, but if the North America APW path was uncertain then such mobilistic interpretations were questionable (RS2) – an unreliability difficulty (RS2).

The paleomagnetic results from the Cretaceous and Eocene may profitably be considered together, inasmuch as large amounts of drift with respect to North America and Europe have been postulated for India, for Japan, and, to some extent, for Australia on the basis of geomagnetic data from rocks of these ages. Such interpretations are based on the conclusion that the Eocene and Cretaceous pole positions for North America and Europe were within 18° of the present geographic north pole and moved up to coincide with it in the Tertiary ... The paucity of paleomagnetic information for these time intervals has already been discussed. In this respect one of the most reliable virtual geomagnetic poles from the few available North American determinations (the Siletz River Volcanic series) falls much closer to the Indian and Australian paleomagnetic poles than to the Cretaceous and Eocene points on the usual polar-wandering curves for North America and Europe. Thus, as for many of the other periods, the North American (and possibly European) field configurations for Cretaceous and Eocene time are far from certain, and, without such a well-defined reference position, the drift interpretations for Australian and Indian Cretaceous and Eocene results may be questioned.

*(Cox and Doell, 1960: 762)*

Despite they themselves, like the Siletz, being situated in the midst of vast currently active orogenic belts (the North American Cordillera) they could not accept the likelihood of local and regional rotations such as had been invoked by Argand (1924), Carey (1958), and others, and already paleomagnetically documented (§3.13, §7.8).

They offered an alternative interpretation of the Cretaceous and Eocene results, which they found as, if not more, plausible than mobilism (RS1 and RS3).

An alternative explanation of the Cretaceous and Eocene measurements, and one that does not require simultaneous rotations of several tens of degrees for Oregon and India, would involve relatively rapid changes in the magnetic-field configuration during this period while maintaining the present continental configuration. The field may have been nondipolar or, alternatively, may have been that of a dipole undergoing somewhat rapid changes in direction. Such changes may or may not have been accompanied by rapid polar wandering. This hypothesis poses many problems. It appears to us, however, to fit the data at least as well as the drift hypothesis and it emphasizes the uncertainties connected with paleomagnetic interpretations at this time.

*(Cox and Doell, 1960: 762–763)*

This shows how heavily their fixist (polar wander only) argument depended on Cox's interpretation of his Siletz result, and how dismissive they were of Irving's well-reasoned and correct tectonic argument (1959a). Also Cox and Doell were not being mindful of the then known age differences when they argued that the tectonic interpretation required the "simultaneous rotation of several tens of degrees" of Oregon and India – the proposed ~65° rotation in Oregon was post-Eocene, whereas that of India was post-Cretaceous; simultaneity was not, as they asserted, a requirement of the mobilistic interpretation of the Siletz and Deccan data.

Cox and Doell thought that the strongest paleomagnetic support for mobilism came from **(f) Carboniferous and Permian** results; there were large significant differences between the positions of Australian poles and their equivalents from Europe and North America.

The Australian Carboniferous and Permian paleomagnetic investigations, although few in number, include concordant results between sediments and volcanic rocks, with an excellent fold test for the sediments. However, the virtual geomagnetic poles are quite significantly different from the poles for Europe and North America. If the axial-dipole hypothesis is valid for these periods, the Australian data constitute evidence for a relative displacement of Australia with respect to North America and Europe which *cannot be ignored*.

*(Cox and Doell, 1960: 762; my emphasis)*

However, they argued that the findings from North America and Europe were not sufficiently reliable to offer strong support for the opening of the North Atlantic since the Permian. They proposed irregular polar wandering as an alternative. Citing the indeterminacy of longitude, and noting that the stratigraphic correlation of the Permian across the Atlantic was not very precise, they offered a consistent but bizarre interpretation involving movement of North America toward Europe.

Even assuming that there is no systematic difference in age of the rocks from North America listed as Permian, and assuming that the field was dipolar, it is important to note that the two sets of Permian data from North America and Europe do not uniquely determine the relative positions of the two continents before a hypothetical displacement.

... This [interpretation of a movement of North America toward Europe] is not intended as a serious interpretation, but points out difficulties that arise in attempting a unique solution for the data now available.

*(Cox and Doell, 1960: 762; my bracketed addition)*

They concluded that collectively the data from North America and Europe offered strong support for polar wandering, just as Doell (1955a, Ph.D. dissertation) (§8.2) and Runcorn (1955b) had done five years earlier (§3.10).

Viewed on a larger scale, however, the consistency of the Permian and Carboniferous results from North America and Europe is most impressive and certainly indicates a magnetic-field configuration vastly different from the present configuration. If the earth's magnetic field was that of an axial dipole, as it probably was from Oligocene to early Pleistocene time and almost certainly was from the late Pleistocene to Recent time, then these results constitute a strong case for polar wandering. This interpretation is also in accord with the axial-symmetry requirements of the dynamo theory.

*(Cox and Doell, 1960: 760)*

Next was **(g) the axial field**. Without mentioning the by then well-documented accord between paleomagnetically determined latitudes and the paleoclimatic evidence from the same regions, they went on to raise a difficulty with the interpretation of apparent polar wandering as necessarily meaning wandering of the geographic pole. They called attention to work of E. J. Öpik (1955) (his brother had attended the Hobart symposium (§6.16)). Öpik had described a geocentric dynamo in which the orientation of core convective currents was controlled not by Earth's rotation but by temperature differences across the core–mantle boundary, calling into question the theoretical support offered for the GAD hypothesis by Elsasser's and Runcorn's appeal to the Coriolis force. They recognized difficulties with Öpik's view, and noted that the principle of uniformitarianism favored the GAD hypothesis, but still maintained, "Öpik's objection should be kept in mind as a possible alternative explanation" (Cox and Doell, 1960: 761).

Finally, there was **(h) statistical weighting**. Cox and Doell raised an unreliability difficulty, arguing that some paleomagnetists had incorrectly applied Fisher's statistics to estimate the accuracy of paleopoles (RS2). Naming no one, their criticism was aimed primarily at Runcorn; he collected oriented sedimentary rock samples from different stratigraphic horizons within a formation, cut several disc-shaped specimens from each, and determined their remanent magnetism. Giving each unit weight as if each were a different point in time, he calculated a mean formation direction and errors. They argued that this approach underestimated errors, claiming that a more realistic estimate would be obtained by averaging results from each specimen to give the mean for each sample, and then by averaging sample means to determine the overall direction and pole formation, and their associated errors. (Their criticism of Runcorn was substantially the same as Irving made of Nairn (§7.2).)

### 8.5 Doell and Cox's earlier attitude to the paleomagnetic case for continental drift

In a moment I shall consider the question of whether the *GSA* 1960 review was a reasonable critique of the paleomagnetic case of mobilism. Before doing so I want to consider if, in the one or two years prior to the appearance of the *GSA* and the *AG* reviews as results of the global paleomagnetic test of continental drift were rapidly accumulating, Cox's and Doell's attitudes changed. During this period they wrote four short reports, in three of which they were strong drift skeptics, at times firmly anti-drift. In the last they tended to be somewhat more favorable to drift.

The four reports, two by Doell alone, were in two abstracts (Doell, 1958, and Doell and Cox, 1959), a short paper (Doell and Cox, 1960), and an evaluation by Doell at the 1960 Helsinki meeting of the IUGG (Anonymous, 1960). I consider Anonymous (1960) later in §8.6. Cox is listed as the first author in the *GSA* review only, but all of them, except perhaps Doell (1958), likely reflect their collective views.

In this light [attributing different views to Allan or me], there is one more thing I wish to relate, and that concerns the authorship in our work – at least until quite late in our association. It was a bit confusing, and it was meant to be that way. We had early come to the conclusion that, as far as paleomagnetism went, our thoughts and knowledge were essentially the same, and hence that we were going to work as a "close team." And for many years this turned out to be a reality. The authorship on our first papers – the reviews – was decided with the flip of a coin, and we alternated back and forth thereafter on all papers of a general nature. Regarding presentations at scientific meetings, whoever was presenting the paper was the 2nd author (we had a lot of fun with this one!). There were some other "rules" regarding studies that we collected separately, but we always wrote papers together during this time, and these latter are not pertinent here. I note this because when you say things about Allan cf. me, or whatever, you are speaking in reality of the same entity.

*(Doell, January 26, 2001 email to author; my bracketed addition)*

The first report was in an abstract Doell prepared while at MIT for the 1958 spring meeting of the AGU. It offered no support for mobilism.

Paleomagnetic Interpretations – Recent paleomagnetic pole determinations combined with previous studies may now indicate the following: (1) No continental drift has taken place since the lower Eocene, with little evidence for such drift since the Permian. (2) The "dipole" hypothesis concerning coincidence of magnetic and axial poles might be questioned. (3) There is a strong need for very careful stratigraphic control in future paleomagnetic studies.

*(Doell, 1958: 513)*

Doell probably wrote this in late 1957 or early 1958. He did not think that the APW paths for Europe, North America, and India were sufficient to tip the scales in favor of mobilism. He did not mention Irving's Australian results, which had begun to appear in 1956; it is very unlikely that he did not know about them as he (and Cox) attended Irving's lecture at Toronto IUGG in September 1957 when he talked about them (§5.5, §5.9, §8.10, §8.12). He also thought the GAD hypothesis questionable, and called, as he would continue to do, for additional data from well-dated rocks.

Doell and Cox talked at a meeting sponsored by the GSA held in Pittsburgh in November 1959. Their abstract, published in December, favored fixed continents and polar wandering.

Conclusions drawn from paleomagnetic data concerning continental drift, continental rotations and polar wandering depend critically upon the statistical treatment used. A recalculation of all available paleomagnetic data, using where possible appropriate homogeneous statistical standards has led to the following general conclusions. As has been pointed out by other workers, the magnetic poles calculated for formations of post-early Tertiary age from all continents rarely differ significantly from the present geographic pole; however, most magnetic-pole positions determined for older formations are significantly different. Paleomagnetic results from Mesozoic and lower Tertiary rocks have recently been interpreted in terms of continental rotations and displacements of thousands of kilometers. An alternative hypothesis – that during this period the earth's magnetic field was a comparatively rapidly changing dipole – correlates these results without requiring displacements of continents. Moreover this interpretation does not require the synchronous rotation of India and Oregon, as is needed for the drift hypothesis.

*(Doell and Cox, 1959: 1590–1591)*

In this report there were no qualifications. They advocated rapid polar wandering during the lower Tertiary and Mesozoic requiring neither displacements of the continents nor Irving's proposed local rotation of the Oregon Coast Range (§7.7); as noted earlier, the mobilistic interpretation of the Oregon and Indian results did not, as they asserted, require their synchronous rotations (§8.4). They also raised the possibility of misuse of Fisher's statistics. Perhaps of more interest is what they did not say. Although claiming that they were considering "all available paleomagnetic data," they said nothing about the very data that they admitted in the GSA and AG reviews offered the strongest support for mobilism, namely, the Australian paleomagnetic data from the Permian and Carboniferous, and they did not mention the Tasmanian dolerite data or the Early Tertiary data from Victoria, both fully described and grossly inconsistent with their Northern Hemisphere equivalents (§3.12, §5.3). Cox, as noted already, had attended Irving's lectures on the Australian data that he gave two years earlier at Berkeley in July 1957, and both he and Doell heard it at the Toronto IUGG meeting in August 1957. Cox and Doell did not include "all paleomagnetic data" and as a consequence, they did not have to temper their rejection of mobilism. Little wonder that Irving became annoyed. I conclude that they were, at this stage, firmly anti-drift.

Their next report, "Paleomagnetism, polar wandering, and continental drift," appeared in a USGS publication, designed to release new findings in a timely manner. In it their criticism of the paleomagnetic support for mobilism was slightly harsher than that they were soon to offer in the *GSA* review. Although their short reports were summaries only, their tactic had become evident: do not consider the paleomagnetic support for mobilism in its entirety, but instead fragment it into short time periods and require each to be complete.

We have concluded, from the paleomagnetic data now available, that the earth's magnetic field had vastly different characteristics during the following periods: post-early Pleistocene, Oligocene to early Pleistocene, Mesozoic to early Tertiary, late Paleozoic, early Paleozoic, and Precambrian. If paleomagnetic results are to be used as evidence supporting or refuting continental drift, it is first essential to determine the configuration of the earth's magnetic field during the time when contemporaneous rocks from different continents were magnetized. For only a few of the above temporal subdivisions has the configuration of the geomagnetic field been established with sufficient certainty to justify application to the problem of continental drift.

*(Doell and Cox, 1960: B426)*

There was no evidence for drift during the late Pleistocene and Recent times, and results from rocks of Oligocene through early Pleistocene allowed for no continental displacements greater than about 10°. They categorized drift interpretations based on Paleozoic data as "extremely hazardous," and said of the Pre-Cambrian results:

Measurements from different continents do not appear to differ significantly, which limits the amount of continental drift since Precambrian time. The North American measurements, however, which far outnumber all the others, do include some widely scattered average pole positions.

*(Doell and Cox, 1960: B426)*

Their opinion of the support for continental drift from rocks of Mesozoic and early Tertiary age was even more harsh.

Although over 50 measurements on rocks of Mesozoic and early Tertiary age have now been reported, no magnetic field configuration has emerged that is comparable in consistency and simplicity with that found in the late Paleozoic and late Tertiary rocks. Some of the virtual geomagnetic poles lie near the present geographical poles, but a significantly large number lie in very low latitudes. Although these rather divergent poles have recently been cited in support of large relative displacement of the continents, the character of the earth's magnetic field during this time has not, in our opinion, been sufficiently well delineated to justify interpretation of the paleomagnetic data as evidence either for or against continental displacement during this time.

*(Doell and Cox, 1960: B427)*

Cox and Doell discounted these well-documented instances as evidence of mobilism: Mesozoic poles from Australia (§5.3), Africa (§5.4), South America (§5.6), and India (§5.2), and Early Tertiary poles from Australia (§5.3) and India (§3.12, §5.2), all of which were strongly and systematically discordant with poles from Europe (§3.6) and

North America (§3.10). Only results from the Permian and Carboniferous offered any support for mobilism. However, they did not consider this support to be as strong as they did somewhat later in their *GSA* review. They emphasized their uneasiness with the paleomagnetic support of North American drift relative to Europe, and addressed the uncertainty of the smaller but not the larger, highly significant discrepancies with Australia.

On the assumption that the earth's field was dipolar, these measurements on late Paleozoic rocks have been interpreted in terms of large drift of Australia relative to the northern hemisphere continents since the Paleozoic, and a smaller westward drift of North America relative to Europe. Although this latter interpretation is suggested by the Permian poles, a simple westward drift of North America does not explain both the Permian and Carboniferous data. An interpretation based on continental drift would require rather improbable relative movements between North America and Europe, and more data from Permian and Carboniferous rocks are highly desirable.

(*Doell and Cox, 1960: B427*)

I think it is fair to say that although they held to no steady line of argument over the period from 1958 through 1960, Doell and Cox gradually came to entertain continental drift as a possible explanation for Late Paleozoic but not for later Mesozoic and Early Tertiary results. I think the same trend continued with the *GSA* review, where they went so far as to say, "Australian [Late Paleozoic] data constitute evidence for a relative displacement of Australia with respect to North America and Europe which cannot be ignored" (Cox and Doell, 1960: 762; my bracketed addition). But even in the *GSA* review, they still were not willing to declare that the support for mobilism was greater than for rapid polar wandering, because as far as the Mesozoic and early Tertiary are concerned (Cox and Doell, 1960: 645), the results "might be explained more plausibly by a relatively rapidly changing magnetic field, with or without wandering of the rotational pole, than by large-scale continental drift."

## 8.6 The *AG* review

Again, let's not kid ourselves that I'm not imagining some of this, paraphrasing some, maybe forgetting much, etc. However, the factual constraints are fixed. Sometime between the 1st of March '59 when I came on duty [at the USGS], and the 1st of Feb '60, when GSA received the manuscript, the above events transpired and we gave up the book and reworked our book text into two reviews. (I just might also mention that Allan was writing up his doctoral thesis during part of this time – he submitted it in Sept of '59.) I don't now know exactly when the *Advances in Geophysics* manuscript had to be submitted, but I recall it wasn't long after the GSA one. We spent a lot of time trying to decide how to split it up and reference the material back and forth between the two audiences but eventually we came to the decision that all our thoughts needed to be in each. We did, of course, emphasize things differently for the two audiences, messed around with the order of the contents, made different looking figures, etc., in an attempt to make the two look a lot different, but the astute reader can easily note that they are essentially the same thing – just with slightly different emphasis. (Or should I admit this at this late date?)

*(Doell, January 26, 2001 email to author; my bracketed addition)*

The *AG* review references the *GSA* review, but not vice versa; evidently the latter was finished first.

Doell is correct about their strong similarity; both raise the same difficulties against mobilism. However, the relationships, for example, between the GAD hypothesis and paleomagnetic testing of polar wandering and continental drift are presented somewhat differently, and mobilism is assessed somewhat more favorably in the *AG* review. Summarizing their opinion of the strength of the support for continental drift and polar wandering, they declared:

To sum up the interpretation of paleomagnetic data in terms of polar wandering [and] continental drift ... there can be little doubt that the magnetic pole has wandered during geologic time; similar wandering of the rotation axis also appears a likely possibility. Moreover, making reasonable assumptions about the earth's field in the past, *it is difficult to explain all of the presently available paleomagnetic data without invoking continental drift.* However, serious problems remain concerning many details of the polar wandering curves, and those shown [for Antarctica, Australia, Europe, India, Japan, and North America] in Fig. 25 may be expected to undergo marked changes in the future. In our opinion, future investigations of these details may well alter some of the current conclusions about continental displacements.[9]

*(Doell and Cox, 1961a: 302; my bracketed addition and emphasis)*

In the *GSA* review, they say that the paleomagnetic evidence in favor of the displacement of Australia relative to North America "cannot be ignored"; in the *AG* review, their declaration is stronger, the displacement is "difficult to explain" without continental drift. Although they thought that future work would markedly change the current APW paths and hence change current views about continental drift, they were not doubting the possibility of drift, and they again singled out the Australian Permo-Carboniferous data as the most significant evidence in its favor.

Perhaps the best documented case for large-scale continental drift is that suggested by Irving for Australia, where Carboniferous and Permian paleomagnetic results from both sediments and lava flows are internally consistent, but have virtual geomagnetic poles displaced more than 70° from those for North America and Europe.

*(Doell and Cox, 1961a: 297)*

Also, in the *AG* review they (and this was, I think an important change) no longer firmly claimed that the Mesozoic and early Tertiary results were more plausibly explained by rapid polar wander than by continental drift. On balance, I think they had become slightly less resistant to mobilism, but still were unwilling generally to favor it over fixism. They (1961a: 300) still maintained a fixist explanation of the Eocene aged Siletz Volcanics, accenting the need for more Eocene data. In fact, the need for more data generally remained one of their major recommendations.

These interpretations of the paleomagnetic data [suggesting large-scale continental drift] are certainly among the most challenging geophysical ideas to be put forward in recent years and are worthy of careful consideration by all who are concerned with the origin and history of the continents. However, a comparison of these curves with the virtual geomagnetic poles … shows that while some parts of the curves are supported by the data now available; for most geologic periods and most continents, the present state of the science is one requiring additional paleomagnetic study of suitable, well-dated rock formations in order to better delineate the configuration of the geomagnetic field. For only a few geologic periods is the geomagnetic field configuration known with sufficient precision to be useful for geologic applications such as dating rocks.

*(Doell and Cox, 1961a: 297–298; my bracketed addition)*

There is one more report of theirs to consider.

Doell discussed paleomagnetic support for mobilism at the 1960 IUGG meeting held during the summer in Helsinki. This was the last and sixth assessment they gave during the late 1950s and early 1960s. He is reported to have said:

Although for the period since the Oligocene the results are consistently in favor of an axially dipolar geomagnetic field which every so often reverses its polarity, the results for older rocks can be plausibly interpreted in terms of a relatively rapid polar wandering. He maintained that the data are not yet sufficiently reliable for any case to be made either for or against the continental drift hypothesis, in contrast to a recent study by Blackett which favours continental drift. Doell voiced the general opinion in calling for more data, combined with a critical approach to the subject.

*(Anonymous, 1960: 475)*

This puts Doell and (by implication since he and Cox held similar views) Cox right back where they were in the *GSA* review, they had not become less resistant to mobilism. However, it is only *a report* of what Doell said. Doell's abstract itself is needed. But, there is no published abstract of Doell's presentation in the volume of the Helsinki proceedings. I asked Doell if he had a copy. He did not. There probably never was an abstract.

This is really puzzling. I can't remember much about this meeting re papers, although I certainly recall being there, spending a lot of time with Jim Balsley, and enjoying the Finns. Now I'd be mighty surprised if I hadn't given a paper (it was a very rare USGS employee that could attend a meeting otherwise!), but my bibliography lists no such abstract either. However, the quote, from wherever it may have come, doesn't seem all that at odds with what one would get from our reviews. As it is, though, it probably doesn't warrant too much consideration.

*(Doell, January 26, 2001 email to author)*

I agree with Doell. The report perhaps does not warrant much consideration, but here it is for the record.

Doell also thinks that the stance in the report does not seem "all that at odds with what one would get from our reviews." Doell believes that there was no change in his and Cox's attitude toward mobilism, and in that he thinks that I am wrong.

Hank, I don't understand why you continue to think there was some change in our attitude from the *GSA* to *Ad. in Geophys.* reviews. As I told you, they were both rewritings of the book we were working on, and were done essentially in tandem. To me, it was simply a matter of trying to find different words to say the same thing. Yes, there was the abstract of the *GSA* effort – but if you look through and read the article you will find, I think, nothing approaching this strong a statement. As I also said before, I can't now understand how we could have written that. And since the Helsinki thing occurred at the same time (and didn't even use my own words) I certainly wouldn't give it any different significance.

*(Doell, February 10, 2001 email to author)*

Perhaps I am imagining differences. Upon his later re-reading of both major reviews, Doell agreed that the *GSA* review's *abstract* favors fixism. In a very forthright declaration, he stated:

I have, as I told you, re-read the two review papers fairly carefully. If I were now to look at just these data reviewed, forgetting all that has happened since, and save for just one sentence that I'll return to, I think I'd say (and by inference Allan too, I think, were he here) pretty much the same things. I dare say I'll be elaborating when I get to your specific questions. The one sentence that puzzles me greatly, though, is item (2) in the abstract for the *GSA* effort. For sure, we note in the text that we find the lower Tertiary and Mesozoic (as well as the pre Carboniferous) significantly more scattered than late Paleozoic, but to use the expression "more plausibly" rather than "possibly" or "alternatively" seems now to be so terribly out of character – especially cf. all the rest that we wrote. I can hardly imagine our using it, but there it sits, big and ugly! Throughout we had been very careful to offer all alternatives to explaining the data and to be judgmental only when we felt very certain of the outcome, for example, on field and self-reversals. I'm really at a loss on this one.

*(Doell, January 26, 2001 email to author)*

Needless to say, it is this very passage in the *GSA* review and its absence in the *AG* review that persuades me that the former was more pro-fixist than the latter.

But retrospectively Doell believed that, except for the *GSA* review's abstract, he and Cox slightly favored mobilism in both reviews. If this were their intended position, they certainly misled Creer, Irving, and Runcorn, and a host of non-paleomagnetists of a fixist persuasion who cited them as a way to set aside the paleomagnetic support for mobilism. Moreover, Verhoogen, their mentor, maintained that both reviews were not favorable toward mobilism.

But American workers [in contrast to British ones] remained skeptical about drift. In their influential review, Cox and Doell (1960) appear unconvinced that discrepancies between poles from different continents could not be explained by inaccuracies in stratigraphic correlation or dating, by changes in the configuration of the magnetic field which may not have remained essentially dipolar through all geologic times, or by episodes of unusually rapid polar wander. In a later review (Doell and Cox, 1961a) these same authors, though impressed by the evidence for drift of Australia (Irving and Green, 1958) still seem unconvinced with regard to India, and to the separation of Europe and North America in Permian times. They do not doubt, however, the reality of polar wander.

*(Verhoogen, 1985: 402; my bracketed addition)*

Verhoogen, like Creer, Irving, and Runcorn, and many non-paleomagnetists, as I shall show, did not consider the *GSA* review at all mobilism friendly.

### 8.7 Was the *GSA* review an unreasonable assessment of the paleomagnetic case for mobilism?

When asked to comment about the *GSA* review, Irving, Runcorn, and Creer did not mince words. Irving in 1981 said:

Probably, the most effective deterrent to the drift point of view ... was the Cox and Doell review. I always regarded that as a cop out. Now I say that and I mean it. It really was to my mind avoiding the main issues in a very, very intelligent and erudite way. That review was enormously useful in listing the data, and the ordering of the data was beautifully done, and I used it and referenced it enormously. It was an enormously useful piece of work. But it was used by people as a means of putting down our case. I think that the one concession, I think, if I remember rightly in that review and I haven't read it for some time, the concession they did make is that there may be a case for continental drift in the Australian data. That for them, for Cox and Doell at that time, who were not really convinced about continental drift and seafloor spreading until 1966, which was six years later, that really was a concession.

*(Irving, 1981 interview with author)*

Runcorn in 1984 said:

Cox and Doell wrote that article really because they had got interested in the field and were anxious, if you like, to establish their credentials as sober critics. In fact, I think I'm right in saying that they were more or less told to review the field as a preliminary to getting a grant for their work.[10] And therefore, they tried to raise all the objections and to weigh the data dispassionately. Of course, as you know they thought the data for some periods was better than for other periods. In our view, as one sees in the 1954 paper, it's the fact that the pole positions all fall on a continuous path that gave them a sort of mutual support. We did not believe that just picking out one geological period for consideration was a very useful way of proceeding at that time.

*(Runcorn, 1984 interview with author)*

Creer wrote in 2000:

We can only assume what went on in the minds of Allan Cox and Dick Doell. Perhaps it was just that the Cambridge group had chosen to interpret the new paleomagnetic results in terms of polar wander and drift and they chose not to follow us but to take another tack, proposing changing complexity of the palaeo-geomagnetic field. They just wanted to be different, and to take the gamble that in due course they would be proven right and we wrong. In my view that would have been reasonable were it not for the work of Wegener and du Toit – the strength of our approach was that the palaeolatitudes given by our poles were consistent with their proposed continental reconstruction. That is what gave us confidence to press on. On the other hand geomagnetic theory had nothing to say about the past structure of the field, so it was reasonable for us to take the view that it had always been essentially dipolar.

As regards the C & D review, I too was rather irritated by the stance they had taken up. I thought briefly about taking them on in the literature, but since Keith was already doing this I felt I could use my own time more constructively – i.e., to give priority to working on my South American collection. I felt that they (C & D) were getting themselves into an untenable position, so I thought, "why not leave them alone to sink into the mud." But in the end as things turned out what saved them was their work on the polarity reversal column, and its application in sea floor spreading: this made their reputations to the extent that their misjudgment of the drift related aspects of the paleomagnetic data was ignored, even forgotten.

*(Creer, September 27, 2000 note to author)*

Irving, Runcorn, and Creer were not alone. Irving also recalled that a year after its appearance Hess agreed it was a "cop-out."

It was at this time on sabbatical from ANU [fall 1961] that at a party at the Hess's house Harry asked me if I really thought drift was real. I have tried to remember my response. I cannot say for sure, but it was along the lines that if GAD was correct and all evidence pointed to it being correct, then yes continents had to have moved. We also discussed the Cox and Doell review which he agreed was a "cop out."

*(Irving, December 2000 email to author; my bracketed addition)*

As Irving said, they found ways to emphasize the weaknesses and downplay the strengths of the paleomagnetic support for mobilism. As Runcorn implied, they found a way to put the paleomagnetic support for mobilism in its worst light. Their piecemeal approach allowed them to de-emphasize the continuity of APW paths, which was the paths' central feature. As Runcorn said, "it's the fact that the pole positions all fall on a continuous path that gave them a sort of mutual support." They chose to focus on short time-slices instead of the long march of poles that defined the paths (Figure 8.1).

The GSA review (Cox and Doell, 1960: 758–759) gave APW paths for Europe, North America, India, Australia, and Japan, all as drawn by other workers (Figures 8.2 and 8.3). They did not include results from Africa, Antarctica, and South America. They did not emphasize their continuity, that they converged to the present geographical poles, that they differed from continent to continent, and that the agreement among individual poles from the same period and continent was much better than that between continents, features that surely had to be indicative of a first-order geological process affecting each continental block differently. Moreover, as Creer said, by downplaying the APW paths, they could ignore the fact that they were consilient within stated errors with the reconstructions of Wegener and du Toit, the very thing that Creer said "gave us the confidence to press on."

Most notably, although Cox and Doell had prepared an unsurpassed list of results, they did not use it either to update these paths with individual poles marked on them, or to determine their limits of error with the statistical rigor that they advocated in their review; they added nothing to the analysis of APW paths which was central to any real assessment of the paleomagnetic case for mobilism.

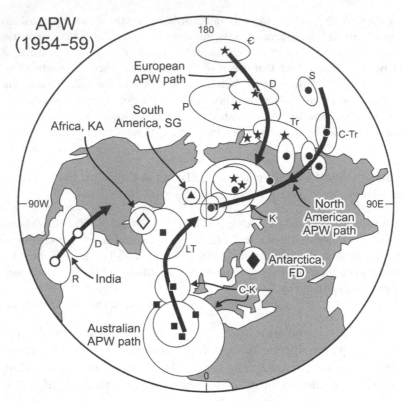

Figure 8.1 APW paths (1954–9). Africa (open diamond), KA – Jurassic Karroo lavas (Nairn, 1956; Graham and Hales, 1957); Antarctica (solid diamond), FD – Jurassic Ferrar dolerites (Turnbull, 1959; Blundell and Stephenson, 1959; Bull and Irving, 1960a); Australia (solid squares), C-K – Late Carboniferous through Cretaceous, LT – Paleogene (Irving and Green, 1958); Europe (solid stars) Є – Cambrian, D – Devonian, P – Permian, Tr – Triassic (Creer, Irving, and Runcorn, 1954, 1957); India (open circles), R – Rahajmahal Traps (Clegg, Radakrishnamurty, and Sahasrabudhe, 1958), D – early Cenozoic Deccan Traps (Irving, 1954; Clegg, Deutsch, and Griffiths, 1956); North America (solid circles), S – Silurian, C-Tr – Carboniferous through Triassic, K – Cretaceous (Runcorn, 1956b; Creer, Irving, and Runcorn, 1957); South America (solid triangle) SG – Jurassic Serra Geral Volcanics (Creer, 1958a).

So far, this section gives instances of the *GSA* review dealing unfairly with the paleomagnetic case for mobilism, casting it in an unfavorable light, unreasonably so. But perhaps it is my arguments that are unsound or incomplete; they have, for example, taken no account of a difficulty raised about the alleged use of Fisher's statistics (§7.2, §8.4). What if the chief culprits, Runcorn's North American APW path and Nairn's European poles are momentarily omitted? Strong paleomagnetic support for mobilism still remains. Hospers', Creer's, Irving's, and Clegg and company's poles (of which more anon) would still provide a firm basis for the British APW path, the paths for the other continents would still be hugely different from one

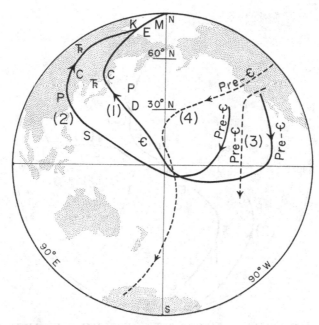

Figure 8.2 Cox and Doell's Figure 33 (1960: 758) captioned "Postulated Paleomagnetic Polar-Wandering Curves for Europe and North America. Pre Є – Precambrian, Є – Cambrian, S – Silurian, D – Devonian, C – Carboniferous, P – Permian, T – Triassic, K – Cretaceous, E – Eocene, M – Miocene." (1) Europe, (2) North America (after Creer, Irving, and Runcorn, 1957); (3) Europe, (4) North America (after Du Bois, 1958).

another, would all converge toward the present geographical poles, would be broadly consilient with Wegener's and du Toit's reconstructions, and, what Cox and Doell hardly mentioned, there would still be the astonishing and by now well-documented intracontinental consilience of the APW paths with the paleoclimatic data. Like other lines of evidence, the paleomagnetic case for continental drift did not stand alone; it had been shown to be consilient with drift reconstructions, with past climatic variations, and with biogeography, particularly of modern plants, all lines of evidence derived by entirely independent ways. These central facts were not addressed in the *GSA* review.

Perhaps Cox and Doell thought their allegation about unreliability and statistical misuse applied equally well to poles that had been used to construct the original British APW path, which they would then have reason to dismiss. Is this so? For brevity I consider only the eleven results available in the mid-1950s, referencing them by bold upper case letter and number, as in the *GSA* review. Cox and Doell first considered Irving's two Precambrian results and found, "From a paleomagnetic point of view the Torridonian sandstone (**B 12** and **B 16**) has been studied as carefully as any formation." They seemed a little unsure about Creer's Precambrian Long-myndian pole (**B 21**) – "Statistical analysis apparently is of directions of individual

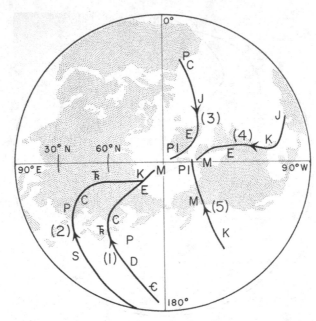

Figure 8.3 Cox and Doell's Figure 34 (1960: 759) captioned "Postulated Paleomagnetic Polar-Wandering Curves for Europe, North America, Australia, India, and Japan. Є – Cambrian, S – Silurian, D – Devonian, C – Carboniferous, P – Permian, T – Triassic, K – Cretaceous, E – Eocene, M – Miocene, Pl – Pliocene." (1) Europe, (2) North America (after Creer, Irving, and Runcorn, 1957); (3) Australia (after Irving and Green, 1958); (4) India (after Clegg, Radakrishnamurty, and Sahasrabudhe, 1958); (5) Japan (after Nagata *et al.*, 1959).

samples, not of mean site directions," but they still accepted the pole, noting that additional tests had been made.

Stability of the late Precambrian Longmyndian Formation (**B 21**) (Creer, 1957b) is suggested by A.C. demagnetization experiments and Thellier's test (that is, there was no change in magnetic directions on remeasurement, after storage in the earth's field). The pole lies between those for the upper and lower Torridonian (**B 12** and **B 16**).

*(Cox and Doell, 1960: 740)*

Creer's Cambrian Caerbwdy Sandstone pole (**C 45**), twelve samples spanning "a stratigraphic interval of 350 feet," presented no difficulties, and they seemed to accept the pole by Clegg and company (**C 2**) and had no problem with Creer's (**C 13**) pole from the Devonian Old Red Sandstone.

Clegg and others ... examined specimens from three oriented samples collected at a single locality (**C 2**) and found that the magnetization was stable in weak D.C. and in strong A.C. magnetic fields. Creer ... sampled much more widely and noted that flat-lying bed gave consistent results (**C 13**) ...

*(Cox and Doell, 1960: 743)*

From the Carboniferous Pennant Sandstone of Gloucestershire (**D 2**), Clegg and company took one huge sample and cut fourteen specimens from it; nevertheless, Cox and Doell appear to have accepted their result. They accepted Creer's Permian pole (**E 4**) of the Exeter Volcanics, based on thirty-four samples from five lava flows, each given unit weight. Clegg and company's pole (**F 12**) from the Triassic Keuper Marls was given high marks, and Creer's supporting results were noted.

The nine sampling sites in the Keuper Marl ... fall into two groups which have directions of magnetization approximately reversed with respect to each other. These sediments are stable in weak D.C. fields and in strong A.C. fields and satisfy a fold test for rather shallow dips. Pole **F 12** is based on the nine mean site directions as listed in the original paper [by Clegg *et al.*, 1954a]. Creer (1957a, p. 136) showed that Keuper Marl samples collected at many other localities have two components of magnetization, one stable, the other unstable; the stable one is essentially parallel to that described by Clegg.

(Cox and Doell, 1960: 749–750; my bracketed addition)

For the Early Tertiary, there was Hospers and Charlesworth's pole (**I 10**) from the Eocene Antrim Basalts of Northern Ireland, based on "6 flows at each of 3 sites and 5 at a fourth site were sampled; 2 to 4 samples were taken from each flow," which they had no problem accepting (p. 728). Finally, there was Hospers' Miocene pole (**A 34**) from Iceland, of which "consistency-of-reversals test indicates stability."

The poles still fell consecutively on a simple path and although use of Fisher's statistics was somewhat varied, that cannot have had an important systematic effect. Analysis of results from other continents would give similar results. Thus I think that there were at the end of the 1950s enough poles to show that the huge differences between the Australian and Indian APW paths and the paths from Europe and North America were real; the paths from the northern and southern continents could hardly have been more different. Indeed, it seems to me that, on the basis of their own criteria that gave them no reason for rejecting these results, they should also have accepted mobilism as the likely cause of these huge intercontinental differences.

The piecemeal approach of the GSA review led to a peculiar contradiction. Their acceptance of very large and systematic differences in Carboniferous and Permian poles from the northern continents and Australia, and their seeming interpretation of these as possibly indicative of continental drift, requires that drift occurred after the Permian, that is, in the Mesozoic or Early Tertiary. But these are the very intervals, as highlighted in their abstract ((a) of §8.4), for which the review strongly favored not drift but rapid polar wander.

The final reason I believe that the GSA review was unreasonable concerns what they did not say. If they had offered strong anti-mobilist arguments from fields other than paleomagnetism, then their refusal to support mobilism might have been reasonable. But they did not discuss work in other fields, and their limited knowledge of geology outside North America did not help them to do so. They said nothing about the fifty years of earlier mobilism history and how paleomagnetic results had, *by entirely independent means*, confirmed the main features of continental drift, which had provided a solution to many intercontinental disjuncts and so many other problems. Nor did they say anything about the excellent correspondence, established by Irving (1956a) and later workers, between independently determined paleomagnetic latitudes and paleoclimatic evidence from the same regions (§3.12). They said nothing about Opdyke and Runcorn's paleowind studies (§5.12–§5.14). They cited, but did not discuss, Irving's paper, noting only (Cox and Doell, 1960: 762) that "Paleobiogeographic studies probably rank first among those in other disciplines that are of interest to paleomagnetism." The biogeographers they did cite, Durham, Chaney, and Stehli, were North American and staunch fixists. They did not mention, as they could have done in the *AG* review, Runcorn's 1959 response to Stehli (§7.10). They did not mention or use pro-drift works by biogeographers, for instance W. G. Chaloner's (1959) excellent review of the paleomagnetic support for mobilism (III, §1.12) and Ronald Good's remarkable *The Geography of the Flowering Plants* (1953) (III, §1.12), which provided readers with a comprehensive botanist's view of continental drift.

## 8.8 Some later reflections on the *GSA* review

Twenty-one years later Irving still regarded the *GSA* review as a cop-out. He was at the time both surprised and angry. He characterized its position as fence-sitting, and thought it unreasonable. As will become apparent shortly from my examination of the correspondence between him and Cox, Irving thought that in writing the *GSA* review they had been swayed by the prevalent anti-mobilist attitude, and also they set their standard of proof so high in order to position themselves above the fray. Doell saw it differently; he agreed that they were anxious that American geologists, their targeted, overwhelmingly fixist audience, would respect their review. He also agreed that he and Cox took the high road. According to Doell, however, they set a high standard of proof because they wanted to make sure that their judgments would not be dismissed as rash or hasty, not because they wanted to remain above the fray. They did not want to shy away from making judgments; they wanted to shy away from making unsure judgments. Here is what he wrote.

And yes, I do think we were taking the "high road" re proof; I think we were very anxious to have our American colleagues in the geological sciences respect our judgment, and thus we only made judgments when we were very sure ... I'm sure that we never had any outside pressures to influence what we wrote. Who would there have been in the country to question our knowledge of views of things paleomagnetic? When we were talking on the phone [in 2000] I certainly was not referring to any specific pressure. The notion was about the general atmosphere in this country and our desire to be very sure of our interpretations. I suppose it is fair to say that the geologic community in the U.S. was less "mobilistic" cf. Europe at that time; and if you mean that as a "pressure" influencing what we wrote, I guess I can accept that, but it was also our own background and stance as well.

(*Doell, January 26, 2001 email to author; my bracketed addition*)

Doell also emphasized that he and Cox took seriously the difficulties they raised.

Perhaps it is difficult now to understand how seriously we took the problems with the data we analyzed. We were definitely puzzled by the Permian/Carboniferous data: relatively many good VGPs [virtual geomagnetic poles] with the Carboniferous showing minimal separation (Americas vs. Europe) and the Permian showing significant separation. Evidence for a reversing dipole axial field in the late Cenozoic, but no reversals in the entire Permian led one to wonder about the axial symmetry for all time. The seemingly rather calm magnetic field in late Paleozoic cf. before and after. Finally, Pre-Cambrian poles all being grouped in a third of the hemisphere.

*(Doell, January 26, 2001 email to author; my bracketed addition)*

Doell and Cox were anxious that North American Earth scientists should not think they had gone too far in supporting mobilism even to the very limited extent that they had. Doell talked about the gutsy move they made in correcting and recalculating the results of other paleomagnetists, which is a different issue, and a skeptic may add that by critically examining the calculations of others, they were making themselves more acceptable to North American geologists who were hostile to continental drift and dubious about paleomagnetism.

I would like you to keep in mind that at that time we were both relatively fresh out of school (definitely true for Allan) and, I further believe, fairly modest persons (that may have changed over the years, but I'm not going to judge that). And I'm speaking here cf. some of the other "actors" who had been well established in the field. I think also that throughout preparing these reviews we were continually anxious about what people would say – especially about the way we had the guts to re-analyze and correct so many of the published data (which we never received a single letter about!). Basically, re-reading these papers now, I'm pretty sure our goal was to offer all the alternatives to explain the assembled data, but to shy away from making conclusions ourselves.

*(Doell, January 26, 2001 email to author)*

Cox actually did get one letter about the recalculation of a pole from Australia, and Irving was not pleased (§8.10). However, examining and reanalyzing data was something careful reviewers were expected to do; most of their corrections were small, in the larger scheme of things, inconsequential arithmetic mistakes made before the use of computers.

There is no question Cox and Doell's accent on examining the paleomagnetic data period by period, instead of as APW paths spanning several periods, put the mobilist interpretation of the paleomagnetic data in its weakest light, taking the data outside their historical context. Unlike mobilist paleomagnetists they did not examine the possible tectonic, paleogeographic, and paleoclimatic consequences of these dramatic broadly sweeping, convergent APW paths from different continents. Nor did they attempt seriously to review what others had done in this regard. There is another possibility. Doell and Cox began working on their analysis of the paleomagnetic results independently, and both adopted a period by period analysis. Doell at MIT started with the Precambrian results, and Cox in California with the Cenozoic. It is possible therefore, as Runcorn maintained, that their piecemeal approach, which effectively put the paleomagnetic support of mobilism in its worst light, may simply have been an artifact of the ordering principle that they initially and independently adopted. This could be so, but there is no evidence that during their extended review process (three years) they did reevaluate their approach, and consider that it might bias their analysis.

In 1959, almost all paleomagnetists thought that drift offered the best solution to the problem of divergent APW poles, all outstanding difficulties having been answered. All British-trained paleomagnetists, who remained active and worked on the problem of divergent APW paths, supported mobilism. The South African group, including the former fixist Hales, had come to accept mobilism. Soviet paleomagnetists, with Khramov leading the way, supported mobilism.

Cox, Doell, and Graham (and Japanese workers as already noted in §5.8 and further discussed in III, §1.15) were the exceptions. They should, I believe, have recognized that the difficulties they had raised were either already answered or were phantom difficulties. Perhaps they were blinded by the intransigent attitudes of North American Earth scientists toward mobilism, as so eloquently documented by Newman (1995), and further discussed by Oreskes (1999). Perhaps, trapped by regionalism's residue, as students they were inhibited by the lack of coverage in their courses and readings in world stratigraphy, which kept them from developing a sympathetic understanding of the classical case for continental drift (Volume I).

Cox and Doell's rejection of mobilism, as a general solution to the problem of the intercontinental dispersion of poles, affords another opportunity to test my hypothesis about theory acceptance according to which scientists *should* change their minds and accept a solution if it is the only one that remains difficulty-free after scrutiny by opponents (I, §1.12). I have argued that by 1960 the mobilist solution to the problem of divergent APW paths as developed by Creer, Irving, Du Bois, Runcorn, Nairn, Blackett, Clegg, Deutsch, Green, and others was the only difficulty-free solution to the problem. When the prospect of such a solution began to be vigorously pursued in 1954, it was based on what was then the questionable assumption that the mean geomagnetic field when averaged over thousands of years was that of the geocentric dipole. To become difficulty-free it required poles from different landmasses, it required further assurances that the paleomagnetic data were reliable, and it required consilience of drift reconstructions of the continents based on paleomagnetism with those based on paleoclimatic data. By 1959, these requirements had been met, and the majority of active paleomagnetists thought the mobilist solution was likely free of potentially fatal difficulties. In addition, as we shall see below, rapid Earth expansion was not supported by paleomagnetic evidence (§8.14). Thus, only the mobilist solution was left standing, and should, I believe, have been accepted as confirmation of continental drift. As I shall later show, some key scientists who were not paleomagnetists accepted this argument.

Cox and Doell raised what they thought were difficulties against mobilism as a general explanation of the intercontinental dispersion of poles; they did not recognize the mobilism solution as difficulty-free, they did not accept it. I have also argued that the difficulties they raised were not legitimate, and that the mobilism solution, if reason had prevailed, would have by 1960 been recognized by many as difficulty-free. If I am correct in claiming that their refusal to accept the mobilist solution was unreasonable, then any hypothesis about what constituted reasonable behavior should categorize their behavior as unreasonable. Of course, this does not show that my hypothesis is correct, only that it implies that Cox and Doell behaved unreasonably.

Cox and Doell did not reject continental drift outright; to the Australian case they were grudgingly favorable, elsewhere there was, they said, no case. They were very far from accepting continental drift as the general solution to the divergent APW paths, even though these were as expected on Köppen and Wegener's paleogeographic reconstructions of the later Phanerozoic, a consilience demonstrated by British, Australian, African, and Soviet workers. I think this grudging acknowledgment of the Australian case and its denial elsewhere was peculiar because it is not possible to have a little bit of continental drift every now and then in one place. It is not something to be accepted in small doses. It can only be a global phenomenon whose acceptance changes one's world view. Cox and Doell, and their predominantly American audience, were, to put it mildly, reluctant to do this, and their *GSA* review reflects that intellectual intransigency of which Newman and Oreskes wrote.

## 8.9 Runcorn's response to the *GSA* review

From the day it was published the *GSA* review became for years one of the most cited articles in paleomagnetism especially for its data compilation, for its discussion of methods, and for its negative assessment of continental drift, to which only Runcorn (1963a, 1964) and Irving (1964) replied publicly. Irving also offered extensive, private replies in letters to Cox in 1960 and 1961. I want to emphasize that although Doell knew that Irving and Cox were corresponding, he knew little about what they were discussing. Indeed, when I showed Doell (some of) the correspondence from Cox's papers forty years later, he was surprised. At the time, Irving knew Doell hardly at all, for instance he had no idea that Doell was senior to Cox in the USGS hierarchy (Doell, February 19, 2001 email to author; Irving, July 2002 email to author).

Commenting many years later on the *GSA* review, Runcorn said:

So we thought that their paper was, you know, somewhat academic – using the word academic to mean a little remote from reality. The same was my feeling about the statistical question. Even today the same question comes up over the [paleomagnetic] moon work: whether you should combine results from nearby before you produce a point on your diagram. I don't believe you can settle statistical questions like that *ab initio*. I think that you just do the best statistics that you can, and eventually, see what happens ... In the end you cannot justify the particular analytical method from pure theory, and you have to just see whether it is adequate when data from different sources and different geological periods are put together. And so

I was trying in my paper more or less to clarify the philosophy one ought to have in using statistics. I felt emboldened about that because I had talked with Fisher about statistics and felt that they were being a bit academic.

*(Runcorn, 1984 interview with author; my bracketed addition)*

Offering a rejoinder to Cox and Doell's charge about the application of Fisher's statistics (RS1), Runcorn (1963a, 1964) sought to justify his procedure, contrasting it with that applicable to lava flows.

The usual procedure [i.e., the one he used] in sampling is to collect hand samples from different stratigraphic positions within one geological formation and to cut a number of discs (usually 4) from each sample. In [our work] magnetization of each disc has been given equal weight on the grounds that one has, as yet, little knowledge of the way in which rocks can be magnetized and even discs in close proximity may have become magnetized at times sufficiently different for one to be sampling uncorrelated values of the geomagnetic field – this means a few hundred years judging from studies of the secular variation. Cox and Doell (1960) however argue that this is an incorrect procedure and that one should determine the mean direction of each sample first, by adding the magnetization vectors of the discs in each sample, giving them equal weight and then proceed with the statistical analysis. There will be general agreement that the latter procedure is absolutely correct where the sample is known to have been magnetized at one time: this would be the case for a sample or samples from a single lava flow in which we know the flow to cool through the Curie point in about a year – in which time the geomagnetic field hardly alters its direction.

*(Runcorn, 1963a: 52; my bracketed additions)*

According to Runcorn, if the conservative method advocated by Cox and Doell[11] had been used when computing confidence limits of poles, it would not have weakened the case for mobilism; the paths for North America and Europe would not change position, would still be continuous, and would still diverge. It did not much matter which procedure was used because he was able to show that for his North American rocks the dispersion between samples was not significantly greater than between specimens from the same sample; moreover, the separation between the APW paths would remain even with the larger limits.

If, however, the angles of confidence had been calculated from the mean direction of each sample, because usually about 4 discs per sample are measured, the angle of confidence would be about doubled. We conclude therefore that a good deal of the scatter ... could be removed by more intensive sampling and smoother polar wandering curves would result. This does not affect the issue of the reality of the discrepancy between the American and British polar wandering paths on which the evidence of continental drift depends, but the use of palaeomagnetism for other purposes, for dating and for the examination of small relative movements within the continents due to tectonics, does depend on the pole moving in a smooth path.

*(Runcorn, 1963a: 54)*

Thus Runcorn demonstrated that Cox and Doell's unreliability difficulty was no impediment to his general drift interpretation (RS2).

## 8.10  Irving's response to the *GSA* review

I met Allan outside the San Francisco Opera House in 1957 when I flew from Australia to San Francisco on my way to Toronto to the IUGG. It would have been in August or late July. I had corresponded with Verhoogen. I did arrive a day early because I had forgotten the dateline. When I rang up Verhoogen, he said, "Irving you're a day early, and Allan Cox will look after you and put you up." He gave me a phone number. I rang Allan, and Allan said, "Do you like ballet." I said, "Sure, I like any of that kind of stuff." So he said, "Meet me outside the Opera House at whatever time it was." And we went to the ballet together, and came to his place, and stayed at his place for several days. I had a very intense time with Allan. He was just beginning to do research. We talked about a lot of things – a lot about statistics.

I actually gave a talk at Berkeley at that time which reviewed the paleomagnetic results. Verhoogen and Turner were there. Turner was quite cool to continental drift.

*(Irving, November 2000 conversation with author)*

In 1960 Irving and Cox corresponded about a variety of issues. There was nothing lukewarm about Irving's reaction to the *GSA* review. He thought their compilation of paleomagnetic data was excellent, but was upset over a few entries. He thought the review well written and would serve as a good introduction to the methods of paleomagnetism, and, not unexpectedly, he agreed with their criticism of Runcorn's use of Fisher's statistics.[11] They corresponded about Irving's critique of Nairn's work (§7.2). Irving questioned their refusal to favor the pro-mobilist interpretation of the paleomagnetic results, and was extremely disappointed in their failure to do so. He essentially told Cox that he worried too much about the views of others. He urged Cox to obtain a new Eocene pole for tectonically stable North America from flat-lying rocks, especially because it was he who complained about the lack of good data for North America and he who continued to cite his pole from the Siletz River Volcanics as inconsistent with the paleomagnetic case for mobilism. Irving raised an internal theoretical difficulty with Cox and Doell's use of the term virtual paleomagnetic poles. Although by introducing the term they were attempting to reduce confusion, Irving argued that it was itself misleading, even philosophically confused. They also discussed attempts to arrange for Irving to spend part of his upcoming sabbatical leave at Menlo Park, one purpose being to help them build a highly sensitive magnetometer.

Unless otherwise noted, I shall consider Irving and Cox's letters in chronological order because I want to document Irving's growing disappointment with the *GSA* review. The surviving letters are in Cox's collected papers at Stanford. There are twenty-three of Irving's letters extant, but only seven of Cox's. Cox wrote the first dated extant letter on February 8, 1960. However, it is not the first letter because it was written in response to two of Irving's earlier letters now lost. The last Cox letter was written on August 12, 1961. Irving's final letter is dated November 9, 1961. Some of Cox's letters appear to be missing. He carbon-copied most of his letters, and kept

Irving's. Irving did not keep the correspondence. Except for one typed letter, Irving wrote in longhand. Cox's surviving letters were typed; the first five appear to be more "official" than the last two because they had been typed on official USGS stationary, for there is sufficient space above the first typed line for a USGS letterhead, and they began with "Branch of Geophysics," centered directly below where the USGS letterhead would have been on the original. These five letters include Irving's address, Cox's salutation followed by a colon, and are signed "Sincerely," with Cox's first and last name and designation "Geophysicist" underneath. The remaining two letters, clearly typed on a different machine, were not on official USGS stationary, don't include Irving's address, a comma follows the salutation, and they are closed with "Best regards," followed by a handwritten "Allan."[12]

### 8.11 Cox–Irving correspondence

Cox first broached the review in a letter dated February 8, 1960, telling Irving, "The review of paleomagnetism for geologists that Dick and I have been working on is finally off to the press, so we can begin to pick up the threads again." He and Doell had talked to J. C. Jaeger, Irving's head of department, about it, and gave him a preview of its general conclusions.

We discussed the general conclusions of our review paper with Jaeger, so you probably know the general position we're taking. The post-Cretaceous drift of India and Australia appears to us a bit shaky, especially in view of the scarcity of data from North America and Europe with good stratigraphic control. On your Australian results we conclude that, given a dipole field, they constitute evidence for continental drift that cannot be ignored. However, for the most part the review is concerned more with the basic techniques and data of paleomagnetism rather than our opinions.

*(Cox, February 8, 1960 letter to Irving)*

Turning to the state of their laboratory, Cox characterized their magnetometer as a "quick and dirty spinner, designed for speed and efficiency of operation rather than ultimate sensitivity." Cox felt they needed a more sensitive astatic magnetometer, and wanted Irving to help them build it.

Replying twelve days later, Irving began with "I'm glad your review has finally gone off – almost as long delayed as the Queen of England's baby," he agreed about the lack of North American data and urged them to remedy the situation by getting more Eocene data from an undeformed area. Otherwise, he would not take seriously the missing-data difficulty they had raised arising from Cox's work on the Siletz Volcanics (§7.6, §7.7).

I agree that the North American work is thin – what I want to see is some work with stability control on rocks which are in an *undeformed* area and I won't believe a word of what you say about your Lower Tertiary field until you provide it.

*(Irving, February 20, 1960 letter to Cox)*

He closed pointedly with "I trust you are madly measuring more Lower Tertiary from undisturbed areas with your quick and dirty spinner."

Irving, having largely devoted his next two letters, July 25 and August 9 (with August 8th crossed out), to his criticism of what he considered to be Nairn's misuse of statistics (§7.2), accused Runcorn of a similar error. Irving, of course, knew from his 1957 discussion with Cox in Berkeley that they were in agreement on this weighting question.

You know we agree really. Anyone with any sense uses site means to get the error. Runcorn and Nairn are out for quick picking and I just won't have it. We did our work properly (Torridonian, Ken's Devonian, etc.) at first and these other people have just slid downhill. My explanation [i.e., what I've done in my paper on Nairn] is physical because this is what a geologist will understand – to talk statistics at him is useless and unnecessary anyway. What we want is a bit of commonsense and to a geologist site means [as opposed to sample or even specimen means] are obviously what you want. Runcorn has sampled quite indiscriminately and his use of statistics is very bad.

*(Irving, August 9, 1960 letter to Cox; my bracketed additions)*

Irving's remark raises the question, why did Irving not directly criticize Runcorn in print when he criticized Nairn for roughly the same mistakes? Forty years later, I asked why he did not do so.

That is a very good question. I really don't know. I think Nairn was worse. Keith was much cleverer ... I think to some extent Keith was more difficult to get at for one thing because he was cleverer at what he did, and the second thing about Keith is that he had been my supervisor. I think I had great respect for Keith even though I thought a lot of his stuff was sloppy. But Keith didn't actively mislead anybody. I don't think. That is correct. Some of the things said about Nairn might have been a bit on the harsh side. But, I was terribly worried. You see there is a time thing here too. I mean Keith had done his damage as it were; had done his thing and got the main answers out. Then Nairn came along, and the horrible thing was that somehow or other the quality was not being improved, and you would have thought that after that initial reconnaissance stage someone like Nairn coming in at a later stage would have been better. I mean he worked with us on the Whin Sill paper in which we had some of the very early a.c. demagnetization and lots of coverage. He knew how to do it, I suppose. He worked with Ken and me, but he seemed somehow or other to have gone off on this retrogression. I suppose that may have been one of the reasons that I thought with the next generation of people coming in, there should have been a systematic increase in the quality of stuff but it was going backwards. And that would have been my reason for attacking Nairn. I just thought that a number of the poles he produced might, in fact, lead us into dangerous water.

*(Irving, September 24, 1999 conversation with author)*

Although not attacking Runcorn in print, Irving had many an oral argument with him over his sampling and weighting techniques.

I argued with Keith about his sampling for ages and ages and ages. Well, yes, Keith thought that each sample represented a spot reading of time, and from my own work of individual sampling individual sites in sedimentary rocks the directions were actually random within a site because I had been unable to find any secular variation within a single site. Therefore, this could mean that we had in fact averaged out the secular variation or we could just have had some fuzzy signal in which the signal was buried in a lot of random noise of one sort or another. I thought it was therefore physically somewhat unsafe to say that if you had four locations from some unit in Arizona and perhaps twenty samples, you should give the twenty samples unit weight. I thought that was wrong. What you had to do was to give the locality, each level, each horizon that you studied unit weight irrespective of how many samples you took from it. And so Keith was able to reduce his error substantially by increasing the number of samples. That was the sleight of hand.

*(Irving, September 20, 1999 conversation with author)*

Irving later talked to Fisher about what he thought of Keith's sampling procedures. Fisher disagreed with Irving. Irving recalled:

Fisher took me to task about this because I was giving him a little lecture about this. He said that what often happens in this instance, and certainly it did to a degree with Keith's data, if you take the average of a series of sites the precision of those sites will be greater than the precision of all the data thrown together irrespective of sites. And consequently the error obtained by grouping the data all together – throwing them all together – may have larger $N$ number but you have a lower precision $k$, and this would correspondingly be somewhat self-correcting. And Fisher is quite right, for the small samples that Keith was dealing with, I think that perhaps I overblew this. I think that it may well be that Fisher was right. Keith may be a bit wrong but he wasn't going to be all that wrong. And that is the reason for it. I did these calculations from test data of one sort or other but I just felt physically that this wasn't the right way to do it, that it was a bit of a kind of statistical wriggle-a-round. I thought that we should take one stratigraphic level and then go to the next stratigraphic level, giving each level unit weight irrespective of how many samples we get from it. That was my rather conservative view. [See for example, Irving (1964): 63–68.]

*(Irving, September 20, 1999 interview with author; my bracketed addition)*

Fisher was right. As shown already, both Runcorn and Nairn redid some of their calculations, and their results were substantially the same. Irving recalled that this lack of any substantial difference was "exactly what Fisher" expected would happen.

You see for a small sample that will be quite valid where your precisions are generally rather low. If there is a good deal of scatter, that will certainly be true. But, on the other hand, physically, it is not a good idea [to give unit weight to samples or even specimens]. That was my point. But, Keith's method was quick and dirty, but it was very smart.

*(Irving, September 20, 1999 interview with author; my bracketed addition)*

Despite Irving's later assessment that he may have been somewhat too hard on Nairn and Runcorn, he, when first reading the Cox and Doell review, clearly agreed with their attack on Runcorn.

Irving also was generally pleased with their compilation – so pleased in fact that he wrote a second letter on August 9 (with no August 8th crossed out) – in which he proposed they join forces with him and publish periodic lists of paleomagnetic data, a process that he had started in the *Geophysical Journal*, the first list appearing in 1960, the same year as the *GSA* review. Such lists would make results accessible to non-paleomagnetists. Irving devoted this second August 9 letter to the proposal, and even underlined "important," which he wrote in large upper case letters across the top of the letter.

I am very impressed by the excellence of your table – a little long-winded perhaps and almost "too fair." I must say it is an excellent job. What are your plans for the future? I suggest that we (that is Cox, Doell and Irving) join forces, Thus: 1) We compile the lists, and we invite further compilers at a later stage when necessary. We will need one from England and I suggest Roy King as a neutral person (you know what I mean). 2) Publish in *Geophysical Journal* [which is where Irving already had begun publishing compilations]. 3) Format ... lists initially with the premise that we will change it if necessary ... 4) We publish an index in due course.

*(Irving, August 9, 1960 letter to Cox; my bracketed addition)*

Irving explained the advantages of his idea, which included wide coverage and close cooperation between groups. He ended the letter like this:

Let me know soon. Your reply does not need to bind you in any way – all I want to know is are you interested? We obviously have much to gain from speaking together rather than separately – certainly some marriage of our lists and methods is desirable.

*(Irving, August 9, 1960 letter to Cox)*

Irving had yet to express any serious reservations about the GSA review, but his apprehension soon became palpable.

### 8.12 Irving becomes critical of the *GSA* review: further Cox–Irving correspondence

The *GSA* review included a finding by Almond, Clegg, and Jaeger (1956) from one Early Tertiary basalt from Tasmania. Irving criticized the review for not including the more recent results from three basalts by Green and Irving (1958). The 1956 result had the wrong inclination because the magnetization was partially unstable, and its "reversal" was an artifact of the core-storage procedure. The matter was important because the 1956 result disagreed with the published directions from Eocene basalts of nearby Victoria whereas the later more complete result was in excellent accord. Cox and Doell were introducing a non-existent conflict, a phantom anomaly difficulty, into the Australian data: fuel for the fixist. No wonder Irving became annoyed.

For your information re. p 754 your review. Tertiary basalt [unoriented bore core] measured by Almond, Clegg & Jaeger unstable. I have [fully oriented] surface samples from near there (Wilson's Creek) and this is what happens. They are a lovely case of partial instability. The reversal in their bore core [which was one of the cores obtained by the Hydroelectric of Tasmania and shipped to Blackett's laboratory] is *not* to be believed – it is due to the drillers turning sections of the core over! Your pole 28 [based on Almond *et al.* (1956)] is worse than meaningless, it means the wrong thing. You would have done better to use the data (p 17 entries C, D, E & F, *Proc. Roy. Soc. Victoria*, 1958, *70*, by Green, R. and Irving). The average Tertiary directions for Tasmania are about −70° inclination not this bloody silly value of −83° you quote. Why don't you ask me about these things? You know my address. I have known about the uselessness of the base core basalt for about 4 years but haven't bothered to correct it. I thought it would not be taken seriously. Let me know if you don't have a reprint of this paper. I'll send one.

*(Irving, August 11, 1960 letter to Cox; my bracketed additions)*

Four days later in his next letter, Irving had become more generally critical of the *GSA* review. He thought it overly influenced by the prevailing anti-mobilism of North America. He seemed more upset about what the review did not do than what it did. It was generally very good. But, it was dead. It had no heart. It harped on the difficulties with the paleomagnetic case for mobilism but almost entirely ignored its merits. Irving believed that the review would harden the case against mobilism and fail to attract new researchers to paleomagnetism. He had come to believe that Cox and Doell's analysis of the paleomagnetic case for mobilism was almost entirely negative; it did not have an eye on the tempting prospects ahead; it was not progressive. Furthermore, although he agreed that Runcorn's work had been sloppy, it was positive for he had worked out an APW path for North America, which was much more than they had done. He did not mince words.

Dear Alan,

Many thanks for the offprint of your review. I hope you will take this that I have to say in good part as I want to disagree with you in a rather fundamental philosophical way. You see, your review read just like Christian apologetics; like the attempt of the medieval schoolman to reconcile their preconceived notions about religion as revealed in the Word with the recently discovered Aristotelean physics. I think you should rename it "paleomagnetic apologetics." I feel that you are too much influenced by contemporary hardness against drift and I feel that your review will tend to harden it …

You seem blinded by techniques and difficulties instead of going to the heart. Of course, I agree that Runcorn should add his vectors up differently but the miracle is that he has vectors to add up at all. If any of you chaps had done half as much work you would be better placed to set standards. (A lot of Keith's mistakes are due to his method of sampling – through not being a geologist.) This leads me to the point that you have done so little yourselves – it is very easy to sit on the very high fence that you do sit on and pontificate about what should be done (it has all been said before anyway) but it is infinitely more difficult to get the right answers. Come down from your ivory tower and go out and show us how to do things by producing half a dozen first class descriptive papers. For, I suppose, a total of about 12 years or so in the subject you really have very little to show for it, and this retarded stunted growth (the review) is not going to bear fruit. I am very much reminded of Gibbon's remark "that Byzantine science was eclipsed by the Arabs due to an over concern with the detection of error rather than the discovery of truth."

For sheer information your review is unparalleled but there is no light shed in dark corners, no concern with what is round the corner, no emergence of generalities without which all is chaos. I am terribly disappointed, for I had expected brightness at last from the Western Hemisphere – I am disappointed not with the review, which is excellent, but that it is only a review – dry bones are not enough. I am mindful of the excuses made to avoid evolution in the past century – you had a great responsibility, which I think you have used wrongly. Time will tell.

Now about the data list, my proposal is ... A coordinated effort on this project would be of immense benefit to the subject and to our groups. I look forward to the time when we see more of each other and can argue more directly. A joint effort would also mean fewer difficulties about preparing lists when on leave, say, or on field work. I hope most sincerely that you will take this offer.

I realize I am not a very soft-spoken person and hope I have not been too harsh about your review. Our approach could not be further apart philosophically at the present time – and I do feel that your arguments about reliability and so on ring rather hollow when you have so little to show for it *in print*. I am trying to goad you to give us some data, particularly on flat-lying un-rotated Eocene of the United States! Any poles from orogenic belts are automatically suspect.

With all the best wishes,

Yours sincerely, Ted

*(Irving, August 15, 1960 letter to Cox)*

"Paleomagnetic apologetics" was strong but, insofar as the GSA review related to continental drift, justified. Irving had hoped that they were going to be more supportive of the drift solution to the problem of intercontinentally dispersed poles. He had thought that they, unlike Graham, were going to be the first American-trained paleomagnetists to show sympathy with mobilism. He was disappointed. Irving believed that they had caved in to anti-mobilist sentiments so prevalent in North America. He also had begun to resent their judgment of Runcorn. Irving, who had collected his share of data, thought they did not have the right to criticize Runcorn because they had obtained very little data, and had limited experience of having done so. Moreover, they had complained about the paucity of data, but as far as Irving knew, they had not yet even tried to obtain a North American pole from truly cratonic Eocene rocks to check for possible rotation of the Oregon Coast Range. Nevertheless, angry as Irving was about what he took to be their caving in to anti-mobilist sentiment, he still wanted to work with them in periodically compiling paleomagnetic data, and offering a critical assessment of new work.

Although Cox wrote his next letter before receiving Irving's "paleomagnetic apologetics" letter, there was plenty to discuss. Regarding Irving's complaint about the Tasmanian Early Tertiary pole, he said that he and Doell had decided that it was not practical to correspond with those whose data they reported, that if asked he would encourage the *Journal of Geology* to publish Irving's critique of Nairn, and that he was still not convinced about Runcorn's use of statistics in his new work even though he had discussed the matter with him during his recent visit. He would talk to Doell about working together on the compilations, but added that working in the laboratory had become more appealing than compiling. Finally, he remarked that many American geologists had doubts about the paleomagnetic case for mobilism.

Dear Ted:
This is in response to your letter of August 11 expressing your displeasure with entry 28, section I in our review.

When Dick and I began assembling data, two decisions faced us at the outset. (1) Should we enter into correspondence with the authors of our source articles to clear up important points that were not clear in the original articles and to inquire about possible relevant unpublished research? (2) Should we be inclusive in our presentation of data, or should we present only the data that looked the most reliable?

Our decision on (1) was not to correspond with all of the workers involved. In part our reason was that this would have been too much work. In part, however, it was because we felt a piece of research should be complete as published and subsequently revised as retraced if necessary, but should not require correspondence with the author. We did try to get everything we could out of what was published, however, even when this meant searching glossaries for obscure place names, to get geographic coordinates and scaling directions from diagrams in order to reconstruct how the statistics had been handled.

As for (2) we decided to be as inclusive in our presentation of data as we could. In part this was in response to the climate of opinion toward paleomagnetism in this country. You may recall a public exchange between yourself and John Graham in Toronto in which Graham stated that much of his own published data which didn't fit the theories didn't get referred to, and strongly implied that there were many more paleomagnetic skeletons in the closet. It's probably hard for you to realize how widespread this type of skepticism toward our field of research actually is in this country. My own experience with the better informed geologist over here is that he's hopeful about the use of paleomagnetism in geology; however, he has read enough to know, for example, that the Cretaceous pole for North America is based on measurements of 3 out of 6 oriented samples from one sampling area, and he really wonders whether there might be something to Graham's oft repeated charge that data at least this good doesn't get published on the various curves. Dick and I felt (and still feel) that it would help clear the air even here if we gathered together all of the data that has been used for poles, together with all the additional data in the literature of comparable quality (however marginal!). We would prefer having a few actual anomalous "skeletons" based possibly on questionable data to contend with in our future work rather than the fear (which wasn't all confined to Graham over here) that there were vast armies of such skeletons hidden away in the literature.

... I hope you feel free to call any of the data in our article into question if you need to do so in your future work, and especially so if you have new information. Our intent was a compilation of the data in the literature and not necessarily an endorsement of it. I'm very sorry that we didn't include mention of the inclination values of 78°, 80°, 69°, and 54° for the four Tasmanian sampling sites listed in Green and Irving, 1958. I seem to have been missed on this reprint, by the way, and would much appreciate one. However, I know I did read the article several years ago, and the inclination data just slipped from my mind. This is the sort of thing that haunts one in preparing a review. My apologies.

If the *J. of Geology* sends your article [criticizing Nairn] around I'll certainly encourage them to publish it complete of course ... I'm still concerned about the statistical treatment. Runcorn dropped by here last week and I had a chance to go over this with him. In his new (GSA, July) article he makes specific use of the argument that the proper vectors to be used in a statistical analysis are those corresponding to individual points in time, and due to slow rates of sedimentation, etc. Specimen disc magnetizations most nearly fit this criterion. I know that we all agree about how lava flows should be treated, but [it] is just exactly in sediments that the physical argument about points in time leads to difficulty. In Runcorn's article there appears to me little doubt that in some of his plots, between-disc scatter is considerably less than that between samples ...

I'll show Dick your proposal about joining forces on some sort of list of data when he returns. Personally I am now in the throes of a strong reaction against too much sterile compiling and writing this past year, and am mainly anticipating more time in the lab now that we have one going.

Sincerely,

Allan Cox

Geophysicist

*(Cox, August 16, 1960 typed letter to Irving; my bracketed additions)*

Cox acknowledged that they had decided to report almost all paleomagnetic data because many North American geologists agreed with Graham that a lot of results did not support the paleomagnetic argument in favor of mobilism. He also told Irving, "It's probably hard for you to realize how widespread this type of skepticism toward our field of research actually is in this country." Cox was very concerned about this widespread skepticism in the United States.

Before turning to Irving's reply to the August 16 letter, I want to go back three years to Irving's presentation of his and Green's APW path for Australia especially at the IUGG meeting in Toronto, in August 1957, the one at which Creer discussed magnetic cleaning and reported his results from South America (§5.5, §5.6). Describing some of the events before the Toronto meeting, Irving recalled:

Ron and I compiled all our data in early 1957 and we drew the first Late Precambrian to present APW path for Australia. We wrote it up and sent it to *JGR*. I submitted it as joint presentation for IUGG meeting Toronto August/September 1957. I presented it first publicly at Sydney University. I then gave it at Berkeley a week or so later, on my way round the world. En route to Toronto and anxious to know the outcome, I visited the editor of *JGR* at Carnegie Institution in Washington in August and learnt that it was rejected. John Graham who had worked on paleomagnetism at Carnegie since the late 1940s was the reviewer. I retrieved the manuscript and while in Newcastle in autumn 1957 I reedited and submitted to the new *Geophysical Journal* of the Royal Astronomical Society who published it in their first volume 1958.

*(Irving, November 2007 email to author)*

At the Toronto meeting Irving argued with Graham and Rutten. Rutten, a prominent Dutch geologist, one of Hospers' former teachers and leaders of the Iceland expedition on which he had made his first collection (§1.12), was helping to organize a group of Dutch paleomagnetists (§5.5, §7.8). Irving also was encouraged by Edward Bullard, who chaired the session. Irving recalled:

Our presentation at Toronto was the first I had given to an international audience. I was essentially unknown. The ANU was essentially unknown. We had documented this huge difference between the APW paths from the northern continents (Europe and North America) and I was wound up for it. Ron and I had worked very hard. I had been working on paleomagnetism for 6 years. As an undergraduate and since I had extensively read the classical case for continental drift. I had thought long and hard about what I was about say for many years. I remember it vividly. The session was chaired by Teddy Bullard. After my talk I was attacked by John Graham, the paleomagnetist from Carnegie, who had just rejected our paper, and Martin Rutten a prominent Dutch geologist, reflecting the universal, adamantine rejection of drift by American and Dutch scientists. Graham brought up a litany of uncertainties especially stress effects. I called his attention to the very good agreement of near vertical directions in Tasmanian dolerites, Permian lavas and Carboniferous glacial strata supported by tilt and conglomerate tests and all with widely differing stress histories; how, I asked, could such uniformity arise if stress was a factor? I argued that field evidence trumped unsupported, speculative theory. Rutten believed the ages of our rocks were insufficiently well known and this could account for our path being so different. This and his condescending manner really got me going. I think he must have been in some sort of shock because the differences between the British and Australian paths was so huge, > 60°, way beyond any possible experimental error. I explained to him that the Australian sequences, because of their economic importance, had been studied by several generations of highly competent Australian geologists who had written highly competent accounts which I had consulted at every step of the way. I had been shown over the outcrops by experts. In any case, the differences between our poles and those from the northern hemisphere were so large that misdating by a geological era (hundreds of millions of years) would not explain them! These guys were talking down to me, as if we hadn't thought about these things, and as if Australian geologists were in some sort of bush league. Rutten never really forgave me. Bullard spoke to me after my talk and remarked on my "spirited defence," and that "God would not have put those magnetizations there just to confuse us." Judging by his frequent later mention of our Australian results, I suspect that it was at this point that Bullard began to take continental drift seriously. Jaeger was sitting near the front and I caught him quietly smiling. Hess was certainly at the Toronto meeting and I think he was in the audience but I am not sure.

*(Irving, September 2007 email to author)*

Elaborating on Bullard's response, Irving also noted:

It was the first time that I got very very solid support from a major figure – someone outside of our own camp as it were – not that Runcorn wasn't a major figure but he was suspect like me because he was responsible [for the paleomagnetic work]. I remember because I felt that Bullard was moving in our direction. You see that was really the first time that I had said more than three words to him.

*(Irving, September 20, 1999 conversation with author; my bracketed addition)*

Cox was right about the Graham/Irving argument. Irving thought that Graham's hard line was wrong and all the more reason Cox and Doell should have come out in favor of mobilism. Their present conciliatory approach would only allow Americans to continue to ignore the paleomagnetic support for mobilism. Irving favored the broadsword approach. Here is his undated reply, this time quoting R. A. Fisher instead of Gibbon.

Dear Alan,

Thanks for letter of Aug. 16. Enclosed is a card with some notes about pole calculations – we agree very well. It may be wise to wait 'till I see you about the data list, but I think we should look to the time when, as compilers, we have *one* person from each country which is producing data. In this way we can save ourselves the terrible trouble geologists have with collecting chemical analyses etc., etc.

About the Tasmanian pole – my feeling is simply that since we correspond so much a note from you would not have been difficult. You did correspond on such points as this with other people. If you had circulated the list it would have helped particularly with your atrocious spelling of Australian aboriginal names!

Your Review will harden American opinion not soften it. Your agnosticism is just the thing that your geologists needed – now they may wipe their hands of the whole business. I would be very much surprised if your review inspires one person to do work in this subject which, what ever is said about it is going to revolutionise geology.

The example you quote from the Cretaceous illustrates the point I made earlier – it is no good criticising early work, what you have to do is to test it – go and see for yourself. It is this preoccupation with error rather than truth which I complained so bitterly about in my last letter. We know Runcorn makes mistakes in detail but he does do something instead of sitting on his backside complaining about the lack of other peoples' data. We can only seek to explain what we have.

Your last paragraph [in which you say that you want to work in the laboratory instead of on another review] is what I have been wanting to hear for years. It may be more difficult in your position than for me but I do feel you are over conscious of the pressure of American scepticism and spend too much time explaining things to them. The truth needs no advocate. Runcorn has oversold things a little but we have to support his effort and not needle him about small things. This your review does and it will delight those who don't particularly like Keith.

The other thing is you overestimate Graham. His papers are not well written or logical and he missed the boat badly through not following rigorously his arguments. E.g., Rosehill stable – right? Now Palaeozoic's of platform areas [are] near present field – right? Thus Rosehill [is] a special effect. Surely he should have said that Rosehill is the one thing we can hang on to, and suspect the rest. His papers are full of this type of argument. Graham has one fault which is greater than all this however – he makes sceptical statements without following them up. His suggestions are full of loose ends. It surprises me that he has been taken so seriously. Graham-worship and fear of CD [continental drift] has been the great trouble. Fisher once said to me, "to do anything really original you need to be a little out of touch with current trends otherwise you let other people' prejudices worry you too much." (These are very wise words.)

LET ME KNOW ABOUT MY STAY WITH YOU. I HAVE TO PLAN THINGS NOW. PLEASE, PLEASE.

Yours, Ted
*(Irving, undated response to Cox's August 16, 1960 letter; my bracketed additions)*

Irving's discontent was manifestly growing. He would have his wife and children with him, so he would need a plan. He found Cox's explanation for not getting in touch with him about entry twenty-eight unsatisfactory. He again told Cox that he and Doell had been too preoccupied with error instead of truth, and favorably contrasted Runcorn's efforts in paleomagnetism with theirs. Runcorn, unlike them, went out and got an APW path for North America, while they continued to sit on their backsides. But there was more. Irving thought they had overestimated the quality of Graham's work, and had mistakenly believed that by taking an agnostic position toward the paleomagnetic case for mobilism they could attract North American Earth scientists to it. Finally, he repeated the charge that they had copped out. Linking his charge with originality in science, he passed on that wonderful advice from Fisher, adding a parenthetical remark of his own, just in case Cox missed the point.

What of Irving's judgment of Graham? It is harsh, but I believe it is correct. As I have already shown (§1.9), Graham even gave up his fold test because he thought it forced him to accept geomagnetic reversals or mobilism. Because of this, he missed both boats. Irving's assessment, "Surely he should have said that Rosehill is the one thing we can hang on to, and suspect the rest" is on the mark. Irving's other complaint that Graham "makes sceptical statements without following them up" that are "full of loose ends" was also fair. Graham's difficulty, for example, about magnetostriction caused much discussion but amounted to very little, especially because paleomagnetists did not determine poles from highly stressed anisotropic rocks (Stacey, 1960; Stott and Stacey, 1959, 1960). For that matter, Doell himself had readily dismissed Graham's magnetostriction difficulty (§8.2).

Irving's predictions that the *GSA* review "will harden American opinion not soften it," and that their "agnosticism" toward the paleomagnetic case for mobilism will allow geologists of fixist persuasion to "wipe their hands of the whole business" also proved correct, for Cox and Doell's review was, as I shall later show, frequently used by fixists as a data source when it was years out of date to excuse their avoidance of paleomagnetism.

Irving's article of faith that paleomagnetism would revolutionize geology regardless of what Cox and Doell said about the strength of its support for mobilism in their review was also correct, although it was for reasons nobody then knew about. In fact, the geomagnetic reversal timescale, which Cox and Doell, and their colleague Brent Dalrymple, played a major role in developing, was one of the reasons paleomagnetism, in one form or another, ended up revolutionizing geology (IV, §5.3). Of course, nobody then knew that the reversal timescale would help bring about the general acceptance of mobilism. I do not think Cox and Doell's agnostic attitude toward mobilism changed until after the confirmation of the Vine–Matthews hypothesis when they accepted it (IV, §5.4–§5.6). Doell's certainly did not.

I think that my best answer [to when I accepted continental drift] is that I never completely accepted the idea in its totality – although I've long thought that some sort of crustal movements were required to explain many phenomena. Then when plate tectonics came along and began explaining so many things so elegantly in the late 60's I considered that the question re continental drift (in the sense of Wegener) had become moot.

*(Doell, January 26, 2001 email to author; my bracketed addition)*

Irving also raised a nomenclatural question about the *GSA* review in two letters written before September 20, 1960. When paleomagnetists first began calculating poles from paleomagnetic data they referred to them as geomagnetic poles or ancient geomagnetic poles. In an attempt to unify terminology, Creer, Irving, and Nairn (1959) introduced the term "paleomagnetic pole" to refer to poles determined from the remanent magnetization of rocks. Cox and Doell coined their own term, "virtual geomagnetic pole," to refer to poles calculated from direct observations of the present field and ancient poles calculated from rock magnetic data regardless of the very different timescales involved. Irving thought that their use of "virtual" was misleading. He raised an internal theoretical difficulty (RS2).

I would very much appreciate clarification on this issue. Could you tell me whether some poles are more virtual than others? If by the use of the word virtual you are expressing some uncertainty then there must be an answer to this question. My view is that it is better to express uncertainty in statistical terms, the truth or otherwise of statements about poles then being a matter of the correctness or otherwise of hypotheses. A pole is not a hypothesis or even a representation but a consequence of hypotheses, and therefore truth in the manner defined, since an early law of logic says that the consequences of hypotheses is truth. In any case one should not make mistakes in solving spherical triangles (although we all do). The uncertainty surely is in the assumptions.

Unless you can answer this question or deny that virtual implies uncertainty then I will have to view your use of the word more suspiciously.

Yours, Ted

PS. Of course virtual may have optical connotations for – "looking through a glass darkly" as it were.

*(Irving, letter to Cox written between August 16 and November 8, 1960)*

Irving thought that a pole is either real or it is not. It exists or it does not. It does not make sense to assert that a pole is virtually real or almost real. Irving thought that Cox and Doell had introduced a level of uncertainty that does not exist. There is a level of statistical uncertainty as calculated by Fisher's method. There is also a level of uncertainty or doubt arising from assumptions such as the GAD hypothesis or the hypothesis that some rocks faithfully record the geomagnetic field. However, their introduction of "virtual" suggested another level of uncertainty where there is none, namely in the calculation of the pole given the hypotheses and data; the move from hypotheses and data to the calculated pole is only a matter of trigonometry.

Irving restated the difficulty in a subsequent letter; yet to hear from Cox, he again pressed Cox about virtual poles.

Sometime you must reply about the polar representation versus polar hypothesis question I asked. That is, my view that given hypotheses A, B, C then poles are true, or your view of virtual poles as a mere representation. Whether your use of the word is general or optical, I don't care, it is still philosophically disagreeable.

*(Irving letter to Cox written after above letter but before November 8, 1960)*

Cox responded on November 8. Cox began by explaining the delay in answering Irving's questions. He also noted that he was not sure if he understood Irving's philosophical objection.

Dick and I have been out of touch while working in Texas and Colorado this past month, hence our delay in answering your queries about our use of the term virtual geomagnetic pole. Now that I have your letters all before me, I find that I'm not entirely clear about the philosophical objections you make. Nonetheless, I'll try to explain our reasoning.

After noting the ambiguity of the term "pole," and the standard uses of "magnetic pole," and "geomagnetic pole," Cox turned to the meaning of "virtual" that he and Doell had in mind.

"Virtual," of course, has nothing to do with virtue, truth, or goodness. It was borrowed from physics, where it is used to describe an event (usually one that is more mathematically tractable than the actual case under consideration) which would give rise to the effect observed.

*(Cox, November 8, 1960 letter to Irving)*

They had not intended to suggest another level of uncertainty; a virtual geomagnetic pole is one that is more mathematically tractable than the actual pole. For example, suppose the present task is to determine the position of the geomagnetic pole. Calculations from direct observations of the present geomagnetic field at different observatories across the world give different virtual geomagnetic poles clustered around the geomagnetic pole, whose position can be determined by taking an average of the virtual geomagnetic poles. However, Cox and Doell also used "virtual geomagnetic pole" to refer to poles determined from remanent magnetizations of ancient rocks – they used it not only to refer to poles derived from single sites or specimens but also to poles derived from averaging many sites or what Irving and others had come to call paleomagnetic poles. Be this as it may, in the remaining correspondence Irving made no further mention of his puzzlement over "virtual geomagnetic pole."[13] Later in his book, he raised a different internal theoretical difficulty with their notion of a virtual geomagnetic pole.[14]

Irving continued to press Cox about getting a new Eocene pole for cratonic North America. Irving became more and more angry on finding out that Cox and Doell were writing a second review. On November 15, 1960, Irving suggested that they stop writing to each other until Cox did some actual paleomagnetic work of his own. In the same letter, Irving returned to the unhealthy state of paleomagnetism in the United States. Alluding to the launching of Sputnik 1 on October 4, 1957, when the United States had been upstaged by the USSR, and impressed by Khramov's work on reversals and progressive views on mobilism (§5.9), Irving told Cox that Soviet paleomagnetists already had pulled ahead of the US counterparts, even though paleomagnetism had begun ten years later than in the United States.

Your country badly needs "experimental" leadership in this work if they are not to be made fools of by the Russians again, starting 10 years later and already well ahead. I regret my impatience with you but we will never understand each other if I keep it down. If I were not very worried by the state of palaeomagnetic work in your country I would not write like this. But research and leadership in research is so very important in your country – more important perhaps than in any other outside Russia – and it is entirely on your shoulders (and those like you).

*(Irving, November 15, 1960 letter to Cox)*

Despite Irving's proposed "embargo" letter, he did not stop writing to Cox; he wrote one more letter in 1960, and nine the following year! Although they no longer discussed specific points about the *GSA* review, they wrote about other disagreements. Irving wrote Cox on July 13, 1961, when on sabbatical leave in Toronto staying with his in-laws. After learning from Jaeger, who had visited Cox and Doell at Menlo Park, that Cox felt that a "situation" had arisen between them, Irving suggested that the "situation" might have arisen from third party gossip. Cox politely and eloquently disagreed. He also suggested that the deterioration of their relationship might have arisen partly because of bureaucratic foul-ups in trying to arrange for Irving to help build an astatic magnetometer during leave at Menlo Park.

Thanks for yours of July 13. Perhaps I did speak too frankly with Jaeger, but I do feel there has been an unfortunate deterioration of our relations, and I don't think saying it isn't so will suddenly change things. I'm not certain how this came about, the first thing was probably our bureaucratic fumbling of your contemplated appointment here. Actually the delays were on several bureaucratic levels above us, but we accept responsibility and are sorry things went as they did.

*(Cox, August 12, 1961 letter to Irving)*

Cox, however, singled out what he described as Irving's "intemperate and hostile letters," as the major reason, and added:

I've also not appreciated your abundant advice on how to run my career. If I want to sit on my butt and write review articles for the rest of my life, or work on mineralogy, or work in seismology, I regard this as my own affair. So when you declared your unilateral "embargo" on further correspondence some months ago, I must confess my reaction was a sigh of relief – a cooling off spell certainly seemed to be a good idea.

*(Cox, August 12, 1961 letter to Irving)*

Irving responded with equal eloquence.

I feel no hostility to you personally. I like arguments and feel we should not be too touchy about firmly stated views (this goes for me more than most people) … There is not much in either of us if we cannot rise out of it.

*(Irving, August 19, 1961 letter to Cox)*

He ended by asking Cox for a reprint of the *AG* review.

Irving also remembered another letter he got from Cox around the same time as the above when he was staying in Toronto. It was handwritten, and there is no copy among Cox's archival papers. According to Irving (January 2001 phone conversation with author), Cox wrote, "Ted, you have no idea of the strength of the objections to continental drift around here." Irving added, "I can remember being so aghast. Of course, it was naive of me to be aghast, but I was aghast because I felt our science was a matter of interpreting observations not of public opinion." Because they, themselves, may have agreed with anti-mobilist sentiment, his comment does not mean that Cox and Doell deliberately tempered any enthusiasm they may have had for mobilism. However, if they knew of reasons outside of paleomagnetism for not preferring mobilism, they certainly did not state them in the *GSA* review. Cox's remark unequivocally implies that he thought that anti-mobilism was very prevalent among North American Earth scientists at the time, that North America was a fortress of fixism.

Irving, Cox, and Doell never ended up spending any time together during Irving's sabbatical year. Talks had already broken down about getting the USGS to hire Irving to help build a highly sensitive astatic magnetometer. Although Irving spent a few days in San Francisco in January 1962 before he and his family returned to Australia, he never met Cox, who left a note saying he had to be out of town. He did remember calling Doell by phone but has no recollection of what was said. Things had definitely cooled (Irving, April 2001 note to author).

Irving and Cox did have enough in them to restore their friendship. They also learned how to handle disagreements better. In 1966, they began to disagree over a statistical issue, and after both made several points, Cox wrote, "Ted, we're getting too old to fight, especially by mail – it's so time consuming. Why don't we carry on over a beer at the AGU?"

In closing this discussion of the Cox–Irving correspondence, I want to mention two additional facts. First, what Cox did not tell Irving. Unbeknownst to Irving, they had attempted to obtain Eocene results from stable North America. Doell recalled what happened.

Our very first collecting trip after we came to Menlo Park was to the Chisos Mts. in Big Bend, TX. With help from our Survey colleagues, we searched the literature to find an Eocene sequence (away from the Siletz) that contained volcanics. The material we found there was unfortunately somewhat altered and we didn't obtain any usable data. We were also aware of a nice Eocene sequence in the Bear Paw Mts. Montana that we wanted to sample (I had actually been a field assistant there for Bill Pecora right after getting my B.Sc.). That was to be high on our list when we got a sensitive magnetometer operating.

*(Doell, January 26, 2001 email to author)*

It took them about eighteen months to get their laboratory started and finish up their *GSA* and *AG* reviews. They made the above field trip during the early fall 1960, stopped off at the GSA meeting in Denver (October 31 – November 2), and returned to Menlo Park before November 8, when Cox wrote his next letter to Irving saying, "Dick and I have been out of touch while working in Texas and Colorado this past month." But he did not tell him that they had collected rocks in an attempt to obtain a North American Eocene pole from stable North America; if he had, their correspondence might have gone differently. Also, if Irving had set up an astatic magnetometer at Menlo Park, they likely would have had success with the Bear Paw Mountains.

Second, Cox kept Doell in the dark. Doell does not recall anything about plans to get Irving to Menlo Park to help build a high sensitivity astatic magnetometer.

I'm beginning to think there must have been a lot in the Irving/Cox correspondence that I don't remember or was never privy to. E.g., the note that Ted proposed coming to the lab in Menlo Park to set up a magnetometer is news to me or at least I sure don't recall it.

*(Doell, February 19, 2001 email to author)*

This is very surprising because Doell was the senior officer and instrumentation was his responsibility. Evidently Cox did not keep Doell fully informed about his correspondence with Irving.

## 8.13 Cox reviews Irving's *Paleomagnetism*

In the years following, Cox and Doell wrote nothing else directly on the paleomagnetic case for mobilism, but in 1965 Cox reviewed Irving's 1964 textbook, indicating that he had not in the meantime changed his mind. He described the book as "an important milestone in the development of a new and rapidly growing field of geophysics," and characterized Irving as a leader in the field, who "with his work in Australia, has produced a virtually unequalled body of paleomagnetic data." He then turned to the paleomagnetic support of polar wandering and continental drift.

One question has often been posed: Is the paleomagnetic evidence for polar wandering and continental drift conclusive? Many geophysicists remain unconvinced, partly because of the increasing complexity and growing numbers of degrees of freedom of the geophysical models needed to explain all of the paleomagnetic observations in terms of polar wandering and continental drift. The present volume avoids the rhetoric of advocacy but, nonetheless, is a book with the central viewpoint that continental drift and polar wandering are supported paleomagnetically. Discrepant data are attributed to inadequate technique or to intracontinental deformation.

*(Cox, 1965: 494)*

He still held no brief for local rotations in orogenic belts. He still was no more supportive of drift than he had been five years before, despite these very active years of paleomagnetic work with numerous new reports of the application of magnetic cleaning (which had now become standard) and of radiometric dating of poles. As previously, he was very impressed with the "compelling evidence for polar wandering."

The most original part of the volume is an extended comparison of ancient paleomagnetic latitudes with various independent indicators of climatic conditions ... The occurrence of both polar wandering and continental drift are inferred from this. Compelling evidence for polar wandering is given by the demonstration that in Australia, as the paleomagnetically determined latitude abruptly increased by more than 50° between the Devonian and the Carboniferous Periods, coral reefs disappeared from Australia and glaciations began.

*(Cox, 1965: 494)*

It is his ambiguous use of the term "polar wandering" and what he omits that is telling. He let the reader know that he believed the paleomagnetic evidence for polar wandering is compelling, implying that the evidence for continental drift is not. What he really meant in the above instance was *apparent* polar wander relative to Australia, polar wander as determined paleomagnetically, not polar wander in the strict classical sense. However, as he accepted polar wander on the basis of what he regarded as "compelling evidence" then continental drift becomes a necessity, because, as Irving's book abundantly demonstrated and was now beyond any possible statistical or physical error, the APW paths from different continents strongly disagreed, and these disagreements were now established using the best modern demagnetization and dating techniques. Despite the industry of paleomagnetists working diligently on samples from all continents, Cox's view remained frozen.

And eight years later there still were no Eocene data from lavas of the Bear Paw Mountains of Montana to check Cox's claim, so central to their case about polar wander based on his Siletz 1957 data.

## 8.14 Cox and Doell on expansion

Cox and Doell (1961a) were the first to apply Egyed's paleomagnetic test of Earth expansion (§6.4). Their paper appeared in January 1961, and they repeated their analysis in their *AG* review (Doell and Cox, 1961a: 301–302). Cox and Doell (1961a: 46–47) concluded that fast expansion theories such as those of Carey and Heezen were unlikely, and that Egyed's theory of slow expansion "is neither confirmed nor rejected by the palaeomagnetic data now available." In reply, Carey (1961) attempted to remove the anomaly difficulty, while they (1961b) argued that he had not done so. I shall examine whether the standards of proof adopted by them in their dismissal of rapid expansion were any less elevated than the standards they demanded of paleomagnetists who argued that their results supported continental drift.

Egyed's method required coeval paleomagnetic results from at least two or more widely separated locations situated on the same paleomeridian at the time the rocks became magnetized. Cox and Doell found what they needed in a timely paper by B. V. Gusev (1959), a Soviet paleomagnetist, who had determined the paleolatitude of the northern Angara region of Siberia based on his study of 1600 samples from five different formations in the Permian Siberian Trap (mafic lavas). They compared them with the now numerous results from well-dated Permian formations (lavas and red beds) from Western Europe. If expansion occurred then the surface distance between the two sites would remain constant but their geocentric angular separation (paleolatitudinal separation) would decrease. Cox and Doell assumed that the time-averaged geomagnetic field had been dipolar during the Permian; they had already argued that this had support (1960: 748). There also had to have been no relative motion between Western Europe and northern Siberia, and they thought this was amply justified because Western Europe and Siberia had been on the same crustal block since the Permian after completion of the deformation of the Ural Mountains. The formations had to be the same age and they argued this was so (1961a: 46): "Although the geological age of the Siberian formations is subject to some uncertainty, Gusev concludes that their radiometrically determined age of 250 million years is probably correct. This would correspond to mid-Permian."[15] Comparing the paleolatitudes of each of the sixteen formations from Western Europe with those of each of the five Siberian formations, they obtained eighty estimates of Earth's radius during the Permian, the average of which was only 60 km less than at present.

The average of the 80 values thus calculated for the Earth's radius during the Permian is:

$R_{Permian}$ = 6,310 km. The standard deviation of individual determinations is 1080 km. and the standard deviation of the mean (21 independent data) is 230 km. This result is subject to an additional uncertainty if, as is quite possible, the average magnetic pole position moved during the Permian and if the time-intervals spanned by the western European data and those from Siberia are not precisely the same.

*(Cox and Doell, 1961a: 46)*

On these grounds, they claimed that Carey's (and Heezen's) very rapid expansion was "unlikely" and were agnostic about Egyed's slow expansion.

Comparing these results with the present Earth radius of 6370 km., we come to the following conclusions. The average Permian magnetic field as seen from the two sampling areas almost 5000 km. apart on the Eurasian land-mass, shows no significant departure from the dipolar configuration and is perfectly consistent with a Permian Earth radius equal to the present one. An increase in the Earth's radius as slow as that suggested by Egyed ... is neither confirmed nor rejected by the paleomagnetic data now available. Although precise statistical confidence limits cannot be assigned to the mean value of 6310 km. for the Permian Earth radius, an increase since the Permian as large as the 1100 km. required to produce a 45 per cent increase in area, as suggested by Carey, appears unlikely.

*(Cox and Doell, 1961a: 46–47)*

Was their assessment that rapid Earth expansion "appears unlikely" consistent with their unwillingness to become favorably disposed toward mobilism because of its paleomagnetic support? Although they (1960: 762) admitted "the Australian data for [the Permian and Carboniferous] constitute evidence for a relative displacement of Australia with respect to North America and Europe *which cannot be ignored*," they did not come out in favor of mobilism. More data were needed from other periods, support through time from at least more than two consecutive periods, yet they also claimed rapid expansionism "appears unlikely," based on only Permian results. Surely, to be consistent, Cox and Doell should have required contrary evidence from other geological periods before concluding that rapid Earth expansion "appears unlikely."

However, there is a difference between having enough support in favor of a hypothesis and having enough support against it. Cox and Doell agreed there was support for mobilism during the Permian and Carboniferous but only from Australia. They wanted more instances. There was not *enough* support. However, to argue that rapid Earth expansion is highly unlikely because it faces a single major anomaly, which is based on reliable data, is a different matter, especially if the same data support a competing hypothesis. Thus, Cox and Doell got what they needed: a Permian radius that was so close to the current radius and 1100 km too large for Carey's rapid expansion, which required a Permian radius of only 5300 km. Because their value was so different from Carey's, they rejected fast Earth expansion. They did not lower their standards in rejecting rapid Earth expansion, and they did act consistently.

Carey's rebuttal and Cox and Doell's reply were published together three months later. Carey argued that their assumption of no tectonic movement since the Permian between Western Europe and Siberia was incorrect.

Cox and Doell state that "in the expansion of the Earth envisaged by Carey, Heezen and Egyed, the continents do not increase in area; hence, the distance between any two points on a stable part of one continent remains constant." When applied to the geotectonic synthesis presented by me in the paper quoted by them, this statement is certainly invalid. In each of the figures for the Asian continent in my paper I show the Asian block as composite and elongation since the Palaeozoic is explicitly shown ... The extensions specified ... fall between the sampling areas used by the authors and involve an extension of some 900 km. This is of the right order for the angular distance measured between these districts to remain unchanged if the surface area of the Earth had increased some 45 per cent since the Palaeozoic.

*(Carey, 1961: 36)*

Carey attempted to turn Cox and Doell's anomaly difficulty on its head; their analysis, he said, actually supported his rapid Earth expansion. Not only was their assumption wrong, but once the increase in distance between Western Europe and northern Siberia postulated by Carey was taken into account, the Earth's Permian radius was not far from that expected by his rapid expansion; they had not studied his maps carefully enough. This they now did. They agreed that he had invoked a 900 km post-Paleozoic elongation of Eurasia, but argued that his proposed line of separation of the two regions was oblique to their common Permian paleomeridian, being off approximately 50°. They argued that this failed to remove the anomaly entirely and therefore that the paleomagnetic test did not support rapid expansion.

The crucial point raised by Prof. Carey concerns proposed separation of the two sampling areas used in our calculation. The 900-km. dilatation suggested by him would certainly mask an increase of 1100 km. in the Earth's radius, provided the elongation took place along the meridian connecting the two sampling areas and the ancient pole. Implicit in Prof. Carey's reconstruction of the Eurasian landmass there is, indeed, a 900-km. displacement of the Siberian sampling area with respect to Europe; its direction, however, is not along the ancient meridian but rather at an angle of some 50° to it. The component of elongation in the ancient meridional direction in Prof. Carey's reconstruction is less than 600 km., not enough to mask an expansion of 1100 km. Thus Prof. Carey's detailed model is not supported by the European and Siberian paleomagnetic results.

*(Cox and Doell, 1961b: 36–37)*

Cox and Doell retreated a little but did not retract; because the anomaly was reduced by almost half they went from saying that Carey's 45 percent increase in area "appears unlikely" to "is not supported by" the paleomagnetic results.

### 8.15 Initiation of the radiometric reversal timescale at the United States Geological Survey and the Australian National University

The development of the radiometric reversal timescale is covered in great detail by Glen in his 1982 *The Road to Jaramillo*, and I shall draw liberally on his account and supplement it with information from the Irving–Cox letters, and from my correspondence with Doell and Tarling. This change of topic at the end of Volume II, to one of such fundamental importance to the later volumes, may surprise the reader, but it is a story that merges historically with what has just transpired.

Their reviews finished, Doell and Cox had several projects in mind as Doell made very clear; developing a reversal timescale with Dalrymple was initially only one of their projects, eventually becoming their outstanding contribution.

No, I would not say that reversals were the main concern once the reviews were finished. We were also interested a lot in paleosecular variation, the accuracy of the paleomagnetic process (see our work on historic Hawaiian lavas), using paleomagnetism for various geologic correlations (see the Icelandic work in conjunction with Dave Hopkins on the opening of the Bering Sea (Allan worked with him on the Pacific side (Pribiloff Ils)), and (of course) the application of paleomagnetism to the expanding earth hypotheses. In spite of what Glen (1982) would have us believe, we were interested in whatever paleomagnetism could be used for – it was just that the reversal part of the work turned out to be the most important.

*(Doell, February 23, 2001 email to author)*

They thought of and worked on several problems before focusing on reversals, for, as remarked already, they devoted their first collecting trip together to attempting, albeit unsuccessfully, to get an Eocene pole from cratonic North America.

While Doell and Cox were launching their research programs, unbeknownst to them, Don Tarling (§7.8), an Irving research student, and Ian McDougall, a new research fellow at ANU, were engaged in separately conceived projects that they would eventually combine into developing a radiometrically dated reversal timescale (IV, §6.4) vital for the acceptance of the Vine–Matthews hypothesis, seafloor spreading and the development of plate tectonics. Tarling's work on the Hawaiian Islands was prompted by his supervisor's (Irving) interest in paleosecular variation of the geomagnetic field, and McDougall wanted to determine radiometrically the age sequence of the islands along the archipelago, information essential to understanding its origin. Irving was also anxious that Tarling should sample the oldest sequences on Pacific oceanic islands to see if relative latitudinal motion of the ocean floor on which they rested could be observed. The contemporaneous development of the reversal timescale at ANU and the USGS in the early 1960s did not begin as single-focus projects; at first, each group also had eggs in other baskets, paleosecular variation being the only one common to both.

The Cox–Irving correspondence touches on the planning by both the ANU and USGS groups in their development of a radiometrically dated reversal timescale. There is also important new information provided by Tarling and by Doell, who had an earlier interest in reversals.

As mentioned earlier, my first interest goes all the way back to my senior thesis on logging reversals in oil wells as a correlation tool. Verhoogen's first graduate seminar on remanent magnetism was heavily on the reversal question and I think that was for my interest (I was then his first and only grad student). John Verhoogen and I tried hard to get financing to collect in Hawaii re the reversal question, as did he and Cox later – both without success. A major part of my thesis eventually revolved around studies of the Neroly Sandstone [see Doell (1955a)]. This also started out as involving the reversal question, but eventually led to proposing a Crystallization Remanent Magnetization phenomenon. My thesis also contains a relatively small amount of work on Cenozoic lavas from the Sierra Nevada – again inspired by reversal interest. And it is of further interest, I think, that when Verhoogen asked me to show Allan the magnetic lab (just before I left for Toronto [in August 1955]) we went to the Sierra Foothills and I showed him about sampling and orienting rocks. And I'm sure we would have talked about the reversal question. Finally, let me note that my first graduate seminars as an instructor at Toronto and MIT were on the reversal problem. Enough, but I just wanted you to know that throughout my career in Paleomagnetism, the reversal question has been very high on my list of interests.

*(Doell, January 26, 2001 email to author; my bracketed additions)*

Coming from the University of California, Berkeley, Doell and Cox knew about the success Garniss Curtis and Jack Evernden had had in using the mass spectrometer of John H. Reynolds to date radiometrically young rocks that are low in radiogenic argon (Glen, 1982: 185–186). Both Curtis (in geology) and Evernden (in geophysics) received their Ph.D.s at Berkeley in 1951. By 1957 both were working in Reynolds' laboratory in the Physics Department. Within a few years, they had assembled their improved apparatus, allowing them to date ever younger rocks (Glen, 1982: 50–71).

Doell and Cox needed dates of their young rocks, and in July 1961 they met Brent Dalrymple, a second year graduate student, supervised by Curtis and Evernden. By the end of 1962, Doell had asked Dalrymple to join Cox and him at the USGS at Menlo Park to work on the reversal timescale (Glen, 1982: 193).

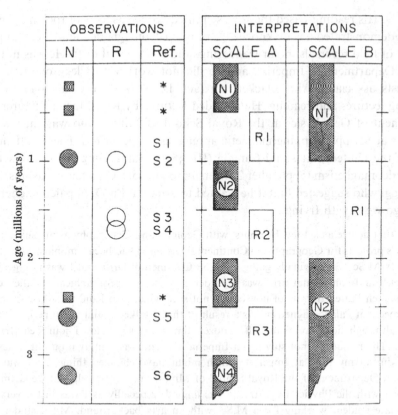

Figure 8.4 Cox, Doell, and Dalrymple's Figure 1 (1963a: 1050). Asterisked data are from Italian lava flows reported by Rutten (1959). Localities S1 through S6 are all from Californian lava flows: S1 is from Owens Gorge; S2, from Big Pine; S3 and S4, from Sutter Buttes; S5, from McGee Mountain; S6, from Owens Gorge.

Hoping to fix the last reversal, Doell and Cox had begun sampling before Dalrymple joined them. Doell (Tarling, 1962: 883) sampled the main island of Hawaii during the first part of 1961, and promptly published on November 18, 1961 (Doell and Cox, 1961b); there was no reversal; all samples were normally magnetized. By summer 1961, they also had begun collecting Pleistocene and late Pliocene lavas in California (Glen, 1982: 189) in which they (Cox et al., 1963a) had found normal and reversed magnetizations. Dalrymple dated them, and they produced their first radiometric reversal timescale, based on their new results from four lava flows in California and three previously published results from Europe (reported in Rutten (1959)) (see Figure 8.4). Their paper was received by Nature on March 19, 1963, and published on June 15, 1963 (Cox et al., 1963a).

Don Tarling received a B.A. in 1957 from the University of Keele, UK, majoring in geology and geography. He then entered the Royal School of Mines, Imperial College, London, and became interested in paleomagnetism, completing his M.Sc. in

May 1959. His reading-based thesis was entitled "Some aspects of rock magnetism." In it, he demonstrated a wide knowledge of the paleomagnetic literature, and argued in favor of field reversals, polar wandering, and continental drift. He was not in the Physics Department at Imperial, and he did not work with Clegg or other paleomagnetists associated with Blackett. In fact, he worked very much in isolation, attending lectures and reading. He attended lectures by Ron Mason, lecturer in the Department of Geophysics at the Royal School of Mines, who was then working summers at Scripps mapping magnetic anomalies in the eastern Pacific off the west coast of the United States and Canada (III, §5.9). But Tarling did not talk to him about paleomagnetism (September 2000 email to author). On a few occasions Tarling met Clegg, who suggested that if he wanted to obtain a Ph.D. in paleomagnetism he should go work with Irving.

I did my first degree as a Joint Honours, with Geology and Geography as my subjects. I did final year's projects for Geography on Continental Drift and had been a mobilist when I was at school. My M.Sc. was originally going to be on Continental Drift, but I was in a geophysics department, and continental drift was considered to be too geographical (or they did not believe in such flat-earth type of nonsense – not sure which!) so I did my Masters thesis on Some Aspects of Palaeomagnetism. As a result of this, I talked mainly with people like John Clegg – although no more than half a dozen times. He suggested I join Ted [Irving] in Australia. The reason for not staying [at Imperial College] was simply no grant. I never met Blackett during my time at Imperial, which might have changed things … I was in the Geophysics Department of the Royal School of Mines – not in the Physics Department. So my contact with the Physics Department was limited, especially as I was just a very junior postgraduate student working for a M.Sc. with an arts background. M.Sc. students were usually people who were not good enough to get on to a Ph.D. in those days – and, in general, still are in the UK! So I was rather a trivial small fry.

*(Tarling, September 2000 email to author; my bracketed additions)*

Tarling applied to ANU and was accepted as a Ph.D. student. Knowing that Tarling's boat was stopping in Aden, Irving suggested that he collect rocks to study the magnitude of secular variation in low latitudes.[16] Irving recognized the Aden Volcanics as a well-exposed target, having been there in the 1940s while in the British Army and going ashore en route to Australia in 1954. Tarling stopped off in Aden, got the samples, and took the next boat. They showed both normal and reversed polarity, so he quickly became familiar at first hand with the phenomenon of reversals. The result turned out to be mainly of interest for tectonics, not secular variation, and were published two years later (§7.8). While Tarling was en route to Australia, Irving, needing to procure funding, enquired by telegram if he wanted to continue working on secular variation elsewhere. Tarling agreed. Recalling what happened, he also noted importantly that the question of reversals was not originally on the agenda.

I contacted Ted [Irving]. The initial work was to collect samples from Devon and Cornwall for "standard paleomagnetic work," but Ted suggested that, as the boat to Australia went *via* Aden, that I collect samples there to look at the amplitude of secular variation in low latitudes. After having collected these and while continuing to Australia, Ted sent a telegram to the boat asking if I would be interested in working on the Pacific Islands – for the same purpose. In other words, the project was secular variation rather than reversals, based on the general assumption that most Pacific Islands to be visited were thought to be less than one million years old.

*(Tarling, September 2000 email to author; my bracketed addition)*

Tarling (1962: 882) began collecting in March 1960 from islands in the south and southwestern Pacific, including importantly, Samoa; thus the ANU were sampling Pacific islands about a year before the USGS. Tarling later sampled most of the major Hawaiian Islands, during June through September or October 1961, some months after Doell had collected from the main island of Hawaii. Although Tarling originally focused, at Irving's behest, on secular variation, once he returned to ANU and found reversed and normal magnetizations, he quickly refocused on reversals and realized that he needed to get his samples accurately dated. Irving agreed.

Serendipitously for Tarling, there was already a potassium–argon facility installed at ANU, cloned under Evernden's supervision from that at Berkeley. Once again Jaeger had, in timely fashion, added to his department a critical, new, developing area of research. This can be traced back to April 1958, when Curtis visited ANU. Staying for two weeks, he lectured on potassium–argon dating. Curtis was interested in improving the geological timescale and was searching for suitable Australian rocks that were well-dated paleontologically. Becoming friendly with Curtis and mindful of his requirements, Irving drove Curtis down to northern Victoria to the Snob's Creek paleontologically well-dated Late Devonian dacite for which, back in Berkeley, Curtis provided an accurate datum for the geological timescale (Irving, January 2011 note to author). Meanwhile, the School of Physics at ANU broke up its Department of Radiochemistry, giving part to Jaeger who then bought a Reynold's mass spectrometer, initiating a long-lasting and very fruitful jointly supported project between ANU and the Bureau of Mineral Resources, the Australian equivalent of the USGS. Jaeger asked Curtis if he or someone from Berkeley would help set up the instrument. Jaeger also said he would send somebody from ANU to Berkeley to learn the techniques.

In late 1959, Jaeger chose Ian McDougall. McDougall had been an undergraduate at the University of Tasmania, where he had taken classes from Carey, and attended the 1956 symposium on continental drift (§6.16). He had just obtained his Ph.D. at ANU on the petrology of the Tasmanian dolerites under Germaine A. Joplin, a classical petrologist. He had met Curtis and was impressed, although unsure whether his physics was strong enough to work in mass spectrometry. Jaeger had no such doubts, and McDougall spent the 1960–1 academic year at Berkeley working with Curtis and Evernden. He met Dalrymple on the first day and they became friends. Glen (1982: 191) quotes Dalrymple:

I went to the K-Ar laboratory one morning and met ... a fellow named Ian McDougall, who was also in there for the very first time. We started out learning about vacuum and argon-extraction systems on the same minute of the same hour of the same day ...

Before McDougall went to Berkeley, Jaeger visited there, and arranged for Evernden to spend August 1960 to January 1961 setting up a potassium–argon laboratory at ANU. Irving described Jaeger's getting Evernden to Australia, and his success in getting the mass spectrometer up and running.

Jaeger went immediately for the potentially geologically important thing, that is, measuring [the ages of] basic rocks and [setting up] the Reynolds mass spectrometer; so he got money from the University to get Jack Evernden over for six months … During his time there Jack had absolute power in the lab; everything was laid out for him, as far as it could be technically, and the preparation of rock samples [was provided]. Evernden had the mass spectrometer going within a few weeks. Jaeger gave him absolute priority in the lab. Evernden [like Curtis] was interested in the [geological] time scale work.

*(Glen 1982: 83; my bracketed additions)*

McDougall spent a successful year at Berkeley. He learned potassium–argon methods, obtained an age for the Tasmanian dolerite, one of the first mafic igneous (often good carriers of geomagnetic field records) bodies to be radiometrically dated (McDougall, 1961). He decided to try to date Hawaiian mafic volcanics, which he planned to collect on his way back to ANU, his motivation being to determine the timing of volcanic activity along the chain of islands. He got to know Cox and Doell and told them of his plans to collect and date Hawaiian rocks (Glen, 1982: 212–214). Doell told McDougall where he had sampled on Hawaii, and explained that they were particularly interested in secular variation; reversals do not seem to have been mentioned. Glen quotes McDougall:

I do not recall great discussions about polarity reversals. They were very interested in hearing about dates on Hawaiian rocks. The interest that they expressed to me was not in polarity studies but in secular variation. They were interested in getting dates.

*(Glen, 1982: 212)*

When McDougall left for Hawaii in August 1961 and returned to ANU, he agreed to keep Cox and Doell informed about his dating of Hawaiian rocks.

Upon arriving in Honolulu, McDougall attended the Pacific Science Conference (August 21 – September 6) where he, by chance, met Tarling; they were, recall, from the same department at ANU. Tarling, who was making his own collection of Hawaiian rocks, was still intent on secular variation studies, studies of the dispersion of the field. Their interests were different, and they did not discuss working together. At the time neither was interested in reversals. McDougall made his own collection, returned to ANU the following month, but, as explained below, did not immediately begin to date his Hawaiian samples. Indeed McDougall became pessimistic about dating the Hawaiian samples after many at the Pacific Science conference meeting in Honolulu told him it could not be done.

By contrast, Tarling began work on his collections immediately upon returning to ANU in September or October 1961. As already noted, he soon found reversed and normally magnetized lavas from the Hawaiian and Samoan Islands, and realized that he probably had located the most recent reversal horizon in both places. His letter to *Nature* appeared on December 1, 1962, and he remarked:

Although it is tempting to suggest a correlation between the level of reversal in the Fagaloa Formation [from Upolu of the Samoan Islands] and the most recent reversal in the Hawaiian Islands, more work will be needed to clarify this. Nevertheless the results outlined here, particularly those from Samoa, make clear the potentialities of this method as an aid in geological correlation within and between the Pacific Islands over very large distances.

*(Tarling, 1962: 883; my bracketed addition)*

Tarling recalled that upon analysis of the Hawaiian rocks "my interest quite quickly switched over to the use of reversals as a relative dating tool, even though if at that stage we couldn't do absolute dating." Tarling also added:

Ted Irving was definitely pressuring Ian McDougall toward the end of this period ... which must have been around about 1962 to try and do radioactive dating and to try and get at least the last event [the Matuyama–Brunhes boundary] as an absolute minimum.

*(Glen, 1982: 216; my bracketed addition)*

McDougall began dating his Hawaiian samples in mid-1962, and he made it clear in his January 1963 paper that Tarling and he now planned to work on the polarity-reversal timescale.

Normal and reversed polarities recently have been reported from the Hawaiian lavas [reference to Tarling (1962)]; by using the potassium-argon method it should be possible to date these reversals.

*(McDougall, 1963: 345: my bracketed addition)*

Tarling and McDougall's first joint paper with their first radiometric reversal time-scale was based entirely on Hawaiian samples, and appeared in October 1963, four months after the publication of the first USGS timescale.

Glen remarked on the sense of urgency Cox and Doell felt about the competition from ANU workers once they saw McDougall's January 26 publication quoted above. Glen quotes Doell, who wrote to a colleague immediately after the appearance of McDougall (1963):

This [the manuscript for Cox, Doell, and Dalrymple, 1963a] is our first summary on the dating of reversals and we're rather anxious to get it out before our competitors in Canberra get something together.

*(Glen, 1982: 227; my bracketed addition)*

Doell and Cox wanted to be first, and they were worried about the ANU group, which they viewed as serious competition. Glen also wonders why ANU workers did not get going earlier, and offered two reasons: McDougall's initial lack of confidence in being able to date such young rocks, and Tarling and McDougall not believing themselves to be in a competition. Glen quotes Tarling.

Ted Irving was trying to pressure McDougall into attempting to date my Hawaiian rocks for more than a year, and in this sense, I think it's quite clear that there was no real race [for production of a polarity-reversal timescale].

*(Glen, 1982: 216; my bracketed addition)*

Glen was very likely correct; ANU workers seemed unaware at the time of Cox and Doell's competitive attitude. But McDougal had another very good reason that Glen does not mention. Beginning with his Ph.D. work on the Tasmanian dolerites, McDougall over the previous years had developed an interest in the vast array of Mesozoic dolerites and basalts of the southern continents and their significance for the proposed fragmentation of Gondwana and continental drift. Commonly thought to be Early Jurassic, the geological constraints on their ages were not tight (§5.7), and obtaining precise dates was a vital requirement. Moreover, he (1961) had already shown that dating of these bodies radiometrically was feasible, whereas dating very young Hawaiian basalts with their lower concentrations of radiometric argon was still problematic. Consequently, upon returning to Australia he gave his work on Mesozoic dolerites first priority. It was a first class project.

So just what progress developing their timescales had both groups made by late 1963? The USGS group presented two timescales in their June 1963 paper based on their California collections. Assuming that reversals were periodic, they (Cox *et al.*, 1963a: 1050) proposed two timescales consistent with their results and those from Europe reported by Rutten. Scale *A* had reversals occurring every 0.5 million years; Scale *B*, every million years. However, they (1963a: 1050), in a note added in proof, remarked that new results from "several lava flows from California also strongly favours scale *B*." They (1963a: 1050) also noted that their assumption that reversals are periodic might not be correct, "Magnetic epochs of unequal length are also consistent with these results – for example, N2 (scale *B*) may be considerably longer than R1. Additional measurements are being made to clarify the important problem of periodicity."

Based on their Hawaiian work of 164 samples from fifty-nine sites on five islands, McDougall and Tarling (1963) proposed that the latest reversed interval began between 2.5 and 2.75 million years ago, and ended about 1 million years ago, being at least 1.5 million years long, prior to which there were several short polarity intervals (Figure 8.5). Going further than Cox and company, they (1963: 56) claimed that reversals in the past few million years had not been periodic: "The present work suggests that models of the Earth's magnetic field should account for irregular periodicity of reversal." McDougall and Tarling were correct; reversals were not periodic but occurred at irregular intervals. They were the first to propose aperiodic reversals rather than merely suggest them; however, the USGS trio were the first to incorporate the idea into a Late Cenozoic timescale, which they did a year later in their June 26, 1964 timescale (IV, §6.4). Meanwhile, Cox, Doell, and Dalrymple (1963b) presented their second reversal timescale in a paper that was

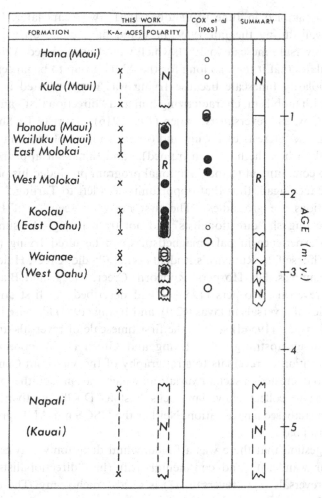

Figure 8.5 McDougall and Tarling's Figure 1 (1963: 55). Open circles are normal polarity; solid circles, reversed polarity.

received in August 1963 and appeared only thirteen days after McDougall and Tarling's. Not having seen the ANU group's timescale, they said nothing about it. With results from ten more California locations, they proposed (1963b: 385) that the last three polarity epochs were not of the same length; they proposed that the current normal and last reversal epoch had lasted for 0.9 to 1 million years, while the previous normal epoch had been "longer by 25 percent or more." Glen (1982, Chapter 6) has described the development of the Late Cenozoic timescale itself in detail and I shall return to it in IV, §6.4.

Now I want to pick up from §8.12, the Cox–Irving correspondence which reveals that Irving originally planned for Tarling to survey the Hawaiian Islands as far back as the fall of 1959. As already shown, the plan was on, off, and then on again until

Tarling finally sampled in June to October 1961, two years later. If Tarling had surveyed Hawaii during the fall 1959, he could have come up with the idea of developing a reversal timescale in 1960, which would have placed ANU well ahead.

Glen speculates that it took so long for the ANU group to begin serious work on the reversal-polarity timescale because Irving was not interested in reversals but preoccupied with mobilism, characterizing him as a "directionalist" and others, such as Doell and Cox, as "reversalists." Glen (1982: 216) contrasts Tarling's interest in reversals with "the interests of Irving, and other members of the English 'directionalists' group (by whom he had been trained)," and thinks "it appears also that his [Irving's] deep commitment to the directional program precluded his possible collaboration with McDougall, thus that opportunity was left to Tarling." I believe both these speculations are groundless. The first makes it seem as if Irving was on autopilot. The English "directionalists" did not train Irving. His supervisor, Runcorn, had no experience in paleomagnetism when he hired Irving (§1.15). Irving largely taught himself, as Runcorn's students generally did; Creer, Hide, and Opdyke being good examples. He, Hospers, Runcorn, Creer, Clegg, and Blackett were all interested in reversals. Hospers (1953–4) had described the first stratigraphically sequential series of reversals in lavas (§2.9), and Irving (1954) likewise in sedimentary rocks (§2.6). Hospers (1954d) set out the first timescale of reversals and the general principles of magnetostratigraphy. Irving and Green (1958) commented on the potential application of reversals to stratigraphy of the Victorian Cenozoic basalts. Irving was also interested in secular variation, which was, in fact, the initial reason he wanted Tarling to collect Hawaiian rocks, just as Doell's motivation to sample Hawaii was to study secular variation. Neither the USGS nor ANU groups confined their interest to poles.

Glen's suggestion that there was a fundamental dichotomy between those interested in polar wandering and continental drift (his "directionalists") and those interested in reversals (his "reversalists") is a false dichotomy. The original Cambridge and Manchester/Imperial College paleomagnetists were interested in both; Hospers was far more interested in reversals than mobilism. Opdyke and Runcorn (1956) studied reversals. Right at this time, Irving and Parry (1963) were establishing the stratigraphic limits and duration of the sixty-million-years-long Kiaman Reversed Polarity Superchron.[17] Doell and Cox were far from being exclusively "reversalists" – they collected rocks from which they determined paleopoles. They attempted to obtain an Eocene cratonic pole for North America (§8.12). They just composed two giant reviews predominantly about "directional" results. They wrote extensively about polar wander. They were very interested in secular variation. They used paleomagnetic directions to determine the Earth's radius in the Permian (§8.14). That they were interested in many things was made very clear by Doell in his February 23, 2001 email to author (§8.15). It is bordering on the absurd to think that the paleomagnetists concerned could be neatly branded as either "reversalists" or "directionalists."

As for his second speculation, Glen does not say how Irving should have collaborated with McDougall and what purpose might have been served by his dropping his own research projects to do so. In the Aden work (§7.8), Irving had introduced Tarling to the practicalities of paleomagnetic work, and the Cox–Irving correspondence shows that he was trying to get Tarling launched on the paleomagnetism of the Pacific Islands to determine what it had to tell about the geomagnetic field and tectonics. To suggest that Irving was in some sort of "directionalist" straightjacket is truly absurd because it was precisely at this time that he was establishing stratigraphic limits of the Kiaman Reversed Superchron, the first polarity superchron to be defined and one of the most important features of the reversal history of the geomagnetic field. The Cox–Irving correspondence confirms, as Glen argued, that ANU workers were unaware of being in a race. It shows that Irving originally hoped to cooperate with Cox and Doell, and coordinate plans; he gave no sign of caring who got the data; everyone just needed to get on with it.

At the beginning of 1960, Irving gave Cox early notice of Tarling's plans.

We are getting interested in oceanic islands. Mr. D. H. Tarling who is a newly arrived research student is hoping to make a study of Pacific islands. The main points are to check the earth's dipole in recent lavas and to study secular variation. Also older sequences may provide information about movements in island arcs, or in the case of the truly oceanic islands about the stability of the ocean floor. I have set him onto this for his Ph.D. problem. He hopes firstly to go to Fiji, the Solomons, New Britain and of course Hawaii which is particularly good for exposures and geological dating.

I do hope this does not interfere with any of your plans and that you will let me know if it does.

Mr. Tarling was hoping to go to Hawaii about September but his wife has elected to have a child about then so he has had to bring his dates forward about six months. The plan now is for him to leave for Hawaii at the end of this month. I am sorry about this short notice and in the normal course of events I would have let you know very much sooner but you can see the reason.[18]

*(Irving, January 1960 letter to Cox and Doell)*

Tarling had originally planned to go to Hawaii in September 1959, almost a year and a half before Doell made his collection, and even longer before he finally sampled there later in 1961.

Cox replied wanting more details, and sounding a note of concern. He let Irving know that the Hawaiian Islands were the only Pacific islands that the USGS would let them survey without having to go through "mountains of red tape." He added that they wanted to find the most recent reversal, although it was just one of three objectives.

On the plans of your student Tarling to study the lavas of oceanic islands, this sounds like an excellent research project. In fact, Dick and I have been thinking along similar lines. It turns out that mountains of red tape and delay are involved in getting Survey permission to work outside of the United States, so the Hawaiian Islands are the only ocean islands readily available to use. Last year, I tried to get Survey personnel on Hawaii to send me some test samples, but this hasn't turned out well. In our budget for next fiscal year Dick and I have submitted plans for the two of us with two collecting crews to spend 45 days each in the islands, and Jim Balsley has approved of this project. (Fortunately there is presently a great deal of interest and activity by Survey geologists, geochemists, and geophysicists in Hawaii.) Our objectives were to investigate the uniformity of the magnetization of some historic lava flows that cooled in known fields; measure the secular variations in a low latitude, and determine the last reversal horizon.

I don't know how Tarling plans to apportion his time between Hawaii and the rest of the Pacific Islands; if he plans an extensive study of the extremely large number of accessible lava flows on Hawaii, then of course there will be no point in our duplicating his work. If, however, he plans a more limited study, then perhaps our respective projects can be planned so that there is only enough overlap to provide some useful points of comparison. Please let us know so we can make the changes in our budget plans if it's necessary.

*(Cox, February 8, 1960 letter to Irving)*

Irving replied on February 20 telling him that Tarling was no longer going to Hawaii because Cox and Doell planned to survey there. Irving got the hint, and doubtless not wanting to put an isolated research student in potential competition with the USGS with far more resources, deferred to them.

Tarling is not now going to Hawaii. He had planned to spend 50–60 days there and make a thoroughly going job of the islands. But as this is clearly better done from your end we now plan to concentrate more on the central Pacific which is rather more crucial for secular variation [because in lower latitudes] but less well known (if at all) geologically. I would be grateful if you would not say too much about Tarling's plans to other people as I want to protect him from the needless worry of having people breathing down his neck – at least not until he gets underway. We have a lot of novel ideas to test (you can no doubt visualize the possibilities) and I want him to get his teeth into this before saying too much to other people.

*(Irving, February 20, 1960 letter to Cox; my bracketed addition)*

Irving also wanted to see if Tarling could track the displacement of Hawaii relative to other Pacific oceanic islands by sampling the oldest sequences there.

In your letter you do not say specifically that you intend sampling the lowest horizons in Hawaii which are Pliocene. One of the points which I am anxious for Tarling to study is the stability of the oceans both in the deep ocean areas and in the island arcs. It is just possible that we can detect movements in the upper Tertiary as the Japanese work has shown, and I want him to get as much of the basement rocks in the islands as he can for this purpose. If you are not interested in sampling these as part of your own studies we would be grateful indeed if you would make some collections and send them to Tarling. The sort of comparisons I wanted him to make (and this was one of the main reasons for going to Hawaii) are between Tertiary or lowest Pleistocene lavas in Hawaii and Tahiti (as ocean islands) and the island arc areas to the north of us, and the mainland work here. From his thesis point of view a comparison with Hawaii is likely to be important.

*(Irving, February 20, 1960 letter to Cox)*

If Cox and Doell worked on such older sequences, could they send their pre-publication results to Tarling? If Cox and Doell did not want to do so, could they send samples to Tarling? It really did not matter who did the work.

From your letter I gather that the secular variation and the last reversal are your first priorities. If this is so may I attempt to prod you into putting the basement work a little higher on the list so that there may be some reasonable chance of it being published or available in pre-publication form within the next three years. Otherwise if your interests are not in that direction we would be grateful if you could get the samples for us. I feel it would be a great pity if this point were not checked on thoroughly.

*(Irving, February 20, 1960 letter to Cox)*

Cox replied on March 16, 1960, writing a short letter reassuring Irving that they would collect Hawaiian samples from lower horizons for Tarling. Cox also told Irving that Balsley had just collected a few samples from Hawaii, but they would not start collecting there themselves until Doell returned from the IUGG conference in Helsinki in early August. Cox predicted that they probably would finish their Hawaiian fieldwork about the time Irving would arrive on sabbatical leave.

This is just a hasty note of reassurance to Tarling through you that we'll be certain to collect at lower horizons in Hawaii. Balsley has been collecting a few reconnaissance samples for us this past week, but Dick and I won't be doing any heavy sampling before he returns from Helsinki. You'll be here in California shortly after we return from Hawaii, so you can make certain that we will get the Hawaiian data of interest ground out in time to be of use to Tarling.

*(Cox, March 16, 1960 letter to Irving)*

Irving never did visit with Cox and Doell. When he eventually passed through San Francisco almost two years later with his family en route back to Australia, Tarling already had his own Hawaiian rocks and Doell and Cox (1961b) had published their first paleomagnetic results (all normal) from Hawaii without locating the last reversal. Moreover, as far as I know, they never did get Tarling his "lower horizons." Irving did not mention Tarling again for about five months, when he told Cox that Tarling had collected many rocks from Tonga and had gotten sick. He also figured (incorrectly) that Doell and Cox had begun to get results from their Hawaiian rocks.

Tarling's rocks are pouring back – he was ill in Tonga but will be back on his feet soon. He has hundreds of specimens – lots of trouble with weathering, but still a lot of fresh stuff. I expect you have your Hawaiian collection now and hope you had good luck with them.

*(Irving, August 15, 1960 letter to Cox)*

Irving had forgotten that Cox already had told him that they would not begin sampling there until Doell had returned from the Helsinki meeting, which ended on August 6, 1960, nine days before Irving wrote his letter.

There is no further mention in the surviving letters on Tarling and the Hawaiian work for a further six months. It seems Cox informed Irving in a missing letter that Doell and he were now "not doing Hawaii," because Irving sent the following:

I'm sorry you are not doing Hawaii. It seems a pity since Tarling could easily have fitted it in, and now, unless others do it (and many are interested as you may know) we shall have to wait to get the results. I trust you will let me know what happens because we still have every intention of going through with it ourselves. Perhaps the cause of sciences would have been better served if I had been less polite and done it ...

*(Irving, undated letter to Cox written before September 20, 1960)*

Just what "not doing Hawaii" means is unclear. In any event Doell began collecting from the main island Hawaii early in 1961 and, as we have seen, found no reversals. Tarling sampled from several older islands along the Hawaiian chain and, as already shown, found a number of reversals.

It is tempting to speculate about what might have happened if Irving had not cancelled Tarling's proposed spring 1960 Hawaiian survey. Evernden could then have played a key role. Evernden arrived at ANU in August 1960 and stayed until January 1961. He soon had the argon extraction line running. Recently Irving wrote:

At this time I was in regular contact with Jack Evernden. He was working through all our rock collections looking for dateable samples. I enjoyed his company. I actually gave him instruction in playing cricket. Evernden knew about the work of Doell and Cox on the California basalts. If Tarling had laid out his reversal sequence along the Hawaiian Chain in later 1960, I am sure I would have called Evernden's attention to it, and I am sure he would have said immediately that it needed dating, and I am equally sure that Tarling would have shouted, "Yippee!" Jaeger had decreed that Evernden should have all the resources that the department could offer during his visit. If Jack had got interested, the dating would have got done – quickly.

*(Irving, January 2011 letter to author)*

In the southern spring 1960, Irving wrote telling Cox about Ian McDougall.

Ian McDougall is over there with Curtis. I suggested he get in touch with you. He has done some excellent work on Dolerite differentiation here. He gives a good colloquium if you want a speaker any time. He is a very good petrologist.

*(Irving, undated letter to Cox written before September 20, 1960)*

Cox and Doell got to know McDougall, and once they learned about his plans to date Hawaiian volcanics, they wanted his results (Glen, 1982: 212).

The Cox–Irving correspondence during the following year concerns other matters (§8.11). It was not until July 3, 1961, in a letter Cox sent to Irving (who was then on sabbatical leave in Toronto), that Tarling and Hawaii are mentioned again. This was well after their original plans for Irving to help build an astatic magnetometer at Menlo Park had fallen apart. Irving still was intending to spend a few days with them en route to Australia. Cox told Irving that Doell had ended up sampling from Hawaii, that they had told Tarling where he had sampled, and had provided him with their preliminary results.

Dick and I certainly hope you have time to visit our lab. I may get over to the Kyoto meeting and possibly may stay a short while in Tokyo [Cox attended the IUGG Kyoto meeting that was held during the last week in September 1961. Runcorn, Creer, and Blackett were also there]. On the way home I may or may not stop off for some more collecting on the Hawaiian Islands, depending somewhat on the extent of Tarling's sampling. We're in good contact with him, by the way; he has a list of all of Dick's sampling localities (980 samples from over 200 flows) and preliminary results. At any rate, I'll probably be back here by the end of the year.

*(Cox, July 3, 1961 letter to Irving; my bracketed addition)*

As far as I know, Doell never did collect any more samples from the Hawaiian Islands in connection with the reversal timescale. As noted previously, they had already collected basalts in California, but Cox made no mention of it.

Irving wrote back ten days later, telling Cox that he hoped their differences would not hurt Tarling.

Regarding Tarling: I hope all works out well in Hawaii – if any difficulties should arise remember they are entirely my fault and not his. I was anxious for him to go to Hawaii years ago for purposes of his Ph.D. Thesis, which is to be on a preliminary survey of Pacific Islands, and Hawaii is an obvious reference point for the last reversal. As his academic supervisor it seemed necessary for me that he should get information from Hawaii to tie with his equatorial islands. The rest you know, and if at times I have been precipitate or impetuous in what I have said then you'll just have to put it down as a black mark against me *not* him.

*(Irving, July 13, 1961 letter to Cox)*

Cox agreed. After telling Irving that he was "glad relations seem to be improving again now" between them, he assured Irving that any ill-feeling that may have developed between them would not affect Tarling. Cox also gave Irving advance notice that they had not located the last reversal horizon on the Hawaiian Islands; all of their samples were normal. Their paper (Doell and Cox, 1961b) appeared just one month later. Still he made no mention of his and Doell's work in California.

There's no reason why any of this should involve Tarling. He has complete lists of all of Doell's sampling localities and knows the general nature of our results to date, so he can benefit a great deal from our disappointments; the main disappointment to date has been to find the main island entirely normal, we'd sampled heavily there in the hope of studying the most recent reversal horizon in detail. We've invited Ian McDougall down from Berkeley to attend our seminars on the Hawaiian results to date; he'll be seeing Tarling in a few weeks, and will be able to give him all the details of our measurements to date. We'll also try to get early copies of our publications into Tarling's hands. He's doing some reconnaissance sampling in areas where Dick collected, which is a very good idea; there's been remarkable little independent checking of data in paleomagnetism compared with other fields, such as radiometric dating.

*(Cox, August 12, 1961 letter to Irving)*

Cox closed with a light-hearted reference to his and Irving's disagreements.

If we all handle our statistics carefully, there should be no disagreements in fact. Any differences in interpretation that may arise must, I suppose, be regarded as healthy signs that our science is still alive.

*(Cox, August 12, 1961 letter to Irving)*

Tarling never got the promised early copies of Cox and Doell's publications. Cox explained the problem in his next letter to Irving.

We've run into some bureaucratic snags on our plans to send Tarling advance copies of manuscripts on the Hawaii results – there's a firm Survey policy forbidding this which we'd Freudianly forgotten. Luckily Ian McDougall had enough time before he left to go over all the results to date with us at several sessions, and he'll be able to pass it all on to Tarling when he visits there – should be past tense by now, I guess.

*(Cox, September 1, 1961 letter to Irving)*

Cox's slip is interesting. Cox wanted McDougall to keep them informed about the dates of his Hawaiian samples. But McDougall does not remember any discussion with them about their objective of defining the polarity-reversal timescale. Perhaps, they forgot to tell him. Also, Warren Hamilton, an employee of USGS at the time, who read through an earlier version of this chapter, disagreed categorically with Cox's characterization of USGS policy.

Cox's claim that USGS policy forbade sending Hawaiian manuscripts to Tarling was false. Cox apparently was ensuring that he got there first. Policy forbade sending info of commercial value to private companies ahead of publication, though this was ignored with regard to common cooperative work with, for example, individual mines. Policy in no way inhibited exchange of academic information with outside colleagues, which many of us did freely all the time.

*(Hamilton, July 24, 2002 email to author)*

Cox was concerned about the ANU group, the members of which were not aware that they were in a race. The mood softened for a while; Cox wrote Irving on September 1, 1961, telling him, "I've just returned from a long tramp in part of our fast dwindling 'out back' to have the pleasure of finding your last two friendly letters waiting." But Cox did not seem to want workers at ANU to get to work on the reversal timescale.

Note that throughout all this, Doell was unaware of the Cox–Irving correspondence; he knew nothing about it until I showed him some of it in September 2001, long after Cox's death (Doell, September 24, 2001 email to author).

## 8.16 Postscript

This chapter has revealed much about the personal side of competition among scientists. Irving and Cox were upset with each other. Irving thought that Cox and Doell's *GSA* review copped out on the drift question. Cox thought that Irving's criticism had become personal. Runcorn and Creer were none too pleased with the *GSA* review. As Creer noted:

I have no doubts that they [Cox and Doell] may well have read our papers in the 1957 Transactions of the Royal Society series, but perhaps not until late that year or early in

1958. Importantly, I would add this comment too: – the scientific research business has always been one of "cut-throat" competition among many. In the fighting, arguing etc., one bad paper can have much more effect in that it can prove infinitely more damaging than half a dozen good ones can be beneficial, in that it provides a "weapon" with which to beat one's opponent or rival. It is convenient to ignore what the "good" papers report, noting that a large number of those "listening in" are not familiar with the literature – too many people form opinions on "hearsay." And I think that this is what happened.

*(Creer, September 2000 email to author; my bracketed addition)*

Finding and exploiting weaknesses in opponents' positions and countering attacks by them on one's own position are at the heart of the mobilism debate and illustrate the three standard research strategies. Give scientists an inch and they will take a mile.

What, however, is heartening about this dispute between Creer, Irving, Runcorn, and Cox and Doell is its pleasant outcomes on the personal level. Irving and Cox began mending their relationship at the end of their correspondence. Irving talked about their later years.

Allan I met first in San Francisco about June 1957 and stayed at his flat for several days. We talked about everything. I appreciated his feeling for music, ballet, his wide reading of many classics, his self-awareness. We corresponded. He came to our pre-nuptial party in Toronto in August. But our friendship slipped probably, as you have now unearthed, in the late 1950s initially because of that cat-and-mouse play with Tarling and then in 1960 with the review. In the 1960s and '70s we connected scientifically and cordially on Cordillera topics. It was however in the citation for the Bucher Medal [which Irving received] in 1979 that he publicly came to terms with our mid-1950s work. Neil [Opdyke] was at the awards table, and I can remember him looking up at me as Allan was reading the citation as if to say, "he's finally said we got it right." Essentially he reviewed in the citation my 1956 paper. Yes he said we had had it right. Some have commented on his omission of that paper or some contemporary paleomagnetic equivalent in his very popular compilation *Plate Tectonics and Geomagnetic Reversals*. A short time before he died he was here in North Saanich, British Columbia, giving a lecture, having lunch at home with Sheila [Irving's wife] and two post-docs I was working with. He said that Freeman wanted a new edition and he wanted to include my 1956 paper. He said it should have been there in the first place, if I remember correctly. For many years we had been on good terms. Not capturing our early closeness perhaps, but good friends.

*(Irving, fall 2000 email to author; my bracketed additions)*

Creer described what happened when he and Cox finally met each other and discussed paleomagnetic issues.

When I finally met up with Allan Cox (after the Berkeley [1963, IUGG] meeting), I found that we got on rather well. Our discussions were mainly about geomagnetic palaeo-secular variations on which subject we were both working actively at that time, rather than on geological aspects of palaeomagnetism. He took me on a tour of San Francisco including Fisherman's Wharf at a time when it was only beginning to take off to be the very big attraction that it is nowadays. I should add that in my view, Allan was never primarily interested in the applications of palaeomagnetism to Earth evolution (drift and polar wander) – but rather to the geomagnetic aspects (reversals and secular variations etc). I cannot recollect him having made any positive contributions to drift via a project involving his own palaeomagnetic measurements on rock formations.

*(Creer, September 2000 email to author; my bracketed addition)*

Finally, I want to identify a tendency among Earth scientists who showed little inclination toward mobilism before its more general acceptance in the mid-1960s to tell and sometimes retell a revisionist history that either obscured the fact that they had not earlier, in light of developments occurring within the Earth sciences, taken mobilism more seriously, or sought to explain their recalcitrance as eminently reasonable. I begin by returning to Cox's citation for Irving's 1979 Bucher Medal, given annually by the AGU for original contributions to the basic knowledge of Earth's crust. It is graceful and correct. He began by describing what Irving had accomplished in the mid-1950s when he (1956a) compared paleomagnetic and paleoclimatic data (§3.12).

In a seminal 1956 paper, Irving determined the paleolatitudes of North America, Europe, and Australia from paleomagnetic data and compared these with the geological records on these continents of paleoclimates. He demonstrated that, except possibly for the Precambrian, the two data sets were in remarkably good agreement: coral reefs and rocks formed in ancient deserts lay at low paleomagnetic latitudes whereas ancient glacial deposits lay at high paleomagnetic latitudes. Using paleomagnetic data, Irving calculated pole paths for North America and Europe and noted that the path for North America lay to the west of the path for Europe. He concluded in his 1956 paper that during the Mesozoic and Paleozoic North America lay closer to Europe and that India, Australia, North America, and Europe had all undergone continental drift.

*(Cox, 1979: 659)*

He then described four central characteristics of Irving's research.

The first was his acceptance, early in his career, of the hypothesis that the geomagnetic field, when averaged over a suitable interval of time, is a dipole field aligned with the earth's rotation axis. The second is his experimental work, which has been both frugal and thorough and has yielded a large body of reliable paleomagnetic data. The third has been his rigor in fully interpreting his data in accord with the dipole hypothesis, even when his interpretations did not conform with accepted ideas. The fourth has been his selection of important, solvable global tectonic problems as the subjects of his research.[19]

*(Cox, 1979: 659)*

Cox concluded the citation by addressing the importance of Irving's work.

The rapid acceptance of plate tectonics during the past decade has been accomplished by a general acceptance of the validity and importance of Irving's work. Today the experimental techniques he helped pioneer and the paleomagnetic data he helped gather are widely recognized as comprising one of the cornerstones of plate tectonics.

*(Cox, 1979: 659)*

I now go back to *Plate Tectonics and Geomagnetic Reversals*, Cox's 1973 compilation of the key papers that led to revolution in the Earth sciences. It was published in 1973, six years before Cox wrote the above citation. Cox's 1973 history differed substantially from what had happened and what he said had happened in his 1979 citation for Irving. He did not include any paper about APW paths. He included no papers by Blackett, Clegg, Creer, Deutsch, Irving, or Runcorn, no example of Hospers' pioneer work on reversals, although he quoted and discussed early work on reversals by others. Thus, it seems that Cox did not think that work done in Britain, Australia, or South Africa seminal. However, he twice mentioned "Cambridge" paleomagnetists.

Interest in continental drift was revived with the publication in 1956 by E. Irving (1956) and S. K. Runcorn (1956) of paleomagnetic data *substantiating* that drift between North America and Europe had occurred since the Paleozoic Era.
*(Cox, 1973: 15; the references correspond to Irving (1956a) and Runcorn (1956b); my emphasis)*

Cox did not say this in the 1960 *GSA* review or in anything else he wrote before the general acceptance of seafloor spreading; though he quoted their papers he did not say that Irving or Runcorn in 1956 first substantiated, on physical grounds, drift between North America and Europe. He did mention British paleomagnetists one other time without, however, citing Blackett, Clegg, Deutsch, and others associated with Blackett.

Cambridge had long been a world center of highly creative geophysical research. The modern school of paleomagnetism with its rigorous statistical approach and its emphasis on the problem of polar wandering and continental drift began at Cambridge with the research of J. Hospers, S. K. Runcorn, K. Creer, and E. Irving, all of whom were graduate students in the early 1950's. Since the time of George Darwin most geophysicists had believed that the earth was too rigid to permit either polar wandering or continental drift. This changed in 1956 with Irving's (1956a) and Runcorn's (1956b) demonstrations using paleomagnetic data that both polar wandering and continental drift had occurred.[20]
*(Cox, 1973: 43)*

It is hard to believe that the Cox of 1960 and 1973 are one and the same.

Paradoxically, Cox in his 1973 compilation included no papers by paleomagnetists in Britain, even though he maintained that the work of the Cambridge paleomagnetists was important and that Irving and Runcorn had demonstrated mobilism. Because he did not include any of their papers (or even mention Creer), he did not have to discuss them, or tell readers what they had actually done. He described them as having demonstrated mobilism, but not his own early reaction to their work.

Cox got it right in his 1979 citation for Irving. Perhaps he had in mind the contrast between his own and Irving's early research when he said that Irving rigorously interpreted his data in accordance with the geocentric axial dipole hypothesis "even when his interpretations did not conform with accepted ideas."

During all this period there was little interaction between Doell and Irving. Later in their lives they saw more of each other. From time to time Doell and his wife visited the Irvings in British Columbia on their way through to Alaska and became good friends, agreeing to differ on the reviews (Irving, October 2008 note to author).

## Notes

1  By contrast, Graham's attack appeared in the *Journal of Geophysical Research*, which, I suspect, was read by substantially fewer geologists than the *Bulletin, Geological Society of America*. Geophysicists read *JGR* and probably still do so more often than the GSA's *Bulletin*, but back then geologists were much less inclined to read papers in *JGR* than in geological journals.

2  There is no question that Verhoogen argued in favor of convection in the mantle, but he did not link mantle convection with continental drift in any of the following: Verhoogen (1946, 1954, 1956, 1959, 1960, 1961) and Turner and Verhoogen (1960). In I, §5.6 and I, §5.10, I describe how support of mantle convection did not necessarily imply support for mobilism.

3  Gilluly's biographer mentions the popularity of *Principles of Geology* (Nolan, 1987: 124).

4  Some of Doell's discussion of this material is also in Glen (1982: 151–154). Doell's recent recollections, however, provide new and quite specific information about the origin of the two reviews.

5  Cox and Doell's compilation appeared in the same years as the first of the "Pole Lists" published periodically by the *Geophysical Journal* in London that were compiled at first by Irving (1960–5) and later by Michael McElhinny, who eventually developed them into a global database (McElhinny, 1973: 282).

6  It is perhaps not surprising that Cox and Doell should make this utopian proposal because, as their record shows, they had little or no experience of the particularities of obtaining reference poles for cratonic blocks; Cox had no experience; Doell had worked briefly on the Colorado Plateau but he set little store by the results (§3.8, §8.2). Moreover, the Plateau, although once part of the North American craton, is now tectonically disconnected, having rotated clockwise relative to it. In early fall 1960 they did attempt to obtain a North American cratonic reference pole for the Eocene of Texas for comparison with the oblique magnetization directions of the Siletz volcanics but were not successful (§8.12).

7  Working backwards, the paleomagnetic results available in 1960 became progressively fewer per unit of time, but there were some results from older rocks that were reliable, and some from younger rocks that were less reliable. Continental drift, as Wegener proposed it, was a later Phanerozoic phenomenon, and only results from Permo-Carboniferous and younger rocks were directly relevant to testing of his hypothesis; Cox and Doell's discussion of results from Devonian and older rocks was therefore strictly not relevant to that test. The requirements specified by Cox and Doell before a discussion of continental drift could proceed to their satisfaction would not be achieved for another twenty or thirty years. In the fixist milieu of North America, theirs was a surefire tactic.

8  Strictly speaking, Cox and Doell were correct. At the time there were no Eocene results from east of the Cordillera. There were, however, available to Cox and Doell results from the Eocene Green River Formation of Colorado (Torreson *et al.*, 1949), which had northerly directions of magnetization and a pole concordant with the APW path for North America as shown in Figure 1 in Creer *et al.*, 1957. Comparison with the east-northeasterly magnetization of the Siletz River Volcanics indicated that the Coast Range of Oregon had rotated ~65° clockwise, as discussed in Irving (1964: 249).

9  Their gloomy prophecy turned out to be wrong. Creer, Irving, Runcorn, Clegg, Du Bois, Green, Blackett, Deutsch, and others got the general form and position of the APW paths back to the Carboniferous essentially correct the first time. Of course there were later

changes because of developments and application of demagnetization techniques and radiometric dating, but the general form and position of the paths survived.

10 I have found no evidence for Runcorn's claim that Cox and Doell did not support drift because they wanted to secure funding and did not want to displease fixists controlling the purse strings. I do not believe it is true. Balsley asked them to write the review but not as a condition for future funding. Moreover, Cox and Doell probably displeased Balsley on another issue; they favored field reversals even though Balsley was a strong advocate of self-reversals (Glen, 1982: 176).

11 This acrimonious debate about how to combine results from sedimentary rock formations between those (e.g., Creer, Irving, Doell, and Cox) who favored the hierarchical approach originally set out by Watson and Irving (1956) of first averaging directions from samples at the lower level (from the same sampling site or horizon) and then averaging site directions taken through the formation, and those (e.g., Runcorn and Nairn supported by R. A. Fisher) who simply averaged all sample directions on the grounds that sediments were deposited slowly and recorded independent points in time, was carried on for a decade. There is now general consensus arrived at on empirical grounds that the former is the better procedure because it does not lead to unrealistically small errors when extensive surveys are carried out (see Irving, 1964: 63–64; Merrill and McElhinny, 1983; McElhinny and McFadden, 2000: 84).

12 The correspondence is found in the Allan Cox Papers in the Stanford University Archives. Unfortunately, the correspondence is not complete, as reference is sometimes made to letters that Cox sent to Irving that are not in the collection. I should like to thank Patricia White, Archives Specialist in the Department of Special Collections at Stanford University Libraries, for sending me this requested material, and the late Richard Doell, then executor of Cox's estate, for giving me permission to quote from the letters.

13 Despite Cox's disclaimer it appears that he and Doell sometimes used "virtual geomagnetic pole" to mean "not real geomagnetic pole." Consider the following passage from Cox and Doell (1961a: 46):

Representative sampling areas in western Europe and Siberia are shown ... together with the virtual geomagnetic poles (fictive geocentric dipoles that would give rise to the palaeomagnetically measured field direction at each locality) ...

Irving perhaps had reason to suppose that Cox and Doell meant to cast another level of uncertainty about paleomagnetic poles through their use of "virtual geomagnetic pole."

14 Irving returned to Cox and Doell's idea of "virtual geomagnetic pole" when writing his 1964 textbook. He raised another internal difficulty. Instead of questioning the "virtual" in virtual geomagnetic pole, he questioned the ambiguous use of the term to refer to poles calculated from direct observations of the present field (a geological instant) and ancient poles calculated from time-averaged rock magnetic data. Irving restricted the use of "virtual geomagnetic pole" to refer to the poles calculated from present field observations, and used "paleomagnetic pole" to refer to poles calculated from time-averaged rock data. Irving's recommendations were generally, but not exactly, followed by later workers. Paleomagnetic poles (or paleopoles or simply poles as they are commonly referred to in this text for brevity's sake where the context is clear) are time averaged and necessarily based on rock data; virtual geomagnetic poles are calculated from spot readings (McElhinny, 1973: 26; McElhinny and McFadden, 2000: 22). Thus, Irving's recommendation about the use of "paleomagnetic pole" to refer to the time-averaged pole based on rock data was adopted by paleomagnetists. Virtual geomagnetic poles are not time-averaged, but, as Irving proposed, are calculated for a single rock outcrop or horizon, what is "geologically speaking, an instant in time." Thus in paleomagnetic studies it takes a number of virtual geomagnetic poles to determine a single paleomagnetic pole, which, I think, captures what Cox originally had in mind when he introduced "virtual geomagnetic pole" as that which is more easily calculated than what is actually sought. But, Cox and Doell also used the term to refer to time-averaged paleopoles, which are not "virtual" in the sense of being more

easily calculated, and understandably led Irving to raise his first difficulty, that of Cox's use of "virtual" as meaning "not real." (See McElhinny (1973: 26) for a discussion of the use of "paleomagnetic pole" and "virtual geomagnetic pole.")

15 An age of 250 Ma for the Siberian Traps is still a good estimate, and is now known to be the age of the top of the Permian System. Hence the Siberian Traps were extruded either at the very end of the Palaeozoic or very beginning of the Mesozoic.

16 The initial motivation for Tarling to collect samples from Aden and the Pacific Islands was to study the dispersion of directions in young lavas and thus estimate the magnitude of secular variation in low latitudes. The overall objective was to describe statistically the recent geomagnetic field for different latitudes. Because the strength of the field of the average geocentric dipole at the pole would be twice that at the equator, simple models of the secular variation would indicate much greater dispersion of the field at the equator than at the poles; both the form and the magnitude of dispersion would vary predictably with latitude, depending on the model employed (Creer *et al.*, 1959; Irving and Ward, 1964). If sufficient dispersion estimates could be obtained over a wide range of latitude, the relationship of dispersion and latitude could be determined empirically providing a test of the GAD hypothesis (Irving, 1964: 146, 243; Gough, Opdyke, and McElhinny, 1964), a matter of critical importance for the paleomagnetic case for continental drift (§5.18). Such studies became widespread and progressively more detailed from the 1970s on (Merrill and McElhinny, 1983).

17 It is noteworthy too that a year earlier, based on contemporary work at ANU, Irving and Parry (1963), working on timescales one or two orders of magnitude greater, had established the stratigraphic limits and duration of the "Kiaman Magnetic Interval" (now called the Kiaman reversed Superchron), the sixty-million-year interval of reversed geomagnetic field at the end of the Paleozoic. This together with the work of McDougall and Tarling established at ANU the full range in variability in reversal frequency during the Phanerozoic.

18 This letter is undated. However, it most likely was written in late January 1961 because Cox responded to Irving's remarks about Tarling in his February 8, 1960 letter to Irving.

19 Cox's citation is, strictly speaking, not quite correct. Irving's observations of high latitude Late Paleozoic glacial beds were first reported in his 1957a paper. He did not have that data in 1955 when he wrote his 1956 paper. He arrived in Australia very late in 1954, and first had to build, with much help from Ron Green, a lab from scratch.

20 Runcorn was no longer a graduate student in 1950, receiving his Ph.D. in 1949 working under Blackett at Manchester. He then moved to Cambridge, where he became Assistant Director of Research in Geophysics in 1950 (§1.2).

# References

Almeida, F. F. M. 1952/1953. Botucatu, a Triassic desert of South America. *Proceedings 19th International Geological Congress*, Algiers, 1952, **VII**, 9–24.

Almond, M., Clegg, J. A., and Jaeger, J. C. 1956. Remanent magnetism of some dolerites, basalts and volcanic tuffs from Tasmania. *Phil. Mag.*, **1**: 771–782.

Almond, M., Davies, J. G., and Lovell, A. C. B. 1951. The velocity distribution of sporadic meteors, part I. *Mon. Not. Roy. Astron. Soc.*, **111**: 585–608.

Almond, M., Davies, J. G., and Lovell, A. C. B. 1952. The velocity distribution of sporadic meteors, part II. *Mon. Not. Roy. Astron. Soc.*, **112**: 21–38.

Amor, K., Hesselbo, S. P., Porcelli, D., Thackrey, S., and Parnell, J. 2008. A Precambrian proximal ejecta blanket from Scotland. *Geology*, **36**: 303–306.

Andrade, E. N. da C. 1910. On the viscous flow in metals, and allied phenomena. *Proc. Roy. Soc. A*, **84**: 1–12.

Andrade, E. N. da C. 1914. On the viscous flow in metals, and allied phenomena. *Proc. Roy. Soc. A*, **90**: 329–342.

Anonymous. 1953. Geophysical discussion. *Observatory*, **73**: 62–69.

Anonymous. 1954. Arizona Arctic. *Time Magazine*, September 27: 40.

Anonymous. 1958. Geophysical discussion. *Observatory*, **78**: 65–68.

Anonymous. 1960. Report of the XII General Assembly of the International Union of Geodesy and Geophysics. *Geophys. J.*, **3**: 475.

Argand, E. 1924. La tectonique de l'Asie. Proceedings of the 13th International Geological Congress, 1, Part 5: 171–372. Translated into English by Carozzi, A.V. 1977. *The Tectonics of Asia*. Hafner Press, New York.

Arkell, W. J. 1956. *Jurassic Geology of the World*. Oliver and Boyd, Edinburgh.

As, J. A. and Zijderveld, J. D. A. 1958. Magnetic cleaning of rocks in palaeomagnetic research. *Geophys. J.*, **1**: 308–319.

Babcock, H. W. 1947a. Remarks on stellar magnetism. *Publ. Astron. Soc. Pac.*, **59**: 112–124.

Babcock, H. W. 1947b. Zeeman effect in stellar spectra. *Astrophys. J.*, **105**: 105–119.

Bagnold, R. A. 1941. *Physics of Blown Sand and Desert Dunes*. Methuen and Co., London.

Bagnold, R. A. 1990. *Sand, Wind and War: Memoirs of a Desert Explorer*. University of Arizona Press, Tucson, AZ.

Baker, H. B. 1912–1914. The origin of continental forms. *Annu. Rep. Michigan Acad. Sci.*, 1912: 116–141; 1913: 26–32, 107–113; 1914: 99–103.

Beck, M. E. 1976. Discordant paleomagnetic pole positions as evidence of regional shear in the Western Cordillera of North America. *Am. J. Sci.*, **276**: 694–712.

Beloussov, V. V. 1954. *Osnovnye Voprosy Geotektoniki (Principal Problems of Geotectonics)*. Gosgeolizdat, Moscow.

Belshé, J. C. 1957. Palaeomagnetic investigations of Carboniferous rocks in England and Wales. *Adv. Phys.*, **6**: 187–191.

Besse, J. and Courtillot, V. 2002. Apparent and true polar wander and the geometry of the geomagnetic field in the last 200 million years. *J. Geophys. Res.*, **107**(B11), 2300, doi:10.1029/2000JB000050.

Billings, M. P. 1960. Diastrophism and mountain building. *Geol. Soc. Am. Bull.*, **71**: 363–397.

Blackett, P. M. S. 1947. The magnetic field of massive rotating bodies. *Nature*, **159**: 658–666.

Blackett, P. M. S. 1952. A negative experiment relating to magnetism and the earth's rotation. *Trans. Roy. Soc. A*, **245**: 309–370.

Blackett, P. M. S. 1956. *Lectures on Rock Magnetism*. Weizmann Science Press of Israel, Jerusalem.

Blackett, P. M. S. 1957. Introductory remarks. *Adv. Phys.*, **6**: 147–148.

Blackett, P. M. S. 1965. Introductory remarks. In Blackett, P. M. S., Bullard, E. C., and Runcorn, S. K., eds., *Symposium on Continental Drift. Philos. Trans. Roy. Soc. London A*, **258**: vii–x.

Blackett, P. M. S., Clegg, J. A., and Stubbs, P. H. S. 1960. An analysis of rock magnetic data. *Proc. Roy. Soc. A*, **256**: 291–322.

Blundell, D. J. 1957. Some geological applications of rock magnetism. Ph.D. Thesis (unpublished). University of London.

Blundell, D. J. and Stephenson, P. J. 1959. Palaeomagnetism of some dolerite intrusions from the Theron Mountains and Whichaway Nunataks, Antarctica. *Nature*, **184**: 1860.

Box, J. F. 1978. *R.A. Fisher, the Life of a Scientist*. John Wiley & Sons, New York.

Bradley, J. 1957. The meaning of paleogeographic pole. *New Zealand J. Sci. Tech.*, **B38**: 354–365.

Bradley, J. 1965. Intrusion of major dolerite sills. *Trans. Roy. Soc. New Zealand*, **3**: 27–55.

Bruckshaw, J. McG. and Robertson, E. I. 1949. The magnetic properties of the Tholeiite Dykes of North England. *Mon. Not. Roy. Astron. Soc., Geophys. Suppl.*, **5**: 308–320.

Brunhes, B. 1906. Recherches sur le direction d'aimantation des roches volcaniques. *J. Phys.*, **5**: 705–724.

Brunnschweiler, R. O. 1958. Indo-Pacific faunal relations during the Mesozoic. In Carey, S. W., Convener, *Continental Drift: A Symposium*. University of Tasmania, Hobart, 128–133.

Brush, S. G. 1996. *A History of Modern Planetary Physics, Volume 1*. Cambridge University Press, Cambridge.

Brynjolfsson, A. 1957. Studies of remanent magnetism and viscous magnetism in the basalts of Iceland. *Adv. Phys.*, **6**: 247–254.

Bucher, W. H. 1954. Opening remarks. *Trans. Am. Geophys. Union*, **35**: 48.

Bull, C. and Irving, E. 1960a. Palaeomagnetism in Antarctica. *Nature*, **185**: 834–835.

Bull, C. and Irving, E. 1960b. The palaeomagnetism of some hypabyssal intrusive rocks from South Victoria Land, *Antarctica. Geophys. J.*, **3**: 211–224.

Bull, C., Irving, E., and Willis, I. 1962. Further palaeomagnetic results from South Victoria Land, *Antarctica. Geophys. J.*, **6**: 320–336.

Bullard, E. C. 1948. The secular change in the earth's magnetic field. *Mon. Not. Roy. Astron. Soc., Geophys. Suppl.*, **5**: 248–257.

Bullard, E. C. 1949. The magnetic field within the earth. *Proc. Roy. Soc. London, A*, **197**: 433–453.

Bullard, E. C. 1974. Patrick Blackett ...: An appreciation. *Nature*, **250**: 370.

Bullard, E. C. and Gaskell, T. F. 1941. Submarine seismic investigations. *Proc. Roy. Soc. A*, **177**: 476–498.

Bullard, E. C., Everett, J. E., and Smith, A. G. 1965. The fit of the continents around the Atlantic. *Phil. Trans. Roy. Soc. London*, **258**: 41–51.

Butler, C. C., Lowell, B., Occhialini, G. P. S., Runcorn, S. K., and Waddington, C. H. 1975. Memorial meeting for Lord Blackett, O.M., C.H., F.R.S. at the Royal Society on 31 October, 1974. *Notes Rec. Roy. Soc. London*, **29**: 135–162.

Butler, C. C., Rochester, G. D., and Runcorn, S. K. 1947. An example of meson production in lead. *Nature*, **159**: 227–228.

Campbell, C. D. and Runcorn, S. K. 1956. The magnetization of the Columbia River basalts in Washington and Northern Oregon. *J. Geophys. Res.*, **61**: 449–458.

Carey, S. W. 1938. Tectonic evolution of New Guinea and Melanesia. D.Sc. Thesis (unpublished). University of Sydney.

Carey, S. W. 1954. The rheid concept in geotectonics. *J. Geol. Soc. Australia*, **1**: 67–117.

Carey, S. W. 1955a. The orocline concept in geotectonics. *Roy. Soc. Tasmania*, **89**: 255–288.

Carey, S. W. 1955b. Wegener's South America–Africa assembly, fit or misfit? *Geological Magazine*, **XCII**: 196–200.

Carey, S. W. 1958. A tectonic approach to continental drift. In Carey, S. W., Convener, *Continental Drift: A Symposium*. University of Tasmania, Hobart, 177–355.

Carey, S. W. 1961. Paleomagnetic evidence relevant to a change in the Earth's radius. *Nature*, **190**: 35.

Carey, S. W. 1963. The asymmetry of the Earth. *Australian J. Sci.*, **25**: 369–384, 479–489.

Carey, S. W. 1970. Australia, New Guinea and Melanesia in the current revolution in concepts of the evolution of the Earth. *Search*, **1**: 178–189.

Carey, S. W. 1976. *The Expanding Earth*. Elsevier, Amsterdam.

Carey, S. W. 1983. Earth expansion and the null universe. In Carey S. W., Convener, *Expanding Earth Symposium, Sydney, 1981*. University of Tasmania, Hobart, 365–372.

Carey, S. W. 1988. *Theories of the Earth and Universe: A History of Dogma in the Earth Sciences*. Stanford University Press, Stanford, CA.

Chaloner, W. G. 1959. Continental drift. In Johnson, M. L., Abercrombie, M., and Fogg, G. E., eds., *New Biology*, **29**. Baltimore, Penguin Books, 7–30.

Chamalaun, F. and Creer, K. M. 1964. Revised Devonian pole for Britain. *J. Geophys. Res.*, **69**: 1607–1616.

Chamalaun, F. H. and Creer, K. M. 1963. A revised Devonian pole for Britain. *Nature*, **198**: 375.

Chamalaun, F. H. and Creer, K. M. 1964. Thermal demagnetization studies of the Old Red Sandstones of the Anglo Welsh cuvette. *J. Geophys. Res.*, **69**: 1607–1616.

Chang, W.-Y. and Nairn, A. E. M. 1959. Some palaeomagnetic investigations on Chinese rocks. *Nature*, **183**: 254.

Chapman, S. (with an added note by Runcorn). 1948a. Variation of geomagnetic intensity with depth. *Nature*, **161**: 52.

Chapman, S. 1948b. The main geomagnetic field. *Nature*, **161**: 462–464.

Chapman, S. and Bartels, J. 1940, 1951. *Geomagnetism*. Clarendon Press, Oxford.

Chevallier, R. 1925. L'aimantation des laves de l'Etna et l'orientation du champ terrestre en Sicile du XII$^e$ au XVII$^e$ siècle. *Ann. Phys.*, **4**: 5–162.

Clark, R. H. and Vella, P. P. 1985. John Bradley, 1910–1985. *Geol. Soc. New Zealand Newsletter*, No. **68**: 33–34.

Clegg, J. A. 1952. The velocity distribution of sporadic meteors. *Mon. Not. Roy. Astron. Soc.*, **112**: 399–413.

Clegg, J. A. 1956. Rock magnetism. *Nature*, **178**: 1085–1087.

Clegg, J. A. 1975. Blackett and rock magnetism. Unpublished, 22 pages.

Clegg, J. A., Almond, M., and Stubbs, P. H. S. 1954a. The remanent magnetization of some sedimentary rocks in Britain. *Phil. Mag.*, **45**: 583–598.

Clegg, J. A., Almond, M., and Stubbs, P. H. S. 1954b. Some recent studies of the prehistory of the Earth's magnetic field. *J. Geomagn. Geoelectr.*, **VI**: 194–199.

Clegg, J. A., Deutsch, E. R., and Griffiths, D. H. 1956. Rock magnetism in India. *Phil. Mag.*, **1**: 419–431.

Clegg, J. A., Deutsch, E. R., Everitt, C. W. R., and Stubbs, P. H. S. 1957. Some recent palaeomagnetic measurements made at Imperial College, London. *Adv. Phys.*, **6**: 219–230.

Clegg, J. A., Radakrishnamurty, C., and Sahasrabudhe, P. W. 1958. Remanent magnetism of the Rajmahal Traps of North-Eastern India. *Nature*, **181**: 830–831.

Cloud, P. and Germs, A. 1971. New Pre-paleozoic nonnofossils frpom the Stoer Formation (Torridonian) north-west Scotland. *Bull Geol Soc Amer.*, **82**: 3469–3474.

Cohen, J. 1998. Uninfectable. *The New Yorker*, July **6**: 34–39.

Collinson, D. W. 1983. *Methods in Rock Magnetism and Paleomagnetism*. Chapman and Hall, London.

Collinson, D. W. 1996. Stanley Keith Runcorn. *Biogr. Mem. Fell. Roy. Soc.*, **42**: 23–38.

Collinson, D. W. 1998. The life and work of S. Keith Runcorn, F.R.S. *Phys. Chem. Earth*, **23**: 697–702.

Collinson, D. W. and Nairn, A. E. M. 1959. A survey of palaeomagnetism. *Overseas Geol. Miner. Resour.*, **7**: 381–397.

Collinson, D. W. and Runcorn, S. K. 1960. Polar wandering and continental drift: evidence from paleomagnetic observations in the United States. *Geol. Soc. Am. Bull.*, **71**: 915–958,

Condon, M. A. and Öpik, A. A. 1956. Summary of continental drift symposium, Hobart, March 1956. Unpublished manuscript from Öpik's papers, Basser Library, Australian Academy of Sciences.

Cook, A. H., Hospers, J., and Paranis, D. S. 1952. The results of a gravity survey in the country between the Clee Hills and Nuneaton. *Q. J. Geol. Soc. London*, **CVII**: 287–302.

Courtillot, V. and Le Mouel, J.-L. 2007. The study of Earth's magnetism (1269–1950). *Rev. Geophys.*, **45**: RG3008, doi: 10.1029/2006RG000198.

Cowling, T. G. 1934. The magnetic field of sunspots. *Mon. Not. Roy. Astron. Soc.*, **94**: 39–48.

Cox, A. 1957. Remanent magnetization of Lower to Middle Eocene basalt flows from Oregon. *Nature*, **179**: 685–686.

Cox, A. 1962. Analysis of present geomagnetic field for comparison with palaeomagnetic results. *J. Geomagn. Geoelectr.*, **13**: 101–112.

Cox, A. 1965. Review of *Paleomagnetism and Its Application to Geological and Geophysica Problems* by E. Irving. *Science*, **147**: 494.

Cox, A. 1973. *Plate Tectonics and Geomagnetic Reversals*. W.H. Freeman and Company, San Francisco.

Cox, A. 1979. Citation for Edward Irving. *Eos*, 659.

Cox, A. 1980. Rotation of microplates in western North America. In Strangway, D. W., ed., *The Continental Crust and its Mineral Deposits*. Geological Association of Canada Special Paper **20**: 305–321.

Cox, A. and Doell, R. R. 1960. Review of paleomagnetism. *Geol. Soc. Am. Bull.*, **71**: 645–768.

Cox, A. and Doell, R. R. 1961a. Palaeomagnetic evidence relevant to a change in the Earth's radius. *Nature*, **189**: 45–47.

Cox, A. and Doell, R. R. 1961b. Reply to Carey on palaeomagnetic evidence relevant to a change in the Earth's radius. *Nature*, **190**: 36–37.

Cox, A., Doell, R. R., and Dalrymple, B. G. 1963a. Geomagnetic polarity epochs and Pleistocene geochronometry. *Nature*, **198**: 1049–1051.

Cox, A., Doell, R. R., and Dalrymple, B. G. 1963b. Radiometric dating of geomagnetic field reversals. *Science*, **140**: 1021–1023.

Creer, K. M. 1955. A preliminary palaeomagnetic survey of certain rocks in England and Wales. Ph.D. Dissertation, Queens' College, University of Cambridge.

Creer, K. M. 1957a. Palaeomagnetic investigations in Great Britain V: the remanent magnetization of unstable Keuper marls. *Phil. Trans. Roy. Soc. London A*, **250**: 130–143.

Creer, K. M. 1957b. Palaeomagnetic investigations in Great Britain IV: the natural remanent magnetization of certain stable rocks from Great Britain. *Phil. Trans. Roy. Soc. London A*, **250**: 111–129.

Creer, K. M. 1958a. Symposium on palaeomagnetism and secular variation. *Geophys. J. Roy. Soc.*, **1**: 99–105.

Creer, K. M. 1958b. Preliminary palaeomagnetic measurements from South America. *Annal. Géophys.*, **14**: 373–390.

Creer, K. M. 1959. A.C. demagnetization of unstable Triassic Keuper marls from S.W. England. *Geophys. J. Roy. Astron. Soc.*, **2**: 261–275.

Creer, K. M. 1962a. Comment on an analysis of the positions of the Earth's magnetic pole in the geological past, by Hibberd, F. H. *Geophys. J.*, **7**: 275–278.

Creer, K. M. 1962b. An analysis of the geomagnetic field using palaeomagnetic methods. *J. Geomagn. Geoelectr.*, **13**: 113–119.

Creer, K. M. 1962c. Palaeomagnetism of the Serra Geral formation. *Geophys. J. Roy. Astron. Soc.*, **7**: 1–22.

Creer, K. M. 1962d. Palaeomagnetic data from South America. *J. Geomagn. Geoelectr.*, **XIII**: 154–165.

Creer, K. M. 1962e. The dispersion of the geomagnetic field due to secular variation and its determination for remote times from paleomagnetic data. *J. Geophys. Res.*, **67**: 3461–3476.

Creer, K. M. 1968. Palaeozoic Palaeomagnetism. *Nature*, **219**: 246–250.

Creer, K. M., Irving, E., and Nairn, A. E. M. 1959. Paleomagnetism of the Great Whin Sill. *Geophys. J.*, **2**: 306–323.

Creer, K. M., Irving, E., Nairn, A. E. M., and Runcorn, S. K. 1958. Palaeomagnetic results from different continents and their relation to the problem of continental drift. *Annal. Géophys.*, **14**: 492–501.

Creer, K. M., Irving, E., and Runcorn, S. K. 1954. The direction of the geomagnetic field in remote epochs in Great Britain. *J. Geomagn. Geoelectr.*, **6**: 163–168.

Creer, K. M., Irving, E., and Runcorn, S. K. 1957. Geophysical interpretation of palaeomagnetic directions from Great Britain. *Phil. Trans. Roy. Soc. A*, **250**: 144–155.

Creer, K. M. and Irving, E. 2012. Testing continental drift: constructing the first palaeomagnetic path of polar wander (1954). *Earth Science History*, **31**: No. 1: forthcoming.

Darabi, M. H. and Piper, J. D. A. 2004. Paleomagnetism of the (late Mesoproterozoic) Stoer Group, nortwest Scotland: implications for diagenesis. age and relationship to the Grenville Orogeny. *Geol. Mag.*, **141**: 15–39.

David, P. 1904. Sur la stabilité de la direction d'animantation dans quelques roches volcaniques. *C. R. Acad. Sci. Paris*, **188**: 41–42.

David, T. W. E. (edited and supplemented by Browne, W. R.) 1950. *The Geology of the Commonwealth of Australia*. Edward Arnold & Co., London.

Day, A. A. and Runcorn, S. K. 1955. Polar wandering: some geological, dynamical and palaeomagnetic aspects. *Nature*, **176**: 422–426.

Debenham, F. 1942. Foreword. In Steers, J. A., ed., *An Introduction to the Study of Map Projections*. University of London Press, London.

Deutsch, E. R. 1963a. Discussion: polar wandering: a phantom event? *Am. J. Sci.*, **261**: 194–199.

Deutsch, E. R. 1963b. Polar wandering and continental drift: an evaluation of recent evidence. In Munyan, A. C., ed., *Polar Wandering and Continental Drift*, Society of Economic Paleontologists and Mineralogists, Special Publication, **10**: 4–46.

Deutsch, E. R. and Watkins, N. D. 1961. Direction of the geomagnetic field during the Triassic period in Siberia. *Nature*, **189**: 543–545.

Deutsch, E. R., Radakrishnamurty, C., and Sahasrabudhe, P. W. 1958. The remanent magnetism of some lavas in the Deccan Traps. *Phil. Mag.*, **3**: 170–184.

Dickins, J. M. and Thomas, G. A. 1958. The correlation and affinities of the fauna of the Lyons Group, Carnarvon Basin, Western Australia. In Carey, S. W., Convener, *Continental Drift: A Symposium*. University of Tasmania, Hobart, 123–127.

Dietzel, G. F. L. 1960. Geology and Permian paleomagnetism of the Merano Region (Province of Balzano, N. Italy). *Geologica Ultraiectina* **4**, 58 pages.

Dirac, P. A. M. 1937. The cosmological constants. *Nature*, **139**: 323.

Dirac, P. A. M. 1938. A new basis for cosmology. *Proc. Roy. Soc. A*, **165**: 199–208.

Doell, R. R. 1955a. Remanent magnetism in sediments. Ph.D. Dissertation. University of California, Geology and Geophysics Department, Berkeley.

Doell, R. R. 1955b. Palaeomagnetic study of rocks from the Grand Canyon of the Colorado River. *Nature*, **176**: 1167.

Doell, R. R. 1956. Remanent magnetism of the Upper-Miocene 'blue' sandstones of California. *Trans. Am. Geophys. Union*, **37**: 156–167.

Doell, R. R. 1958. Paleomagnetic interpretations. *Trans. Am. Geophys. Union*, **39**: 513.

Doell, R. R. 1974. Memorial to John Warren Graham. *Geol. Soc. Am. Mem.*, **III**: 105–108.

Doell, R. R. and Cox, A. 1959. Analysis of paleomagnetic data (Abstract). *Geol. Soc. Am. Bull.*, **70**: 1590–1591.

Doell, R. R. and Cox, A. 1960. Paleomagnetism, polar wandering, and continental drift. In *Geological Survey Research 1960*, US Geological Survey Professional Paper, **400**: B426–B427.

Doell, R. R. and Cox, A. 1961a. Paleomagnetism. *Adv. Geophys.*, **8**: 221–313.

Doell, R. R. and Cox, A. 1961b. Paleomagnetism of Hawaiian lava flows. *Nature*, **192**: 645–646.

Du Bois, P. M. 1955. Palaeomagnetic measurements of the Keweenawan. *Nature*, **176**: 506–507.

Du Bois, P. M. 1957. Comparison of palaeomagnetic results for selected rocks of Great Britain and North America. *Adv. Phys.*, **6**: 177–186.

Du Bois, P. M. 1959a. Late Tertiary geomagnetic field in northwestern Canada. *Nature*, **183**: 1617–1618.

Du Bois, P. M. 1959b. Palaeomagnetism and rotation of Newfoundland. *Nature*, **184**: 63–64.

Du Bois, P. M., Irving, E., Opdyke, N. D., Runcorn, S. K., and Banks, M. R. 1957. The geomagnetic field in Upper Triassic times in the United States. *Nature*, **180**: 1186–1187.

Dunlop, D. J. and Özdemir, Ö. 2001. *Rock Magnetism: Fundamentals and Frontiers*. Cambridge University Press, Cambridge.

Durham, J. W. 1952. Early Tertiary marine faunas and continental drift. *Am. J. Sci.*, **250**: 321–343.

Durham, J. W. 1959. Palaeoclimates. *Phys. Chem. Earth*, **3**: 1–15.

du Toit, A. L. 1927. *A Geological Comparison of South America with South Africa*. Carnegie Institution, Washington.

du Toit, A. L. 1937. *Our Wandering Continents: An Hypothesis of Continental Drifting*. Oliver and Boyd, London.

du Toit, A. L. 1954. *The Geology of South Africa* (3rd edition). Oliver and Boyd, London.

Edwards, W. N. 1955. The geographical distribution of past floras. *Rep. Br. Assoc. Adv. Sci.*, **46**: 165–176.

Egyed, L. 1956a. The change of the Earth's dimensions determined from paleogeographical data. *Geofis. Pura Appl.*, **33**: 42–48.

Egyed, L. 1956b. Determination of changes in the dimensions of the Earth from palaeogeographical data. *Nature*, **178**: 534.

Egyed, L. 1956c. A new theory on the internal constitution of the Earth and its geological-geophysical consequences. *Acta Geol. Acad. Sci. Hung.*, **4**: 43–78.

Egyed, L. 1957. A new dynamic conception of the internal constitution of the Earth. *Geol. Rundschau*, **46**: 101–121.

Egyed, L. 1959. The expansion of the Earth in connection with its origin and evolution. *Geophysica*, **7**: 1–22.

Egyed, L. 1960a. Some remarks on continental drift. *Geofis. Pura Appl.*, **45**: 115–116.

Egyed, L. 1960b. Dirac's cosmology and the origin of the solar system. *Nature*, **186**: 621–622.

Egyed, L. 1961. Paleomagnetism and the ancient radii of the Earth. *Nature*, **190**: 1097–1098.

Egyed, L. 1963. The expanding Earth? *Nature*, **197**: 1059–1060.

Egyed, L. 1969. The slow expansion of the Earth. In Runcorn, S. K., ed., *The Application of Modern Physics to the Earth and Planetary Interiors*. Wiley-Interscience, London, 65–75.

Einarsson, T. and Sigurgeirson, T. 1955. Rock magnetism in Iceland. *Nature*, **175**: 892.

Elliston, J. 2002. Professor S.W. Carey's struggle with conservatism. *Aust. Geologist*, **125**: 17–23.

Elsasser, W. M. 1939. On the origin of the Earth's magnetic field. *Phys. Rev.*, **55**: 486–498.

Elsasser, W. M. 1946a. Induction effects in terrestrial magnetism. Part I. Theory. *Phys. Rev.*, **69**: 106–116.

Elsasser, W. M. 1946b. Induction effects in terrestrial magnetism. Part II. The secular variation. *Phys. Rev.*, **70**: 202–212.

Elsasser, W. M. 1947. Induction effects in terrestrial magnetism. Part III. Electric modes. *Phys. Rev.*, **72**: 821–833.

Elsasser, W. M. 1950. The Earth's interior and geomagnetism. *Rev. Mod. Phys.*, **22**: 1–35.

Elsasser, W. M. 1954. Remarks. *Trans. Am. Geophys. Union*, **35**: 74.

Elsasser, W. M. 1959. Acceptance of the William Bowie Medal. *Trans. Am. Geophys. Union*, **40**: 91–94.

Encarnación, J., Fleming, T. H., Elliot, D. H., and Eales, H. V. 1996. Synchronous emplacement of Ferrar and Karroo dolerites and the early breakup of Gondwana. *Geology*, **24**: 535–538.

Evans, M. E. 1976. Test of the dipolar nature of the geomagnetic field through the Phanerozoic time. *Nature*, **262**: 276–277.

Evans, M. E. 2005. Testing the geomagnetic dipole hypothesis: palaeolatitudes sampled by large continents. *Geophys. J. Int.*, **161**: 266–267.

Evans, M. E. and Hoye, G. S. 2007. Testing the GAD throughout geological time. *Earth Planets Space*, **59**: 697–701.

Fisher, R. A. 1953. Dispersion on a sphere. *Proc. Roy. Soc. London A*, **217**: 295–305.

Fisher, N. L., Lewis, T., and Embleton, B. J. J. 1987. *Statistical Analysis of Spherical Data*. Cambridge University Press, Cambridge.

Folgerhaiter, G. 1899. Sur les variations cularies de l'inclinaison magnetique dans antiguit. *J. Phys.*, **8** (3rd series): 5–16.

Frankel, H. R. 1981. The paleobiogeographical debate over the problem of disjunctively distributed life forms. *Stud. Hist. Phil. Sci.*, **12**: 211–259.

French, A. N. and Van der Voo, R. 1979. The magnetization of the Rose Hill Formation at the classical site of Graham's fold test. *J. Geophys. Res.*, **48**: 7688–7696.

Frenkel, J. 1945. *C.R. Acad. USSR*, **49**: 98–101.

Gelletich, H. 1937. Uber magnetitfuhrende eruptive Gänge und Gangesysteme im mittleven teil des südlichen Transvaals. *Beitr. Angew. Geophys.*, **6**: 337–406.

Gilbert, C. 1956. Dirac's cosmology and the general theory of relativity. *Mon. Not. Roy. Astron. Soc., Geophys. Suppl.*, **6**: 684–690.

Gill, E. D. 1958. Australian Lower Devonian palaeobiology in relation to the concept of continental drift. In Carey, S. W., Convener, *Continental Drift: A Symposium*. University of Tasmania, Hobart, 103–122.

Gilluly, J., Waters, A. C., and Woodford, A. O. 1951. *Principles of Geology*. W.H. Freeman and Co., San Francisco.

Gilluly, J., Waters, A. C., and Woodford, A. O. 1959. *Principles of Geology* (2nd edition). W.H. Freeman and Co., San Francisco.

Glen, W. 1982. *The Road to Jaramillo*. Stanford University Press, Stanford, CA.

Gold, T. 1955. Instability of the Earth's axis of rotation. *Nature*, **175**: 526–529.

Good, R. 1953. *The Geography of the Flowering Plants* (2nd edition). Longmans, Green & Co., London.

Gough, D. I. 1956. A study of the palaeomagnetism of the Pilansberg Dykes. *Mon. Not. Roy. Astron. Soc., Geophys. Suppl.*, **7**: 196 213.

Gough, D. I. 1989. Landmarks of a life in geophysics. *S. Afr. Geophys. Assoc. Yearbook*, **1988**, 22–29.

Gough, D. I., Opdyke, N. D., and McElhinny, M. W. 1964. The significance of paleomagnetic results from Africa. *J. Geophys. Res.*, **69**: 2509–2519.

Graham, J. W. 1949. The stability and significance of magnetism in sedimentary rocks. *J. Geophy. Res.*, **54**: 131–167.

Graham, J. W. 1952. Note on the significance of inverse magnetizations of rocks. *J. Geophys. Res.*, **57**: 429–431.

Graham, J. W. 1953. Changes of ferromagnetic minerals and their bearing on magnetic properties of rocks. *J. Geophys. Res.*, **58**: 243–260.

Graham, J. W. 1954. Rock magnetism and the Earth's magnetic field during Paleozoic time. *J. Geophys Res.*, **59**: 215–222.

Graham, J. W. 1955. Evidence of polar shift since Triassic time. *J. Geophys. Res.*, **60**: 329–317.

Graham, J. W. 1956. Paleomagnetism and magnetostriction. *J. Geophys. Res.*, **61**: 735–739.

Graham, J. W. 1957. The role of magnetostriction in rock magnetism. *Adv. Phys.*, **6**: 362–363.

Graham, J. W. and Torreson, O. W. 1951. Contrasting magnetizations of flat-lying and folded Paleozoic sediments (abstract). *Trans. Am. Geophys. Union*, **32**: 336.

Graham, J. W., Buddington, A. F., and Balsley, J. R. 1957. Stress-induced magnetizations of some rocks with analyzed magnetic minerals. *J. Geophys. Res.*, **62**: 465–474.

Graham, J. W., Buddington, A. F., and Balsley, J. R. 1959. Magnetostriction and palaeomagnetism of igneous rocks. *Nature*, **183**: 1318.

Graham, K. W. T. 1961. Palaeomagnetic studies on some South African Rocks. Ph.D. dissertation. University of Capetown, South Africa.

Graham, K. W. T. and Hales, A. L. 1957. Palaeomagnetic measurements on Karroo dolerites. *Adv. Phys.*, **6**: 149–161.

Graham, K. W. T., Helsley, C. E., and Hales, A. L. 1964. Determination of the relative positions of continents from paleomagnetic data. *J. Geophys. Res.*, **69**: 3895–3900.

Green, D. H. 1998. Alfred Edward Ringwood. *Biogr. Mem. Fell. Roy. Soc.*, **44**: 349–362.

Green, R. 1958. Polar wandering, a random walk problem. *Nature*, **182**: 382–383.

Green, R. 1961a. Study of the palaeomagnetism of some Kainozozic and Palaeozoic rocks. Ph.D. thesis, Australian National University, Canberra.

Green, R. 1961b. Palaeomagnetism of some Devonian rock formations in Australia. *Tellus*, **13**: 119–124.

Green, R. and Irving, E. 1958. The palaeomagnetism of the Cainozoic basalts from Australia. *Proc. Roy. Soc. Victoria*, **70**: 1–17.

Griffiths, D. H. and King, R. F. 1954. Natural magnetization of igneous and sedimentary rocks. *Nature*, **173**: 1114–1117.

Gusev, B. V. 1959. Vozrast shchelochno-ultraosnownykh porod maymecha-kotuyskogo rayona po paleomagnitnym dannym. *Inf. Byull. Inst. Geol. Arktiki*

*Leningr.*, **4**: 30–33. (1960 translation: Age of alkaline-ultrabasic rocks of Maymechakotuy region according to paleomagnetic data. *Int. Geol. Rev.*, **2**: 327–329.)

Gustafson, J. K., Burrell, H. C., and Garretty, M. D. 1950. Geology of the Broken Hill ore deposit, N.S.W. Australia. *Geol. Soc. Am. Bull.*, **61**: 1369–1438.

Gutenberg, B. 1940. Geotektonische hypothesen. *Handbuch der Geophysik*, **3**, part 4: 442–547.

Gutenberg, B. 1951a. Introduction. In Gutenberg, B., ed., *Internal Constitution of the Earth* (2nd edition). Dover Publications, New York, 1–7.

Gutenberg, B. 1951b. Hypotheses on the development of the Earth. In Gutenberg, B., ed., *Internal Constitution of the Earth* (2nd edition). Dover Publications, New York, 178–226.

Hales, A. L. 1986. Geophysics on three continents. *Ann. Rev. Earth Planet. Sci.*, **14**: 1–20.

Hales, A. L. and Gough, D. I. 1947. Blackett's fundamental theory of the Earth's magnetic field. *Nature*, **160**: 746.

Halls, H. C. 1976. A least-squares method to find a remanence direction from converging remagnetization circles. *Geophys. J. Roy. Astron. Soc.*, **45**: 297–304.

Halm, J. K. E. 1935. An astronomical aspect of the evolution of the earth. *Astron. Soc. S. Afr.*, **4**: 1–28.

Haslett, A. W. 1954. Research report. *Science News*, **34**: 115–123.

Hess, H. H. 1962. History of ocean basins. In *Petrologic Studies: A Volume to Honor A. F. Buddington*. Geological Society of America, New York, 599–620.

Hess, H. H. and Maxwell, J. C. 1953. *Geol. Soc. Am. Bull.*, **64**: 301–316.

Hibberd, F. H. 1962. An analysis of the positions of the Earth's magnetic pole in the geological past. *Geophys. J.*, **6**: 221–244.

Hide, R., 1953. Some experiments on thermal convection in a rotating liquid. Ph.D. dissertation (unpublished), University of Cambridge.

Hide, R., 1996. Stanley Keith Runcorn FRS (1922–1995). *Q. J. Roy. Astron. Soc.*, **37**: 463–465.

Hide, R. 1997. Response to winning the William Bowie Medal of the American Geophysical Union. AGU website.

Hilgenberg, O. C. 1933. *Vom wachsenden Erdball*. Giessmann and Bartsch, Berlin.

Hill, M. N. and King, W. B. R. 1953. *Q. J. Geol. Soc. London*, **109**: 1–20.

Hill, M. N. and Laughton, A. S. 1954. *Proc. Roy. Soc. A*, **222**: 348–356.

Hills, G. F. S. 1947. *The Formation of the Continents by Convection*. Arnold, London.

Holmes, A. 1929. A review of the continental drift hypothesis. *Mining Mag.*, **40**: 205–209, 286–288, 340–347.

Holmes, A. 1944. *Principles of Physical Geology* (1st edition). Thomas Nelson & Sons, Edinburgh.

Holmes, A. 1947. *Principles of Physical Geology* (Reprint of 1st edition). Thomas Nelson, London.

Holmes, A. 1965. *Principles of Physical Geology* (2nd edition). Thomas Nelson, London.

Hospers, J. 1951. Remanent magnetism of rocks and the history of the geomagnetic field. *Nature*, **168**: 1111–1112.

Hospers, J. 1953a. Reversals of the main geomagnetic field, part I. *Proc. Roy. Netherlands Acad. Sci. Amsterdam, Series B*, **56**: 467–476.

Hospers, J. 1953b. Reversals of the main geomagnetic field, part II. *Proc. Roy. Netherlands Acad. Sci. Amsterdam, Series B*, **56**: 477–491.

Hospers, J. 1953c. Palaeomagnetic studies of Icelandic rocks. Ph.D. thesis (unpublished). University of Cambridge.

Hospers, J. 1954a. Reversals of the main geomagnetic field, part III. *Proc. Roy. Netherlands Acad. Sci. Amsterdam, Series B*, **57**: 112–121.

Hospers, J. 1954b. De natuurlijke magnetiztie van Ijslandse Gesteenten. *Geol. Mijnbouw*, **16**: 48–51.

Hospers, J. 1954c. Rock magnetism and polar wandering. *Nature*, **173**: 1183–1184.

Hospers, J. 1954d. Magnetic correlation in volcanic districts. *Geol. Mag.*, **XCI**: 352–360.

Hospers, J. 1955. Rock magnetism and polar wandering. *J. Geol.*, **63**: 59–74.

Hospers, J. and Charlesworth, H. A. K. 1954. The natural permanent magnetization of the lower basalts of Northern Ireland. *Mon. Not. Roy. Astron. Soc., Geophys. Suppl.*, **7**: 32–43.

Hospers, J. and Willmore, P. L. 1953. Gravity measurements in Durham and Northumberland. *Geol. Mag.*, **XC**: 117–126.

Hull, D. L. 1988. *Science As a Process: An Evolutionary Account of the Social and Conceptual Development of Science*. University of Chicago Press, Chicago.

Irving, E. 1954. The palaeomagnetism of the Torridonian sandstone series of North-Western Scotland. Ph.D. dissertation (unpublished), University of Cambridge.

Irving, E. 1956a. Palaeomagnetic and palaeoclimatological aspects of polar wandering. *Geofis. Pura Appl.*, **33**: 23–48.

Irving, E. 1956b. The magnetisation of the Mesozoic dolerites of Tasmania. *Pap. Proc. Roy. Soc. Tasmania*, **90**: 157–168.

Irving, E. 1957a. Directions of magnetization in the Carboniferous glacial varves of Australia. *Nature*, **180**: 280–281.

Irving, E. 1957b. Rock magnetism: a new approach to some palaeogeographic problems. *Adv. Phys.*, **6**: 194–218.

Irving, E. 1957c. The origin of the palaeomagnetism of the Torridonian sandstones of north-west Scotland. *Phil. Trans. Roy. Soc. London A*, **250**: 100–110.

Irving, E. 1958a. Rock magnetism: a new approach to the problems of polar wandering and continental drift. In Carey, S. W., Convener, *Continental Drift: A Symposium*. University of Tasmania, Hobart, 24–57.

Irving, E. 1958b. Palaeogeographic reconstruction from palaeomagnetism. *Geophys. J. Roy. Astron. Soc.*, **1**: 224–237.

Irving, E. 1959a. Palaeomagnetic pole positions. *Geophys. J.*, **2**: 51–79.

Irving, E. 1959b. Magnetic 'cleaning'. *Geophys. J.*, **2**: 140–141.

Irving, E. 1960. Preface to English translation by A. J. Lojkine of N. J. Khramov's *Palaeomagnetism and Stratigraphic Correlation*. Geophysics Department, Australian National University, iii–iv.

Irving, E. 1961. Paleomagnetic methods: a discussion of a recent paper by A.E.M. Nairn. *J. Geol.*, **69**: 226–231.

Irving, E. 1962. An analysis of the positions of the Earth's magnetic pole in the geological past. *Geophys. J.*, **7**: 279–283.

Irving, E. 1963. Paleomagnetism of the Narrabeen Chocolate shales and Tasmanian dolerite. *J. Geophys. Res.*, **68**: 2283–2287.

Irving, E. 1964. *Paleomagnetism and Its Application to Geological and Geophysical Problems*. John Wiley & Sons, New York.

Irving, E. 1977. Drift of the major continental blocks since the Devonian. *Nature*, **270**: 304–309.

Irving, E. 1988. The paleomagnetic confirmation of continental drift. *Eos*, **69**: 994–1014.

Irving, E. 2004. The case for Pangea B, and the intra-Pangean Megashear. In Channell, J. E. T., Kent, D. V., Lowrie, W., and Meert, J. G., eds., *Timescales of the Paleomagnetic Field*, AGU Geophysical Monograph, **145**: 13–27.

Irving, E. and Banks, M. R. 1961. Paleomagnetic results from the Upper Triassic lavas of Massachusetts. *J. Geophys. Res.*, **66**: 1935–1939.

Irving, E. and Green, R. 1957a. Palaeomagnetic evidence from the Cretaceous and Cainozoic. *Nature*, **179**: 1064–1065.

Irving, E. and Green, R. 1957b. The palaeomagnetism of the Kainozoic basalts of Victoria. *Mon. Not. Roy. Astron. Soc., Geophys. Suppl.*, **7**: 347–359.

Irving, E. and Green, R. 1958. Polar movement relative to Australia. *Geophys. J. Roy. Astron. Soc.*, **1**: 64–72.

Irving, E. and Opdyke, N. D. 1965. The paleomagnetism of the Bloomsburg red beds and its possible application to the tectonic history of the Appalachians. *Geophys. J. Roy. Astron. Soc.*, **9**: 153–166.

Irving, E. and Parry, L. G. 1963. The magnetism of some Permian rocks from New South Wales. *Geophys. J. Roy. Astron. Soc.*, **7**: 395–411.

Irving, E. and Runcorn, S. K. 1957. Palaeomagnetic investigations in Great Britain II. Analysis of the paleomagnetism of the Torridonian sandstone series of Northwest Scotland I. *Phil. Trans. Roy. Soc. London A*, **250**: 83–99.

Irving, E. and Tarling, D. H. 1961. The paleomagnetism of the Aden volcanics. *J. Geophys. Res.*, **66**: 549–556.

Irving, E. and Ward, M. A. 1964. A statistical model of the geomagnetic field. *Geofis. Pura Appl.*, **57**: 25–30.

Irving, E., Robertson, W. A., Stott, P. M., Tarling, D. H., and Ward, M. A. 1961. Treatment of partially stable sedimentary rocks showing planar distribution of directions of magnetization. *J. Geophys. Res.*, **66**: 1927–1933.

Jaeger, J. C. 1956. Palaeomagnetism. *Australian J. Sci.*, **19**: 100–102.

Jaeger, J. C. and Irving, E. 1957. *Palaeomagnetism and the Reconstructions of Gondwanaland, C. R. 3rd Congress Pacific Indian Ocean Science*. Imprimerie Officielle, Tananarive, Madagascar, 233–242.

Jaeger, J. C. and Joplin, G. A. 1955. Rock magnetism and the differentiation of dolerite sill. *J. Geol. Soc. Australia*, **2**: 1–19.

Jardetsky, W. A. 1949. On the rotation of the Earth during its evolution. *Trans. Am. Geophys. Union*, **30**: 797–817.

Jeffreys, H. 1929. *The Earth, Its Origin, History and Physical Constitution* (2nd edition). Cambridge University Press, Cambridge.

Jeffreys, H. 1950. *Earthquakes and Mountains* (2nd edition, revised). Methuen and Co., London.

Jeffreys, H. 1952. *The Earth: Its Origin, History and Physical Characteristics* (3rd edition). Cambridge University Press, Cambridge.

Jeffreys, H. 1959. *The Earth: Its Origin, History and Physical Characteristics* (4th edition). Cambridge University Press, Cambridge.

Johnson, E. A., Murphy, T., and Torreson, O. W. 1948. Pre-history of the Earth's magnetic field. *Terrestrial Magnetism*, **53**: 349–372.

Kanasewich, E. R. J., Havskov, J., and Evans, M. E. 1978. Plate tectonics in the Phanerozoic. *Can. J. Earth Sci.*, **15**: 919–935.

Kawai, N., Ito, H., and Kume, S. 1961. Deformation of the Japanese Islands as inferred from rock magnetism. *Geophys. J.*, **6**: 124–130.

Kennedy, W. Q. 1946. The Great Glen Fault. *Q. J. Geol. Soc. London*, **102**: 41–76.

Kent, D. V. and Irving, E. 2010. Influence of inclination error in sedimentary rocks on the Triassic and Jurassic apparent pole wander path for North America and implications for Cordilleran tectonics. *J. Geophys. Res.*, **115**, B10103, doi: 10.1029/2009JB007205.

Kern, J. W. 1961a. Effects of moderate stresses on directions of thermoremanent magnetization. *J. Geophys. Res.*, **66**: 3801–3805.

Kern, J. W. 1961b. The effect of stress on the susceptibility and magnetization of a partially magnetized multidomain system. *J. Geophys. Res.*, **66**: 3807–3816.

Kern, J. W. 1961c. Stress stability of remanent magnetization. *J. Geophys. Res.*, **66**: 3817–3820.

Khramov, A. N. 1958. *Palaeomagnetism and Stratigraphic Correlation.* Gostoptechizdat, Leningrad. English translation by Lojkine, A. J. 1960. Geophysics Department, Australian National University.

Khramov, A. N. 1994. Acceptance of Bucher Award. *Eos, Trans. AGU*, **75**: 106.

Khramov, A. N. and Sholpo, L. Ye. 1967. Synoptic tables of U.S.S.R. paleomagnetic data. Appendix I of *Paleomagnetism.* Nedra Press, Leningrad, 213–233. 1970. Translation by Hope, E. R., Irving, E., ed., Report T510R, Directorate of Scientific Information Services, DRB Canada.

King, L. 1953. Necessity for continental drift. *Bull. Am. Assoc. Petrol. Geol.*, **37**: 2163–2177.

Knight, S. H. 1929. The Fountain and Casper formations of the Laramie Basin. *Univ. Wyoming Pub. Sci., Geol.*, **1**: 64–66.

Köppen, W. and Wegener, A. 1924. *Die Klimate der Geologischen Vorzeit.* Gebrüder Borntraeger, Berlin.

Köppen, W. (and Wegener, A.) 1940. *Die Klimate der Geologischen Vorzeit* (2nd edition). Gebrüder Borntraeger, Berlin.

Kreichgauer, D. 1902. *Die Aquatorfrage in der Geologie.* Missionsdruckerie, Steyl (2nd edition, 1926).

Krishnan, M. S. 1949. *Geology of India and Burma* (2nd edition). Madras Law Journal Office, Madras.

Kristjansson, L. 1982. Paleomagnetic research on Icelandic rocks: a bibliographic review 1951–1981. *Jokull*, **32**: 91–106.

Kristjansson, L. 1984. Paleomagnetic research on Icelandic rocks 1951–81, additional notes and references. *Jokull*, **34**: 77–79.

Kristjansson, L. 1993. Investigations on geopolarity reversals in Icelandic lavas, 1953–78. *Terra Nova*, **5**: 6–12.

Laj, C., Kissel, C., and Guillou, H. 2002. Bruhnes' research revisited: magnetization of volcanic flows and baked clays. *Eos, Trans. AGU*, **83**: 381–382.

Laming, D. J. C. 1954. Sedimentary processes in the formation of the New Red Sandstone of South Devonshire. Ph.D. thesis (unpublished), University of London.

Laming, D. J. C. 1958. Fossil winds. *J. Alberta Soc. Petrol. Geol.*, **6**: 179–183.

Larmor, J. 1919. How could a rotating body such as the Sun become a magnet? *Br. Assoc. Adv. Sci. Rep.*: 159–160.

Lee, J. S. 1952. Distortion of continental Asia. *Palaeobotanist*, **1**: 298–315.

Lees, G. M. 1953. The evolution of a shrinking Earth. *Q. J. Geol. Soc. London*, **109**: 217–257.

Le Grand, H. E. 1988. *Drifting Continents and Shifting Theories.* Cambridge University Press, Cambridge.

Le Grand, H. E. 1989. Conflicting orientations: John Graham, Merle Tuve, and paleomagnetic research at the DTM, 1938–1958. *Earth Sci. Hist.*, **8**: 55–65.

Longwell, C. R. 1958a. My estimate of the continental drift concept. In Carey, S. W., Convener, *Continental Drift: A Symposium*. University of Tasmania, Hobart, 1–12.

Longwell, C. R. 1958b. Epilogue. In Carey, S. W., Convener, *Continental Drift: A Symposium*. University of Tasmania, Hobart, 356–358.

Lovell, B. 1968. *The Story of Jodrell Bank*. Harper & Row, New York.

Lovell, B. 1975. Patrick Maynard Stuart Blackett. *Biogr. Mem. Fell. Roy. Soc.*, **2**: 1–115.

Lowes, F. J. 1998. Keith's early work in geomagnetism. *Phys. Chem. Earth*, **23**: 703–707.

Ma, T. Y. H. 1953. The sudden total displacement of the outer solid Earth shell by sliding relative to the fixed rotating core of the Earth. Research of the past climate and continental drift, VI. Privately published, Taiwan.

Manley, H. 1949. Paleomagnetism. *Sci. News*, **12**: 43–64.

Markowitz, W. 1945. Redeterminations of latitude and longitude. *Trans. Am. Geophys. Union*, **26**: 197–199.

McDougall, I. 1961. Determination of the age of a basic igneous intrusion by the potassium-argon method. *Nature*, **190**: 1184–1186.

McDougall, I. 1963. Potassium-argon ages from western Oahu, Hawaii. *Nature*, **197**: 344–345.

McDougall, I. and McElhinny, M. W. 1970. The Rajmahal traps of India: K–Ar ages and palaeomagnetism. *Earth Planet. Sci. Lett.*, **9**: 371–378.

McDougall, I. and Tarling, D. H. 1963. Dating polarity zones in the Hawaiian Islands. *Nature*, **200**: 54–56.

McElhinny, M. W. 1973. *Paleomagnetism and Plate Tectonics*. Cambridge University Press, Cambridge.

McElhinny, M. W. 1994. Citation for Khramov receiving Bucher Award. *Eos, Trans. AGU*, **75**: 106.

McElhinny, M. W. and McFadden, P. L. 2000. *Paleomagnetism: Continents and Oceans*. Academic Press, San Diego, CA.

McKee, E. D. 1933. *The Coconino Sandstone: Its History and Origin*. Carnegie Institution, Washington, Publication No. 440: 77–115.

McKee, E. D. 1938. Structures in modern sediments aid in interpreting ancient rocks. Carnegie Institution, Washington, Publication No. 501: 688–690.

McKee, E. D. 1940. Three types of cross-lamination in Paleozoic rocks of northern Arizona. *Am. J. Sci.*, **238**: 811–824.

Menard, H. W. 1986. *The Ocean of Truth: A Personal History of Global Tectonics*. Princeton University Press, Princeton, NJ.

Mercanton, P. L. 1926a. Inversion de l'inclinasion magnetique terrestre aux ages geologiques. *Terr. Magn. Atmos. Electr.*, **31**: 187–190.

Mercanton, P. L. 1926b. Aimantation des basaltes groenlandais. *C.R. Acad. Sci. Paris*, **182**: 859–860.

Merrill, R. T. and McElhinny, M. W. 1983. *The Earth's Magnetic Field: Its History, Origin and Planetary Perspective*. Academic Press, New York.

Miknulic, D. J. and Kluessendorf, J. 2001. Gilbert O. Raasch, student of Wisconsin's ancient past. *Geoscience Wisconsin*, **18**: 75–93.

Milankovitch, M. 1933. Säkulare Polverlagreungen. *In Handbuch der Geophysik*, **1**, Part 7, Gebrüder Borntraeger, Berlin, 438–500.

Milankovitch, M. 1934. Der Mechanismus der Polverlagerungen und die darasus sich ergebenden polbahnukurven. *Gerlands Beitr. Z. Geophys.*, **42**: 70–97.

Morel, P. and Irving, I. 1981. Paleomagnetism and evolution of Pangea. *J. Geophys. Res.*, **86**: 1858–1887.

Munk, W. H. 1956. Polar wandering: a marathon of errors. *Nature*, **177**: 551–554.

Munk, W. H. 1958. Remarks concerning the present position of the pole. *Geophysica*, **6**: 335–355.

Munk, W. H. and MacDonald, G. J. F. 1960. *The Rotation of the Earth*. Cambridge University Press, Cambridge.

Muttoni, G., Kent, D. V., Garzanti, E., Branck, P., Abrahamsen, N., and Gaetani, M. 2003. Early Pangea 'B' to Late Permian 'A'. *Earth Planet. Sci. Lett.*, **215**: 379–394.

Nagata, T. 1952. Reverse thermo-remanent magnetism. *Nature*, **169**: 704–705.

Nagata, T. and Shimizu, Y. 1959. Natural remanent magnetization of the Pre-Cambrian gneiss of Ongul Islands in the late Antarctic. *Nature*, **184**: 1472–1473.

Nagata, T., Akimoto, S., and Uyeda, S. 1951. Reverse thermo-remanent magnetization. *Proc. Imp. Acad. Japan*, **27**: 643–645.

Nagata, T., Akimoto, S., Shimizu, Y., Kobayashi, K., and Kuno, H. 1959. Paleomagnetic studies on Tertiary and Cretaceous rocks in Japan. *Proc. Japan Acad.*, **35**: 378–383.

Nairn, A. E. M. 1956. Relevance of palaeomagnetic studies of Jurassic rocks to continental drift. *Nature*, **178**: 935–936.

Nairn, A. E. M. 1957a. Palaeomagnetic collections from Britain and South Africa illustrating two problems of weathering. *Adv. Phys.*, **6**: 162–168.

Nairn, A. E. M. 1957b. Observations paleomagnetiques en France: Roches Perminnes. *Bull. Soc. Geol. France*, **7**: 721–727.

Nairn, A. E. M. 1959. A palaeomagnetic survey of the Karroo system. *Overseas Geol. Min. Res.*, **7**: 398–410.

Nairn, A. E. M. 1960. Paleomagnetic results from Europe. *J. Geol.*, **68**: 285–308.

Nairn, A. E. M. 1961. The scope of palaeoclimatology. In Nairn, A. E. M., ed., *Descriptive Palaeoclimatology*. Interscience, New York, 45–59.

Nakashima, H., Shioji, Y., Kobayashi, T., Aoki, S., Shimizu, H., and Miyasaka, J. 2011. Determining the angle of repose of sand under low-gravity conditions using discrete element method. *J. Terramechanics*, **48**: 17–26.

Néel, L. 1949. Théorie du traînage magnétique des ferromagnétiques en grains fins avec application aux terres cuites. *Ann. Géophys.*, **5**: 99–136.

Néel, L. 1951. Inversion de l'aimantation permenente des roches. *Ann. Géophys.*, **7**: 90–102.

Néel, L. 1952. Confirmation expérimentale d'un mécanisme d'inversion de l'aimantation thermorémanente. *C. R. Acad. Sci.*, **234**: 1991–1993.

Newman, R. P. 1995. American intransigence: the rejection of continental drift in the great debates of the 1920s. *Earth Sci. Hist.*, **14**: 62–83.

Nolan, T. B. 1987. James Gilluly, 1896–1980. *Biogr. Mem.*, **56**: 118–132.

Nye, M. J. 1999. Temptations of theory, strategies of evidence: P.M.S. Blackett and the earth's magnetism, 1947–52. *Br. J. Hist. Sci.*, **32**: 69–92.

Nye, M. J. 2004. *Blackett*. Harvard University Press, Cambridge, MA.

Officer, C. D. 1954. South-west Pacific crustal structure. *Trans. Am. Geophys. Union*, **35**: 356.

Oldham, R. D. 1906. The constitution of the Earth. *Q. J. Geol. Soc. London*, **62**: 465–472.

Opdyke, N. D. 1958a. Palaeoclimates and palaeomagnetism. *20th Congr. Geol. Intern. C. R.*, Mexico, 1956, Section IX, 193–199.

Opdyke, N. D. 1958b. Palaeoclimatology and palaeomagnetism in relation to polar wandering and continental drift. Ph.D. dissertation, Durham University.

Opdyke, N. D. 1959. The impact of paleomagnetism on paleoclimatic studies. *Int. J. Bioclimatol. Biometeorol.*, **3**, Part VI, Section A: 1–11.

Opdyke, N. D. and Runcorn, S. K. 1956. New evidence for reversal of the geomagnetic field near the plio-Pleistocene boundary. *Science*, **123**: 1126–1127.

Opdyke, N. D. and Runcorn, S. K. 1959. Paleomagnetism and ancient wind directions. *Endeavour*, **18**: 26–34.

Opdyke, N. D. and Runcorn, S. K. 1960. Wind direction in the western United States in the late Palaeozoic. *Geol. Soc. Am. Bull.*, **71**: 959–972.

Öpik, E. J. 1955. Paleomagnetism, polar wandering and continental drift. *Irish Astron. J.*, **3**: 191–236.

Oreskes, N. 1999. *The Rejection of Continental Drift*. Oxford University Press, New York.

Paterson, M. S. 1982. Jaeger, John Conrad. *Biogr. Mem. Fell. Roy. Soc.*, **28**: 163–203.

Peach, B. N., Horne, J., Gunn, W., Clough, C. T., Hinxman, L. W., and Teall, J. J. H. 1907. *The Geological Structure of the NW Highlands of Scotland*. Memoirs of the Geological Survey of Great Britain.

Phillips, J. D. and Forsyth, D. W. 1972. Plate tectonics, paleomagnetism and the opening of the Atlantic. *Geol. Soc. Am. Bull.*, **83**: 1579–1600.

Piper, J. D. A. and Darabi, M. H. 2005. Paleomagnetic study of the (late Mesoproterozoic) Torridon group, NW Scotland: age, magnetostratigraphy, tectonic setting and partial magnetic overprinting by Caledonian Orogeny. *Precambrian Research*, **142**: 45–81.

Poole, F. G. 1957. Paleo-wind directions in late Paleozoic and early Mesozoic time on the Colorado plateau as determined by cross-strata. *Geol. Soc. Am. Bull.*, **68**: 1870.

Raasch, G. O. 1958a. Editorial foreword. *J. Alberta Soc. Petr. Geol.*, **5**: 139.

Raasch, G. O. 1958b. The Baraboo (Wis.) Monadnock and palaeo-wind direction. *J. Alberta Soc. Petr. Geol.*, **6**: 183–187.

Rainbird, R. H., Hamilton, M. A., and Young, G. M. 2001. Detrital zircon geochronology and provenance of the Torridonian, NW Scotland. *J. Geol. Soc. London*, **158**: 15–27.

Ramsey, W. H. 1949. On the nature of the Earth's core. *Mon. Not. Roy. Astron. Soc., Geophys. Suppl.*, **9**: 409–426.

Reiche, P. 1938. An analysis of cross-lamination: the Coconino sandstone. *J. Geol.*, **46**: 905–932.

Roche, A. 1950. Sur les caracteres magnétiques du système éruptif de Gergovie. *C. R. Acad. Sci. Paris*, **230** (1951): 113–115.

Roche, A. 1951. Sur les inversions de l'aimantation rémanente des roches volcaniques dans les monts d'Auvergne. *C. R. Acad. Sci. Paris*, **233**: 1132–1134.

Roche, A. 1953. Sur l'origine des inversions d'aimantation constatées dans les roches d'Auvergne. *C. R. Acad. Sci. Paris*, **236**: 107–109.

Roche, A. 1957. Sur l'aimantation de laves Miocènes d'Auvergne. *C. R. Acad. Sci. Paris*, **250**: 377–379.

Runcorn, S. K. 1948. (with an appendix by Chapman, S.) The radial variation of the Earth's magnetic field. *Proc. Phys. Soc. London*, **61**: 373–382.

Runcorn, S. K. 1954. The Earth's core. *Trans. Am. Geophys. Union*, **35**: 49–63.

Runcorn, S. K. 1955a. Rock magnetism: geophysical aspects. *Adv. Phys.*, **4**: 244–291.

Runcorn, S. K. 1955b. Palaeomagnetism of sediments from the Colorado plateau. *Nature*, **176**: 505–506.

Runcorn, S. K. 1955c. The Earth's magnetism. *Sci. Am.*, **192**: 152–162.

Runcorn, S. K. 1956a. Paleomagnetic survey in Arizona and Utah: preliminary results. *Geol. Soc. Am. Bull.*, **87**: 301–316.

Runcorn, S. K. 1956b. Palaeomagnetic comparisons between Europe and North America. *Proc. Geol. Assoc. Canada*, **8**: 77–85.

Runcorn, S. K. 1956c. Paleomagnetism, polar wandering and continental drift. *Geol. Mijnbouw*, **18**: 253–258.

Runcorn, S. K. 1957. The sampling of rocks for palaeomagnetic comparisons between the continents. *Adv. Phys.*, **6**: 169–176.

Runcorn, S. K. 1958. Address by professor S. K. Runcorn on palaeomagnetism. *J. Alberta Soc. Petrol. Geol.*, **5**: 140–144.

Runcorn, S. K. 1959a. On the theory of the geomagnetic secular variation. *Ann. Geophys.*, **15**: 87–92.

Runcorn, S. K. 1959b. On the hypothesis that the mean geomagnetic field for parts of geological time has been that of a geocentric axial multipole. *J. Atmos. Terr. Phys.*, **14**: 167–174.

Runcorn, S. K. 1959c. Discussion on the Permian climate zonation and paleomagnetism. *Am. J. Sci.*, **257**: 235–237.

Runcorn, S. K. 1959d. Rock magnetism. *Science*, **129**: 1002–1012.

Runcorn, S. K. 1961. Climatic change through geological time in the light of the palaeomagnetic evidence for polar wandering and continental drift. *Q. J. Roy. Meteorol. Soc.*, **87**: 282–313.

Runcorn, S. K. 1963a. Palaeomagnetic methods of investigating polar wandering and continental drift. In Munyan, A. C., ed., *Polar Wandering and Continental Drift*. Society of Economic Paleontologists and Mineralogists, Special Publication No. 10: 47–54.

Runcorn, S. K. 1963b. Satellite gravity measurements and convection in the mantle. *Nature*, **200**: 628–630.

Runcorn, S. K. 1964. Paleomagnetic results from Precambrian sedimentary rocks in the western United States. *Geol. Soc. Am. Bull.*, **75**: 687–704.

Runcorn, S. K., Benson, A. C., and Moore, A. F. 1950. The experimental determination of the geomagnetic radial variation. *Phil. Mag.*, **41**: 783–791.

Runcorn, S. K., Benson, A. C., Moore, A. F., and Griffiths, D. H. 1951. Measurements of the variation with depth of the main geomagnetic field. *Phil. Trans. Roy. Soc. A*, **244**: 113–151.

Rutten, M. G. 1959. Paleomagnetic reconnaissance of mid-Italian volcanoes. *Geol. Mijnbouw*, **21**: 373–374.

Rutten, M. G. and Veldkamp, J. 1958. Paleomagnetic research at the Utrecht University. *Ann. Géophys.*, **14**: 519–521.

Rutten, M. G., Van Everdingen, O. R., and Zijderveld, J. D. A. 1957. Palaeomagnetism in the Permian of the Oslo Graben (Norway) and of the Estérel (France). *Geol. Mijnbouw*, **19**: 193–195.

Sahni, B. 1944. Microfossils and problems of Salt Range geology. *Proc. Natl. Acad. Sci. India*, **14**, Section B: i–xxxii.

Scalera, G. and Braun, T. 2003. Ott Christoph Hilgenberg in twentieth century geophysics. In Scalera, G. and Jacob, K.-H., eds., *Why Expanding Earth? A Book in Honour of Ott Christoph Hilgenberg, Proc. 3rd Lautenthaler Montanistisches Colloquium.* Mining Industry Museum, Lauthenthal, Germany, May 26, 2001, 25–41.

Schmidt, P. W. and McDougall, I. 1977. Palaeomagnetic and potassium-argon dating studies of the Tasmanian dolerites. *J. Geol. Soc. Aust.*, **24**: 321–328.

Schmidt, P. W., Williams, G. E., and Embleton, B. J. J. 1991. Low paleolatitude of Late Proterozoic glaciation: early timing of remanence in hematite of the Elatina Formation, South-Australia. *Earth Planet. Sci. Lett.*, **105**: 355–376.

Schuster, A. 1891. Recent total solar eclipses. *Proc. Roy. Inst.*: 273–276.

Schwartz, E. J. 1963. A paleomagnetic investigation of Permo-Triassic red beds and Andesites from the Spanish Pyrennes. *J. Geophys. Res.*, **68**: 3265–3271.

Schwarzbach, M. 1955. *Climates of the Past. Introduction to Palaeoclimatology (Russian translation).* Izd-vo Inostrannol literatury, Moscow.

Sclater, J. G. 2001. Heat flow under the oceans. In Oreskes, N., ed., with Le Grand, H., *Plate Tectonics*. Westview Press, Boulder, CO, 128–147.

Sclater, J. G. and Francheteau, J. 1970. The implications of terrestrial heat flow observations on current tectonic and geochemical models of the crust and upper mantle of the earth. *Geophys. J. Roy. Astron. Soc.*, **20**: 509–537.

Shotton, F. W. 1937. The Lower Bunter sandstones of North Worcestershire and East Shropshire. *Geol. Mag.*, **74**: 534–553.

Shotton, F. W. 1956. Some aspects of the New Desert in Britain. *Liverpool and Manchester Geol. J.*, **1**: 450–466.

Smith, A. G., Briden, J. C., and Drury, G. E. 1973. Phanerozoic world maps. In Hughes, N. F., ed., Organisms and Continents through Time. Special Paper *Palaeontology*, **12**: 241–269.

Smith, R. L. and Piper, J. D. A. 1984. Paleomagnetic study of the (Lower Cambrian) Longmyndian sediments and tuffs. *Geophy. J. Roy. Astro. Soc.*, **79**: 875–892.

Smith, R. L., Stearn, J. E. F., and Piper, J. D. A. 1983. Paleomagnetic study of sediments of NW Scotland. *Scott. J. Geol.*, **19**: 29–45.

Snavely, P. D., Jr. and Baldwin, E. M. 1948. Siletz River Volcanic series, northwestern Oregon. *Am. Assoc. Petr. Geol. Bull.*, **32**: 805–812.

Special Correspondent. 1954. Drift of polar axis. *The Times*, September 9.

Stacey, F. D. 1958. Effect of stress on the remanent magnetism of magnetite-bearing rocks. *J. Geophys. Res.*, **63**: 361–368.

Stacey, F. D. 1960. Magnetic anisotropy of igneous rocks. *J. Geophys. Res.*, **65**: 2429–2442.

Stacey, F. D. 1963. Physical theory of rock magnetism. *Adv. Phys.*, **12**: 45–133.

Steers, J. A. 1942. *An Introduction to the Study of Map Projections* (with a Foreword by F. Debenham). University of London Press, London.

Stehli, F. G. 1957. Possible Permian climatic zonation and its implications. *Am. J. Sci.*, **255**: 607–618.

Stehli, F. G. 1959. Reply. *Am. J. Sci.*, **257**: 239–240.

Stevens, G. 1988. John Bradley: a New Zealand pioneer in continental drift studies. *Geol. Soc. New Zealand Newsletter*, No 17: 30–38.

Stewart, A. D. 2002. *The Later Paleozoic Torridonian Rocks of Scotland: Their Sedimentology, Geochemistry and Origin.* Memoir 24. Geological Society of London, London.

Stewart, A. D. and Irving, E. 1974. Palaeomagnetism of the Precambrian rocks from Northwest Scotland and the apparent polar wandering path for Laurentia. *Geophys. J. Roy. Astron. Soc.*, **37**: 51–72.

Stewart, A. D. and Irving, E. 1974. Paleomagnetism of sedimentary rocks from NW Scotland and the apparent polar wander path for Laurentia. *Geophys. J. Roy. Soc.*, **37**: 51–72.

Stott, P. M. and Stacey, F. D. 1959. Magnetostriction and palaeomagnetism of igneous rocks. *Nature*, **183**: 384–385.

Stott, P. M. and Stacey, F. D. 1960. Magnetostriction and palaeomagnetism of igneous rocks. *J. Geophys. Res.*, **65**: 2419–2424.

Strahow, N. M. 1948. *Outlines of Historical Geology* (in Russian). Moscow.

Sullivan, W. 1974. *Continents in Motion: The New Earth Debate*. McGraw-Hill, New York.

Swann, W. F. G. and Longacre, A. 1928. An attempt to detect a magnetic field as the result of the rotation of a copper sphere at high speed. *J. Franklin Institute*, **206**: 421–434.

Tarling, D. H. 1962. Tentative correlation of Samoan and Hawaiian Islands using reversals of magnetization. *Nature*, **196**: 882–883.

Tarling, D. H. 1985. Paleomagnetic studies of the Orcadian Basin. *Scott. J. Geol.*, **21**: 261–273.

Termier, H. and Termier, G. 1952. *Histoire Géologique de la Biosphere*. Masson, Paris.

Thompson, W. O. 1937. Original structures of beaches, bars, and dunes. *Geol. Soc. Am. Bull.*, **48**: 723–752.

Torreson, O. W., Murphy, T., and Graham, J. W. 1949a. Magnetic polarization of sedimentary rocks and the Earth's magnetic history. *J. Geophys. Res.*, **54**: 111–129.

Torreson, O. W., Murphy, T., and Graham, J. W. 1949b. Rock magnetization as a clue to Earth's magnetic history. *Phys. Rev.*, **75**: 208–209.

Turnbull, G. 1959. Some palaeomagnetic measurements in Antarctica. *Arctic*, **12**: 151–157.

Turnbull, M. J. M., Whitehouse, M. J., and Moorbath, S. 1996. New isotopic age determination for the Torridonian, NW Scotland. *J. Geol. Soc. London*, **153**: 955–964.

Turner, F. J. and Verhoogen, J. 1960. *Igneous and Metamorphic Petrology* (2nd edition). McGraw-Hill, New York.

Tuve, M. A. 1949. Department of Terrestrial Magnetism. In *Carnegie Institution of Washington, Year Book No. 48: 1948–1949*. Lord Baltimore Press, Baltimore, MD, 57–80.

Tuve, M. A. 1950. Department of Terrestrial Magnetism. In *Carnegie Institution of Washington, Year Book No. 49*. Lord Baltimore Press, Baltimore, MD, 61–81.

Twenhofel, W. H. 1950. *Principles of Sedimentation*. McGraw-Hill, New York.

Urey, H. C. 1953. On the origin of continents and mountains. *Proc. Natl. Acad. Sci.*, **39**: 933–946.

van Bemmelen, W. 1883. De isogonen in de 16de en 17de eeuw. Dissertation, Utrecht.

Vandamme, D., Courtillot, V., Besse, J., and Montigny, R. 1991. Paleomagnetism and the age determinations of the Deccan Traps (India): results of a Nagpur-Bombay traverse and review of earlier work. *Rev. Geophys.*, **29**: 159–190.

van Hilten, D. 1962. Presentation of paleomagnetic data, polar wandering, and continental drift. *Am. J. Sci.*, **260**: 401–426.

van der Lingen, G. J. 1960. Geology of the Spanish Pyrennes, north of Canfranc, Huesca Province. *Estud. Geol.*, **16**: 205–242.

van der Voo, R. 1967. The Rotation of Spain: paleomagnetic evidence from the Spanish Meseta. *Palaeogeogr., Palaeoclimatol., Palaeoecol.*, **3**: 393–416.

van der Voo, R. 1969. Paleomagnetic evidence for the rotation of the Iberian Penisula, *Tectonophysics*, **7**: 5–56.

van der Voo, R. 1990. Phanerozoic paleomagnetic poles from Europe and North America and comparisons with continental reconstructions. *Rev. Geophys.*, **28**: 167–206.

van der Voo, R. and Torsvik, T. H. 2001. Evidence for Late Paleozoic and Mesozoic non-dipole fields provide an explanation for Pangea reconstruction problems. *Earth Planet. Sci. Lett.*, **187**: 71–81.

Veldkamp, J. 1984. *History of Geophysical Research in the Netherlands and Its Former Overseas Territories*. North-Holland, Amsterdam.

Verhoogen, J. 1946. Volcanic heat. *Am. J. Sci.*, **244**: 745–771.

Verhoogen, J. 1954. Petrological evidence on temperature distribution in the mantle of the Earth. *Am. Geophys. Union Trans.*, **35**: 50–59.

Verhoogen, J. 1956. Ionic ordering and self-reversal of magnetization in impure magnetites. *J. Geophys. Res.*, **61**: 201–209.

Verhoogen, J. 1959. The origin of thermo-remanent magnetization. *J. Geophys. Res.*, **64**: 2441–2449.

Verhoogen, J. 1960. Temperatures within the Earth. *Am. Sci.*, **48**: 134–159.

Verhoogen, J. 1961. Heat balance of the Earth's core. *Geophys. J. Roy. Astron. Soc.*, **4**: 276.

Verhoogen, J. 1985. North American paleomagnetism and geology. In Drake, E. T. and Jordan, W. M., eds., *Geologists and Ideas: A History of North American Geology, GSA Centennial Special Volume I*. Geological Society of America, Boulder, CO.

Vincenz, S. A. 1952. Remanent magnetism of some igneous rocks of Great Britain and its geophysical significance. Ph.D. thesis, University of London.

Voisey, A. H. 1958. Some comments on the hypothesis of continental drift. In Carey, S. W., Convener, *Continental Drift: A Symposium*. University of Tasmania, Hobart, 161–171.

Voisey, A. H. 1991. *Sixty Years on the Rocks: The Memoirs of Professor Alan H. Voisey*. The Earth Sciences History Group, Geological Society of Australia, Sydney.

Walker, F. 1956. The magnetic properties and differentiation of Dolerite Sills: a critical discussion. *Am. J. Sci.*, **254**: 433–443.

Wanenmacher, J. M., Twenhofel, W. H., and Raasch, G. O. 1934. The Paleozoic strata of the Baraboo areas, Wisconsin. *Am. J. Sci.*, **28**: 1–30.

Wang, T., Teng, H., Li, C., and Yeh, S. 1960. *Acta Geophys. Sinica*, **9**: 125–138.

Watson, G. S. and Irving, E. 1957. Statistical methods in rock magnetism. *Mon. Not. Roy. Astron. Soc., Geophys. Suppl.*, **7**: 289–300.

Wegener, A. 1915. *Die Entstehung der Kontinente und Ozeane*. Friedrich Vieweg & Sohn, 1st edition, 1915; 2nd edition, 1920, 3rd edition, 1922; 4th edition, 1924; 4th revised edition, 1929; 5th edition, revised by K. Wegener, 1936.

Wilson, H. A. 1923. An experiment on the origin of the Earth's magnetic field. *Proc. Roy. Soc. A*, **104**: 451–455.

Wood, R. M. 1985. *The Dark Side of the Earth*. George Allen & Unwin, London.

Zijl, J. S. V. van, Graham, K. W. T., and Hales A. L. 1962. The palaeomagnetism of the Stormberg lavas of South Africa, II. *Geophys. J.*, **7**: 169–182.

# Index

Printed in the United States
By Bookmasters